A Synopsis of
Regional Anatomy

Contents

NECK 175

UPPER LIMB 221

THORAX 333

PLEURA, LUNGS AND DIAPHRAGM 375

ABDOMEN 391

PELVIS   467

LOWER LIMB   535

BACK   641

Head

The head contains the brain, cavities for the special sensory organs, openings for passage of air and food and teeth for mastication. The body framework of the head is the skull. The scalp is the soft tissue covering the calvaria.

Chapter 1

The Scalp

Chapter Outline

1. Layers and Muscles
2. Arteries
3. Veins
4. Nerves
5. Lymphatics

The head contains the brain, cavities for the special sensory organs, openings for the passage of air and food and teeth for mastication. The bony framework of the head is the skull. The scalp is the soft tissue covering the calvaria.

The scalp covers the cranium. It extends from the supraorbital margins in front to the superior nuchal lines behind and from one temporal line to the other laterally. It may be regarded as a prolongation of the facial skin and muscles over the vertex of the skull.

1. Layers and Muscles

The scalp comprises five layers which may be recalled by the acronym S (skin), C (close connective tissue), A (aponeurosis of the epicranius muscle), L (loose connective tissue) and P (pericranium):

Skin, which is very thin and contains numerous long hairs (except over the forehead) sweat glands and sebaceous glands (it is the most common site for the occurrence of sebaceous cysts). It is firmly attached to the subcutaneous layer.

The subcutaneous layer is a dense vascularized and innervated connective tissue which firmly binds the skin to the underlying aponeurotic layer. It contains fat enclosed in lobules. The amount of these lobules decrease with age.

Wounds involving this layer bleed profusely because the adventitia of blood vessels is firmly anchored to the dense connective tissue of this layer and are prevented from retracting.

Epicranial aponeurosis (galea aponeurotica) is a strong sheet of fibrous tissue extending between the frontal and occipital bellies of the occipitofrontalis (epicranius) muscle. It is attached to the external occipital protuberance and highest nuchal line behind. It extends down as a thin membrane over the temporal fascia to the zygomatic arch.

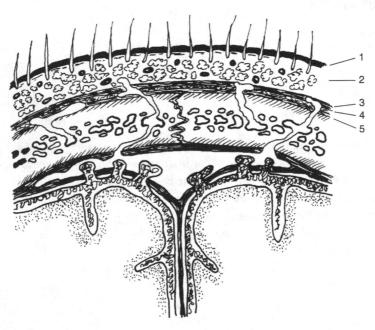

FIG. 1-1 Layers of scalp. (1) Skin; (2) Subcutaneous tissue; (3) Galea aponeurotica; (4) Loose subaponeurotic tissue; (5) Pericranium.

Wounds of the scalp may not gape unless the epicranius or its aponeurosis is divided transversely.

MUSCLES OF THE SCALP AND EXTERNAL EAR

These muscles belong to the same group as the muscles of facial expression and are similarly supplied by the facial nerve (VII).

Epicranius muscle consists of two bellies of the occipitofrontalis muscle that are connected by the intervening epicranial aponeurosis.

Frontal Bellies (Frontalis)

Frontal bellies are long, wide and united. They have no bony attachments.

- **Origin:** Anterior part of epicranial aponeurosis.
- **Insertion:** Skin and dense connective tissue near the eyebrows and root of the nose.
- **Action:** Wrinkles the forehead transversely and pulls the scalp forward. Their antagonists are the orbital part of orbicularis oculi.
- **Nerve supply:** Temporal branch of the fazcial nerve (VII).

Occipital Bellies (Occipitalis)

Occipital bellies are small and separate. They arise from bone.

- **Origin:** Lateral two-thirds of the superior nuchal line.
- **Insertion:** Posterior part of epicranial aponeurosis.
- **Action:** Draws the scalp backwards and fixes the scalp allowing

frontalis to pull on the skin of the forehead.

- **Nerve supply:** Posterior auricular branch of the facial nerve (VII).

Auricularis muscles are three small muscles inserted into the deep (cranial) aspect of the auricle (pinna). Auricularis anterior and superior are thin and fan shaped. They arises from the temporal fascia. Their nerve supply is the temporal branch of the facial nerve (VII). Auricularis posterior is a narrow muscle that arises from the mastoid process. It is supplied by the posterior auricular branch of the facial nerve (VII).

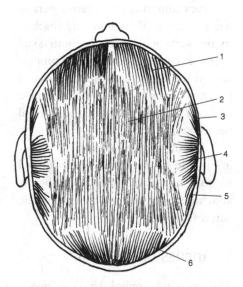

FIG. 1-2 Musculoaponeurotic layer of scalp. (1) Frontal belly of occipitofrontalis; (2) Galea aponeurotica; (3) Auricularis anterior; (4) Auricularis superior; (5) Auricularis posterior; (6) Occipital belly of occipitofrontalis.

Loose subaponeurotic layer consists of loose connective tissue that allows movement of the first three layers and forms a subaponeurotic space. It contains emissary veins that connect the venous sinuses of the skull with the veins of the subcutaneous layer.

The subaponeurotic layer is sometimes called the **"danger space of the scalp"** because blood and pus can accumulate and infections spread easily in it limited only by the attachments of the epicranius and aponeurosis. Spread occurs into the periorbital connective tissue or via emissary veins to the intracranial region and cranial venous sinuses. This layer is the plane of separation in an injury that tears the scalp.

Pericranium is the external periosteum of the skull. It is loosely attached to the surface of the squamous skull bones and has poor osteogenic properties. At the suture lines, it dips between the skull bones as the suture membrane becomes continuous with the internal periosteum of the skull.

A skull fracture may cause bleeding in the subperiosteal plane limited to the individual skull bone by its attachments of the pericranium at the sutures.

2. Arteries of the Scalp

The scalp is supplied by five arteries, three anterior and two posterior to the coronal plane of the ear on each side. The anterior three are branches of the ophthalmic branch of the internal carotid artery; and the posterior two are branches of the external carotid artery. They anastomose freely with one another artery with the arteries of the opposite side, they contribute to the rapid healing of scalp wounds.

Supratrochlear artery is a terminal branch of the ophthalmic artery (from the internal carotid artery). It leaves the upper medial angle of the orbital margin piercing the superior palpebral fascia to

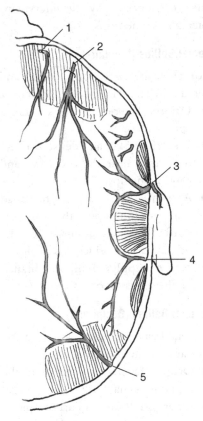

FIG. 1-3 Arteries of scalp. (1) Supratrochlear a; (2) Supraorbital a; (3) Superficial temporal a; (4) Posterior auricular a; (5) Occipital a.

supply the anterior and medial scalp of the forehead. It anastomoses with its opposite artery and the supraorbital artery.

Supraorbital artery is also a branch of the ophthalmic artery (from the internal carotid). It leaves the supraorbital foramen or notch of the orbital margin ascending over the frontal bone to supply the forehead and vault of the skull. It anastomoses with the angular branch of the facial artery and frontal branch of the superficial temporal artery.

Superficial temporal artery is a terminal branch of the external carotid artery. It divides in front of the auricle and supplies the parotid gland, adjacent muscles and glands of the face, anterior part of the external ear and scalp of the frontal and parietal regions. It anastomoses with the artery of the opposite side, frontal and supraorbital arteries.

Posterior auricular artery is a branch of the external carotid artery. It passes under the parotid gland, along the upper border of sternocleidomastoid then divides near the mastoid process and supplies the scalp above and behind the back of the ear. Its important stylomastoid branch supplies the facial nerve, middle ear and mastoid antrum and air cells. It anastomoses with the artery of the opposite side and with the occipital artery.

Occipital artery is a branch of the external carotid artery. It passes along the lower border of sternocleidomastoid, then it passes into the occipital groove deep to the mastoid process of the temporal bone and reaches the scalp of the occiput. It supplies the surrounding muscles, dura of the posterior cranial fossa and skin over the occiput and anastomoses with the opposite occipital artery, posterior auricular and superficial temporal arteries.

3. Veins of the Scalp

Veins of the scalp parallel the arteries. They anastomose with each other, form a network within the subcutaneous layer, and receive emissary veins connecting with intracranial venous sinuses.

Supratrochlear and supraorbital veins drain the forehead, communicate with the superior ophthalmic vein, and unite at the medial angle of the eye to form the angular vein.

The angular vein communicates with the superior ophthalmic vein and then the cavernous sinus.

Superficial temporal veins descends from the temple to the upper part of the parotid gland, where it unites with the maxillary vein to form the retromandibular vein.

Posterior auricular vein forms from a plexus on the side of the head and back of the ear and descends behind the auricle. Below the parotid gland, it joins the posterior division of the retromandibular vein to form the external jugular vein.

Occipital vein drains the posterior parietal and occipital regions of the scalp. It pierces the trapezius, and

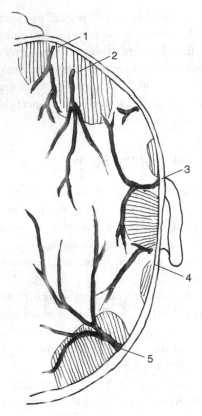

FIG. 1-4 Veins of scalp. (1) Supratrochlear v;
(2) Supraorbital v; (3) Superficial temporal v;
(4) Posterior auticular v; (5) Occipital v.

terminates in the suboccipital plexus
communicating with the vertebral vein.

4. Nerves of the Scalp

Sensory nerves of the scalp are sequen-
tial branches of all three divisions of the
trigeminal nerve and have a distribution
similar to that of the blood vessels that
accompany them. In addition, the area
behind the ears is supplied by the
ventral rami of cervical spinal nerves
arising in the neck from the cervical
plexus (C2 and 3).

Supratrochlear nerve, a branch
of the frontal nerve (of V1), leaves the
orbit at the medial end of the supraorbital margin between the trochlea of
the superior oblique muscle and supraorbital foramen or notch and pierces
the frontal belly of occipitofrontalis to
supply the medial part of the upper
eyelid and forehead.

Supraorbital nerve is the continuation of the frontal nerve (of V1). It
leaves the orbit through the supraorbital
foramen or notch, pierces the frontal
belly of the occipitofrontalis, then it
supplies the forehead scalp to the vertex, upper eyelid and frontal sinus.

Zygomaticotemporal nerve is a
branch of the zygomatic nerve (of V2).
It leaves the orbit through the zygomatico-temporal foramen and supplies
the anterior part of the temple.

Auriculotemporal nerve is a
branch of the mandibular division of
the trigeminal. It (V3) leaves the
infratemporal fossa by winding behind
the capsule of the temporo-mandibular
joint. It then divides on the surface at
the upper end of parotid gland and supplies the side of the scalp, auricle, and
acoustic meatus (outer surface of the
tympanic membrane).

Lesser occipital nerve (C2 ventral ramus from the cervical plexus)
ascends along the posterior border of

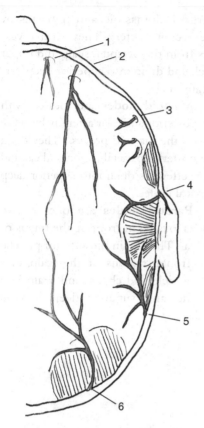

FIG. 1-5 Nerves of scalp. (1) Supratrochlear n;
(2) Supraorbital n; (3) Zygomaticotemporal n;
(4) Auriculotemporal n; (5) Lesser occipital n;
(6) Greater occipital n.

sternocleidomastoid muscle between the ear and occipital artery. It supplies the skin of the head and cranial surface of the auricle.

Greater occipital nerve (C2 medial branch of dorsal ramus) also winds around the posterior border of sternocleidomastoid and ascends with the occipital artery to pierce semispinalis capitus and trapezius. It

supplies the posterior part of scalp, skin over the parotid and both surfaces of the auricle.

Third occipital nerve (C3 medial branch of dorsal ramus) ascends to pierce the trapezius and supply the skin of the back of the head.

5. Lymphatic Drainage of the Scalp

Lymphatic vessels of the head and neck drain into the venous system at the junction of the internal jugular vein and subclavian vein directly from the tissues or indirectly after traversing outlying lymph nodes.

Lymph vessels of the scalp drain into superficial lymph nodes (superficial to the investing layer of deep cervical fascia) consisting of a few small nodes along the external jugular vein in line with the superficial parotid nodes. These drain into the deep cervical nodes (lie deep to the investing layer of deep cervical fascia and sternocldiedomastoid muscle) and have been named according to their location.

In relation to lymphatic drainage of the scalp, some of the deep nodes are located in a collar at the junction of the head and neck on the stem of named arteries ("collar chain" nodes). These are:

Occipital nodes are one to three nodes overlying the superior nuchal line between the attachments of

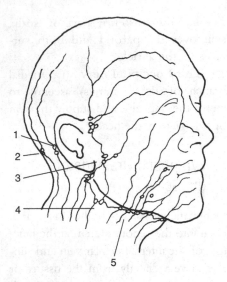

FIG. 1-6 Lymph nodes of head. (1) Retro-auricular; (2) Occipital; (3) Parotid; (4) Superficial cervical; (5) Submandibular.

sternocleidomastoid and trapezius by the occipital artery. They receive vessels from the occipital scalp and upper neck and drain into superior deep cervical nodes.

Mastoid nodes are located with the posterior auricular artery behind the ear on the mastoid process. They drain the posterior parietal region and ear and their efferents drain into superior deep cervical nodes.

Parotid nodes are one or two nodes located in front of the tragus of the ear. They drain the anterior parietal and frontal regions of the scalp, eyelids, auricle and cheek; and drain into parotid and superior deep cervical nodes.

Chapter *2*

Skull

Chapter Outline

The skull is the total bony structure of the cranium and the mandible. It is a composite of 22 bones, some single and some paired. In addition to these bones, an unattached bone in the neck, the hyoid bone is associated with the skull for developmental reasons. In addition, there are three ossicles found in the middle ear region.

1. Introduction

The skull can be divided into an upper part, the cranium, and a lower part, the facial skeleton. **Cranium** is the skull minus the mandible. It forms the walls of the cranial cavity that houses and protects the brain. Facial skeleton supports the eyes and provides paths of entry into the respiratory and digestive systems.

It provides teeth and jaws for mastication and cavities for the special organs of hearing and equilibrium, smell and sight.

Calvaria is the cranium or vault without the facial bones.

Neurocranium is the part of the skull forming a protective case around the brain. It comprises parts that are developed from membrane (flat bones that surround the brain as a

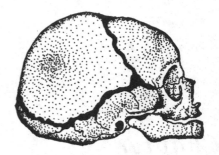

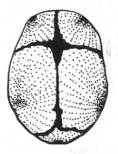

FIG. 2-2   Skull of newborn infant.

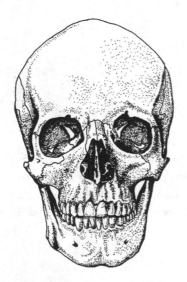

FIG. 2-1   Skull — norma frontalis.

vault) and parts that are developed from cartilage (the base of the skull or chondrocranium).

Chondrocranium is the base of the skull that develops in the cartilage.

Viscerocranium (splanchnocranium) comprises bones of the face formed mainly by the first two pharyngeal arches. It has membranous parts (bones or parts of bones that develop from intramembranous ossification) and cartilaginous parts (bones or parts of bones that develop from intracartilaginous ossification).

Cranial bones have external and internal tables of compact bone separated by a venous sinus containing spongy bone layer, the diplöe. **Pericranium** (periosteum) covers the outer layer and endosteum (endosteal layer of dura mater) the inner layer.

Bones of the skull are connected by immovable joints or sutures. Connective tissue between bones (the sutural ligament) allows bone growth primarily in response to the expansion of the brain. The mandible and auditory ossicles are the only bones joined by movable joints (of the synovial type).

2. Suture Lines

Frontal suture separates the two developing halves of the frontal bone until about six years of age. Metopic suture is a persistent part of the frontal suture in some adults.

Coronal (frontal) **suture** exists between the frontal and parietal bones.

Cruciform suture unites the two palatine processes of the maxillae in front and the horizontal processes of the palatine bones behind.

Table 1. Developmental Origin of the Skull

Part	Origin	Definitive Structure
Viscerocranium (cartilaginous)	Maxillary process of first arch	Maxilla, zygomatic, part of temporal bone
	Mandibular process of first arch	Meckel's cartilage (replaced by membrane bone in mandible)
		Sphenomandibular ligament
Viscerocranium (membranous)	Maxillary process	Incus, malleus
	Mandibular process	
	Second arch	Stapes
Neurocranium (membranous)	Condensed mesenchyme	Frontal bone (squamous)
		Parietal bone (squamous)
		Occipital bone (squamous)
		Temporal bone (squamous)
Neurocranium (cartilaginous)	Parachordal cartilage	Base of occipital
	Occipital sclerotomes (3)	
	Hypophysial cartilages	Body of sphenoid
	Trabeculae cranii	Ethmoid
	Ala orbitalis	Lesser wings of sphenoid
	Ala temporalis	Greater wings of sphenoid
	Periotic capsule	Petrous and mastoid parts of temporal bone

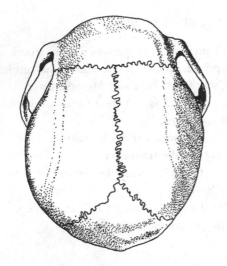

FIG. 2-3 Superior aspect of skull (norma verticalis).

Lambdoid (parieto-occipital) suture is found between the occipital and two parietal bones. In the fetus, it is occupied by the posterior fontanelle that closes soon after birth.

Occipitomastoid suture separates the occipital bone from the mastoid process of the temporal bone. It runs upward to meet the lambdoid suture.

Parietomastoid suture separates the parietal bone from the mastoid process of the temporal bone. It meets the lower end of the lambdoid suture.

Sagittal suture separates the two parietal bones in the median plane.

Squamous suture is found between the parietal bone and squamous part of the temporal bone.

Pterion is the point of junction between the frontal, sphenoid and parietal bones.

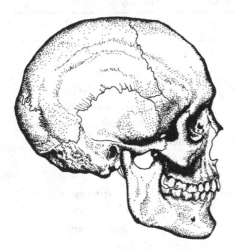

FIG. 2-4 Lateral aspect of skull (norma lateralis).

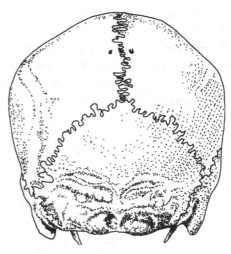

FIG. 2-5 Posterior aspect of skull (norma occipitalis).

Intermaxillary suture is found in the median plane between the two maxillae.

Internasal suture is found in the median plane between the nasal bones.

Beneath pterion is the anterior branch of the middle meningeal artery and the stem of the lateral sulcus of the brain. In the fetus and infant, it is the site of the anterolateral fontanelle.

Bregma is the point of intersection between the sagittal and coronal sutures. In the fetus, it is the site of the fibrous anterior fontanelle (Fr little fountain) that bridges the gap between the margins and angles of ossifying bones. The fontanelle closes during the second postnatal year.

Asterion is the point of junction of the temporal, parietal and occipital bones. It is the site of the posterolateral fontanelle in the fetus and infant.

Lambda is the point of junction of the occipital and two parietal bones. In the fetus and infant, it is the site of a membranous area, the posterior fontanelle.

3. Views of the Skull (External Surface)

The skull can be viewed from various aspects and these are given names as follows:

Norma frontalis: A view of the anterior aspect of the skull.

Norma lateralis: A view of the lateral aspect of the skull.

Norma occipitalis: A view of the posterior aspect of the skull.

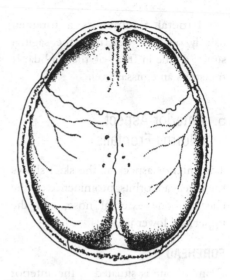

FIG. 2-6 Internal (endocranial) surface of calva.

Norma verticalis: A view of the superior aspect of the skull.

Norma basalis: A view of the base of the skull.

In addition, the calvaria is removed to view the internal aspect of the skull.

4. Superior Aspect (Norma Verticalis)

Viewed from above, the skull is ovoid in shape and shows four bones. The frontal in front, occipital behind, and right and left parietal bones between. It presents the following landmarks.

Vertex is the highest point on the skull lies on the sagittal suture a few centimeters behind bregma.

Parietal eminence is the most convex portion of each parietal bone.

Parietal foramen is a foramen usually on each side near the sagittal suture a little in front of the lambda. It transmits an emissary vein.

5. Anterior Aspect (Norma Frontalis)

The anterior aspect of the skull shows the forehead, orbits, prominence of the cheeks, bony external nose, and the upper and lower jaws.

FOREHEAD

Frontal bone is situated in the anterior part of the cranium that comprises a vertical plate, the squama that forms the skeleton of the forehead and a horizontal plate that forms the greater part of the roof of the orbits.

Frontal eminence is the fullness at the centre of each half of the squama. It is the site where (membranous) ossification of the frontal bone began.

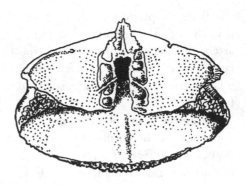

FIG. 2-8   Inferior aspect of frontal bone.

Zygomatic process of the frontal bone is a stout projection of the frontal bone at the lateral end of the supraorbital margin. It articulates with the frontal process of the zygomatic bone and forms the superolateral part of the orbital margin.

Temporal line is a sharp line curving upwards and backwards from the zygomatic process of the frontal bone. It separates the temporal fossa below from the scalp above. The temporal fascia is attached to this line.

Glabella is a smooth depression above the nasion between the supraciliary arches. Its covering skin is smooth and devoid of hair *in vivo* hence its name.

Superciliary arches are the curved ridges extending on each side laterally from the glabella. They are so named because of they lie beneath the eyebrows *in vivo*.

Metopic suture is the remnant of a frontal suture that separated the two halves of the frontal bone until about six years. It may persist into adulthood.

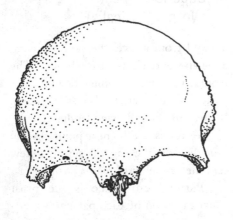

FIG. 2-7   External aspect of frontal bone.

ORBITAL ADITUS

Each orbit is a bony four-sided pyramid housing the eye. Its base, the aperture (aditus) on the front of the skull is slightly constricted which tends to prevent proptosis and protect the eyeball from trauma.

Supraorbital margin the superior margin of the orbit that is formed by the frontal bone. It has a supraorbital notch or foramen medially.

Lateral margin is formed by the zygomatic bone and zygomatic process

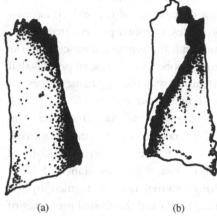

(a)   (b)

FIG. 2-11 Right nasal bone — (a) external surface; and (b) internal surface.

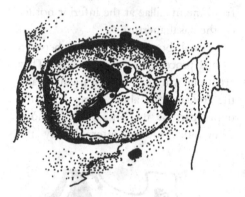

FIG. 2-9 Right orbit — anterior aspect.

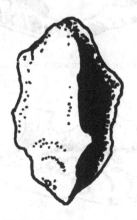

FIG. 2-10 Right lacrimal bone — lateral surface.

of the frontal bone (and greater wing of the sphenoid posteriorly).

Medial margin is not continuous. The upper half of the margin lies more posteriorly and is in line below with the lacrimal crest on the lacrimal bone. The lower half of the margin is formed by the lacrimal crest of the maxilla. The lacrimal sulcus lies between the crests lodges the lacrimal sac.

Inferior margin is formed by the maxilla and the zygomatic bone.

Supraorbital foramen or notch lodges the supraorbital nerve (from the frontal branch of V1), supraorbital artery (from the ophthalmic artery) and an accompanying vein.

PROMINENCE OF THE CHEEK

The prominence of the cheek is formed by the zygomatic bone and the zygomatic process of the maxilla.

Zygomatic bone is located in the lower lateral side of the orbit. It has two processes, a frontal process that articulates with the zygomatic process of the frontal bone and a temporal process that articulates with the zygomatic process of the temporal bone.

Zygomaticofacial foramen on the lateral surface of the zygomatic bone transmits the zygomaticofacial nerve (from zygomatic branch of V2). Bony external nose is formed by the nasal bones and the frontal processes of the maxillae.

Piriform aperture is the anterior opening of the bony nose. The nasal cavity can be seen through the aperture to be divided by the vertical nasal septum. *In vivo*, we can see the vertical plate of the ethmoid, the vomer and septal cartilage. Along each lateral wall of the nasal cavity are curled bony plates, the conchae (turbinates), each separated from the other by a space or meatus.

Anterior nasal spine is a median sharp bony spur that projects forward from the maxillae at the inferior border of the nasal aperture.

Nasal bones are located between frontal processes of the maxillae. They articulate with each other medially and the frontal bone above. They form the upper part of the bridge of the nose. They are grooved internally by the

FIG. 2-12　Lateral aspect of right zygomatic bone.

FIG. 2-14　Right inferior concha — lateral surface.

FIG. 2-13　Medial aspect of right zygomatic bone.

FIG. 2-15　Right inferior concha — nasal surface.

anterior ethmoidal nerves (from V1) that become the external nasal nerve.

Nasion is the point of intersection of the internasal and frontonasal sutures.

ANTERIOR ASPECT OF THE MAXILLA

The two maxillae (and the palatine bones) form the upper jaw. Each maxilla has a **body** (containing a maxillary sinus) from which four processes project — **a frontal process** (which articulates with the frontal bone); **a zygomatic process**; (that articulates with the zygomatic bone); a **palatine process** (that projects medially and forms most of the hard palate); and an **alveolar process** which is a thick ridge that carries the maxillary teeth.

The body of the maxilla presents five surfaces and so contributes to the walls of the orbital cavity, the nasal

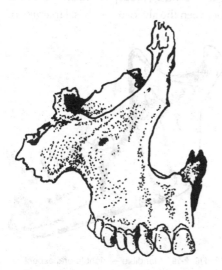

FIG. 2-16 Right maxilla — anterior aspect.

cavity, front wall of the infratemporal fossa and the oral cavity.

Nasal surface or base forms the lateral surface of the nasal cavity and articulates with the middle and inferior conchae.

Orbital surface forms most of the floor of the orbit. It articulates with the medial border of the lacrimal, ethmoid and orbital process of the palatine bone.

Infratemporal surface forms the front wall of the infratemporal fossa.

Anterior surface is covered with facial muscles.

Palatine surface is the palatine process which forms most of the roof of the palate and floor of the nasal cavity.

Canine eminence is a vertical ridge corresponding to the root of the canine teeth on the surface of the alveolar process. This ridge separates the incisive fossa in front from the canine fossa behind. These fossae are landmarks for surgical approaches to the maxillary sinus.

Premaxilla is the part of the maxilla supporting the incisor teeth. It is separated from the palatine process by a suture passing from the incisive fossa between the lateral incisor and canine teeth. It is formed from a medial downgrowth of the medial nasal process that also forms the lower part of the nasal septum and median part of the upper lip.

Infraorbital foramen is the anterior orifice of the infraorbital canal in the canine fossa below the infraorbital

infraorbital nerve after entering the infraorbital fissure in the orbit. It also transmits the infraorbital artery (from the main trunk of the maxillary artery).

ANTERIOR ASPECT OF THE MANDIBLE

Anterior aspect of the mandible comprises a body and pair of rami. Each ramus is surmounted by a condylar process and a coronoid process.

Alveolar part of the mandible is the part of the bone carrying the mandibular teeth.

Mandibular symphysis, in the midline anteriorly, is the region where two halves of the body of the mandible united in development.

Incisor fossa is a depression below the ridges produced by incisor teeth. It gives origin to the mentalis and a part of orbicularis oris muscles.

Mental foramen is located below the second bicuspid tooth midway between the alveolar and basal margins of

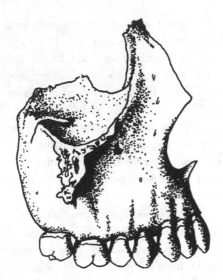

FIG. 2-17   Right maxilla — lateral aspect.

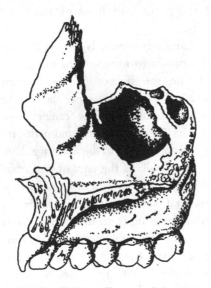

FIG. 2-18   Right maxilla — medial aspect.

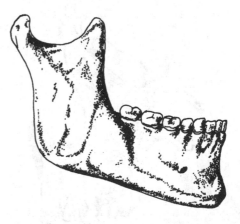

FIG. 2-19   Mandible — right lateral aspect.

margin. It transmits the infraorbital nerve — the terminal branch of the maxillary nerve (V2) that becomes the

the bone. It transmits the mental nerve (terminal branch of V3) and vessels (terminal branches of the maxillary artery).

Mental protuberance is a forward triangular projection of the lower anterior border of the mandible.

6. Lateral Aspect of the Skull (Norma Lateralis)

The lateral aspect of the skull includes some parts of some bones comprising the base of the skull, the lateral aspect of the maxilla, mandible, and the temporal and infratemporal fossae.

FEATURES OF THE LATERAL ASPECT OF THE TEMPORAL BONE

The temporal bone lies at the side and base of the skull and has a **squamous part** above and anteriorly, a **petrous part** medially, a **mastoid part** posteriorly and a **tympanic part** below.

Squamous part of the temporal bone forms the lateral cranial wall (cerebral surface), a large part of the temporal fossa (temporal surface) and a small part of the infratemporal surface. It articulates at the squamosal suture with the parietal bone and in front with the greater wing of the sphenoid.

Zygomatic process of the temporal bone projects forward to articulate with the zygomatic bone. Its lower border gives origin to the masseter muscle.

The complete zygomatic arch (zygoma) is composed of the zygomatic bone, the temporal process of the zygomatic bone and the zygomatic process of the temporal bone.

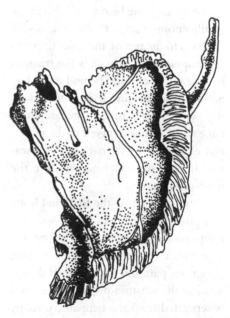

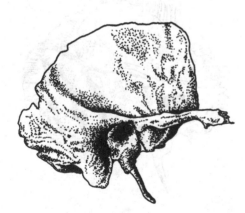

FIG. 2-20 Right temporal bone — lateral aspect.   FIG. 2-21 Right temporal bone — superior aspect.

Mandibular fossa is a depression behind the articular tubercle for articulation with the head of the mandible (through a cartilaginous disc) in mandibular retrusion or when the teeth are in occlusion.

Articular tubercle on the lower border of the zygomatic arch attaches the lateral ligament of the temporomandibular joint. The head of the mandible (through a cartilaginous disc) articulates with the tubercle in mandibular protrusion.

External acoustic meatus (medial or osseous part) is a canal that leads *in vivo* to the tympanic membrane. Located behind the mandibular fossa, it is bounded by the root of the zygomatic arch above, and by the tympanic part of the temporal bone below. (The lateral or cartilaginous part of the meatus is attached to the rim of the osseous part.)

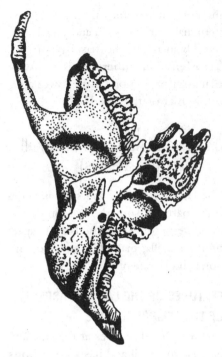

FIG. 2-22 Right temporal bone — inferior aspect.

Supramental triangle is the triangular area just above and behind the external acoustic meatus. It is bounded by the suprameatal crest, posterosuperior edge and a vertical tangent to the posterior edge of the external acoustic meatus. It is a surgical landmark for the mastoid antrum.

Mastoid part of the temporal bone comprises the posterior part of the temporal bone which lies below the suprameatal crest. It is fused to the squamous part and is prolonged downward as the mastoid process. The latter is separated from the tympanic plate by the tympanomastoid suture.

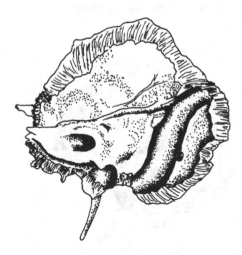

FIG. 2-23 Right temporal bone — medial aspect.

The mastoid process is poorly developed at birth and its growth is gradual during childhood. It may fail to protect the facial nerve leaving the skull through the stylomastoid foramen from damage during a forceps delivery.

Mastoid foramen located posterior to the base of the mastoid region transmits an emissary vein.

Tympanic part of the temporal bone is a curved plate of bone forming the floor and anterior wall of the external acoustic meatus. It is fused with the petrous part medially and separated from the squamous part anteriorly by the squamotympanic fissure and mastoid part posteriorly by the tympanomastoid fissure.

Styloid part of the temporal bone (styloid process) is a downward and forward projection of variable length from beneath the tympanic plate. It is a remnant of the second pharyngeal arch cartilage and connects with the hyoid bone through the stylohyoid ligament.

LATERAL ASPECT OF THE MANDIBLE

Oblique line extends from the mental tubercle to the ramus. It gives origin to the depressors of the lower lip and angle of the mouth.

Lateral surface of the ramus gives attachment to the masseter muscle.

Coronoid process is flat and triangular pointing up and forward. It gives attachment to the temporalis, the attachment extends down on the anterior border of the ramus.

Condyloid process is divided into a head and a neck.

Head is convex from before backwards and less so lateral to medial. It articulates with an articular disc of the temporomandibular joint lying in the articular fossa when the mouth is shut and below the articular eminence when the mouth is open.

Neck is the compressed segment below the head. It gives attachment to the capsule of the joint and the lateral pterygoid muscle is attached to its anterior surface (in the pterygoid fovea).

Inferior border of the mandible gives attachment to the fascia of the neck.

TEMPORAL FOSSA

Temporal fossa is the space on the side of the skull occupied by the temporalis muscle. Its boundaries are as below.

Superior boundary: The temporal line is usually double. Temporal fascia is attached to the superior line and temporalis muscle to the parallel inferior line. The boundary arches backward from the zygomatic process of the frontal bone across the frontal and parietal bones and becomes continuous with the supramastoid crest and posterior root of the zygomatic process.

Inferior boundary: The infratemporal crest, a ridge of bone that extends across the greater wing of the sphenoid, is continuous with a ridge that extends across the squamous part of the temporal bone to the anterior root of the zygomatic process.

Medial wall: The medial wall consists of parts of the frontal and parietal bones, the temporal squama, and the greater wing of the sphenoid bone.

Lateral wall: The lateral wall is formed by the zygomatic arch.

Pterion: The pterion is the junction of four bones in the floor of the temporal fossa, frontal, greater wing of the sphenoid, the parietal and temporal bones. It is the site of the anterolateral fontanelle in the fetus.

Contents: Temporalis muscle.

The pterion overlies the anterior branch of the middle meningeal artery and the point where the lateral sulcus of the cerebrum divides into three rami. The artery lies in a groove on the inner surface and is frequently damaged by blows to the side of the head.

Zygomaticotemporal foramen is in the anterior wall of the temporal fossa. It transmits the zygomaticotemporal branch of V2, a cutaneous nerve that supplies the temple.

INFRATEMPORAL FOSSA

Infratemporal fossa is the irregular region behind the maxilla, below the temporal fossa and deep to the zygomatic arch and ramus of the mandible.

Its boundaries are as below.

Lateral wall: Ramus and coronoid process of the mandible.

Anterior wall: Posterior surface of the maxilla presenting maxillary tuberosity behind the third molar and alveolar foramina for the posterior superior alveolar vessels and nerves (of V2) to the molar and premolar teeth.

Medial wall: Lateral pterygoid plate of the sphenoid (pterygoid process), and tubercle of palatine bone which wedges between the tubercle and the maxilla.

Superior boundary (infratemporal crest of): The greater wing of the sphenoid.

Posterior boundary: Articular eminence laterally and spine of the sphenoid medially.

Contents: Mandibular nerve (V3); chorda tympani nerve (branch of VII); pterygopalatine ganglion; otic ganglion; terminal branches of the maxillary artery; and accompanying veins; pterygoid venous plexus; fat and fibrous tissue; medial and lateral pterygoid muscles; and the otic ganglion.

The fossa communicates with the orbit through the infraorbital fissure; with the pterygopalatine fossa through the pterygomaxillary fissure; and with the temporal fossa directly under the zygomatic arch.

LATERAL ASPECT OF MAXILLA

The convex lateral surface of the maxilla has the following boundaries:

Anterior: The origin of the zygomatic process of the maxilla and a ridge that descends from the zygomatic process toward the roots of the first molar tooth.

Above: The infraorbital fissure is a horizontal cleft through which the infratemporal fossa communicates with the orbit. It transmits the infraorbital artery and vein, zygomatic nerve, orbital branches of the pterygopalatine ganglion and veins from the orbit to the pterygoid plexus.

Behind: Pterygopalatine fossa.

Posteroinferior: The maxillary tuberosity which is a rounded eminence at the end of the alveolar process. It gives attachment to the upper head of the lateral pterygoid muscle.

Posterior superior alveolar foramina (usually two): On the posterior surface of the maxilla. They transmit the posterior superior alveolar vessels (from the maxillary artery) and nerves (from V2) that supply the upper teeth and maxillary sinuses.

Alveolar process is an arch of spongy bone surrounding the roots of teeth and forming the sockets (alveoli).

Canine eminence is a bony elevation over the root of the canine tooth.

Canine fossa is the depression behind the canine eminence. Levator anguli oris is attached in the fossa.

The pterygomaxillary fissure is a triangular interval formed by the divergence of the maxilla from the pterygoid process of the sphenoid behind. It leads to the pterygopalatine fossa and transmits the terminal part of the maxillary artery as it leaves the infratemporal fossa.

LATERAL ASPECT OF LOWER JAW (MANDIBLE)

Body (or horizontal ramus): The tooth bearing horizontal segment of the mandible is joined with its opposite at the symphysis menti.

Vertical rami: The posterior segment (paired) extend upward from the mandibular angle. They end posteriorly in the condylar process which articulates with the temporal bone and anteriorly in the coronoid process which gives insertion to the temporalis muscle.

Alveolar part is spongy bone of the horizontal ramus that carries the mandibular teeth.

Mental foramen lies between the roots of the first and second bicuspid teeth or below the second bicuspid tooth. It transmits the mental nerve (from V3) and vessels.

Oblique line is a bony ridge that passes upwards from below the mental foramen becoming continuous with the anterior border of the ramus. Buccinator and depressor anguli oris are attached to the posterior part of the line.

Coronoid process is a flattened upward triangular projection of the anterior vertical ramus. It gives attachment to the temporalis muscle.

Mandibular notch lies between the coronoid and condylar processes.

It transmits the masseteric nerve (from V3) and vessels (from the maxillary artery) passing laterally to reach the masseter muscle.

Condyloid process is divided into head and neck. The head is enlarged above in a transverse direction (that articulates with the mandibular fossa and articular tubercle of the temporal bone). Below the condyloid process is a narrower part, the neck. The lateral pterygoid muscle is attached to the pterygoid fovea on the anteromedial part of the neck.

7. Posterior Aspect of the Skull (Norma Occipitalis)

The back of the skull is convex and consists of the parietal bones above, occipital squama below, and mastoid temporal bones on each side.

External occipital protuberance is a projection near the center of the squamous part of the occipital bone. It gives attachment to trapezius muscle and to the ligamentum nuchae. Its centre is known as *inion*.

External occipital crest is a crest of bone extending from the external occipital protuberance towards the foramen magnum. It gives attachment to the ligamentum nuchae.

Superior nuchal line is a curved ridge extending laterally from the external occipital protuberance. It gives attachment to trapezius, sternocleidomastoid, splenius and occipital belly of occipitofrontalis muscles.

Highest nuchal line is a ridge above the superior nuchal line. It gives attachment to epicranial aponeurosis.

Lambda is the point of intersection of the sagittal and lambdoid sutures.

Asterion is the point of intersection of the parietomastoid and occipitomastoid sutures.

Mastoid foramen is an opening on each side near the occipitomastoid suture. It transmits an emissary vein.

8. Inferior Aspect

This aspect can be described in areas separated by transverse lines.

Anterior transverse line passes through the articular tubercle, foramen ovale, auditory tube, foramen lacerum and spheno-occipital synchondrosis.

Posterior transverse line passes through the stylomastoid foramen, between the styloid and mastoid processes, posterior margin of the jugular foramen, hypoglossal canal and junction of the anterior third and posterior two thirds of the occipital condyle.

Anteromedial area comprises the following bones: maxillae, palatine bones, body and pterygoid processes of sphenoid, and the vomer. The palatine bone has a perpendicular and a horizontal plate joined at a right angle.

Bony (hard) palate forms the roof of the mouth (within the dental arches) and the floor of the nasal cavity. It comprises the palatine process of the maxillae in front and the horizontal plates of the

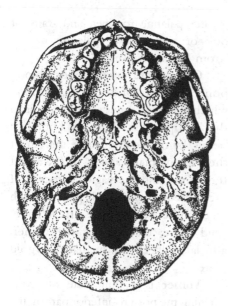

FIG. 2-24 Inferior surface of base of skull.

FIG. 2-26 Right palatine bone — lateral aspect.

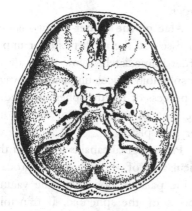

FIG. 2-25 Internal (endocranial) surface of cranial base.

FIG. 2-27 Right palatine bone — posterior aspect.

palatine bones behind. Anterior and lateral boundaries of the bony palate are the alveolar processes that support the teeth and the maxillary tuberosity behind. The posterior margin of the bony palate is free and gives attachment to the aponeurosis of the soft palate.

Median palatal suture is found in the midline between the palatine processes of the maxillae in front and horizontal plates of the palatine bones behind.

Transverse palatine suture is between the palatine processes of the

FIG. 2-28 Right palatine bone — medial aspect.

maxillae and horizontal plates of the palatine bones.

Posterior nasal spine projects from the middle of the free posterior border of the bony palate. It gives attachment to musculus uvulae.

Incisive fossa is a depression behind the incisor teeth. Four incisive foraminae open into the fossa transmitting the long sphenopalatine nerves (from V2) and terminal branches of the greater palatine vessels from the nose.

Greater palatine forarninae are located at the posterolateral angle of the horizontal plate of the palatine bone. The greater palatine nerves (from V2) and vessels run forwards from the foraminae in the grooves near the alveolar process.

Lesser palatine forarninae may be double. They are located behind each greater palatine foramen and transmit one or more lesser palatine nerves (from V2) and vessels.

Choana are the posterior oblong bony apertures of the nose, through which the nasal cavities continue into the nasopharynx. Each is bounded; below, by the posterior free margin of the palatine plate; above, by the ala of the vomer and the vaginal process of each medial pterygoid plate, which meet and cover the base of the sphenoid bone; medially, by the vomer; and laterally, by the medial pterygoid plate.

Vomer is a thin plate of bone forming the postero-inferior part of the nasal septum. It has a free posterior border.

Alae of the vomer are two lateral processes of the most posterior upper border immediately below the body of the sphenoid articulating with the vaginal process of the the sphenoid (medial extension of upper end of the medial pterygoid plate).

Palatovaginal canal is formed at the articulation of the sphenoidal process of the palatine bone with the vaginal process of the sphenoid. It transmits the pharyngeal nerve and artery from

FIG. 2-29 Vomer (right surface).

the pterygopalatine ganglion backwards to the pharynx.

Sphenoid bone lies in the base and sides of the skull in front of the occipital and temporal bones and behind the frontal and ethmoid bones. It is composed of a body which enclose a pair of large sinuses and attaches three pairs of processes, greater wings, lesser wings, and pterygoid processes. The anterior and inferior surfaces of the body of the sphenoid form part of the roof of the nose and pharynx, respectively. In the adult, the posterior end of the body is fused with the basilar part of the occipital bone.

Greater wings extend sidewards from the lateral surfaces of the body. They present cerebral, orbital, temporal, and infratemporal surfaces.

Lesser wings extends sidewards from the upper and front parts of the body by two roots between which is the optic canal. The anterior margin articulates with the orbital plate of the frontal bone but the posterior border is free ending medially as the anterior clinoid process.

Pterygoid processes are wing-like vertical structures on either side of the posterior nasal aperture. They

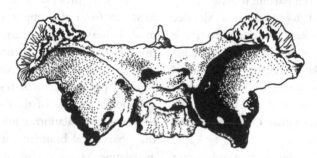

FIG. 2-30 Superior aspect of sphenoid bone.

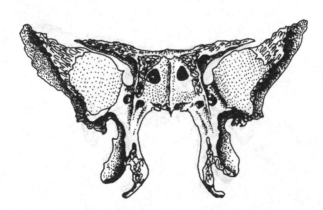

FIG. 2-31 Anterior aspect of sphenoid bone.

consist of narrow medial plates and broad lateral plates. The plates are fused anteriorly at their upper parts enclosing the pterygoid fossa and are free behind and below. Lateral pterygoid muscles are attached to the lateral surfaces of the lateral pterygoid plate and medial pterygoid muscles are attached to the medial surface of the medial pterygoid plate.

Pterygoid fossa is a wedge shaped area enclosed by the fused parts of the lateral and medial pterygoid plates. Above it is a small depression, the scaphoid fossa which gives attachment to the tensor veli palatini muscle.

Pterygoid hamulus is a slender hook-like process projecting from the lower posterior border of the medial pterygoid plate. Tensor veli palatini muscle hooks around the hamulus passing from its origin at the base of the medial pterygoid plate to insert into the palatine aponeurosis.

Pterygoid tubercle is a projection from the upper end of the posterior border of the medial plate. It covers the posterior opening of the pterygoid canal which transmits the nerve of the pterygoid canal to the pterygopalatine fossa.

Pterygoid canal is located above and lateral to the pterygoid tubercle. It traverses the root of the pterygoid process to reach the foramen lacerum. It transmits the vessels and nerve of the pterygoid canal (formed by the union of the **great petrosal nerve** carrying secretomotor fibres to the lacrimal and nasal glands) and the **deep petrosal nerve** (from sympathetic plexus around internal carotid artery).

Vaginal process is an extention medially from the upper end of the medial pterygoid plate inferior to the body of the sphenoid to articulate with the ala of the vomer. Anteriorly, the vaginal process articulates with the sphenoidal process of the palatine bone to form the palatinovaginal canal (for the pharyngeal branches of the pterygopalatine ganglion and of the maxillary artery).

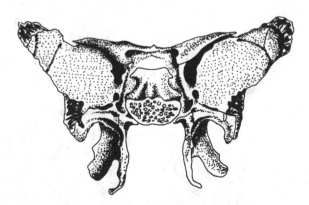

FIG. 2-32   Posterior aspect of sphenoid bone.

Palatinovaginal canal is located on the under surface of the vaginal process. It transmits a pharyngeal ramus of the pterygopalatine ganglion and pharyngeal branch of the third part of the maxillary artery.

ANTEROLATERAL AREA

Foramen ovale is located on the infratemporal surface of the greater wing of the sphenoid at the base of the lateral pterygoid plate posterior to foramen rotundum and medial to the spine of the sphenoid. It transmits the mandibular nerve (V3), accessory meningeal artery (from the maxillary artery) and lesser petrosal nerve (from the tympanic branch of the glossopharyngeal nerve).

Foramen spinosum is smaller than the foramen ovale and posterolateral to it, it is immediately in front of the spine of the sphenoid in the greater wing of the sphenoid. It transmits the middle meningeal artery and vein and a recurrent branch of V3.

Spine of the sphenoid is a sharp spur of bone projecting down in front of the tympanic temporal bone medial to the root of the styloid. It gives attachment to the sphenomandibular and pterygospinous ligaments and tensor veli palatini.

The auriculotemporal nerve (secretomotor to the parotid gland) is lateral to the spine of the sphenoid while the chorda tympani nerve (secretomotor to the other. Large salivary glands and gustatory nerves from the anterior two thirds of the tongue) is medial to it. Fracture of the sphenoid spine may therefore interrupt the appreciation of taste and saliva production.

Inferior orbital fissure lies between the sphenoid bone and the maxilla. It leads anterolaterally from the infratemporal fossa to the orbit and medially to the pterygopalatine fossa. It transmits the maxillary nerve (V2) that becomes the infraorbital nerve, infraorbital vessels, zygomatic nerve, rami from the pterygopalatine ganglion and veins connecting orbital veins with the pterygoid plexus.

POSTEROMEDIAL AREA

Occipital bone placed at the back and base of the cranium, develops in four parts around the foramen magnum: the basilar part anteriorly (fused with the sphenoid), the squamous part posteriorly, and two lateral or condylar parts.

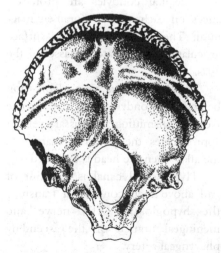

FIG. 2-33 Occipital bone — internal aspect.

Separate at birth, all are fused by five years of age.

The **basilar part** is a bar of bone that, in the adult, joins the occipital bone with the sphenoid. Near its center externally is an elevation, the pharyngeal tubercle for attachment of the pharyngeal raphe, a median seam between right and left pharyngeal muscles (the pharyngeal constrictors) and suspending the pharynx from the skull.

The squamous part is a large curved plate of bone forming the posterior part of the vault of the skull and only the middle of the posterior edge of foramen magnum. On each side, the upper margin articulates with the parietal bone and the lower part with the mastoid part of the temporal bone.

The condylar part forms most of the edges of foramen magnum and in the adult consists of the occipital condyles and jugular process.

Occipital condyles are protuberances on each side of foramen magnum. Their convex articular surface articulates with the lateral masses of the atlas.

Condylar fossa is a depression behind each condyle which may become a canal transmitting an emissary vein. It approximates the posterior margin of the atlas when the head is extended.

Hypoglossal canal is in front of and above each condyle. It transmits the hypoglossal (XII) nerve and meningeal branch of the ascending pharyngeal artery.

Jugular process is a shelf of bone that extends laterally from each condyle to articulate with the petrous part of the temporal bone.

Jugular notch is the anterior hollowed-out part of the jugular process. It contributes to the posterior boundary of the jugular foramen.

Jugular foramen is located between the petrous part of the temporal bone and occipital bone. Posteriorly, it accommodates the jugular bulb of the internal jugular vein and anteriorly the inferior petrosal sinus. Between these are the glossopharyngeal nerve (IX), vagus nerve (X), and accessory nerve (XI).

The posterolateral area consists mostly of the parts of the temporal bone.

The temporal bone consists of five parts that can be seen from outside the skull. The squamous part, mainly on the lateral aspect of the skull, extends into the skull base. The styloid process is derived from the second (hyoid pharyngeal arch) and projects downwards and forwards from the lower surface. The tympanic part is a curved plate fused behind with the mastoid and petrous parts. It forms a sheath of the styloid process. The mastoid part, the most posterior part is a blunt downward projection fused with the petrous part a pyramidal shaped part that contains the internal and middle ear. It extends forward and medially between the occipital bone and sphenoid bone.

Squamous part has temporal and small infratemporal surfaces that can be seen from outside the skull. The inferior surface articulates with the greater wing of the sphenoid bone and upper part overlaps and articulates with the parietal bone.

Zygomatic process extends forward from the squamous part to articulate with the temporal process of the zygomatic bone completing the zygomatic arch.

Mandibular fossa is for articulation with the head of the mandible (with a disc intervening) when the teeth are in occlusion (jaw retruded).

Articular eminence is a smooth convex surface in front of the mandibular fossa which continues into the anterior root of the zygomatic process. In protrusion of the mandible, the head of the mandible and disc move onto the eminence.

Styloid process is a slender projection which is directed downward and forward between the mastoid process and external acoustic meatus. A derivative of the second pharyngeal arch cartilage, the stylohyoid ligament also a derivative, projects from its apex and suspends the hyoid bone. Stylopharyngeus is attached to the base of the styloid process medially; and attached to the stylohyoid posteriorly at approximately its middle; and styloglossus externally near its tip.

Base of the petrous part is shaped like a three-sided pyramid (fused with the mastoid bone). Its apex is directed anteromedially. The petrous part is the hardest bone in the body.

The inferior surface presents the following features:

Jugular foramen is between the petrous part of the temporal bone and the jugular notch of the occipital bone. It transmits the inferior petrosal sinus medially; the sigmoid sinus (becoming the internal jugular vein) laterally; cranial nerves IX, X and XI between the two sinuses; the posterior meningeal artery; a branch of the ascending pharyngeal artery; and an occasional meningeal branch of the occipital artery.

Carotid canal is a large canal in the petrous temporal bone posterior to the spine of the sphenoid and in front of the jugular foramen. It transmits the internal carotid artery and associated carotid sympathetic nerve plexus.

Tympanic canaliculus is a small passage opening on the ridge of bone between the carotid canal and the jugular foramen. It transmits the tympanic branch of the glossopharyngeal (IX) nerve to the tympanic cavity.

Quadrate area is an area of bone between the carotid canal and foramen lacerum. It gives origin to the levator veli palatini muscle. The groove between the quadrate area medially and the sphenoid laterally is filled by the cartilaginous part of the auditory tube.

Foramen lacerum is an irregular opening between the apex of the petrous temporal bone, greater wing of the sphenoid and basilar part of the occipital bone. The gap is closed by fibrocartilage in the living subject. The pterygoid canal opens into the anterior wall of foramen lacerum above the cartilage and several meningeal arteries and emissary veins traverse the foramen.

Groove for the auditory tube *in vivo* is the auditory tube that connects the pharynx with the middle ear (tympanic cavity) and consists of cartilaginous as well as bony parts. The medial cartilaginous part runs in the groove of the sphenopetrosal fissure between the petrous temporal bone and greater wing of the sphenoid and anteromedially from the pterygoid tubercle to the spine of the sphenoid.

Tympanic part is a curved ring like plate of bone that forms the anterior, inferior and posterior walls and the floor of the external acoustic meatus. It extends down the anterior aspect of the styloid process forming a sheath for the process.

Squamotympanic fissure separates the upper border of the tympanic plate from the squamous part of the temporal bone. It extends from the anterior margin of the external acoustic meatus, passes behind the mandibular fossa to the groove for the auditory tube.

Tegmen tympani is a downward projecting sliver of bone that belongs to the petrous temporal bone. It divides the squamotympanic fissure into two parts:

Petrosquamous fissure the anterior part that transmits the tympanic branch of the middle meningeal artery (a branch of the maxillary artery).

Petrotympanic fissure the posterior part that transmits the chorda tympani nerve (branch of the facial (VII) nerve), and anterior tympanic artery (branch of the maxillary artery).

Mastoid part lies behind the external acoustic meatus and forms the posterior part of the temporal bone.

Mastoid process is a downward bony projection immediately posterior to the external auditory meatus. It develops after birth and is filled in the adult with mastoid air cells that communicate with the middle ear through the mastoid antrum. The lateral side the mastoid process gives origin to several muscles that move the neck, sternocleidomastoid, splenius capitus and the longissimus capitis, mastoid notch. The medial side of the mastoid process gives origin to the posterior belly of the digastric.

Occipital groove located medial to the mastoid notch is occupied by the occipital artery (a branch of the external carotid artery).

Mastoid canaliculus is a small passage that opens on the lateral wall of the jugular fossa. It transmits the auricular branch of the vagus (X) nerve to the tympanomastoid fissure.

Mastoid foramen lies in the groove for the sigmoid sinus. It transmits a mastoid emissary vein to an occipital vein.

Tympanomastoid fissure on the side of the skull separates the front of the mastoid process from the tympanic plate. It transmits the auricular branch of the vagus (X) nerve.

Stylomastoid foramen is located between the styloid and mastoid processes. It transmits the facial nerve (VII) and stylomastoid artery.

9. Floor of Cranial Cavity

The cranial surface of the base of the skull is divided into three terraces, the *anterior*, *middle*, and *posterior cranial fossae* by bony ridges at the posterior border of the lesser wing of the sphenoid (the "sphenoidal ridge") in front and the superior border of the petrous temporal bone (the "petrous ridge") behind.

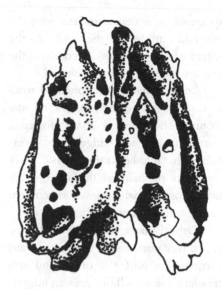

FIG. 2-34   Ethmoid bone — superior surface.

ANTERIOR CRANIAL FOSSA

The floor of the anterior cranial fossa is formed by portions of the frontal bone, cribriform plate, crista galli of the ethmoid and the lesser wings and anterior part of the body of the sphenoid bone (the jugum).

Frontal crest is a midline ridge on the cerebral surface of the frontal bone. It gives attachment to the falx cerebri.

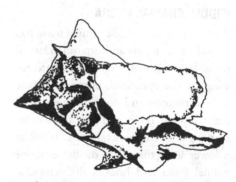

FIG. 2-35   Ethmoid bone — right lateral surface.

Crista galli is the median process of the ethmoid bone behind the foramen caecum that, with the frontal crest, gives attachment to the falx cerebri.

Foramen cecum is a blind pit immediately behind the crista galli. It may transmit a vein from the nasal mucosa to the superior sagittal sinus.

Cribriform plates of the ethmoid at each side of the crista galli form part of the roof of the nose and are perforated by small apertures which transmit filaments of the olfactory nerves (I) from the nasal mucosa to the olfactory bulbs.

Sphenoidal jugum is the anterior upper surface of the body of the sphenoid bone that articulates with the ethmoid and forms the roof of the sphenoidal air sinus(es). It connects the two lesser wings of the sphenoid.

Lesser wings of the sphenoid articulate with the orbital plates of the frontal. They have a sharp posterior border (sphenoidal ridge), that overhangs the middle cranial fossa. They terminate medially in the anterior clinoid processes which gives attachment to the tentorium cerebelli.

Orbital plates of the frontal bone are convex portions of the frontal bone forming the roof the orbits and ethmoidal air sinuses. They present impressions of the cerebral gyri and sulci.

MIDDLE CRANIAL FOSSA

The floor of the middle cranial fossa has a small median part and expanded lateral parts. It consists of the body and greater wing of the sphenoid, anterior surface of the petrous and squamous parts of the temporal bone and the anteroinferior angle of the parietal bone. It lies at a lower level than that of the anterior cranial fossa and houses the temporal lobes of the cerebral hemispheres laterally and the hypophysis medially.

MEDIAN PART OF MIDDLE CRANIAL FOSSA

Body of the sphenoid is cuboidal in shape and houses the sphenoidal air sinus(es). On its superior surface, it lodges the hypophysis. Its anterior surface contributes to the roof of the nose, posterior surface is attached to the

occipital bone and its inferior surface contributes to the roof of the pharynx.

Hypophyseal fossa is located on the upper surface of the body of the sphenoid bone immediately behind the chiasmatic groove. It is bounded in front by the tuberculum sellae, posterolateral to which are small tubercles, the middle clinoid processes. Behind by an upright plate of bone, the dorsum sellae, which has the posterior clinoid processes at its upper angles. The sella turcica (turkish saddle) includes the hypophyseal fossa, tuberculum in front and dorsum sellae behind.

Chiasmatic sulcus is a shallow, transverse, furrow parallel to the sphenoidal limbus and just behind and above which lies the optic chiasma. On each side, this sulcus leads to the optic canal that transmits the **optic nerve** (II) and **ophthalmic artery** (a branch of the internal carotid artery).

Carotid sulcus is located on the side of the body of the sphenoid bone, lateral to the hypophysial fossa. It begins at the foramen lacerum, runs upward then forward then upward again medial to the anterior clinoid process. It lodges the **internal carotid artery** and **carotid sympathetic plexus**.

LATERAL PART OF THE MIDDLE CRANIAL FOSSA

Superior orbital fissure is located anteriorly is the slit between the body medially, greater wing below and lesser wings of the sphenoid bone above leading into the orbit. Transmits the oculomotor (III), trochlear (IV), abducent (VI) nerves and branches of the ophthalmic division of the trigeminal (V) nerve, superior and inferior ophthalmic veins.

Foramen rotundum is located below and behind the medial end of the superior orbital fissure. It transmits the maxillary nerve (V2) from the trigeminal ganglion to the pterygopalatine fossa.

Foramen ovale is a large oval shaped foramen located posterolateral to the foramen rotundum. It transmits the mandibular nerve (V3) **accessory meningeal artery** (branch of the maxillary artery), **venous plexus of foramen ovale** and the **lesser petrosal nerve** (a branch of the glossopharyngeal nerve).

Foramen spinosum is a small foramen posterolateral to the foramen ovale. It transmits **middle meningeal vessels** (branch of the maxillary artery) and the **meningeal branch of the mandibular (V3) nerve**. A groove for the vessels extends laterally and forward on the inner surface of the skull before they divide into anterior and posterior grooves.

Foramen lacerum is a jagged opening that extends from the exterior base of the skull to the middle cranial fossa. It is located between the medial part of the greater wing of the sphenoid, the petrous temporal bone posteriorly and laterally to the base of dorsum sellae. *In vivo*, it is closed by fibrous tissue and related below to the cartilaginous part of the auditory tube. The carotid canal opens into the foramen and the internal carotid artery crosses the foramen to reach the carotid groove on the sphenoid. The internal carotid lies above the cartilage normally bridging the foramen. The lateral wall of foramen lacerum is formed by a spur of the sphenoid, the lingula. The foramen contains the small **deep petrosal** and **greater petrosal nerves** which combine to form the nerve of the pterygoid canal.

ANTERIOR SURFACE OF THE PETROUS TEMPORAL BONE

Trigeminal impression is a shallow concavity near the apex of the petrous temporal bone. The trigeminal ganglion lies in the impression extending forward over the lateral part of foramen lacerum and posteroinferior in vivo to the cavernous sinus.

Arcuate eminence is a rounded elevation on the anterior surface of the petrous temporal bone lateral to the sulci for the petrosal nerves. It indicates the position of the underlying anterior semicircular canal of the internal ear.

Sulcus for the greater petrosal nerve lies in front of the arcuate eminence. It is a hiatus which leads into a groove running toward the foramen lacerum. It houses the **greater petrosal nerve** which carries secretomotor fibres for the lacrimal and nasal mucous glands from nervus intermedius (VII).

Sulcus for the lesser petrosal nerve lies lateral to the groove for the greater petrosal nerve. It runs from the hiatus

Table 2.   Content of Opening in Skull

Opening in Skull	Content of Opening
Anterior ethmoidal foramen	Anterior ethmoidal artery, vein, and nerve (a branch of nasociliary nerve)
Aqueduct of vestibule	Endolymphatic duct
Carotid canal	Internal carotid artery and carotid plexus
Cochlear canaliculus	Aqueduct of the cochlea (perilymphatic duct)
Condylar canal (inconstant)	Emissary vein
Foramen lacerum	Internal carotid artery, carotid plexus, nerve of pterygoid canal
Foramen lacrimale (inconstant)	Anastomosis between middle meningeal and lacrimal arteries
Foramen ovale	Mandibular nerve (V3), accesaory meningeal artery, lesser petrosal nerve
Foramen rotundum	Maxillary nerve (V2)
Foramen spinosum	Meningeal branch of V3, middle meningeal artery and vein
Greater palatine canal	Greater palatine artery, vein, and nerve
Hiatus for greater petrosal	Greater petrosal nerve, petrosal branch of nerve middle meningeal artery
Hypoglossal canal	Hypoglossal nerve, meningeal branch of ascending pharyngeal artery
Incisive foramina and canal	Nasopalatine nerve
Infraorbital foramen	Infraorbital nerve (continuation of maxillary nerve (V2), artery and vein, zygomatic nerve, orbital branch from pterygopalatine ganglion, veins from orbit to pterygoid plexus
Internal acoustic meatus	Facial nerve (VII), vestibulocochlear nerve (VIII), and labyrinthine artery and vein
Jugular foramen	Internal jugular vein, glossopharyngeal nerve (IX), vagus nerve (X), accessory nerve (XI), inferior petrosal sinus, sigmoid sinus
Lesser palatine foramina	Lesser palatine nerves
Mandibular foramen	Inferior alveolar artery, vein, and nerve (from V3)
Mastoid canaliculus	Auricular branch of vagus nerve
Mastoid foramen	Emissary vein
Optic canal	Optic nerve (II), ophthalmic artery, sympathetic fibers of internal carotid plexus

(Continued)

Table 2. (*Continued*)

Opening in Skull	Content of Opening
Palatinovaginal canal	Pharyngeal branch of pterygopalatine ganglion
Petrosquamous fissure	Tympanic branch of middle meningeal artery
Petrotympanic fissure	Chorda tympani nerve, anterior tympanic, Tympanic branch of maxillary artery
Posterior ethmoidal foramen	Posterior ethmoidal artery vein and nerve.
Posterior superior alveolar	Posterior superior alveolar artery, vein, and nerve (V2) foramen
Pterygoid canal	Artery, vein, and nerve of pterygoid canal
Pterygomaxillary fissure	Maxillary artery and vein
Tympanosquamous fissure	Part of tegmen tympani
Stylomastoid foramen	Facial nerve (VII), stylomastoid branches of posterior auricular artery
Superior orbital fissure	Oculomotor nerve (III), trochlear nerve (IV), abducent nerve (VI), ophthalmic nerve (V1), frontal nerve, Iacrimal nerve, sympathetic plexus from internal carotid plexus, ophthalmic vein
Supraorbital foramen	Supraorbital artery, vein, and nerve.
Tympanic canaliculus	Tympanic branch of glossopharyngeal nerve
Tympanomastoid fissure	Auricular branch of vagus nerve
Vestibular aqueduct	Endolymphatic duct
Zygomaticofacial foramen	Zygomaticofacial artery, vein, and nerve
Zygomaticoorbital foramen	Zygomatic nerve (V2)
Zygomaticotemporal foramen	Zygomaticotemporal nerve

for the lesser petrosal nerve to foramen ovale. It houses the lesser petrosal nerve which carries secretomotor fibres toward the otic ganglion from the glossopharyngeal (IX) nerve.

Tegmen tympani is the smooth thin bone between the sulci for the two petrosal nerves posterolateral to the arcuate eminence. It forms the roof of the tympanic antrum and cavity, mastoid antrum and auditory tubes. The anterior part turns downwards into the squamotympanic fissure (observed from the base of the skull).

Sulcus for the superior petrosal sinus grooves the upper margin of the petrous temporal bone lateral to the notch for the **trigeminal nerve**.

POSTERIOR CRANIAL FOSSA

The posterior cranial fossa houses the cerebellum, pons and medulla. The floor of the fossa consists of the dorsum sellae, clivus of the sphenoid bone and basilar part of the occipital bone.

Clivus is the sloping inner surface of the basilar part of the occipital bone in front of the foramen magnum. It is closely related to the pons and medulla.

Foramen magnum is bounded by the basilar part of the occipital in front, and lateral and squamous parts behind and through which the posterior cranial fossa is continuous with the vertebral canal (brain continuous with the spinal cord). It also transmits the spinal roots of tha accessory nerve (XI), the two vertebral arteries and sympathetic plexuses, apical ligament of the dens, and tectorial membrane (continuation of the posterior longitudinal ligament).

POSTERIOR SURFACE OF THE PETROUS TEMPORAL BONE

Internal acoustic meatus (porus) opens directly medial to the external acoustic meatus. It transmits the facial (VII), vestibulocochlear (VIII) nerves and labyrinthine artery (branch of the basilar artery) and vein.

Aperture of the aqueduct of the vestibule is a depression or slit located behind and lateral to the internal acoustic meatus. It transmits the **endolymphatic duct** from the internal ear.

Sulcus for the superior petrosal sinus runs along the upper border of the petrous temporal bone. The superior petrosal sinus reaches the transverse sinus where the latter bends into

the sigmoid sinus. Tentorium cerebelli is attached to its margins.

Internal occipital crest is a median ridge behind foramen magnum leading upward from the foramen magnum to the internal occipital protuberance. It gives attachment to a median fold of dura, the **falx cerebelli**. On each side of the crest is a **cerebellar fossa**.

Sulcus for the transverse sinus is occupied in vivo by the transverse sinus, it extends laterally on each side from the internal occipital protuberance turning downward to become the sulcus for the sigmoid sinus.

The transverse sinus becomes the sigmoid sinus. The corresponding sulcus for the transverse sinus turns medially on the mastoid temporal bone becoming the sulcus for the sigmoid sinus, which then runs forward and medial on the jugular process of the occipital to the lateral end of the jugular foramen (where the sigmoid sinus becomes the **internal jugular vein**).

Groove for the sigmoid sinus is a continuation of the groove for the transverse sinus at the posteroinferior angle of the parietal bone. The mastoid foramen transmitting an emissary vein opens into the groove.

Confluence of the sinuses is the region where the superior sagittal and straight sinuses end in a common pool and the right and left transverse sinuses begin. Symmetry is variable.

Hypoglossal canal opens on each side above the anterior margin of the foramen magnum. It transmits the **hypoglossal nerve** (XII).

Jugular tubercle is an ovoid elevation of the occipital bone above the hypoglossal canal between the jugular foramen and foramen magnum. It may be grooved posteriorly by cranial nerves IX, X, and XI.

10. Roof of Cranial Cavity

The bones of the roof of the skull (or vault) have outer and inner plates or laminae of compact bone with a layer of spongy bone (the diploë) in between (except in bones covered with muscle). The diploë houses an extensive anastomotic network of diploic veins that connect the external and internal (dural) vessels.

The inner aspect of the calvaria exhibits the same sutures as those described on its external surface. It also presents digital impressions, sagittal sulcus, parietal foramen and vascular grooves.

Digital impressions corresponds to convolutions of the orbital gyri and sulci.

Sagittal sulcus runs in the midline from the frontal crest and crista galli to the internal occipital protuberance. It lodges the superior sagittal sinus. It becomes wider as it extends from the frontal crest along the sagittal suture (as blood draining into the sinus increases from anterior to posterior) to the internal occipital protuberance, where it usually joins the sulcus for the right transverse sinus. A number of depressions (granular pits) along this sinus lodge arachnoid granulations, enlargements of arachnoid villi concerned with the return of cerebrospinal fluid to the venous system.

Parietal foramen found on both side of the groove for the sagittal sinus transmit an emissary vein.

Vascular grooves are most evident on the surface of the parietal bones. They carry branches the meningeal vessels.

Chapter 3

Contents of the Cranial Cavity

Chapter Outline

1. Meninges and Venous Sinuses
2. Cranial Nerves
3. Arteries of the Brain

The cranial cavity contains the brain, meninges, blood vessels and parts of the cranial nerves. The roof is the skull cap and the floor is the upper surface of the skull base.

1. Meninges and Venous Sinuses

The central nervous system is enclosed in three membranes or meninges: the dura mater, arachnoid mater and the pia mater. Between the arachnoid and pia is a space filled with cerebrospinal fluid (CSF) produced in the ventricular system percolating to the subarachnoid space. The system acts as a shock absorber to the brain.

THE DURA MATER

The dura mater the outermost tough fibrous protective layer is subdivided into outer and inner layers. The outer layer of dura mater (endosteum or endocranium), which is the most external membrane, adheres closely to the inner aspect of cranial bones and continuous at sutures and through foraminae with the external periosteum (pericranium). The middle meningeal artery and corresponding veins run in this layer. The fibrous tissue between bones of the skull is known as the sutural ligament. The inner layer of dura mater (the meningeal layer) adheres to the outer layer except where it is reduplicated into four inwardly projecting folds that subdivide the cranial cavity into compartments and whose edges contain venous sinuses.

The **epidural space** is a virtual space external to the dura (outside of the endosteum). Intracranial bleeding creates a space occupying haematoma. General increase in intracranial pres-

sure results in depressed consciousness, headache, nausea and papilloedema. If the hematoma displaces the temporal lobe through the tentorial notch, the oculomotor nerve is compressed depressing pupillary response. Compression of the medulla may depress vital functions.

The **subdural space** is a capillary space located between the deep surface of dura and the arachnoid mater over the brain (and spinal cord).

PROCESSES OF DURA

The internal or meningeal layer of dura sends four reduplicated folds which project inwardly between major subdivisions of the brain and partly subdivides the cranial cavity.

The **falx cerebri** is a sickle shaped partition. It lies in the median plane between the cerebral hemispheres. It extends from the crista galli and frontal crest anteriorly to the interior occipital protuberance posteriorly

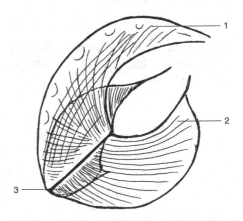

FIG. 3-1   Dural folds. (1) Falx cerebri; (2) Tentorium cerebelli; (3) Falx cerebelli.

joining the tentorium cerebelli posteriorly. Superiorly, the two layers of the falx split and attach to the lateral edges of the sagittal groove to enclose the superior sagittal sinus and to the posterior clinoid process. The free inferior border of falx cerebri follows the corpus callosum and encloses the inferior sagittal sinus.

The **tentorium cerebelli** is a crescent shaped fold lies in the transverse plane between the occipital lobes of the cerebrum and cerebellum. It is attached around the occipital bone enclosing the transverse sinus and along the ridge of the petrous part of the temporal bone enclosing the superior petrosal sinus where it attaches to the falx cerebri, and encloses the straight sinus. Medially, the edge of the tentorium twists so that the free edge continues anteriorly to attach to the anterior clinoid process and the attached border continues medially to the posterior clinoid process.

The **tentorial notch** is formed by the free border of the tentorium and dorsum sellae. It closely grips the midbrain and divides the posterior cranial fossa from the remainder of the cranial cavity.

Space occupying lesions of the temporal lobe of the brain may result in transtentorial or uncal **herniation** with compression of the midbrain and the oculomotor nerve.

The **cavernous sinus** is formed by the triangular piece of dura between the free superior border of the tentorium cerebelli (passing to the anterior clinoid

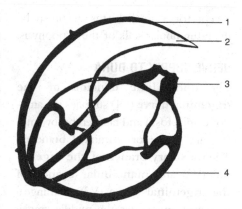

FIG. 3-2 Dural sinuses. (1) Superior sagittal sinus; (2) Inferior sagittal sinus; (3) Cavernous sinus; (4) Transverse sinus.

process) above and lateral border (reaching the posterior clinoid process) below. The triangular area created is pierced by the oculomotor and trochlear nerves.

The **trigeminal cave** is a recess formed by an evagination of the meningeal layer of dura at the apex of the petrous temporal bone beneath the superior petrosal sinus and meningeal layer of the middle cranial fossa. It is fused anteriorly with the lateral wall of the cavernous sinus. It houses the trigeminal ganglion.

The **falx cerebelli** is a sickle shaped evagination of dura lying in the median plane below the tentorium and attached to the internal occipital crest. Its posterior border encloses the occipital sinus.

The **diaphragma sellae** is a horizontal duplication of the meningeal dura that covers the hypophyseal fossa except for a small hiatus occupied by the infundibular stalk of the hypophysis.

NERVE SUPPLY TO DURA

The **ophthalmic division of the trigeminal nerve** (V1) supplies the anterior cranial fossa and falx cerebri by anterior and posterior ethmoidal branches. The **maxillary branch of the trigeminal** (V2) and **mandibular branch of the trigeminal nerve** (V3) (meningeal branches) innervates the middle cranial fossa. The **vagus** (X) and **hypoglossal nerves** (XII) (recurrent meningeal branches) carries branches of C1–3 that innervate the posterior cranial fossa.

Dural Venous Sinuses

Venous sinuses occur between the outer (endosteal) and inner (meningeal) layers of dura except in the free edges of dura where they lie within reduplications of the inner layer. They have an endothelial layer but no valves. Tributaries arise from neighboring parts of brain, middle meningeal veins and diploic veins. All drain ultimately to the internal jugular vein. Emissary veins connect sinuses with extracranial veins. The structure of sinuses prevents their collapse. If a sinus is ruptured by a skull fracture or during surgery, air can enter the venous system and produce a fatal air embolism.

The **superior sagittal sinus** runs backward in the entire length of the attached margin of the falx cerebri. It begins in front of crista galli and nasal crest (where it might receive a small vein from the nasal cavity through foramen caecum), receives superior cerebral veins and communicates with arachnoid granulations. Near the internal occipital protuberance, the superior sagittal sinus is joined by the straight sinus, and occipital sinuses and divides into left and right transverse sinuses at the confluence of sinuses.

The **lateral lacunae** are invaginations of arachnoid into the lumen usually of the superior sagittaal sinus through which cerebrospinal fluid is returned to the venous system.

The **inferior sagittal sinus** (unpaired) runs in the free lower border of the falx cerebri and ends in the straight sinus.

The **straight sinus** (unpaired) runs in the junction of the falx cerebri and tentorium cerebelli. It receives the great cerebral vein anteriorly and joins the confluence of sinuses posteriorly.

The **occipital sinus** (unpaired) lies in the attached margin of the falx cerebelli. It begins near the margin of the foramen magnum and ends at the confluence of the sinuses. It communicates with the vertebral venous plexus.

The **confluence of the sinuses** is the point of union of the superior sagittal, straight and occipital sinuses at the internal occipital protuberance. The left and right transverse sinuses drain from the confluence. The superior sagittal sinus most commonly drains into the right transverse sinus while the right transverse, straight and occipital sinuses most commonly drain into the left transverse sinus.

The **left and right transverse sinuses** lie in the peripheral edge of the tentorium cerebelli and drain the confluence of sinuses. Commonly, the superior sagittal sinus drains into the right transverse sinus and the straight and occipital sinuses drain into the left transverse sinus.

The **sigmoid sinuses** are continuations of the transverse sinuses on each side. They curve downward sharply in deep grooves on the temporal bone and in the jugular foramen, then become continuous with the internal jugular veins. Many small sinuses (e.g. sphenoparietal and inferior petrosal) are not associated with dural folds.

The **cavernous sinus** (paired) is a flattened expanded venous sinus formed between meningeal and periosteal layers of dura. It is broken into multiple small venous channels by a honeycomb of fibrous septa. It lies between the superior

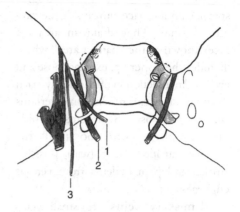

FIG. 3-4 Relationships of structures in the cavernous sinus from above. (1) Oculomotor n; (2) Abducent n; (3) Trochlear n.

orbital fissure anteriorly and the apex of the petrous temporal behind on each side of the body of the sphenoid. Each sinus receives ophthalmic veins, central vein of the retina, middle and inferior cerebral veins and the sphenoparietal sinuses and emissary veins (from the pterygoid plexus via foramen ovale). They drain into the superior and inferior petrosal sinuses. Flow through the sinuses is a slow percolation. The internal carotid artery and its sympathetic plexus and the abducent nerve (VI) run through the cavernous sinus medially within sheaths of endothelium. The oculomotor (III), trochlear (IV) and ophthalmic (V1) nerves run in the lateral wall of the sinus. The maxillary nerve (V2) lies in the dura immediately lateral to the sinus.

Ophthalmic veins (through angular veins from the face) and deep facial veins (from the pterygoid plexus and veins passing through foramen ovale,

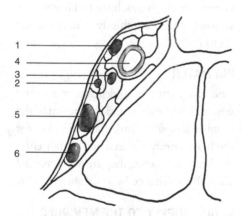

FIG. 3-3 Cavernous sinus — coronal section. (1) Oculomotor n; (2) Trochlear n; (3) Abducent n; (4) Internal carotid a; (5) Maxillary n; (6) Mandibular n.

spinousum and lacerum) reach the cavernous sinus. They drain an area (of face) known as the "danger area" which includes the upper lip, external nose and eyelids. An infection in this area may reach the cavernous sinus resulting in cavernous sinus thrombosis. Symptoms include ophthalmplegia, thrombosis of the internal carotid artery (with possible stroke) and papilloedema from venous engorgement of the retina.

Emissary veins are small veins passing through small foraminae or large apertures in the bones of the skull and connect the dural venous sinuses inside the skull and deep veins of the scalp or veins below the base of the skull. They have no valves but the flow is usually away from the brain. The superior ophthalmic vein is an emissary vein connecting the angular vein of the face with the cavernous sinus. Mastoid emissary veins connect the posterior auricular vein with the sigmoid sinus. Parietal emissary veins connect veins of the scalp with the superior sagittal sinus. An emissary vein may connect veins of the nasal cavity with the superior sagittal sinus. An emissary vein traversing the condylar canal may connect the sigmoid sinus and veins of the suboccipital triangle.

Diploic veins are lodged in channels in the diploe of the cranial vault. They are divided into frontal joining the supraorbital vein, *anterior parietal* joining the deep temporal vein, *posterior parietal* joining the transverse sinus and *occipital* joining the occipital vein (outside) and transverse sinus (inside the skull).

ARACHNOID MATER

The arachnoid mater is a delicate membrane lying beneath the dura from which it is separated by the subdural space, which is actually a cleft. Fine connective tissue trabeculae comprising the arachnoid cross between outlying arachnoid layer and the innermost pia mater creating the **subarachnoid space** which contains cerebrospinal fluid.

Cerebrospinal fluid (CSF) is produced by the choroid plexuses which project into the ventricles of the brain. It is a clear, colorless aqueous solution containing sodium chloride and other salts and traces of sugars, proteins, urea and other organic substances and a sprinkling of red blood cells. It flows through the ventricular system entering the subarachnoid space of the spinal cord and brain through the median and lateral apertures in the roof of the fourth ventricle. It re-enters the circulatory system through arachnoid villi into the venous sinuses (chiefly the superior sagittal sinus).

PIA MATER

The pia mater is a delicate membrane which closely invests and is adherent to the brain and spinal cord. Blood vessels for the brain ramify within it and as they enter the brain substance, they are accompanied for a short distance by a pial sheath.

BLOOD SUPPLY TO THE MENINGES

The anterior meningeal branch of the anterior ethmoidal artery is a branch of the **ophthalmic artery** in the orbit that enters the anterior ethmoidal canal entering the anterior cranial fossa. It crosses the cribriform plate of the ethmoid reaching the nasal slit beside crista galli and descends into the nose.

The anterior meningeal branch of internal carotid artery is a small branch of the **internal carotid** as it crosses the cavernous sinus.

The posterior ethmoidal branch of the ophthalmic artery enters the anterior cranial fossa between the ethmoid and orbital plate of the frontal bone.

The anterior branch of middle meningeal artery branches from the middle meningeal near pterion.

The middle meningeal branch of the **maxillary artery** enters the middle cranial fossa through foramen spinosum and divides into anterior and posterior branches. Branches are embedded in dura and occupy grooves in the bone. They cross the floor of the middle cranial fossa and ascend in the lateral wall to reach the calvaria.

The posterior meningeal arteries are branches of the **vertebral artery** before it pierces the dura near foramen magnum. They supply the dura of the posterior cranial fossa.

The meningeal branches of the occipital branch of the **external carotid** (via the mastoid canal to the sigmoid sulcus) ascends to the pharyngeal (via the jugular foramen) and the vertebral artery (through the foramen magnum).

The posterior branch of the **middle meningeal artery** also supplies the posterior cranial fossa.

Epidural hemorrhage commonly results from a fracture of the temporal or parietal bones when meningeal vessels are torn. Symptoms result from brain compression. There is commonly transient loss of consciousness followed by a lucid interval then unconsciousness and death.

Subdural hemorrhage results from the tearing of cerebral veins as they cross the subdural space to join the dural sinuses. Tearing may result from a blow to the head or sudden jarring. Symptoms depend on the extent of tearing and may fluctuate if the haemorrhage is slow.

Subarachnoid hemorrhage follows the rupture of a weakened cerebral artery. It may follow strenous exercise. Blood is found in the cerebrospinal fluid.

2. Cranial Nerves (as seen during removal of the brain)

Cranial nerves differ from spinal nerves in several ways (after Elliott 1963):

1. They are not formed regularly with dorsal sensory and ventral motor roots.
2. Some are entirely motor nerves while some are mixed nerves.
3. In mixed nerves, sensory and motor roots are attached to the same or almost the same part of the brainstem.
4. Some contain visceral sensory and visceral motor components more

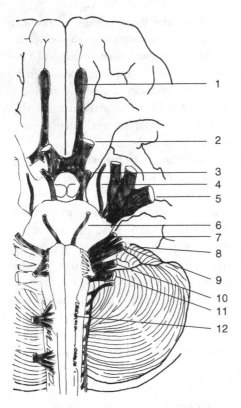

FIG. 3-5 Superficial origin of cranial nerves. (1) I;
(2) II; (3) III; (4) IV; (5) V; (6) VI; (7) VII; (8) VIII;
(9) IX; (10) X; (11) XI; (12) XII.

distinctly represented than in spinal
nerves.

5. Some crainial nerves contain taste
fibers.

6. Some cranial nerves contain
branchial motor components distributed to muscles that originate
from branchial arches. These are
not found in spinal nerves.

7. Most have peculiarities of central
connection, course and peripheral
distribution.

OLFACTORY NERVE (I)

The oflactory nerves arise from receptor cells in the epithelial lining of the
nose. Processes of these cells pass
through the cribriform plate of the ethmoid and enter the olfactory bulbs.
Here, they synapse with neurons whose
axons pass in olfactory tracts posteriorly
to the piriform cortex and several other
regions on the medial surface of the
temporal lobes. Olfactory nerves mediate the sense of smell.

OPTIC NERVE (II)

The optic nerve originate as axons of
ganglion cells in the retina. The axons
pass through the lamina cribrosa of the
sclera to form the optic nerve and travel
in the optic foramen. After entering
the cranial cavity, they join in the **optic
chiasm** immediately in front of the
hypophysis where fibers from the medial
half of the retina cross to the opposite
side while those from the lateral half
remain on the same side. Fibers then
continue in the optic tracts to the **lateral
geniculate bodies** of the thalamus and
superior colliculi. Optic nerves are actually
a fibre tract within the central nervous
system and not a peripheral nerve. They
mediate sight and visual reflexes.

OCULOMOTOR NERVE (III)

The oculomotor nerve arises on the
medial side of the cerebral peduncle. It
pierces the dura to enter the lateral wall of
the cavernous sinus above and medial to
the trochlear nerve. It then passes forward crossed laterally by the trochlear and

frontal branch of the trigeminal before reaching the superior orbital fissure and entering the orbit. The oculomotor nerve contains somatic motor fibers to certain **extrinsic ocular muscles** — superior rectus, levator palpebrae superioris, inferior and medial recti, and inferior oblique muscles.

Oculomotor nerves also carry preganglionic parasympathetic nerves to the **ciliary ganglion** where they relay and postganglionic fibers then influence the ciliary muscle (and therefore lens shape — a process called accomodation) and sphincter muscle of the iris (and therefore pupil adjustment to light intensities).

TROCHLEAR NERVE (IV)

The trochlear nerve arises posteriorly on the brainstem below the inferior colliculi. Trochlear nerves decussate on the dorsal aspect of the midbrain and curve around the lateral surface of the cerebral peduncle. They then pass in the lateral wall of the cavernous sinus below the oculomotor nerve and traverse the superior orbital fissure lateral to the tendinous ring and enter the orbit. The trochlear nerve is the motor nerve to the superior oblique muscle.

TRIGEMINAL NERVE (V)

The trigeminal nerve arises from the side of the pons by a small motor root and a large sensory root. It arches over the most medial upper border of the petrous temporal bone below the attached margin of the tentorium cerebelli. Cell bodies of the sensory neurons are located in the trigeminal ganglion located in an invagination of the dura, the trigeminal cave near the apex of the petrous part of the temporal bone. Peripheral axons of the sensory part of the trigeminal nerve reach the trigeminal ganglion through branches that arrive at the anterior edge of the ganglion.

The **ophthalmic nerve** (V1) runs along the lateral wall of the cavernous sinus below the oculomotor and trochlear nerves and enters the orbit through the superior orbital fissure.

The **maxillary nerve** (V2) leaves the middle of the anterior surface at the trigeminal ganglion then leaves the middle cranial fossa through foramen rotundum becoming the infraorbital nerve.

The **mandibular nerve** (V3) leaves the lateral surface of the trigeminal ganglion and enters the foramen ovale. It is the largest division and is joined on its deep aspect by the motor root of the trigeminal nerve.

The trigeminal nerve supplies motor fibers to the muscles of mastication, somatic sensory fibers from the skin from the vertex to the chin, mucous membrane of the mouth, paranasal sinuses, teeth and cranial cavity above the tentorium cerebelli. In addition, it carries parasympathetic fibers to the lacrimal and salivary glands, intraocular muscles and taste fibers from the anterior two thirds of the tongue.

ABDUCENT NERVE (VI)

The abducent nerve arises from the lower border of the pons in line with the oculomotor nerve. It pierces the dura over the inferior petrosal sinus to reach the cavernous sinus where it lies lateral to the internal carotid. The abducent nerve enters the superior orbital fissure between the heads of the lateral rectus muscle and supplies the lateral rectus muscle in the orbit.

FACIAL NERVE (VII)

The facial nerve arises from the brainstem between the olive and pons at the lower border of the pons. It has larger part which is a motor root that supply the muscles of facial expression (derived from the second branchial arch) and a smaller sensory root (nervus intermedius) which contains taste fibers (from the **anterior two thirds of the tongue**) and preganglionic parasympathetic fibers supplying the lacrimal and salivary glands and some pain fibers. Sensory fibers arise from neurons in the geniculate ganglion within the petrous temporal bone. Both roots pass forwards and laterally enter the internal acoustic meatus and join at the base of the meatus. The nerve enters the canal for the facial nerve.

VESTIBULOCOCHLEAR NERVE (VIII)

The vestibulocochlear nerve is actually two separate nerves, the **vestibular nerve** concerned with equilibrium and the **acoustic nerve** concerned with hearing which arise from the brainstem between the pons and inferior cerebellar peduncle. Nerves pass into the internal acoustic meatus with the facial nerve. Neurons contributing to the acoustic part are located in the spiral ganglion in the cochlea while those contributing to the vestibular part are located in the vestibular ganglion.

GLOSSOPHARYNGEAL NERVE (IX)

The glossopharyngeal nerve emerges from the upper medulla in the groove between the olive and inferior cerebellar peduncle. It passes laterally over the flocculus of the cerebellum to the jugular foramen piercing the dura separately and leaving the skull in front of the vagus and accessory nerves. The nerve is a mixed nerve with sensory, motor and autonomic components. It supplies motor fibers to the **stylopharyngeus** muscle (derived from the third branchial arch). Sensory fibers arise from the pharyngeal mucosa, tonsillar region and **posterior third of the tongue**, auditory tube and middle ear, carotid sinus and carotid body as well as **taste** from the posterior third of the tongue. Parasympathetic fibers in cranial nerve IX supply the parotid gland.

VAGUS NERVE (X)

The vagus nerve emerges from the brainstem in the groove between the inferior cerebellar peduncle and the olive, it also passes laterally over the flocculus to the jugular foramen passing in the same dural sheath as the accessory nerve behind the glossopharyngeal nerve. The vagus nerve carries **sensory** fibers from the dura of the posterior cranial fossa, part of the external acoustic meatus, part of the external surface of the tympanic membrane, mucosa of the pharynx and larynx and **taste fibers** from the epiglottic region. **Motor fibers** supply most of the muscles of the soft palate, pharyngeal muscles and muscles of the larynx. **Autonomic** (preganglionic parasympathetic fibers) fibers supply the viscera of the thorax and abdomen.

ACCESSORY NERVE (XI)

The accessory nerve is formed by the union of cranial (internal) and spinal (external) portions. The spinal part arises from the lateral part of the spinal cord from the upper five cervical nerve segments. The cranial part emerges from the side of the medulla in line with the glossopharyngeal and vagus nerves. They unite to form a trunk that ascends in the vertebral canal through foramen magnum behind the ligamentum denticulatum then passes out of the jugular foramen in the same sheath as the vagus. The cranial part of the accessory supplies motor fibers to the soft palate, pharyngeal constrictors, larynx and cardiac branches to the vagus. The spinal part supplies motor fibers to the sternomastoid and trapezius muscles.

HYPOGLOSSAL NERVE (XII)

Hypoglossal nerve arises as bundles of roots from the groove between the inferior olive and pyramid (in line with the oculomotor and adbucent nerves).

Bundles come together and the combined nerve pierces the dura then passes to the hypoglossal canal. The nerve supplies motor fibers to the extrinsic muscles of the tongue (except palatoglossus) and intrinsic muscles of the tongue as well as thyrohyoid and geniohyoid (from the first cervical nerve muscles.)

3. Arteries of the Brain

The **internal carotid** and **vertebral arteries** alone supply the brain.

INTERNAL CAROTID ARTERY

The internal carotid artery from the carotid canal, passes over the foramen lacerum medial to the lingula of the sphenoid entering the cavernous sinus. It ascends in the subarachnoid space in the angle between the optic nerve and tract and ends by dividing into the **anterior** and **middle cerebral arteries.**

Ophthalmic artery, one of the terminal branches, enters the orbit through the optic canal lateral to and below the optic nerve. It crosses over the nerve to reach the superomedial angle of the orbit.

Posterior communicating artery, the second terminal branch, arises from the posterior part of the artery. It runs backward to join the posterior cerebral artery (branch of the vertebral artery).

VERTEBRAL ARTERY

The vertebral artery pierces the dura behind the occipital condyle and grooves the margin of the foramen

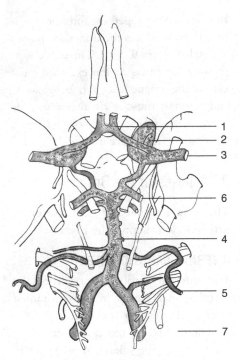

form the basilar artery at the lower border of the pons.

The **posterior inferior cerebellar artery,** a terminal branch of the vertebral artery, arises near the pons. It courses between the roots of XII then XI and X to reach the undersurface of the cerebellum.

The **basilar artery** is formed by the union of right and left vertebral arteries (both branches of the subclavian artery). The terminal branch extends from the base of the pons to upper border of the pons where it bifurcates into right and left posterior cerebral arteries. The basilar artery gives off three paired branches: anterior inferior cerebellar; superior cerebellar; posterior cerebral.

FIG. 3-6   Arteries supplying the brain. (1) Anterior cerebral a; (2) Internal carotid a; (3) Middle cerebral a; (4) Basilar a; (5) Anterior inferior cerebellar a; (6) Posterior cerebral a; (7) Vertebral a.

The **arterial circle** is a polygonal shaped anastomosis between the vertebral and carotid arteries. It is formed by the posterior cerebral, posterior communicating, internal carotid, anterior cerebral and posterior communicating arteries. Variations are common and the anastomosis may be important in establishing collateral circulation in cases of obstruction of any of the major contributing arteries.

magnum. It winds around the medulla over the ligamentum denticulatum, passes between the hypoglossal and first cervical nerve uniting with its fellow to

Chapter 4

Face

Chapter Outline

1. Muscles
2. Arteries
3. Veins
4. Nerves
5. Lymphatics

The face is the part of the front of the head between the ears and from the chin to the hairline. The skin of the face is thick with underlying superficial fascia containing blood vessels and nerves. On the outer surface of buccinator muscle but deep to the anterior margin of masseter, the buccal fat pad lies between the superficial fascia and the buccinator muscle. This pad prevents the cheeks from collapsing in suckling. The submucous layer contains many small mucous glands in the region of the lips and cheek known as the labial and buccal glands, respectively. Their ducts open into the vestibule of the mouth.

1. Muscles

During development, the mesoderm of the second branchial arch forms a thin sheet of muscle, that spreads over the neck, face and scalp. These are all supplied by the facial (7th cranial) nerve. Some of the muscle fibers organize around the apertures (eye, nose and mouth) and each aperture comes to possess a sphincter and a dilator mechanism. As a group, these muscles affect facial expression. Fibers are skeletal muscle but they are inserted into the skin. They may be grouped according to their name or location.

MUSCLES OF THE SCALP
See Chapter 1.

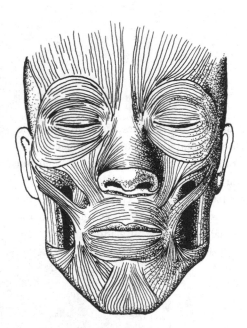

FIG. 4-1 Muscles of facial expression.

MUSCLES OF THE EYELIDS
ORBICULARIS OCULI

Orbicularis oculi is the orbital sphincter muscle and consists of three parts.

(1) *Orbital part* (surrounding orbit)
 - **Origin:** Frontal and maxillary bones at the medial orbital margin.
 - **Insertion:** Skin only; encircles orbit and has no bony insertion laterally.
 - **Action:** The orbital part closes the lids forcibly, decreasing the volume of the conjunctival sac. If the sac is brimful of tears, they spill over down the cheeks. The opponent of this part is occipitofrontalis, a scalp muscle.

(2) *Palpebral part* (slightly arched fibers in eyelids)
 - **Origin:** Medial palpebral ligament. Palpebral ligament and raphe are attached to medial and lateral bony margins of orbit, and split to go to edges of eyelids.
 - **Insertion:** Lateral palpebral raphe.
 - **Action:** The palpebral part closes the lids gently edge to edge, with no diminution in volume of conjunctival sac, so that tears are not extruded. Blinking wipes tears medially toward the lacrimal sac on to the nasal cavity. The palpebral part of orbicularis oculi is opposed by levator palpebrae superioris.

(3) *Lacrimal part* attached to lacrimal sac
 - **Origin:** Posterior lacrimal crest, forming posterior edge of lacrimal fossa.

- **Insertion:** Posterior aspect of medial ends of tarsal plates.
- **Nerve** of the entire muscle is supplied by the facial nerve (VII).
- **Action:** Expands the lacrimal sac or expel its contents.

Levator Palpebrae Superioris

- **Origin:** Roof of the orbit anterior to the origin of superior rectus (an extrinsic muscle of the eyeball).
- **Insertion:** Is into the skin, tarsal plate and conjunctiva of the upper lid. It is an antagonist of the orbicularis oculi.
- **Nerve:** Oculomotor (3rd cranial) nerve.

The inferior part of levator palpebrae superioris is a smooth muscle, supplied by sympathetic fibers from the superior cervical ganglion via the carotid plexus. The upper lid droops (ptosis), if the sympathetic pathway is interrupted.

Corrugator Supercilli

Corrugator supercilli acts on the skin of the forehead, but developmentally, it is a detached portion of the orbicularis oculi.

- **Origin:** Frontal bone at the medial end of superciliary ridge.
- **Insertion:** Fibers are directed upwards and laterally through orbicularis oculi to the skin of the forehead.
- **Action:** Produces frown of vertical wrinkles above nose.

MUSCLES OF THE NASAL REGION
Compressor Naris (the sphincter of the nose)

- **Origin:** Maxilla at side of lower part of bony anterior nares.
- **Insertion:** Aponeurosis continued across the nose to its fellow on the opposite side.
- **Action:** Compresses the upper part of the cartilaginous nasal aperture.

Dilatator Naris (the dilator of the nose)

- **Origin:** Maxilla below compressor naris (above the lateral incisor).
- **Insertion:** Side of alar cartilage.
- **Action:** Pulls on lateral wing of cartilage to widen the nostril.

These two muscles constitute the nasalis muscle. The compressor is called the transverse part and the dilatator the alar part.

Procerus

Procerus is a continuation of median fibers of frontalis down on to the nose.

- **Origin:** Fascia over bridge of nose.
- **Insertion:** Skin of the forehead between the eyebrows.
- **Action:** Produces small transverse wrinkles at root of nose by slight elevation of external nose.

MUSCLES OF THE MOUTH REGION
Orbicularis Oris

Orbicularis oris encircles the lips and has its own proper fibers attached to bone above and below in midline, the superior and inferior incisive slips. Its bulk is much increased by fibers received from buccinator and the dilatator muscles described below.

- **Origin:** Muscles converging on the mouth.
- **Insertion:** Fibers form an ellipse.
- **Action:** Closes the mouth into a small circle, as in whistling.

Dilatators of mouth are arranged radially around the lips. They consist of elevators of the upper lip and angle of mouth and depressors of the lower lip and angle of mouth.

Levator Labii Superioris

- **Origin:** Arises from the infraorbital margin above infraorbital margin.
- **Insertion:** Skin of the upper lip.

Levator Labii Superioris Alaeque Nasi

- **Origin:** Arises from the upper part of the frontal process of the maxilla.
- **Insertion:** The labial part blends with levator labii superioris. The nasal part is inserted into alar cartilage of nose.
- **Action:** The labial part elevates lip. The nasal part dilates nostril.

Zygomaticus Major

- **Origin:** Temporall process of the zygomatic bone.
- **Insertion:** Skin of the angle of the mouth.
- **Action:** Lifts the angle of the mouth upwards and laterally. Some

of its fibers decussate with those of depressor anguli oris.

Zygomaticus Minor

- **Origin:** Outer surface of the zygomatic bone in front of the origin of zygomaticus major.
- **Insertion:** Upper lip at the nasolabial groove.
- **Action:** Elevates the upper lip.

Levator Anguli Oris

- **Origin:** From canine fossa of maxilla where it lies under the cover of levator labii superioris and the infraorbital nerve.
- **Insertion:** Angle of the mouth at the modiolus.
- **Modiolus:** Is the point of decussation of central fibers of buccinator which receives elevators and depressors of angle of mouth. The crossing of so many fibers makes a palpable nodule inside angle of mouth opposite first upper premolar tooth. The three muscles, zygomaticus major and minor and levator anguli oris descend to be inserted into the angle of the mouth at the modiolus.

Depressor Labii Inferioris

- **Origin:** From the front of the mandible below the mental foramen.
- **Insertion:** Skin of the lower lip.
- **Action:** Depresses the lower lip.

Depressor Anguli Oris

- **Origin:** Anterior part of external oblique line of mandible.

- **Insertion:** Angle of mouth at the modiolus.
- **Action:** Pulls the angle of the mouth inferolaterally as in an expression of sadness.

Buccinator

Buccinator forms the muscular plane of the cheek.

- **Origin:** A deep posterior origin from the pterygomandibular raphe that is attached above to the hamulus of the medial pterygoid plate and below to the posterior end of the mylohyoid line of the mandible. It also arises from the adjacent areas of the outer surfaces of the maxilla and mandible near the three molar teeth.
- **Insertion:** Into the orbicularis oris-upper and lower fibers into respective lips, central fibers decussate at modiolus.
- **Action:** Obliterates the space between cheek and teeth. It is used in sucking and chewing when it pushes the food between the teeth (balanced by the tongue).

The pterygomandibular raphe is attached above to the hamulus of the medial pterygoid plate and below to bone beside posterior border of third molar tooth at the posterior end of the mylohyoid line of the mandible. It unites the buccinator and the superior constrictor, in the same plane.

Buccinator is pierced by the parotid duct, buccal branch of mandibular nerve and ducts of molar glands. The tendon of tensor veli palatini passes above its border between tuberosity of maxilla and hamulus.

Risorius

Risorius is the smiling or grinning muscle that appears to be a prolongation on to the face of some fibers of the platysma.

- **Origin:** Fascia over masseter. Horizontal bundles converge on the angle of the mouth passing through the modiolus.
- **Insertion:** Skin over the angle of the mouth.
- **Action:** Pulls the mouth laterally and helps to widen the mouth and with zygomaticus major, it produces the nasolabial folds (laughing muscles).

MUSCLE OF THE SUPERFICIAL NECK REGION
Platysma

Platysma muscle fibers migrate over the anterior surface of the neck and may extend onto the upper part of the thoracic wall.

- **Origin:** From the fascia over the upper part of the chest and shoulder. Platysma forms a wide thin sheet covering front and side of the neck with the fibers directed upwards and medially.
- **Insertion:** Into lower border of mandible and spreads up on to face where it may attach to the angle of the mouth replacing risorius.

- **Nerve:** Cervical branch of VII.
- **Action:** Platysma pulls the corner of the mouth down and to the side.

The platysma lies in the deep part of the superficial fascia of the neck; it covers the external jugular vein and the cutaneous branches of the cervical plexus.

Facial (Bell's) palsy results from damage to the facial nerve. The face is pulled to the unaffected side and the patient cannot wrinkle the forehead, frown, close the eye, twitch the nose, smile or whistle on the affected side. Food collects between the cheek and gums.

2. Arteries

These are principally branches of the external carotid artery with a small contribution from the internal carotid.

Facial artery is a branch of the external carotid artery. It reaches the face at the anterior inferior angle of the masseter muscle, ascends tortuously terminating as the angular artery by anastomosing with dorsalis nasi of ophthalmic artery. It crosses the mandible, buccinator, and levator anguli oris and lies deep to platysma, risorius, zygomaticus major, and usually levator labii superioris. The facial vein is straighter than and posterior to the artery.

The branches of the facial artery in the face are as follows:

Inferior labial artery arises near angle of mouth. It has a tortuous course;

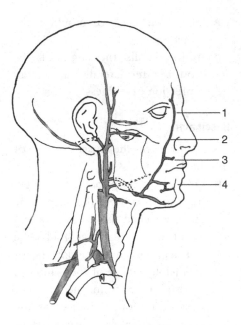

FIG. 4-2 Arteries of the face. (1) Angular a; (2) Lateral nasal a; (3) Superior labial a; (4) Inferior labial a.

penetrates the lower lip and orbicularis oris; and anastomoses freely with the opposite artery.

Superior labial artery arises with or near the inferior labial, having corresponding course in the upper lip. It anastomoses with the opposite artery and supplies a branch to septum of nose.

Lateral nasal artery supplies ala and side of nose and anastomoses with the opposite artery, nasal branch of ophthalmic, and infraorbital branch of maxillary.

Angular artery terminal branch of the facial at the medial angle of the eye. It anastomoses with the dorsal nasal and palpebral branches of the ophthalmic artery (from the internal carotid artery).

3. Veins

Veins of the face form a network that is drained by the **common facial vein.** They communicate with the ophthalmic veins (via the nasofrontal vein) and the pterygoid plexus (via the deep facial vein.)

The supratrochlear and supraorbital veins unite to form the angular vein, which continues as the facial vein at the lower margin of the orbit. It passes downward and backward behind the facial artery and receives the tributaries that correspond to branches of the facial artery. It leaves the face at the anterior inferior angle of the masseter muscle, pierces the deep cervical fascia, and joins the anterior division of the retromandibular vein to form the **common facial vein.**

4. Nerves

The nerves of the face are derived from three sources.

FACIAL NERVE (VII)

This cranial nerve leaves the cranial cavity via the internal acoustic meatus, facial canal, and stylomastoid foramen it exchanges fibers with almost all sensory cutaneous branches of the trigeminal nerve (V).

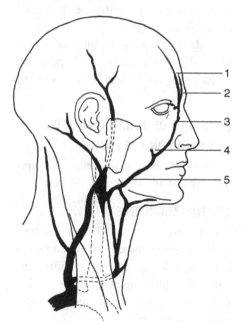

FIG. 4-3   Veins of the face. (1) Supratrochlear v; (2) Supraorbital v; (3) Angular v; (4) Deep facial v; (5) Facial v.

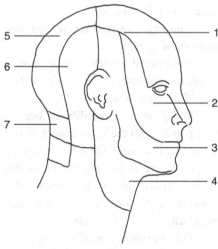

FIG. 4-4   Cutaneous innervation of the head. (1) Ophthalmic n; (2) Maxillary n; (3) Mandibular n; (4) Great auricular n; (5) Greater occipital n; (6) Lesser occipital n; (7) Third occipital n.

The following branches of the facial nerve exit from stylomastoid foramen:

Posterior auricular nerve passes up behind the external acoustic meatus and supplies the auricular muscles and occipitalis (occipital belly) of occipitofrontalis. It also carries sensory fibers to the auricle. It supplies to the stylohyoid muscle and posterior belly of the digastric muscle digestion.

Branches the facial nerve on the face enter the parotid gland and form a **parotid plexus**. Terminal branches emerge from the anterior border of the parotid gland and communicate with branches of trigeminal nerve.

The **temporal branch** emerges from the parotid gland at its upper pole in front of the superficial temporal artery. It supplies orbicularis oculi, frontalis of occipitofrontalis and corrugator supercilli muscles.

The **zygomatic** branch supplies muscles of eyelid.

The **buccal** branch supplies buccinator, risorius, orbicularis oris and muscles of external nose.

The (marginal) **mandibular** branch runs parallel with the lower border of the mandible. It crosses the inferior border of jaw where it lies superficial to the submandibular gland. It supplies the muscles of the lower lip and chin and ends below angle of mandible

The **cervical** branch leaves the parotid at its inferior pole. It perforates cervical fascia beneath mandible to supply platysma.

FACIAL BRANCHES OF GREATER AURICULAR NERVE

Facial branches of greater auricular nerve are from the second and third cervical nerves. These sensory branches innervate the skin over the lower part of the parotid gland and the masseter muscle.

Great auricular nerve (C2, 3) winds round the posterior margin of sterno-mastoid to reach parotid gland. It gives off the following branches:

Facial branches supplies the skin of the face over the parotid and parotid fascia.

Auricular nerve supplies the posterior surface of the pinna and lateral surface below external auditory meatus. It communicates with the posterior auricular nerve.

Mastoid nerve supplies skin over mastoid process joining with posterior auricular of facial and lesser occipital nerve.

BRANCHES OF TRIGEMINAL NERVE (SENSORY) ON THE FACE

These branches arise from all three divisions of the trigeminal nerve.

OPHTHALMIC NERVE (V1)

Frontal nerve enters the orbit lateral to IV passing forward on levator palpebrae superioris. It divides into the supratrochlear nerve, supraorbital nerve, nasocillary nerve and the anterior ethmoidal nerve.

Supratrochlear nerve is a terminal branch of the frontal nerve. It is directed forwards in the orbit and leaves the orbit medially between the trochlea and supraorbital notches. It supplies the skin of the forehead and upper eyelid and communicates with the infratrochlear nerve.

Supraorbital nerve arises as a continuation of the frontal nerve as it passes through the supraorbital notch or foramen and supplies palpebral filaments. It ends on the forehead by dividing just outside the orbit into two branches which supply the anterior half of the scalp, upper eyelid and frontal sinus.

Nasociliary nerve enters the orbit between the heads of lateral rectus.

Anterior ethmoidal nerve is a continuing branch of the nasociliary nerve, that passes through the anterior ethmoidal foramen. It re-enters the cranium (as the internal nasal nerve) and passes downwards by the side of the crista galli; dividing into terminal branches, a medial branch, supplying the mucous membrane of the adjacent part of the septum, and a lateral branch that runs along a groove on the internal surface of the nasal bone and supplies the lateral wall of the nose. This branch passes between the bone and lateral cartilage to supplies skin of the ala and tip of nose (external nasal nerve).

Infratrochlear nerve is a branch of the nasociliary nerve that passes forwards on the medial wall of the orbit and supplies the lacrimal sac, skin of upper eyelid and root of nose.

MAXILLARY NERVE (V2)

The maxillary nerve emerges on the face from the infraorbital foramen as the **infraorbital nerve** under the levator labii superioris and joins with branches of the facial nerve to form the **infraorbital plexus**. Its branches supply the skin of the lower eyelid, medial part of cheek, lateral edge of nose, and upper lip.

Zygomaticotemporal nerve is a branch of the zygomatic nerve (branch of V2), which passes along the lateral wall of the orbit, enters a foramen in the zygomatic bone, and becomes cutaneous on the temple.

Zygomaticofacial nerve passes to lower and lateral angle of orbit, goes through zygomatic foramen and becomes cutaneous to the skin over the bony cheek.

MANDIBULAR NERVE (V3)

Buccal nerve emerges between two heads of lateral pterygoid; passes on to buccinator and is sensory to mucous membrane and skin of cheek.

Auriculotemporal nerve runs backwards inferior to lateral pterygoid muscle then behind mandibular joint capsule; turns upwards with superficial temporal artery to temporal fossa where it becomes cutaneous. It lies posterior to superficial temporal artery as they cross the zygomatic arch. It gives off branches that supply the temporomandibular joint, skin of upper half of outer surface of auricle, upper part of external acoustic meatus, tympanic membrane and scalp above meatus. Postganglionic parasympathetic fibers from the otic ganglion pass with the auriculotemporal nerve and go to the parotid gland.

Mental nerve is a terminal branch of the inferior alveolar nerve, it passes in a backward direction from the mental foramen and supplies the skin of the lower lip from angle of mouth to midline, as well as the mucosa and gum of the anterior part of the vestibule.

5. Lymphatics

All lymphatics from the head and neck drain into deep cervical lymph nodes

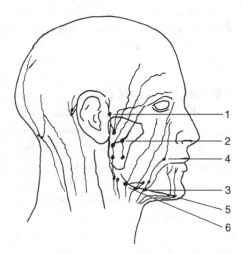

FIG. 4-5 Lymphatic drainage of the face. (1) Preauricular nodes; (2) Parotid nodes; (3) Submandibular nodes; (4) Buccal nodes; (5) Submental nodes; (6) Superficial cervical nodes.

directly from the tissues or indirectly after traversing outlying groups of nodes.

Several groups of outlying nodes form a "pericervical collar" at the junction of the head and neck. These are in five groups, *occipital, mastoid, parotid* (some superficial and some deep) *submandibular* (related to small nodes along the course of the facial artery) and *submental* nodes. They receive lymph from the skin of the face (and nose, mouth, tongue and salivary glands) and drain into upper deep cervical nodes (related to the internal jugular vein).

Lymph drainage of the superficial head is usually as follows:

Scalp will drain to occipital, mastoid and parotid groups of pericervical collar nodes.

Forehead will drain to parotid nodes.

Eyelids and orbicularis oculi will drain to parotid and submandibular nodes.

Cheek will drain to parotid and submandibular nodes.

External nose and vestibule of nose will drain to submandibular nodes.

Auricle and external auditory meatus will drain to mastoid, parotid and superficial cervical nodes.

Lips (skin and mucous membrane) will drain to submental and submandibular nodes.

Chapter 5

Orbital Region

Chapter Outline

1. Superficial Features
2. Eyelids
3. Lacrimal Apparatus
4. Orbit and Contents
5. Eyeball

The orbits are the bony cavities that contain the eyeballs and their associated muscles, nerves, blood vessels, fat and most of the lacrimal apparatus. Each orbit is shaped like a four-sided pyramid with its apex posteriorly and base (aditus) anteriorly. The eyeballs occupy the anterior third or less of the orbits. They are enveloped in fat and protected from trauma by the eyelids and from dehydration by tear secreting lacrimal glands.

1. Superficial Features

Palpebral fissure is the opening bounded by the upper and lower eyelids.

Lateral and the medial angles (canthi) are the junctions of the upper and lower lids. The lateral angle is more acute than the medial angle.

Lacrimal lake is the shallow triangular depression at the medial angle separating the eyelids. It is mostly filled with the lacrimal caruncle.

Lacrimal caruncle is a red mass of connective tissue lying in the lacrimal lake. It contains some small sebaceous glands.

Semilunar fold is the fold of conjunctiva that forms the lateral margin of the lacrimal lake. It is a remnant of the nictitating membrane in some animals.

Lacrimal punctum is a pit on the medial end of the lower eyelid. It is the external opening of the lacrimal canaliculus on the summit of a small elevation, the lacrimal papilla.

2. Eyelids

The eyelids (palpebrae) are movable, musculofibrous folds in front of each orbit. They protect the eye and control excessive light entry. The upper eyelid is more movable than the lower and by closure, the lids protect the eye from injury. Blinking spreads a film of tears on the cornea from the superior and laterally placed lacrimal glands and pumps away excessive tears to the medially placed lacrimal duct.

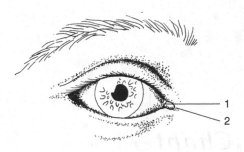

FIG. 5-1 Superficial features of the eye. (1) Lacrimal lake; (2) Lacrimal caruncle.

Eyelids consist of a number of layers from without inward skin, areolar tissue, orbicular muscle, tarsal plate, and orbital fascia, tarsal glands, and conjunctiva. The upper lid also contains the aponeurosis of the levator palpebrae muscle that is attached along the upper margin of the tarsal plate.

The skin of the eyelid is very thin and hairless except for the eyelashes or cilia. It is continuous onto the inner surface of the eyelid with the conjunctiva (palpebral conjunciva) that is in turn reflected over the front of the eyeball (bulbar conjunctiva) at the fornices.

Subcutaneous layer comprises areolar tissue that contains large sebaceous glands associated with the cilia and modified sweat glands (ciliary glands) that open onto the skin or ducts of sebaceous glands. There is no fat in the eyelids.

Muscular layer is the palpebral part of the orbicularis oculi and part of levator palpebrae superioris. The lower eyelid has a poorly developed depressor muscle.

Fibrous layer comprises the tarsal plate and the orbital septum.

Tarsal plates are condensed connective tissue laminae supporting each eyelid. The larger superior plate is half oval in shape and the lower is a narrow oblong strip. Embedded in each plate are the sebaceous tarsal glands the ducts of which open by minute foraminae at the edge of the eyelids. The medial and lateral ends of the tarsal plates are anchored to the margin of the orbit by the *medial* and *lateral palpebral ligaments*. The superior tarsal plate receives the main insertion of the *levator palpebrate superioris* muscle.

Orbital septum is a fascial sheet attached to the periosteum of the orbital margin and tarsal plates. It consists of superior and inferior palpebral fascia. These membranes are continuous with each other at the medial angle of the eye behind the lacrimal sac and at the lateral angle by means of the lateral palpebral

ligament which is fused with the fascia. The fascia is also continuous with the periosteum of the orbit. The orbital septum acts as a barrier preventing the spread of infection from eyelid into the orbit. Orbital fat is retained by the orbital septum.

Fascia bulbi (or bulbar sheath) is a connective tissue socket enclosing the eyeball. It is attached to the eyeball at the corneoscleral junction and envelopes the extrinsic muscles.

Check ligaments are fibrous expansions of the fascia bulbi most pronounced running from the lateral rectus muscle to the zygomatic tubercle of the zygomatic bone (lateral check ligament) and from the medial rectus muscle to the posterior lacrimal crest of the lacrimal bone (medial check ligament).

Suspensory ligament of the eyeball is a thickening of the inferior portion of fascia bulbi. It forms a hammock attached medially to the posterior lacrimal crest and laterally to the zygomatic tubercle.

Mucous membrane of the eyelids is known as the **palpebral part of the conjunctiva.** It forms the posterior layer of both eyelids and at the free edges of the eyelids joins the skin. It is firmly attached to the deep surface of each tarsal plate (to prevent wrinkling). It is reflected on to the eyeball, the lines of reflection being known as the superior and inferior fornices, of which the superior is the deeper; and into which some fibers of the levator palpebrate superioris are inserted.

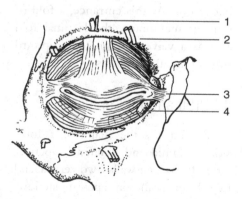

FIG. 5-2 Tarsofascial layer of the right orbit. (1) Supraorbital n; (2) Supratrochlear n; (3) Medial palpebral ligament; (4) Lacrimal sac.

3. Lacrimal Apparatus

The lacrimal apparatus consists of the tear-secreting lacrimal gland, lacrimal passages (the lacrimal canaliculi and sac and the nasolacrimal duct).

Lacrimal gland is a serous gland located in a fossa in the superolateral angle of the orbit. The gland is divided into a larger, orbital part and a smaller, palpebral part by the lateral border of the aponeurosis of the levator palpebrae superioris muscle. The ducts of the gland (12 or 14) open into the superior conjunctival fornix.

Lacrimal secretion flows down medially under pressure provided by the eyelids. About half of the secretion evaporates; the rest is drained by the lacrimal canaliculi into the inferior meatus of the nasal cavity. Excess secretion flows down the cheeks as tears.

The lacrimal gland is supplied by the **lacrimal artery** (branch of the ophthalmic artery), **lacrimal nerve** (sensory branch of the ophthalmic division of V), and **parasympathetic fibers** (secretory from the pterygopalatine ganglion).

Lacrimal canaliculi are short, paired narrow tubes beginning medially at the lacrimal punctum arching in the free edge of the lid and passing medially to open into the lacrimal sac.

Lacrimal sac is placed in a groove (the lacrimal fossa) formed by the lacrimal bone and the frontal process of the maxilla. It is placed behind the medial palpebral ligament and in front of the lacrimal part of the orbicularis oculi muscle. The sac is the dilated upper end of the nasolacrimal duct.

Nasolacrimal duct descends through the nasolacrimal canal formed by the lacrimal, maxilla and inferior nasal concha and enters the nasal cavity (in the inferior meatus). It is about two centimeters long, and is directed downwards and slightly laterally and backwards. At this entrance, a fold of mucous membrane, the lacrimal fold acts as a valve that prevents upward flow of nasal secretions.

4. Orbit and Its Contents

Each orbit has the shape of a four-sided pyramid and is lined by periosteum. It has four walls (two vertical and two horizontal), an apex posteriorly near the optic canal and a base. The orbital aditus is bounded by the orbital margin. The orbits contain the eyeballs,

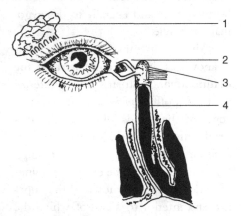

FIG. 5-3 Lacrimal apparatus. (1) Lacrimal gland; (2) Lacrimal canaliculus; (3) Lacrimal sac; (4) Nasolacrimal duct.

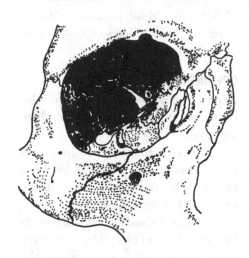

FIG. 5-4 Right bony orbit.

related muscles, blood vessels, and nerves embedded in a large amount of (periorbital) fat.

ANATOMICAL RELATIONSHIPS OF THE ORBITS

These related structures are frequently involved in diseases of or damage to the orbit.

Above: Anterior cranial fossa (and frontal lobes of the brain) and frontal paranasal sinus.
Laterally: Temporal fossa in front and middle cranial fossa behind.
Below: Maxillary (paranasal) sinus.
Medially: (paranasal) Ethmoidal sinuses and sphenoidal sinus.

Margins of the orbital aditus comprise three bones separated by three sutures.

SUPRAORBITAL MARGIN

The supraorbital notch (occasionally a foramen) is near the junction of middle and medial thirds of the supraorbital margin. It transmits the supraorbital nerve (continuation of the frontal branch of the ophthalmic nerve, V1) and supraorbital artery (branch of the ophthalmic artery).

LATERAL MARGIN

Orbital tubercle is located on the zygomatic bone. It gives attachment to the lateral palpebral ligament.

INFRAORBITAL MARGIN

Infraorbital foramen opens about one centimeter below the margin. It transmits the **infraorbital nerve** (terminal branch of the maxillary, V2) and infraorbital vessels (branches of the maxillary artery).

MEDIAL (ORBITAL) MARGIN

The margin is not continuous. The upper half formed by the frontal bone lies more posterior and is continued below by the lacrimal crest of the lacrimal bone. The lower half lies more anteriorly and is formed by the lacrimal crest of the maxilla. Between is the lacrimal sulcus for the lacrimal sac.

Lacrimal crest of the maxilla is a continuation of the inferior orbital margin onto the frontal process of the maxilla.

Lacrimal crest of the lacrimal bone is the continuation of the superior orbital margin down onto the lacrimal bone.

Lacrimal sulcus formed by the expanded medial margin of the orbit (maxilla and lacrimal) sbetween the lacrimal crests.

Nasolacrimal canal is formed by the maxilla laterally and lacrimal bone and inferior nasal concha medially. A downward continuation of the fossa, it transmits the nasolacrimal duct from the nasolacrimal sac to the inferior meatus of the nasal cavity.

There is also a feature of interest at each angle of the aditus.

Superomedial: Trochlear fovea or spine for attachment of the trochlea traversed by the tendon of superior oblique.

Superolateral: Fossa for the lacrimal gland.

Inferomedial: Nasolacrimal canal continuous with the lacrimal sulcus.

Inferolateral: The inferior orbital fissure about two centimeters behind the orbital margin.

The palpebral fascia is attached to the posterior lacrimal crest so that the lacrimal sac is not located within the orbit. The orbital aditus is slightly constricted which helps to prevent proptosis of the eyeball.

WALLS OF THE ORBIT

Roof: Orbital plate of frontal bone and lesser wing of sphenoid bone.

Floor: Maxilla, zygomatic bone, and orbital process of palatine bone.

Lateral wall: Zygomatic anteriorly and greater wing of sphenoid bone posteriorly.

Medial wall: Orbital plate of ethmoid, lacrimal, frontal process of maxilla part of body of sphenoid bone.

OPENINGS INTO THE ORBIT

There are nine openings into the orbit of which three are paired and three unpaired.

Superior orbital fissure (transmits cranial nerves III, IV and VI, branches of the ophthalmic nerve, sympathetic nerves and ophthalmic veins).

Inferior orbital fissure (transmits the maxillary nerve, V2), zygomatic nerve, orbital branches of pterygopalatine ganglion and infraorbital artery.

Supraorbital foramen transmits the supraorbital artery and nerve.

Infraorbital foramen transmits the infraorbital nerve and vessels.

Anterior ethmoidal foramen is a small opening at the junction of the medial wall and the roof. It transmits the anterior ethmoidal nerve (from the ophthalmic V1) and vessels.

Posterior ethmoidal foramen is also at the junction of the medial wall and roof behind the anterior ethmoidal foramen. It transmits the posterior ethmoidal branch of the nasociliary nerve and vessels (when present).

The optic canal is situated between the two roots of the lesser wing of the sphenoid above and medial to the superior orbital fissure. Through it, passes the optic nerve and ophthalmic artery.

A common tendinous ring gives origin to some of the orbital muscles. It is attached above, medial to and below the optic foramen and to the lateral side of the bend in the superior orbital foramen.

The zygomatico is the orbital foramen on the orbital surface of the zygomatic bone divides within the bone to emerge as the zygomatico-temporal and zygomatico-facial foramina.

The zygomatico-facial foramen transmits the zygomatic nerve (branch of the maxillary (V2) nerve) that branches into the zygomatico-facial and zygomatico-temporal nerves to the face and temporal regions respectively.

The nasolacrimal canal is formed by the maxilla laterally and lacrimal bone and inferior nasal concha medially. It transmits the nasolacrimal duct from the lacrimal sac to the inferior meatus of the nasal cavity.

EXTRINSIC OCULAR MUSCLES

Of the seven extrinsic ocular muscles, one elevates the eyelid and the remainder move the eyeball. With the exception of inferior oblique, the extrinsic muscles arise from a circumscribed area at the back of the orbit, and diverge forwards to surround the eyeball forming the "cone of muscles". The four recti arise from a fibrous band surrounding the optic canal and crossing the medial part of the superior orbital fissure — the **common tendinous ring**. The lateral rectus muscle extends above and below the bend in the superior orbital fissure. Each passes forwards and is inserted into the eyeball just behind the corneoscleral junction.

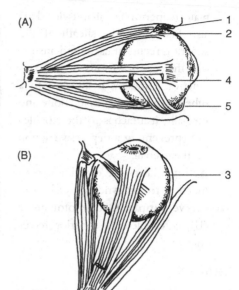

FIG. 5-5 Extrinsic ocular muscles (A) Lateral view (B) from above. (1) Levator palpebrae superioris; (2) Superior rectus; (3) Superior oblique; (4) Lateral rectus; (5) Inferior oblique; (6) Inferior rectus.

Levator Palpebrae Superioris

Levator palpebrae superioris may be considered as split from superior rectus to ensure that the upper eyelid is raised with the eyeball when the latter is directed upward.

- **Origin:** Roof of orbit (lesser wing of sphenoid) anterior to the optic canal. It overlies the superior rectus and widens anteriorly to an aponeurosis.
- **Insertion:** Pierces the orbital septum to become attached to the skin of the upper lid. Part attaches to the upper border of the superior tarsal

plate (superior tarsal muscle). Part also fuses with the sheath of the superior rectus muscle and inserts into the superior fornix of the conjunctiva. Laterally, it attaches to a tubercle on the zygomatic bone and medially, it attaches to the trochlea thus preventing a depressive action on the eyeball.

- **Action:** Raises the eyelid and conjunctival fornix synchronously.
- **Nerve supply:** Oculomotor nerve (III) after piercing superior rectus muscle.

Recti

Four straight muscles (recti) are named according to their location with reference to the eyeball: *superior, inferior, medial,* or *lateral recti.* Lateral and medial rectus are in the same horizontal plane and superior and inferior rectus in the same vertical plane. All four recti arise medial to the vertical axis of the eyeball so superior and inferior recti adduct the eyeball in addition to elevating and depressing the centre of the cornea.

- **Origin:** Common tendinous ring surrounding the optic canal and medial part of the superior orbital fissure. The lateral rectus arises from the part of the ring that spans the superior orbital fissure and is described as having two heads of origin, from the upper and lower parts of the lateral portion of the tendinous ring.

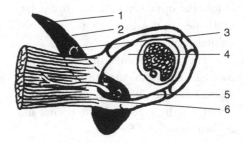

FIG. 5-6 Fibrous ring and superior orbita fissure — anterior aspect. (1) Lacrimal n; (2) Frontal n; (3) Trochlear n; (4) Oculomotor n; (5) Nasociliary n; (6) Abducent n.

- **Insertion:** On the eyeball by wide tendons, 0.5 cm behind the sclerocorneal junction (corneal limbus).
- **Nerve supply:** Lateral rectus muscle by the abducent nerve (VI); the other recti by the oculomotor nerve (III). Innervation of the extrinsic ocular muscles can be recalled by the mnemonic formula $LR_6SO_4 \cdot (AO_3)$ — Lateral rectus by the sixth nerve, superior oblique by the fourth nerve, all others by the third nerve.

Superior Oblique

- **Origin:** Orbital roof (body of sphenoid) adjacent to optic canal and medial to origin of levator palpebrae superioris. The tendon runs forward along the medial wall and as a slender tendon, passes through the trochlea, a fibrocartilaginous angled tube attached to the trochlear fossa, and directs the tendon posterolaterally under the superior rectus muscle.

- **Insertion:** Superolateral quadrant of the eyeball behind its equator.
- **Nerve supply:** Trochlear nerve (IV).

Inferior Oblique

- **Origin:** Front of the orbital floor form a depression on the upper surface of the maxilla.
- **Insertion:** The muscle passes laterally under the inferior rectus, but between the lateral rectus and the eyeball, to be inserted into the upper and lateral quadrant of the eyeball behind its equator.
- **Nerve supply:** Oculomotor nerve (III).

ACTION OF EXTRINSIC OCULAR MUSCLES

The eyeball is balanced in the orbit by fascial attachments and muscle action. Note that while the medial walls of the orbit are parallel, the lateral walls are perpendicular to one another. The visual or optic axis lies in the sagittal plane when looking forward. The orbital axis, the long axis of the eyeball, is the axis around which the recti are arranged (from optic canal to center of orbital aditus). When both eyes look directly forward, the visual axis does not correspond to the orbital axis.

Movements of the eyeball are resolved into those taking place around the primary axes at right angles to one another.

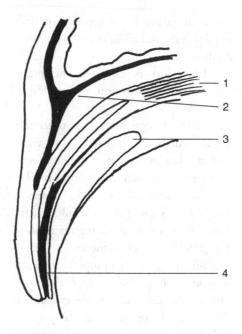

FIG. 5-7 Sagittal section through the upper eyelid. (1) Levator palpebrae superioris; (2) Orbital septum; (3) Superior conjunctival fornix; (4) Tarsus.

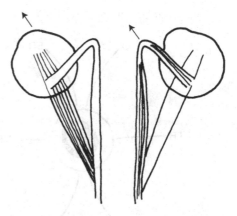

FIG. 5-8 Action of extrinsic ocular muscles. When the eye looks medially, superior oblique directs the eye downwards. The other eye is rolled (intorted) by superior oblique.

Abduction is movement of the pupil laterally and adduction is movement medially.

Elevation is directing the pupil upward and **depression** is directing it downward.

All four recti arise medial to the vertical axis ot the eyeball so that when looking directly forwards, superior and inferior recti adduct, in addition to elevate and depress the orbit. Superior rectus rotates the globe medially, (intorsion) and inferior rectus rotates it laterally (extorsion). When the eye is turned medially, the (adducted) superior oblique is positioned to depress the eyeball and inferior oblique to elevate the eyeball. In the abducted eye, superior oblique can only intort and inferior oblique extort.

To test the extrinsic ocular muscles, the eye must be placed in a medial or lateral gaze and elevation and depression tested. With the eye looking medially, the obliques are most suited to elevate and depress. With the eye looking laterally, it is the superior and inferior recti that are most suited to be the elevators and depressors.

ARTERIES

Ophthalmic artery is a large and important branch of the intracranial part of the internal carotid. It arises after internal carotid has pierced the dura mater on the medial side of anterior clinoid process of the sphenoid and enters orbit through optic canal lateral to and below optic nerve. It then crosses above the nerve to its medial side and runs forward between superior oblique and medial rectus to the front of the orbit and divides into its terminal branches, the supratrochlear and dorsalis nasi arteries.

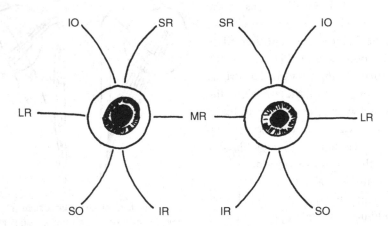

FIG. 5-9 Action of extrinsic ocular muscles from in front. The eye is directed directly downward by the action of the obliques, superior and inferior recti. IO; SR; SO; IR; LR, MR.

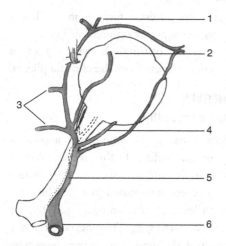

FIG. 5-10 Arteries of the orbit. (1) Supratrochlear a; (2) Supraorbital a; (3) Ethmoidal a;
(4) Long ciliary a; (5) Ophthalmic a; (6) Internal
carotid a.

Branches of ophthalmic artery:

Lacrimal branch accompanies the
lacrimal nerve over the lateral rectus
muscle to reach the lacrimal gland. It
gives off lateral palpebral and zygomatic
branches and a branch which re-enters
the cranial cavity through the superior
orbital fissure.

Central artery of the retina
begins before the ophthalmic artery
becomes lateral to the optic nerve into
which it sinks one centimeter behind
eyeball. It runs in the substance of the
optic nerve to the retina. Note that it is
a virtual end artery i.e. there is no effective collateral circulation to the retina so
that occlusion results in blindness.

Short posterior ciliary arteries
are several small vessels that perforate
sclera and supply the choroid.

Long posterior ciliary arteries
pierce the sclera and pass forward
between the choroid and sclera to supply the iris and ciliary body.

Supra orbital artery accompanies
the supraorbital nerve ascending into
the orbit between levator palpebrae
superioris and the periosteum. It passes
through the supraorbital notch; and
ascends over the frontal bone to supply
the forehead and scalp.

Muscular branches superior and
inferior supply to the muscles of the
orbit. These give off several *anterior ciliary arteries* that pierce sclera behind the
cornea and supply the iris.

Ethmoidal branches separates
into a posterior and an anterior branch.
The posterior ethmoidal artery passes
through the posterior ethmoidal foramen to supply the mucoperiosteum of
the sphenoidal and posterior ethmoidal
air sinuses and the mucosa of the nose.
The anterior ethmoidal artery passes
through the anterior ethmoidal foramen
and supplies the mucoperiosteum of
the anterior ethmoidal air cells, dura
mater and frontal sinus. It accompanies
the external nasal branch of anterior
ethmoidal nerve to supply the skin of
the nose (anterior nasal artery).

Medial palpebral branches supply each eyelid, and arise near the pulley
for the superior oblique. They form an
arch in each eyelid and supply lacrimal
apparatus.

Supratrochlear branch (terminal
branch) leaves the orbit superomedially

and with the supratrochlear nerve supplies the medial scalp and forehead.

Dorsal nasal branch (terminal branch) passes between medial palpebral ligament and trochlea to supply the root of the nose.

VEINS

Superior ophthalmic vein is formed near the root of the nose by the union of the supraorbital and angular veins where it anastomoses with the facial vein. It leaves the orbit through the superior orbital fissure and drains into the cavernous sinus.

Inferior ophthalmic vein begins on the floor of the orbit, it connects

with the anterior facial vein. It drains into the superior ophthalmic vein.

Both ophthalmic veins receive vorticose veins from the choroid plexus.

NERVES
Optic Nerve (II)

Optic nerve (II) comprises axons of ganglion cells of the retina. Axons pierce the lamina cribrosa of the sclera to become myelinated (but they have no neurilemma) and are enclosed by the cranial meninges. The "nerve", really a tract of the brain, and its meninges pass backward and medially to leave the orbit through the optic canal.

Oculomotor Nerve (III)

Oculomotor nerve (III) is the chief motor nerve to ocular muscles that enters the orbit as two branches through the superior orbital fissure between the heads of the lateral rectus muscle.

Superior Branch

Superior branch passes *across* the optic nerve and terminates by innervating the superior rectus and levator palpebrae superioris muscles.

Inferior Branch

Inferior branch passes forward on the floor of the orbit and supplies medial and inferior recti and terminates in the inferior oblique. The nerve to inferior oblique gives preganglionic parasympathetic fibers to the ciliary ganglion, where they synapse. Postganglionic fibers pass via the short ciliary nerve to the eyeball,

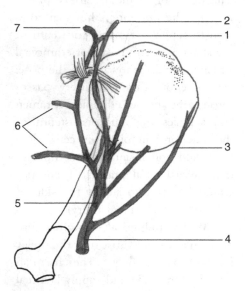

FIG. 5-11   Nerves of the orbit. (1) Supratrochlear n; (2) Supraorbital n; (3) Lacrimal n; (4) Ophthalmic n; (5) Nasociliary n; (6) Ethmoidal n; (7) External nasal n.

to innervate the ciliary and sphincter pupillae muscles.

Trochlear Nerve (IV)

Trochlear nerve (IV) enters the orbit through the superior orbital fissure lateral to (outside of) the common tendinous ring. It passes above levator palpebrae superioris and innervates the superior oblique muscle.

Abducent Nerve (VI)

Abducent nerve (VI) enters the orbit through the superior orbital fissure passing through the common tendinous ring. It passes forward on the surface of lateral rectus that it innervates.

Sensory nerves of the orbit are the branches of the ophthalmic nerve (V_1). They arise nearby and pass through the superior orbital fissure individually.

Lacrimal Nerve (a branch of V1)

Lacrimal nerve (a branch of V1) enters the orbit lateral to (outside of) the common tendinous ring. It follows the upper border of lateral rectus muscle, and innervates the lacrimal gland, conjunctiva, and upper eyelid. A sensory nerve, it also carries postganglionic secretory fibers originating in the pterygopalatine ganglion. These fibers are carried in the zygomatic nerve and then by a communicating branch to the lacrimal nerve.

Frontal Nerve (a branch of V1)

Frontal nerve (a branch of V1) is the largest branch of the ophthalmic nerve.

It lies on levator palpebrae superioris beneath the roof of the orbit and divides into two branches:

Supraorbital Nerve

Supraorbital nerve passes between levator palpebrae superioris and roof of the orbit. It leaves the orbit through the supraorbital foramen and innervates the skin of the forehead and scalp.

Supratrochlear Nerve

Supratrochlear nerve leaves the orbit above the trochlea for superior oblique to supply the conjunctiva and skin of the upper eyelid and the skin of the forehead.

Nasociliary Nerve (branch of V1)

Nasociliary nerve (branch of V1) is the sensory nerve of the eye. It enters the orbit through the common tendinous ring and passes towards the medial wall of the orbit. It branches as follows:

Communicating Branch

The communicating branch carries sensory fibers of the eyeball (cell bodies in the trigeminal ganglion). Fibers pass through the ciliary ganglion to reach the nasociliary nerve.

Long Ciliary Nerves

Long ciliary nerves pass along the medial side of the optic nerve, join the short ciliary nerves pierce the sclera with their corresponding arteries and distribute to the ciliary body and iris. The nerves carry afferent fibers from the uvea and cornea and sympathetic (motor) fibers from the superior cervical ganglion (via the internal carotid plexus) to innervate the dilator pupillae muscle.

Posterior Ethmoidal Nerve

Posterior ethmoidal nerve (may be absent) enters the posterior ethmoidal foramen, to innervate the mucoperiosteum of the sphenoidal and posterior ethmoidal sinuses.

Anterior Ethmoidal Nerve

Anterior ethmoidal nerve is the continuation of the nasociliary nerve. It passes through the anterior ethmoidal foramen into the anterior cranial fossa. It travels on the cribriform plate and passes through a nasal slit by the side of crista galli into the nose. An internal nasal branch supplies the mucous membrane of the nasal walls while an external nasal branch passes between the nasal bone and lateral cartilage to supply the skin of the ala and tip of the nose.

Infratrochlear Nerve

Infratrochlear nerve is given off after the anterior ethmoidal branch, this nerve passes forward on the medial wall of the orbit and passes beneath the trochlea to emerge above the medial angle of the eye. It innervates root of the nose, the skin of the upper eyelid, and the lacrimal sac.

Ciliary Ganglion

Ciliary ganglion is a peripheral parasympathetic ganglion located lateral to the optic nerve near the apex of the orbit. Its connections (roots) entering or leaving posteriorly are:

Sensory (Long) Root

Sensory (long) root passes to the eyeball (choroid, iris and cornea) from the nasociliary and (of the ophthalmic) join the posterior angle of the ganglion. Its cell bodies are located in the trigeminal ganglion.

Parasympathetic (Short) Root

Parasympathetic (short) root arises from the inferior branch of the oculomotor nerve to inferior oblique. Preganglionic cell bodies are located in the anterior part of oculomotor nucleus (accessory oculomotor nucleus). Their axons synapse in the ciliary ganglion and postganglionic axons pass in short ciliary nerves to supply the sphincter pupillae and ciliary muscles.

Sympathetic Root

Sympathetic root consists preganglionic cell bodies are located in the superior cervical ganglion and their axons join the carotid plexus. They frequently enter with the sensory root of the ciliary ganglion passing through the ganglion and travel in the short ciliary nerves to reach the eyeball. The fibers are vasoconstrictor to intrinsic vessels of eyeball.

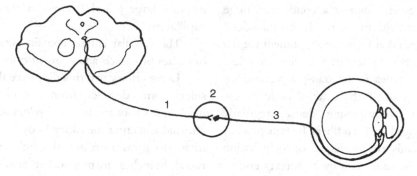

FIG. 5-12 Autonomic innervation of the bulb. (1) Oculomotor n; (2) Ciliary ganglion; (3) Short ciliary n.

Nerves leaving the ciliary ganglion are the short ciliary nerves, ten or twelve in two bundles, large inferior and small superior. They pass forwards above and below the optic nerve, with long ciliary nerves of the nasociliary. The nerves separate and pierce the sclera around the optic nerve. They run in grooves on the internal surface of the eyeball to end in the ciliary muscle, sphincter pupillae (parasympathetic) and cornea (sensory).

5. Eyeball

The eyeball is a sphere occupying the anterior third of the orbit that bulges anteriorly. It is enveloped by the bulbar fascia (except the cornea), which separates the eyeball from the periorbital fat. It has three concentric coverings (tunics) that enclose three refractive media.

The anterior and posterior poles are the central points of the cornea and sclera in the anteroposterior axis of the eye.

Equator is the imaginary line equidistant between the poles.

Meridian is a line on the globe from pole to pole cutting the equator at right angles.

The anteroposterior diameter of the eyeball may be normal (eumetropic), greater (myopic) or less (hypermetropic).

OUTER (FIBROUS) TUNIC

Sclera is the opaque dense connective tissue capsule that covers approximately five sixths of the eyeball. It is continuous with the **cornea** at the sclerocorneal junction (limbus). The sclera is covered anteriorly by the transparent **bulbar conjunctiva** which contains many small blood vessels and nerves. It receives the insertions of the extrinsic ocular muscles and is pierced posteriorly by vessels and nerves.

Lamina cribrosa is a meshwork of scleral connective tissue fibers formed where the sclera is perforated posteriorly by the fiber bundles of the optic nerve.

Scleral spur is a continuous ridge around the inside of the corneoscleral juntion that forms an attachment for the meridional fibers of the ciliary muscle.

Cornea is the transparent, anterior one sixth of the external tunic of the eyeball comprising dense, regularly arranged collagen fibers. Its transparency depends of its degree of dehydration that is maintained by its limiting epithelia. The anterior surface is covered by bulbar conjunctiva. The cornea is avascular and richly supplied with sensory nerve endings. Most of the refraction of light by the eye occurs at the surface of the cornea.

Middle (vascular) tunic (uvea) comprises the choroid, ciliary body and iris.

Choroid extends from the optic nerve as far forward as the ora serrata of the retina. It is situated between the sclera and the retina, and is the vascular tunic of the eyeball. It is continued anteriorly into the ciliary body, forming a number of inward foldiing projections known as the ciliary processes that are arranged in a circle. The choroid is composed of a pigmented meshwork of areolar tissue in which are embedded nerve plexuses, small arteries (derived from the short ciliary arteries) and draining veins (draining to vorticose veins).

Externally, the choroid is connected to the sclera by loose connective tissue (suprarchoroid lamina) traversed by vessels and nerves. The intermediate layer (vascular layer) contains capillaries and veins. The most internal layer (basal lamina) contains capillaries.

The vascular lamina contains small branches of vessels and pigment cells.

Long ciliary arteries (2) pierce the sclera some distance from the optic nerve, pass forward between sclera and choroid and enter the ciliary body. They divide to form a greater iris circle and radial branches form another smaller circle (the lesser iris circle).

Short ciliary arteries pierce the sclera at the periphery of the optic disc, pass forwards and send branches inwards to end in the innermost layer. They are the major source of nutriment to the pigmented layer of the retina. Terminal branches reach a rich capillary plexus in the ciliary body.

Anterior ciliary arteries pierce the sclera behind the corneoscleral juntion contribuing to a plexus in the ciliary body and anastomose with the greater iris circle.

Veins lie external to the arteries and join together into four or five principal trunks, which pierce the sclera (as vorticose veins) midway between the cornea and the optic nerve. These veins drain into superior and inferior ophthalmic veins.

Ciliary body is the elevated anterior continuation of the choroid. It is a wedge-shaped ring that connects the choroid with the iris. It suspends the lens by zonular fibers which insert into the capsule of the lens.

Ciliary processes are about 70–80 radiating folds of the ciliary body which project into the posterior chamber. They have a role in secretion of aqueous humor and serve as an insertion for zonular fibers.

Ciliary muscle is smooth muscle running from the base of the ciliary body to the scleral spur. It is arranged in two sets:

Meridional (radial) fibers, which arise from the sclerocorneal junction and pass backwards to attach to the ciliary processes and scleral spur;

Oblique fibers which form a ring internal to the meridional fibers.

Ciliary muscle acts to accommodate for near vision by drawing the choroid and ciliary processes forward. This relaxes the suspensory ligament of the lens (zonular fibers), so the lens becomes more convex. The ciliary muscle is innervated by the parasympathetic fibers of the short ciliary nerves.

Iris is a contractile, annular plate, continuous with the ciliary body located vertically between the cornea and the lens. The circular aperture formed by the free ends of the iris is the **pupil**. The iris divides the space in front of the lens into anterior and posterior chambers.

The anterior surface of the iris lacks an endothelial cover so the connective tissue core of the iris is exposed. It is colored and marked by wavy lines converging towards the free edge of the pupil.

The posterior surface is lined with a double layer of darkly pigmented epithelium and is marked with folds prolonged from the ciliary processes. The framework of the iris is a delicate stroma of connective tissue, containing blood vessels, nerves, pigment cells, and two groups of involuntary muscular fibers.

Within the loose stroma of the iris are two involuntary muscles: sphincter pupillae and dilator pupillae.

Sphincter pupillae comprises circular fibers located in the posterior stroma near the margin of the pupil. It is innervated by parasympathetic fibers by the short ciliary nerves. Contraction results in constriction of the pupil (miosis).

Dilator pupillae comprises radiating smooth muscle fibers located near the posterior surface of the iris. These are innervated by sympathetic fibers of the long ciliary nerves. Contraction causes dilatation of the pupil (mydriasis).

Blood vessels of the iris consist of the two *long posterior ciliary* and the *anterior ciliary* arteries; the former pierce the sclera close to the optic nerve, and pass forwards in the space between the lamina fusca of the sclera and the suprachoroid lamina of the choroid to enter the outer surface of the iris. Ciliary arteries anastomose with the corresponding vessels of the opposite side, and with those from the vascular zone of the sclera, formed by the anterior ciliary arteries. These form the *greater arterial circle* of iris. Small branches from this circle converge

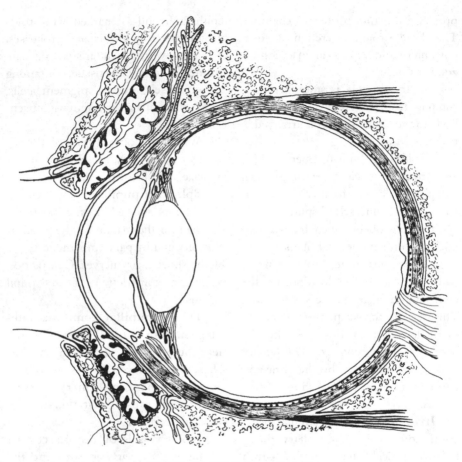

FIG. 5-13   Sagittal section of the bulb.

towards the pupil, and freely anastomose with one another, forming the
lesser arterial circle.

Veins follow the same arrangement
as the arteries, and drain into veins of
choroid coat.

Nerves of the choroid and iris are
about fifteen in number, and are the
ciliary nerves from the ciliary ganglion
and the nasociliary branch of the

trigeminal. They follow the course of
the blood vessels very closely and,
reaching the ciliary body, form a plexus,
which sends twigs to the ciliary muscle,
iris, and cornea.

The iris divides the chamber of the
orbit into two parts, known as the anterior and posterior chambers.

The *anterior chamber* is bounded in
front by the cornea, behind by the iris,

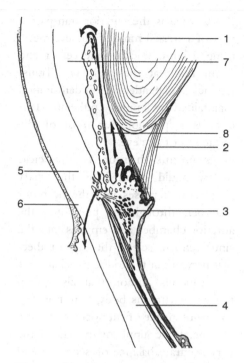

FIG. 5-14 Sagittal section of ciliary region and iridocorneal angle. (1) Lens; (2) Ciliary zonule; (3) Ciliary muscles; (4) Sclera; (5) Scleral spur; (6) Sinus venosus; (7) Anterior chamber; (8) Posterior chamber.

and opposite the pupil by the anterior part of the lens.

The *posterior chamber is* the triangular interval at the circumference of the lens between the ciliary processes, the iris, and the ciliary zonule.

INNER (NERVOUS) TUNIC

Detailed structure of the retina is best studied in programs in Neuroscience.

Retina is the internal nervous tunic of the eye that contains special receptors to light onto which an inverted image of objects is cast by the refractive media of the eye It is disposed in two layers:

Pigmented stratum comprises a single layer of pigment containing cuboidal cells that developed from the outer layer of the optic cup.

Cerebral stratum comprises five types of neurons arranged in three layers of neurons developed from the inner layer of the optic cup. Photoreceptors occupy the outermost layer adjacent to and interdigitated with the pigmented layer. A microscopic subretinal space containing proteoglycan separates the two layers.

The retina is divided into a functional posterior optic part up to the ora serrata and anterior non-functional ciliary part covering the ciliary body and posterior part of the iris.

The optic part of the retina contains the photoreceptors, supporting cells and interneurons. It extends from the optic disc to the ora serrata, a jagged abrupt line where the number of cell layers reduces to two.

Optic disc (blind spot) is the region devoid of receptors where the nerve fibers (axons of ganglion cells) leave the eyeball. It is about three millimeter to the medial side of the yellow spot (macula lutea). Coursing over the optic disc may be seen the central vessels of the retina.

Macula lutea, or yellow spot, is situated in the axis of the globe. Fovea centralis is a depression in the middle of the macula lutea in which vision is most acute. The retina is very thin at the fovea and some layers of the retina are absent. It is avascular and nourished by the choroid.

Ciliary part of the retina is a double layer of cuboidal cells that extends from the ora serrata over the ciliary body to the iris. It is insensitive to light.

The iridial part of the retina extends as a bilayer of cuboidal epithelium on the posterior surface of the iris extending to its pupillary margin: it is continuous with the ciliary part.

BLOOD VESSELS OF THE RETINA

The **central artery of the retina** passes through the optic nerve, and reaches the inner surface of the retina through the optic disc. Here, it divides into two branches, a superior and inferior, and each of these branches into a lateral or temporal division, and a medial or nasal division.

The lateral branches give small branches to end in the capillaries round the fovea. The remainder of the branches are distributed, as capillaries, to the retina, as far as the ora serrata, but the smaller branches do not anastomose with one another or with any other vessels. The veins follow the same distribution as the arteries.

Light entering the eye is bent, or refracted, at the interfaces of the various media through which it passes. The media are cornea, aqueous humor, vitreous body and lens.

Cornea is the anterior transparent part of external tunic of the eye measuring 0.5 mm thick at the center and 1 mm thick at the periphery. Transparency is dependent upon dehydration maintained by limiting epitheium. This layer is responsible for most of the refraction of the eye.

Aqueous humor is the clear watery liquid secreted by the ciliary processes. It passes into the posterior chamber, through the pupil into the anterior chamber and empties into the sinus venosus sclerae through a trabecular network at the iridocorneal angle.

Aqueous humor, that also penetrates the vitreous body, is responsible for maintenance of intraocular pressure and metabolic supply of the lens. If the very accurate balance of secretion and resorption of aqueous humor is disturbed, intraocular pressure increases leading to glaucoma.

Vitreous body (humor) is a transparent gelatinous mass that fills four fifths of the eyeball between the lens and retina. It not only transmits light rays but maintains the position of the lens and retina (in contact with the pigment layer).

Lens lies behind the iris and pupil and is held in place by its suspensory ligament (ciliary zonule) consisting of a series of fibers that extend from the capsule of the lens to the ciliary body. The lens is biconvex, transparent, and composed of: a structureless membrane — the capsule of the lens; epithelium of the lens — a layer of cuboidal cells lining its anterior surface. At the equator, cells divide by occasional mitoses, elongate and differentiate into; and lens fibers which are elongated ribbon like cells which form the cortex of the lens.

Chapter **6**

Nasal Region

Chapter Outline

1. External Nose
2. Nasal Cavity
3. Arteries
4. Veins
5. Nerves
6. Paranasal Sinuses

The nose consists of the external nose and the nasal cavities. The most essential element, however, is the olfactory mucous membrane. It contains the organ of smell through its olfactory mucous membrane. It also conditions inspired air by humidifying, filtering and controlling temperature of inspired air.

1. External Nose

The external nose presents:

Tip or apex: Free lower angle of the pyramidal shaped external nose.
Root or bridge: Upper attachment of external nose to the forehead.
Dorsum: Rounded border between the apex and root.
Nostrils (or nares): Inferior perforation of the external nose.
Alae: Form the lateral boundaries of the nostril.
Septum: Is the medial boundary of the nostril.

The framework of the upper external nose includes the nasal and frontal bones and the frontal processes of the maxillae. The lower part is cartilaginous and comprises five major hyaline cartilages:

A septal cartilage contributes to the septum between the nasal cavities. It has two lateral nasal cartilages (upper lateral nasal cartilages), triangular expansions of the septal cartilage.

Two greater alar cartilages placed below each lateral cartilage. They have lateral and medial crurae (lower lateral nasal cartilages).

Several smaller plates, the lesser alar cartilages embedded in the fibrous membrane connecting the lateral nasal cartilages and the alar cartilages.

MUSCLES OF THE EXTERNAL NOSE

All the muscles of the external nose are supplied by the facial (VII) nerve. Two muscles constitute the nasalis muscle — the compressor is called the transverse part and the dilatator the alar part.

Compressor Naris (the Sphincter)

- **Origin:** Maxilla at the side of the lower part of the bony anterior nares.
- **Insertion:** Aponeurosis continued across the nose to its fellow of the opposite side.

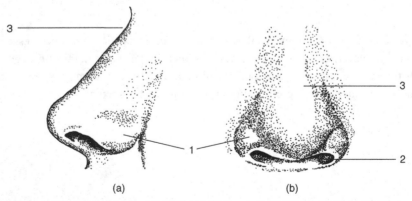

(a)   (b)

FIG. 6-1   External nose (a) lateral (profile) and (b) from in front. (1) Ala; (2) Naris; (3) Dorsum.

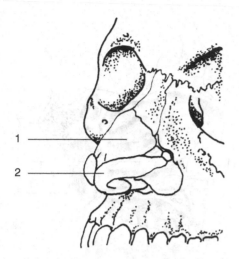

FIG. 6-2 Cartilage plates of the nose. (1) Lateral nasal cartilage; (2) Greater alar cartilages.

- **Action:** Compresses the upper part of the cartilaginous nasal aperture.

Dilatator Naris

- **Origin:** Maxilla below compressor naris.
- **Insertion:** Side of the alar cartilage.
- **Action:** Pulls on lateral wing of cartilage to open up nostril.

Procerus

Procerus is the continuation of median fibers of frontalis down onto the nose.

- **Origin:** Fascia over the bridge of the nose.
- **Action:** Produces small transverse wrinkles at root of nose by slight elevation of the external nose.

Depressor Septi

- **Origin:** Incisive fossa of the maxilla.
- **Insertion:** Septum and posterior part of the alae.
- **Action:** Depresses the septum of the nose.
- **Blood supply:** Branches of the facial (lateral nasal) and ophthalmic (dorsal nasal) arteries.
- **Nerve supply:** Branches of the ophthalmic (V1) (external nasal and supratrochlear and infratrochlear nerves) and the maxillary (V2) (infraorbital) nerves.

2. Nasal Cavity

The nasal cavities, separated from each other by the nasal septum, are open in front at the nostrils and behind at the posterior apertures (choanae) where they open into the nasopharynx.

Nares are the anterior apertures of the external nose.

Piriform aperture is the bony opening bounded by the nasal bones and laterally and maxillae below.

Choanae are the oblong posterior apertures leading from the nose to the nasopharynx. They are bounded medially by the vomer, inferiorly by the horizontal plate of the palatine bone, laterally by the medial pterygoid plate, and superiorly by the body of the sphenoid.

BOUNDARIES OF THE NASAL CAVITIES

The floor of the nasal cavities consists of the palatine process of the maxilla and the horizontal plate of the palatine bone.

The roof of the nasal cavities consists of nasal cartilages and nasal bone, anteriorly; the anterior and inferior surfaces of the body of the sphenoid, posteriorly; and cribriform plate of the ethmoid between.

Medial Wall (Septum)

Medial wall (septum) or the nasal septum is formed from before backward; by the septal cartilage, perpendicular plate of ethmoid and the vomer. The columella is the mobile part of the septum near the apex of the nose.

Lateral Wall

Lateral wall is formed by parts of the nasal bones, frontal process of the

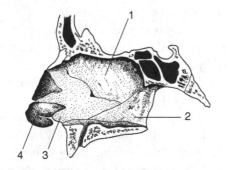

FIG. 6-3 Nasal septum. (1) Perpendicular plate of ethmoid; (2) Vomer; (3) Septal cartilage; (4) Greater alar cartilage.

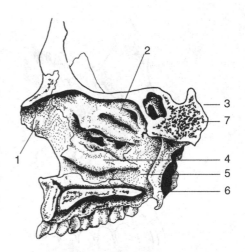

FIG. 6-4 Bony lateral nasal wall. (1) Maxilla; (2) Superior concha; (3) Middle concha; (4) Inferior concha; (5) Perpendicular plate of palatine bone; (6) Medial pterygoid plate; (7) Body of sphenoid.

maxilla and the lacrimal bone, ethmoid (labyrinth and conchae), maxilla, inferior nasal concha, vertical plate of the palatine and medial pterygoid plate of the sphenoid, posteriorly. The lateral wall is irregular because of three medial projections, the superior, middle and inferior conchae. The superior and middle conchae are processes of the ethmoid. The inferior concha is a separate bone.

The conchae subdivide the nasal cavity into grooved passageways named meatuses. Meatuses are named by the concha above — the superior meatus lies below the superior concha, middle meatus below middle concha and the inferior meatus is below the inferior concha (and above the roof of the mouth).

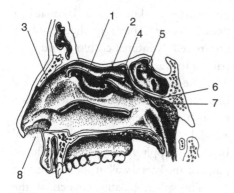

FIG. 6-5 Lateral walls of nose (conchae removed). (1) Hiatus semilunaris; (2) Bulla ethmoidalis; (3) Atrium of middle meatus; (4) Sphenoethmoidall recess; (5) Superior meatus; (6) Middle meatus; (7) Inferior meatus; (8) Vestibule of nose.

Sphenoethmoidal recess is the space above and behind the superior concha. It receives the opening of the sphenoidal sinus.

Superior meatus lies lies beneath the superior concha. It receives the posterior ethmoidal air cells (and in the dried skull, the sphenopalatine foramen).

Middle meatus is the region bounded by attachments of the middle concha above and inferior concha below. There are several features in the meatus:

Bulla ethmoidalis beneath the middle concha is an elevation caused by the underlying middle ethmoidal air cells. Two ostia usually open onto the bulla.

Hiatus semilunaris is a curved trench in front of and below the bulla ethmoidalis (beneath the ethmoidal bulla and the uncinate process of the ethmoid). It receives usually two openings of anterior ethmoidal air cells and the opening of the maxillary sinus.

Ethmoidal infundibulum is a narrow passage upward and in front of the hiatus semilunaris. It receives secretion from the frontal sinus.

Uncinate process is a curved ridge lying inferior to the bulla ethmoidalis.

Ostium of the maxillary sinus is located below the bulla ethmoidalis at the end of the hiatus semilunaris.

Inferior nasal concha is a separate bone attached to the lateral wall of the nose. It articulates with the maxilla, lacrimal, ethmoid and palatine bones.

Inferior meatus is the passage bounded by the maxilla, lacrimal and inferior nasal concha. It receives the nasolacrimal duct. Nasal reflux is prevented by a lacrimal fold of mucous membrane.

Vestibule is the dilated part of the nasal cavity above the nostril. It is lined with skin bearing sebaceous glands, sweat glands and hair.

Atrium is the wide space in front of the middle meatus (above the vestibule). It is limited posteriorly by a ridge, the **agger nasi.**

The nasal cavity is lined with a thick, vascular mucous membrane (respiratory area) closely adherent to underlying periosteum and continuous with that of the pharynx, vestibule and adjoining sinuses. The epithelium of the posterior two thirds of the nasal cavity is pseudostratified ciliated with goblet cells and including intraepithelial glands. Mucus secreted by goblet cells traps particulate matter and cilia of columnar cells beat towards the pharynx expelling the mucus.

Bipolar chemoreceptors are located interspersed between supporting cells in the epithelium of the olfactory region that extends over the superior nasal concha and upper third of the nasal septum. Their dendrites reach the surface and their unmyelinated axons collect into bundles that penetrate the ethmoid and enter the olfactory bulb.

3. Arteries

The mucosa of the nasal cavity is highly vascular especially over the middle and inferior conchae. Arteries feed a subepithelial capillary plexus and then a plexus of large thin walled veins with enlargements known as lacunae or "swell bodies". Flow in the vessels is controlled by intimal cushions and contribute to warming of inspired air.

The anterior and posterior ethmoidal branches of the **ophthalmic artery** pass through the anterior and posterior ethmoidal foraminae and their nasal branches supply the upper and anterior parts of the lateral wall of the nasal cavity.

The sphenopalatine branch of the **maxillary artery** passes through the sphenopalatine foramen to enter the nasal cavity. It provides posterior lateral nasal branches to supply the posterior lateral wall of the nasal cavity.

Sphenopalatine artery continues beneath the body of the sphenoid bone to terminate as the posterior septal branches on the nasal septum.

Greater palatine artery descends from the pterygopalatine fossa through the greater palatine canal (with the

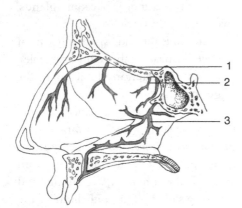

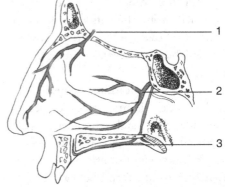

FIG. 6-6 Arteries of nasal septum. (1) Anterior ethmoidal a; (2) Posterior ethmoidal a; (3) Septal branch of sphenopalatine a.

FIG. 6-7 Arteries of lateral nasal wall. (1) Anterior ethmoidal a; (2) Sphenopalatine a; (3) Greater palatine a.

greater palatine nerve) emerging at the posterolateral angle of the hard palate. It passes forward in a groove on the hard palate to supply the hard palate. Lesser palatine branches given off in the lesser palatine canal, a branch of the greater palatine artery, and supplies the soft palate.

Epistaxis or bleeding from the nose is commonly from the region at the junction of the area supplied by the septal branch of the superior labial artery and septal branch of the sphenopalatine artery.

4. Veins

Veins drain an extensive plexus beneath the mucous membrane. The principal draining veins correspond approximately to the arteries:

Anterior ethmoidal veins follow the ethmoidal arteries through the nasal slit to the anterior cranial fossa. Here, it receives meningeal veins, traverses the anterior ethmoidal foramen to join the superior ophthalmic vein that then drains into the cavernous sinus.

Posterior ethmoidal veins traverse holes in the cribriform plate to enter the anterior cranial fossa then posterior ethmoidal foramen. They enter the superior ophthalmic vein and reach the cavernous sinus.

Nasal veins drain to the ophthalmic veins and the cavernous sinus

Sphenopalatine vein through the sphenopalatine foramen to reach the pterygoid plexus in the pterygopalatine fossa. Blood reaches the maxillary

vein or the cavernous sinus via emissary veins.

Tributaries drain to the greater palatine vein (including through the incisive foramen) enter the greater palatine canal to reach veins in the pterygopalatine fossa and connect with the pterygoid plexus. From here, blood drains to the maxillary vein, then retromandibular vein (or cavernous sinus via emissary veins).

Septal veins drain to the superior labial vein join the anterior facial vein.

Small channels from the vestibule of the nose join the tributaries of the anterior facial vein. A small vein may be present that passes through foramen caecum to join the superior sagittal sinus.

5. Nerves

In general, the anterior parts of the medial and lateral walls of the nasal cavity are supplied by the **anterior ethmoidal nerve.** An area near the cribriform plate and lateral walls are supplied by the **olfactory nerve** and the remainder are supplied by branches of the **maxillary nerve.** Secretory parasympathetic nerves to mucous cells are derived from the **pterygopalatine ganglion.**

SPECIAL SENSORY NERVES

Olfactory nerve (I) innervates the mucosa of the olfactory area of the nasal cavity. Olfactory receptor cells are bipolar neurons located in mucosa. The olfactory nerve consists of a number of bundles of unmyelinated nerve fibers of the receptor cells covered by meninges. Olfactory mucosa is that part of the nasal mucous membrane receptive to odors. It is yellowish in color and confined to the superior nasal concha, roof and upper third of the nasal septum. Fiber bundlespass through the cribriform plate of the ethmoid to reach the olfactory bulb.

GENERAL SENSORY NERVES

Anterior ethmoidal nerve (terminal branch of the nasociliary branch of ophthalmic division of the trigeminal nerve) enters the cranial cavity from the orbit through the anterior ethmoidal foramen. It runs along the lateral margin of the cribriform plate and through the ethmoidal slit at the side of crista galli, and enters the nasal cavity. It descends beneath the nasal bone, provides the internal nasal branches, and leaves the nasal cavity between the nasal bone and lateral nasal cartilage as the external nasal nerve that supplies the skin of the dorsum and the tip of the nose.

Posterior superior nasal nerves from the pterygopalatine ganglion, pass through the sphenopalatine foramen to supply the upper and posterior parts of the lateral wall of the nasal cavity.

Nasopalatine nerve also arises from the pterygopalatine ganglion. It passes through the sphenopalatine foramen, crosses the roof of the nasal cavity beneath the body of the sphenoid, then descends along the nasal

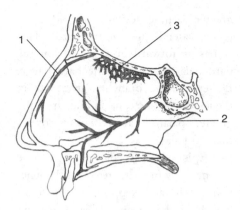

FIG. 6-8   Nerves of nasal septum. (1) Anterior ethmoidal n; (2) Nasopalatine n; (3) Olfactory n.

septum to the palate, and reaches the roof of the mouth through the incisive foramen. The nasopalatine nerve innervates the nasal septum and anterior hard palate.

Greater palatine nerve descends from the pterygopalatine ganglion through the greater palatine canal. It emerges at the greater palatine foramen to reach the hard palate. The posterior inferior nasal branches arise in the canal and innervate the lower part of the nasal cavity.

Lesser palatine nerve descends from the pterygopalatine ganglion. It passes in the greater palatine canal (with the greater palatine nerve) then enters the lesser palatine canal emerging from the lesser palatine foramen to innervate the tonsil and soft palate.

Pharyngeal nerve arises from the pterygopalatine ganglion. It passes backward through the palatinovaginal canal to supply the mucous membrane

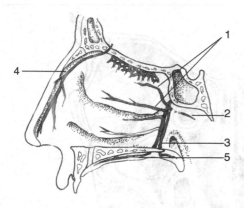

FIG. 6-9 Nerves of lateral nasal wall. (1) Lateral posterior superior nasal n; (2) Lateral posterior inferior nasal n; (3) Posterior ethmoidal n; (4) Anterior ethmoidal n; (5) Palatine n.

of the roof and posterior wall of the nasopharynx.

6. Paranasal Sinuses

Air sinuses located within the maxilla, frontal, ethmoid and sphenoid are small at birth but enlarged greatly at puberty. They develop as diverticula from the nasal cavity and are therefore lined with mucous membrane continuous with respiratory mucosa of the nasal cavity. They drain by ciliary action and suction. They serve to decrease the weight of the skull bones and act as resonating chambers for the voice.

Frontal sinus is paired, vary greatly in size and symmetry and are separated by a bony septum. They are located between the vertical and orbital plates of the frontal bone and are probably anterior ethmoidal sinuses that invade the

frontal bone after birth. Each sinus usually opens by a narrow passage, the frontonasal duct, into the upper anterior part of the hiatus semilunaris.in the nasal cavity.

Ethmoidal sinuses are numerous, small, thin-walled (ethmoidal) cells located within the ethmoidal labyrinth between the upper parts of the nasal cavities and orbits. They are divided into three groups: (1) *anterior* that open into the hiatus semilunaris, (2) *middle* that open onto bulla ethmoidalis; and (3) *posterior* that open into the superior meatus.

Sphenoidal sinus is located in the body of the sphenoid bone, and paired, divided by a bony septum. Each part varies in size and opens into the sphenoethmoidal recess of it own side.

The sphenoiddal sinus has important relationships in surgical approaches to the base of the brain. It is related posteriorly to the pons and basilar artery, superiorly to the optic chiasma and hypophysis, laterally to the cavernous sinus, internal carotid artery and ophthalmic and maxillary divisions of the trigeminal nerve.

Maxillary sinus is the largest paranasal sinus and occupies the body of the maxilla. It is a four-sided hollow pyramid that opens from its upper medial wall into the middle meatus via the posterior part of the hiatus semilunaris. The roof of the sinus is directly related to the floor of the orbit, the floor is the alveolar process of the maxilla, the anterior wall is the surface of the face

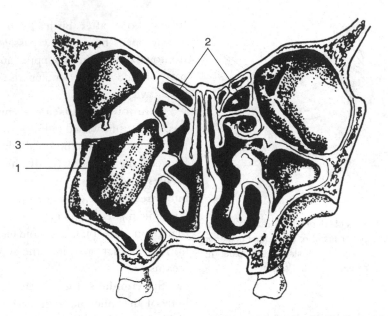

FIG. 6-10   Paranasal sinuses. (1) Maxillary sinus; (2) Ethmoidal air cells; (3) Maxillary ostium.

and the posterior wall is the infratemporal and pterygopalatine fossae.

In upright posture, the opening of the maxillary sinus is close to the roof of the sinus therefore limiting natural drainage. Because of the close relationship of the openings of the frontal and ethmoidal sinuses, infection may spread between the sinuses via the hiatus semilunaris.

Infection can spread from maxillary molar teeth to the maxillary sinus while infection of the sinus can result in toothache.

Chapter 7

Ear

Chapter Outline

1. External Ear
2. Middle Ear
3. Internal Ear

The ear is composed of three parts: external, middle and internal ears. Each part has a separate embryological origin. The external and middle ears conduct sound waves in air from the tympanic membrane, across a system of bony ossicles to receptor cells in the fluid filled internal ear. Energy is not lost in the transfer because of an impedance match between the large tympanic membrane and the small area of the foot plate of the stapes. Force is increased but amplitude is decreased. There is a small amplification resulting from the configuration of the ossicles. The inner ear contains sense organs for balance as well as for hearing. The vestibulocochler nerve (VIII) conducts impulses from both receptors to the brainstem.

1. External Ear

The external ear comprising the auricle and external acoustic meatus serves to direct sound toward the middle ear.

Auricle (Pinna) is a plate of elastic cartilage covered by tightly adherent skin. Cartilage is absent from the lobule that is composed of fat and fibrous tissue. The depressions and elevations of the lateral surface of the auricle are produced by the underlying cartilage.

Helix is the folded rim of the auricle. It begins at the concha and ends in the lobule.

Antihelix is the elevation in front of the helix (opposite the helix).

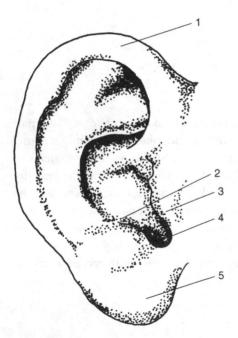

FIG. 7-1   Auricle; (1) Helix; (2) Antihelix; (3) Tragus; (4) Antitragus; (5) Lobule.

Tragus overlaps the anterior part of the concha and is separated from the helix by the anterior notch.

Antitragus projects over the concha from behind and below (opposite the tragus). It is separated from the tragus by the intertragic notch.

The important anterior relationships of the tragus is the auriculotemporal nerve and superficial temporal artery.

Fractures of the head of the mandible can be diagnosed by palpation through the external auditory meatus. The pulp of the examiner's little finger is directed forward while the patient opens and closes the mouth.

ARTERIES

Superficial temporal and **posterior auricular** arteries (branches of the external carotid artery).

NERVES

Great auricular nerve (C2,3) supplies the skin of the facial and cranial aspects of the auricle below the meatus.

Auriculotemporal nerve (V_3) supplies the skin of the facial aspect of the auricle above the meatus and the anterior and superior walls of the external auditory meatus and adjacent tympanic membrane.

Lesser occipital nerve (C2, 3) supplies the skin of the cranial aspect of the auricle above the external meatus.

Auricular branch of the vagus (X) supplies skin of the posterior concha at its root and the remainder of the tympanic membrane.

Great auricular nerve (C2,3) supplies the lower part of the auricle on both surfaces.

Lesser occipital nerve (C2,3) supplies the upper portion of the cranial surface of the auricle.

Vagal innervation of the external ear explains why irritation of the meatus may cause reflex coughing or vomiting.

Auriculotemporal innervation of the ear (a branch of the mandibular nerve that also supplies teeth of the lower jaw) explains why a diseased tooth may give clinical symptoms of earache.

EXTERNAL ACOUSTIC MEATUS

External acoustic meatus extends from the bottom of the concha to the tympanic membrane. The lateral third, or cartilaginous part is continuous with the auricle. The medial two-thirds is the bony part. Fibrous tissue connects the two parts. The firmly adherent skin of the meatus contains hairs and ceruminous glands that secrete wax.

TYMPANIC MEMBRANE

Tympanic membrane is an elliptical sheet of radially and circularly arranged collagenous fibres separating the external from the middle ears. Its lateral surface is covered with epidermis and is concave. Its medial surface is covered with mucous membrane and is convex. The thickened margin, known as the fibrocartilaginous ring, is set in the tympanic sulcus of the tympanic part of the temporal bone. The ring is incomplete

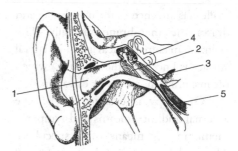

FIG. 7-2 Coronal section through right ear; (1) External acoustic meatus; (2) Tympanic membrane; (3) Tympanic cavity; (4) Internal ear; (5) Auditory tube.

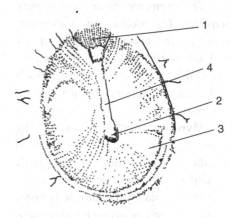

FIG. 7-3 External view of right tympanic membrane; (1) Pars flaccida; (2) Umbo; (3) "Cone of light" (4) Handle of malleus.

above at the tympanic notch. Below the notch is the V-shaped pars flaccid part of the membrane where connective tissue fibers are thin. It is bounded by the anterior and posterior malleolar folds, which meet where the lateral process of the malleus is attached to the membrane. The remainder of the membrane is the rigid tense part. The handle of the malleus is attached to the membrane and draws the membrane medially. The point of maximal concavity on its lateral surface is known as the umbo (L) which means, in Latin, a boss or elevation.

Otoscopy is examination of the external auditory meatus and tympanic membrane by means of a speculum. The auricle is pulled backwards and upwards to straighten the cartilaginous part. A normal tympanic membrane reflects a "cone of light" in its anteroinferior quadrant.

The tympanic membrane is very sensitive. The medial surface is innervated by the tympanic branch of IX and the lateral surface is supplied by the auricular branch of X posteriorly and auriculotemporal branch of V_3 anteriorly.

Myringotomy (incision of the tympanic membrane usually to drain middle ear infections) is preferably located in the posterior inferior quadrant thus avoiding the chorda tympani nerve that crosses between the malleus, the incus, and nerves and blood vessels that enter the membrane from above.

2. Middle Ear

Comprises the tympanic cavity in the temporal bone. It communicates with the nasopharynx by the auditory tube and the mastoid air cells and mastoid antrum by means of the **aditus**. It is traversed by a chain of bones which transmit vibrations from the tympanic membrane to the internal ear.

TYMPANIC CAVITY

This air space lined with mucous membrane is located within the temporal bone between the tympanic membrane and the internal ear. It consists of the tympanic cavity proper and the epitympanic recess. The latter is that part of the cavity above the level of the tympanic membrane that contains the greater part of the incus and the upper half of the malleus. The tympanic cavity communicates with the mastoid air cells through the aditus and with the nasopharynx through the auditory tube.

BOUNDARIES OF THE TYMPANIC CAVITY

Lateral (membranous) wall: Is formed by the tympanic membrane and the bone around it.

Roof (tegmental) wall: Comprises a portion of the petrous temporal bone (tegmen tympani). It separates the middle ear from the middle cranial fossa.

Floor (jugular) wall: Is thin sheet of bone that separates the cavity from the jugular fossa, posteriorly; and ascending part of the carotid canal, anteriorly. Inferiorly, the following openings are seen.

Squamotympanic fissure, through which the anterior ligament of the malleus and anterior tympanic branch of maxillary artery pass.

Posterior canaliculus from which the chorda tympani, a branch of the facial nerve emerges.

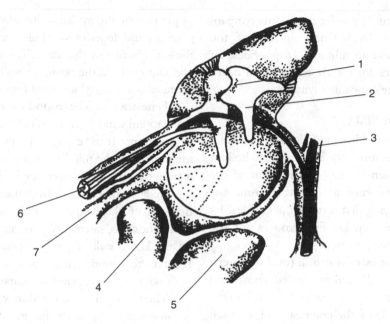

FIG. 7-4 Lateral wall of middle ear; (1) Malleus; (2) Incus; (3) Chorda tympani n; (4) Carotid canal; (5) Jugular fossa; (6) Tensor tympani m; (7) Auditory tube.

The anterior canaliculus through which the chorda tympani passes to join the lingual nerve.

ANTERIOR (CAROTID) WALL

The anterior wall contains the openings to two semi-canals. A thin bony shelf, the cochleariform process divides this tube into an upper compartment occupied by tensor tympani muscle (semicanal for the tensor tympani), and a lower compartment occupied by the pharyngotympanic (Eustachian) tube.

Part of the auditory tube is elastic cartilage and part is bone. The medial or cartilaginous part is trumpet-shaped, and terminates in an oval opening on the lateral wall of the nasopharynx. The lateral osseous part lies along the angle of union of the squamous and petrous parts of the temporal bone and is about 1.5 cm long. Below this a thin plate of bone separates the tympanum from the carotid canal.

POSTERIOR (MASTOID) WALL

Posterior (mastoid) wall contains the following structures.

Aditus ad antrum which connects the epitympanic recess with a cavity, the mastoid antrum.

Pyramidal eminence (pyramid) which is a small hollow projection below the aditus. It contains the stapedius muscle, its tendon projecting through the apex of the pyramid.

Canaliculus for the chorda tympani nerve is located near the top of the tympanic membrane. It provides an entrance for the chorda tympani nerve into the tympanic cavity.

MEDIAL WALL

Medial wall has the following features.

Promontory is a rounded, hollow, prominence below and in front of the fenestra vestibuli resulting from the underlying first turn of the cochlea. Its surface contains fine grooves for the tympanic plexus.

Fenestra vestibuli (oval foramen) is an oval window located above and behind the promontory. It leads into the vestibule of the inner ear and is closed by the base of the stapes.

Prominence of the facial nerve canal indicates the path of the facial nerve as it courses backward in the upper part of the medial wall above the pyramid and fenestra vestibuli towards the lower border of the aditus. The canal continues behind the posterior wall and terminates at the stylomastoid foramen.

Fenestra cochleae (round window) is the round window situated behind the promontory. It is closed by mucous membrane of the middle ear separating it from the scala tympani of the cochlea.

Auditory tube extends from the anterior wall of the tympanic cavity downwards, forwards, and medially, terminating on the lateral wall of the nasopharynx. It serves to equalize the air pressure on both sides of the tympanic membrane.

The bony part of the auditory tube lies in the sphenopetrosal fissure of the temporal bone below the semicanal for the tensor tympani muscle.

The cartilaginous part of the auditory tube is triangular in shape, with

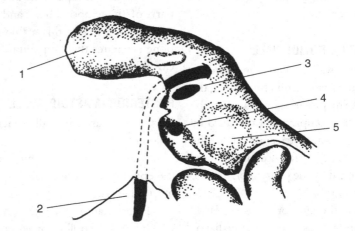

FIG. 7-5 Medial wall of middle ear; (1) Antrum; (2) Facial n; (3) Fenestra vestibulae; (4) Fenestra cochleae; (5) Promontory.

the apex attached to the medial end of the bony part. The free base of the cartilage is curved and forms the tubal elevation behind the pharyngeal opening of the tube.

In section, the tube is C-shaped folded into two laminae. The medial lamina forms the posteromedial wall and is continuous above with the narrow lateral lamina. The tube is completed inferolaterally by the membranous lamina.

The cartilaginous tube widens at its anteromedial end so that it is trumpet shaped. Tensor veli palatini arises from the membranous lamina (as well as the adjacent spine of sphenoid and scaphoid fossa). Levator veli palatini arises from the medial lamina (and adjacent petrous bone) while salpingopharyngeus arises from the anterior edge of the medial lamina. During swallowing, contraction of the tensor tends to pull the auditory tube open and help equilibrate pressures between the pharynx and middle ear.

CONTENTS OF THE TYMPANIC CAVITY
Muscles

Two muscles are associated with the middle ear.

TENSOR TYMPANI

- **Origin:** Mostly from the cartilaginous part of the auditory tube and greater wing of the sphenoid. It enters the semicanal above the auditory tube. Its tendon turns laterally around the processus cochlerformis.

- **Insertion:** Handle (manubrium) of the malleus near its root.
- **Action:** Draws the tympanic membrane medially tightening the tympanic membrane damping down over vibration of membrane and ossicles.
- **Nerve supply:** Via otic ganglion by mandibular division of trigeminal nerve.

STAPEDIUS

- **Origin:** Interior of the pyramid on the posterior wall of the tympanic cavity.
- **Insertion:** Neck of the stapes.
- **Action:** Draws the anterior end of the base of the stapes laterally damping down over vibration of membrane and ossicles
- **Nerve supply:** Branch of the facial nerve (VII).

Paralysis of stapedius or of tensor tympani result in inability to dampen vibration in the ossicles with resultant excessive sensitivity to sound (hyperacusis).

Tympanic Ossicles

These bones are enveloped by the mucous membrane of the middle ear.

MALLEUS (L, HAMMER)

- *Head* which has on the upper part of its posterior surface. This presents a facet for articulation with the body of the incus.

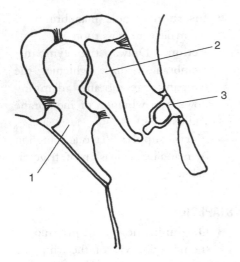

FIG. 7-6 Auditory ossciles (Diagramatic); (1) Malleus; (2) Incus; (3) Stapes.

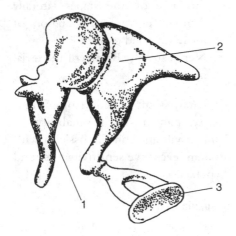

FIG. 7-7 Auditory ossciles (Medial aspect); (1) Malleus; (2) Incus; (3) Stapes.

- *Neck* is a constriction below the head. Both head and neck are located in the epitympanic recess of the middle ear.

- *Lateral (short) process* which arises from the root of the handle and projects laterally to be attached to the tympanic membrane.
- *Manubrium* (handle), a long tapering process that passes downwards and backwards and is also attached laterally to the tympanic membrane.
- *Anterior (long) process,* a slender spicule connected by the anterior ligament of the malleus to the petrotympanic fissure and then to the spheno-mandibular ligament (remnants of the first pharyngeal arch).

Incus (L, anvil)

- The body incus is shaped likea molar teeth. *Body* articulates in front by a saddle-shaped facet with the head of the malleus.
- *Short process* attached to the margin of the aditus on the posterior wall.
- *Long process* passes downwards behind and parallel to the handle of the malleus with its tip terminating medially in a knob, the lenticular process that articulates with the head of the stapes.

Stapes (L, stirrup)

- *Head* faces laterally and articulates with the incus.
- *Base* fixed at its periphery by the annular ligament thereby closing the fenestra vestibuli (oval window).
- *Limbs (crura)* arise from a constricted part, the neck, and pass medially to the extremities of the base.

Joints

Incudomallear joint

- *Type:* Synovial saddle.
- *Articulating elements:* Body of incus with head of malleus.

Incudostapedial joint

- *Type:* Synovial ball and socket.
- *Articulating* elements: Lenticular process of the incus with the head of the stapes.

Tympanostapedial joint

- *Type:* Syndesmosis.
- *Articulating elements:* Membrane of fenestra vestibuli with base of the stapes.

Otosclerosis is a result of overgrowth of bone around the foot plate of stapes and around the fenestra vestibuli resulting in deafness.

Neurovascular Supply of the Middle Ear

Arteries and Veins

Anterior tympanic artery (branch of the maxillary artery) enters the tympanosquamous fissure to reach the tympanic membrane.

Inferior tympanic artery (from the ascending pharyngeal) enters the cavity through the typanic canaliculus.

Stylomastoid artery (branch of the posterior auricular artery) enters through the facial canal.

Posterior tympanic artery (from the stylomastoid artery) enters through the canaliculus for the chorda tympani.

Middle meningeal artery gives two branches to the middle ear

Caroticotympanic artery (from the internal carotid artery in its canal) supplies the anterior wall.

Veins parallel the arteries and terminate in the superior petrosal sinus and pterygoid venous plexus.

Nerves

Sensory innervation is by the auriculotemporal nerve (V_3), tympanic branch of the glossopharyngeal nerve (IX) and auricular branch of the vagus (X).

Tympanic branch of IX arises from the inferior ganglion of IX, passes through the tympanic canaliculus (between the jugular foramen and carotid canal) to reach the tympanic cavity. It divides to contribute to the tympanic plexus on the promontory on the medial wall of the tympanic cavity. It gives sensory branches to the mucous membrane of the cavity, mastoid air cells and auditory tube. The tympanic branch of IX also contains preganglionic parasympathetic (secretomotor) fibers for the parotid gland (cell bodies in the inferior salivatory nucleus in the brainstem).

Caroticotympanic nerves (sympathetic) derived from the carotid plexus (preganglionic cell bodies in the superior cervical ganglion), enter the middle ear through the wall of the carotid canal and ramify, together with the tympanic branch of the IX, in the **tympanic plexus.**

The **chorda tympani nerve** merely passes through the middle ear, crossing from the posterior wall, over the upper part of the manubrium of the malleus, to the anterior wall.

Mastoid (Tympanic) Antrum

The mastoid antrum is a large hollowed recess behind and above the tympanic cavity with which it is connected by an opening, the aditus on the upper part of the posterior tympanic wall. The antrum developed with the nasal cavity and is lined by a continuation of its mucous membrane. **Mastoid air cells** open into the antrum.

The mastoid antrum lies deep to the suprameatal triangle bounded by the suprameatal crest, posterosuperior border of the external acoustic meatus and a vertical tangent to the posterior edge of the meatus. Infections of the mastoid antrum commonly result from spread of infection from the tympanic cavity (otitis media) that also results from infection in the nasopharynx.

3. Internal Ear

The internal ear is located within the petrous temporal bone. It is divided lnto bony and membranous labyrinths, the former enclosing the latter. **Endolymph** is the fluid within the membranous labyrinth and the **perilymph** the fluid between the membranous and osseous labyrinths.

The internal ear contains the organs of hearing and equilibrium and

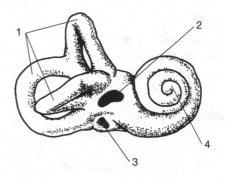

FIG. 7-8 Bony labyrinth; (1) Semilunar canals; (2) Fenestra vestibulae; (3) Fenestra cochleae; (4) cochlea.

within are the fibres of the vestibulo-cochlear (eighth cranial) nerve.

Bony labyrinth consists of the vestibule, semicircular canals and cochlea. It encloses sacs and ducts of the membranous labyrinth.

VESTIBULE

Vestibule is the middle part of the bony labyrinth.

Lateral wall corresponds with the medial wall of the middle ear. On it is the fenestra vestibule closed by the base of the stapes.

Medial wall has a depression perforated by several holes for the fibres of the eighth nerve. Behind is a ridge, the vestibular crest; and still further back is the medial opening of the aqueduct of the vestibule, for transmission of the endolymphatic duct.

Posterior wall has five openings of the semicircular canals and anteriorly is an opening which leads to the scala vestibuli of the cochlea.

The cavity of the vestibule has the following boundaries:

Medial: Fundus of the internal acoustic meatus.

Lateral: Fenestra vestibuli.

Anterior: Communicates with the cochlea.

Posterior: Communicates with the semicircular canals.

A narrow canal, the vestibular canaliculus (aqueduct), runs vestibule to the posteroinferior surface of the petrous temporal bone (posterior cranial fossa). It transmits the endolymphatic duct.

SEMICIRCULAR CANALS

Semicircular canals are three arched osseous canals, placed above and behind the vestibule that open into that chamber by five rounded apertures (two adjacent canals have a common opening). Each canal forms about two thirds of a circle, and has at one end a dilated part, the ampulla.

There are three canals:

Anterior (superior) canal lies in a vertical plane transverse to the long axis of the tempral bone (about 45° to the sagittal plane). It produces the arcuate eminence in the middle cranial fossa. Its lateral end is ampullated; the medial end joins the posterior canal to form the crus commune.

Posterior canal lies in the vertical plane (at right angles to the anterior canal). It arches parallel to the long axis

of the petrous temporal bone vestibule. Its lower end is ampullated and has a separate opening into the vestibule.

Lateral canal is the smallest canal. It arches laterally in a horizontal plane and has separate openings into the vestibule.

COCHLEA

The cochlea is the organ of hearing. It is shaped like a snail's shell lying on its side so that the base is turned to the internal auditory meatus, and the apex is opposite the canal for the tensor tympani. The cochlea consists of a tapering spiral canal of two and a half turns, around a bony axis or modiolus. The canal is divided into three compartments by partitions of bone (spiral laminae) and membranes. The first turn of the canal bulges into the middle ear and forms the promontory. The bony spiral lamina ends at the apex of the cochlea and between it and the modiolus is a small opening, the helicotrema, by which the upper and lower compartments communicate. The modiolus is pierced by small canals for the fibres of the cochlear nerve.

A small canal, which winds round the modiolus contains the spiral ganglion of the cochlear nerve. The peripheral processes of the ganglion cells go to the **spiral organ (of Corti)** and the central processes form the cochlear part of the vestibulocochlear (VIII) nerve.

The three compartments in the cochlea are the scala tympani (lower), the cochlear duct (middle) and scala vestibuli (upper). The scala tympani commences at the fenestra cochleae (round window). Near the fenestra cochleae is the opening of the aqueduct of the cochlea (perilymphatic duct) that conveys perilymph from the basal turn of the cochlea through the petrous temporal to open at the anterior margin of the jugular foramen where it communicates with the cerebrospinal fluid in the subarachnoid space.

MEMBRANOUS LABYRINTH

The membranous labyrinth is a complex fibrous sac lined with epithelium. It is filled with a clear fluid, the endolymph. It lies in the bony labyrinth and is bathed by perilymph. The fibres of the vestibulocochlear nerve are distributed on its wall. Parts of the membranous labyrinth are:

Utricle is a sac in the larger of two sacs (utricle and saccule) in the upper part of the vestibule. The membranous semicircular ducts open into its posterior part. Vestibular nerves enter the anterior part.

The interior of the utricle has a thickened area called the macula on which there is a small mass of calcareous grains known as otoliths. Inferiorly, there is a canal, the utriculosaccular canal that joins the utricle with the saccule. This canal gives off the slender endolymphatic duct, which ends in a dilated pouch, the endolymphatic sac on the posterior surface of the petrous temporal bone.

Saccule is a smaller sac located in front of the utricle and below it.

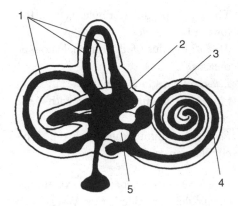

FIG. 7-9 Membranous labyrinth; (1) Semilunar canals; (2) Utricle; (3) Saccule; (4) Cochlear duct; (5) Ducts reuniens.

The ductus reuniens is the short tube connecting the saccule with the cochlear duct.

Maculae signal the position of the head in space as well as tilting movements of the head. If stimulation is prolonged, motion sickness may result.

Semicircular ducts are about one-third the size of the canals, except at the ampullae, where each dilates to almost fill the bony canal. Each duct is free on its concave aspect and the convexity is fixed to the osseous canal. On the surface of the ampulla where it is attached to the bony canal, there is a transverse projection, the ampullary crest, in which some filaments of the vestibular nerve end.

Ampullae contain a crista ampullaris with a cupula that spans the entire height of the ampulla. Vestibular cells have hair-like processes embedded in the cristae which are deformed by relative movement of the cristae in the fluid filled canal when the head rotates leading to nerve impulse generation.

Cochlear duct is triangular in section. It (cochlear duct) separates the scala vestibuli from the scala tympani. The duct contains the spiral organ to which the cochlear nerve is distributed and is separated from the scala vestibuli by the vestibular membrane that passes from the spiral lamina upwards and laterally to the wall of the bony cochlea. The duct is bounded below by the basilar membrane to which the spiral organ is attached and is connected to the saccule by the ductus reuniens. Above, it terminates in a blind cone-shaped extremity, partly bounding the helicotrema.

Pressure changes in the perilymph from sound waves cause vibrations to be transferred to endolymph of the cochlear duct and the basilar membrane upon which special sensory (hair) cells are placed. Only a specific area of the membrane vibrates in response to a particular frequency. The hair cells are displaced relative to the tectorial membrane in which the hairs of the hair cells are embedded. The shearing movements produce ionic changes in the hair cells and lead to nerve impulse generation.

The neurovascular system of the membranous labyrinth is supplied by arteries: Labyrinthine artery, a branch of the basilar artery or anterior inferior cerebellar artery that enters the inner ear through the internal acoustic meatus; veins: labyrinthine vein originates from branches similar to those of the artery and drains into the inferior petrosal sinus; nerves: vestibulocochlear nerve (VIII) consisting of two functionally distinct parts.

Cochlear nerve originates from the central process of bipolar neurons in the spiral ganglion in a canal in the modiolus. Its peripheral processes extend to the spiral organ (of Corti) by way of the osseous spiral lamina.

Vestibular nerve originates from the central processes of bipolar neurons in the vestibular ganglion located in the internal acoustic meatus. Its peripheral fibers that enter the inner ear and break up into three branches distributed to the maculae of the saccule, utricle, and the ampullary crests of the semicircular ducts.

Internal acoustic meatus is the opening in the petrous temporal bone that leads into a short canal located medial to the internal ear. It transmits the facial (VII) and vestibulocochlear (VIII) nerves and the labyrinthine vessels.

Chapter 8

Parotid Region

Chapter Outline

1. Introduction
2. Fascia
3. Parotid Gland
4. Related Structures

The parotid region consists of the parotid gland and its bed. It is part of the side of the neck above the limits of the anterior triangle.

1. Introduction

The parotid gland is a salivary gland producing saliva which is a predominantly serous secretion contributing to a mixed serous and mucous secretion. The gland is superficially triangular in shape (apex pointing downward) but pyramidal in cross section. The surface shows fibrous septa that divide the gland into lobules. The gland is traversed by nerves, veins and arteries in separate planes

The parotid bed is the space occupied by the parotid gland. It extends medially to the pharynx and superficially it overflows onto the face.

The boundaries of the parotid bed are:

Anterior: Ramus of the mandible with the masseter laterally and medial pterygoid medially.

Posterior: Mastoid process with sternocleidomastoid muscle laterally and posterior belly of the digastric medially.

Superior: External acoustic meatus and back of the capsule of the temporomandibular joint.

Medial: Styloid process (with styloglossus attached to its tip, stylohyoid attached to the posterolateral middle third and stylopharyngeus attached to its medial base).

2. Fascia

SUPERFICIAL

This fascia covers the gland. It is a part of the subcutaneous tissue of the face. It contains lymph glands and facial branches of the great auricular nerve. (C2, 3).

DEEP

The fascia enclosing the parotid gland is part of the investing layer of deep cervical fascia. It splits between the angle of the mandible and tip of the mastoid process to enclose the parotid gland as the **parotid fascia**. It is attached above from the tip of the mastoid, across the cartilaginous auditory tube to the lower border of the zygoma. Parotid fascia extends anteriorly to blend with areolar tissue on the surface of masseter.

The deep layer of parotid fascia is attached in a line to the tip of the styloid process, lower edge of the tympanic temporal bone and tympanosquamous fissure. The part of the fascia between the styloid process and angle of the mandible is thickened to produce the stylomandibular ligament which separates the parotid gland from the submandibular gland.

3. Parotid Gland

The parotid gland is the largest of the three major paired salivary glands. Connective tissue septa subdivide the gland into lobules and the gland has several extensions as follows: superficial part (lobe) is connected by a narrow bridge with the deep part of the gland.

Facial process is an anterior extension of the superficial part onto the

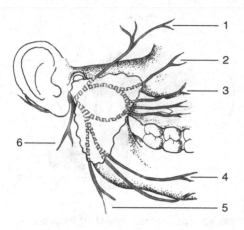

FIG. 8-1 Nervous plane of the parotid gland. (1) Temporal branch of VII; (2) Zygomatic branch of VII; (3) Buccal branch of VII; (4) Marginal branch of VII; (5) Cervical branch of VII; (6) Branch of VII to digastric and stylohyoid m.

surface of masseter. It may be detached from the bulk of the gland, a small, separate **accessory parotid** is located above the parotid duct and below the zygomatic arch.

Deep part (lobe) is the largest part of the parotid gland extends medially to the styloid process and may reach the wall of the pharynx (the pharyngeal process). The pterygoid process is an anterior extension of the deep process between the ramus of the mandible and medial pterygoid muscle.

The neurovascular system of the parotid gland is supplied by arteries and veins from traversing vessels; and the sensory nerves supplying the gland are the greater auricular (C2,3) and auriculotemporal nerve (from V3).

Parasympathetic secretomotor fibers reach the parotid gland by a complex route. Preganglionic neurons are located in the **superior salivatory nucleus** in the brainstem. Their axons travel in the glossopharyngeal (IX) nerve then leave the main trunk as the **tympanic nerve** that enters the middle ear cavity through a small canal in the bone between the jugular foramen and carotid canal. Axons then form a plexus on the surface of the promontory on the medial wall of the tympanic cavity. Axons then leave the plexus as the **lesser petrosal nerve** through a canal for the lesser petrosal nerve and enter the middle cranial fossa. The lesser petrosal nerve courses briefly across the petrous temporal bone and exits via the foramen ovale. These (preganglionic) axons finally synapse in the **otic ganglion**. Postganglionic fibers pass from the otic ganglion to the **auriculotemporal nerve** (branch of V3) to secretory cells of the parotid gland.

Preganglionic sympathetic cell bodies are located in the upper thoracic spinal cord which synapse with postganglionic neurons in the superior cervical ganglion. Postganglionic axons then envelop the external carotid and leave the middle meningeal artery to reach the parotid.

Parotid duct is formed by union of lobular ducts of the gland and emerges from the anterior border of the gland and running forward across masseter muscle. It turns medially around the anterior border of masseter, pierces the buccal fat pad, buccopharyngeal fascia, buccinator muscle, and mucous membrane, to open into the vestibule of the mouth at the parotid papilla opposite the second upper molar tooth.

4. Related Structures

Nerves, veins and arteries traverse the parotid gland in separate planes from superficial to deep. Since nerves are most superficial, they are particularly vulnerable to injury in facial injuries.

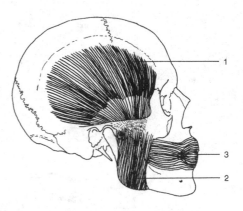

FIG. 8-2   Lateral aspect of head. (1) Temporalis m; (2) Masseter m; (3) Buccinator and orbicularis oris.

NERVOUS PLANE

The facial nerve (VII) leaves the stylomastoid foramen, runs forward crossing the root of the styloid process and gives small branches such as **posterior auricular branch** supplies the posterior auricular muscles and intrinsic muscles of the auricle; **occipital branch** supplies the occipital branch of occipitofrontalis; **digastric branch** supplies the posterior belly of the digastric and the stylohyoid muscle (a migrated anterior part of the digastric muscle).

The part of the facial nerve in the parotid gland crosses the posterior edge of the mandibular ramus dividing into upper and lower divisions on each side of a bridge separating superficial from deep parts of the gland. The upper division divides into two branches (**temporal** and **zygomatic** branches) while the lower division divides into three branches (**buccal, marginal** and **cervical** branches).

Auriculotemporal nerve (of V3) may be a content of the gland. It passes between the temporomandibular joint and external acoustic meatus supplying the surroundings, the meatus, joint and gland as well as the skin of the upper lateral side of the auricle and temple.

Anterior branch of great auricular nerve (C2) may be a content of the parotid. It supplies the skin over the angle of the jaw, the only part of the face not supplied by the trigeminal (V) nerve.

VENOUS PLANE

Retromandibular vein is formed below the zygoma by the union of the superficial temporal and maxillary veins. It descends in the parotid gland deep to the facial nerve. A posterior branch of the retromandibular vein joins the posterior auricular vein forming the external jugular vein (external to sternocleidomastoid).

The main trunk emerging from the apex of the parotid joins the facial vein and drains into the internal jugular vein (internal to sternocleidomastoid).

ARTERIAL PLANE

External carotid artery ascends into the parotid gland from the carotid triangle at the upper border of the posterior belly of the digastric muscle. It ends at the neck of the mandible by dividing into two terminal branches:

Maxillary artery is a terminal branch arises in the parotid gland opposite the neck of the mandible. It runs forward on a deep plane in the infratemporal fossa.

Superficial temporal artery which is the other terminal branch, gives branches to the parotid gland, temporalis muscle and external ear.

Transverse facial artery is a branch of the superficial temporal artery arising within the parotid. It lies just above the parotid duct. It leaves the gland at its superior border and supplies the surrounding gland and muscle.

Occipital artery runs posterosuperiorly along the lower edge of posterior belly of digastric and grooves the skull medial to the mastoid notch. It supplies the local muscles, trapezius, digastric, stylohyoid and sternocleidomastoid, meninges and muscles of skin of the occiput.

Posterior auricular artery runs along the upper border of the posterior border of the digastric. A small but impotant branch (stylomastoid artery) supplies the middle and internal ear.

MASSETER

Masseter muscle is a powerful muscle of mastication that elevates the mandible. A product of the mandibular arch, it is supplied by nerve of the mandibular arch, the mandibular division ot the trigmeninal nerve (V3).

- **Origin:** Anterior two thirds of the lower border of the zygomatic arch (superficial head) and entire medial surface of the zygomatic arch (deep head).
- **Insertion:** Lateral surface of the ramus of the mandible.
- **Action:** Muscle of mastication; raises the mandible. The superficial part of the muscle assists in protraction of the mandible while the deep part assists in retraction of the protracted mandible.
- **Nerve supply:** Masseteric branch of V3 crosses the infratemporal fossa and mandibular notch to enter the deep aspect of the muscle.

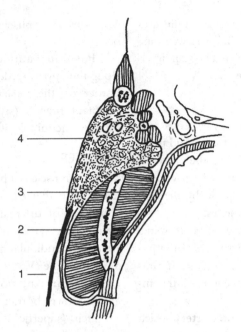

FIG. 8-3 Horizonal section through parotid gland. (1) Facial n; (2) Masseter m; (3) Medical pterygoid m; (4) Parotid gland.

Chapter 9

Temporal Region

Chapter Outline

1. Boundaries
2. Contents

The temporal region consists of the temporal fossa and its contents. The temporal fossa contains the temporalis muscle.

1. Boundaries

Above: Temporal lines.
In front: Frontal process of the zygomatic bone.
Laterally: Zygomatic arch.
Below: Infratemporal crest of the sphenoid.

2. Contents

Superficial fascia contains fat, extrinsic muscles of the ear and superficial blood vessels and nerves.

    Deep fascia is the strong and dense temporal fascia which covers the temporalis muscle above the zygomatic arch. It is attached to the superior temporal line above but below, it divides into two layers a superficial lamina attached to the upper border of the zygomatic arch and a deep lamina attached to the inner surface and blending with fascia lining the deep surface of masseter.

Temporalis Muscle

Temporalis muscle is a fan-shaped muscle occupying the temporal fossa.

- **Origin:** Superficial head from the floor of the temporal fossa below the inferior temporal line and deep surface of the temporal fascia. The deep head arises from the roof of the infratemporal fossa and infratemporal crest.
- **Insertion:** Tendinous to the apex of the coronoid process and medial surface and anterior border of the ramus of the mandible.
- **Action:** Anterior fibers elevates the mandible and posterior fibers retract the protruded mandible in closure of the mouth.
- **Nerve supply:** Deep temporal branches of the anterior division of the mandibular nerve (V3).

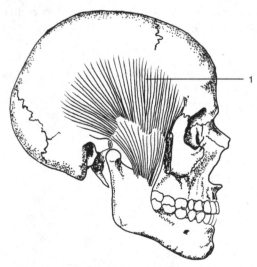

FIGURE 9-1 (1) Temporalis m.

Chapter **10**

Infratemporal Fossa

Chapter Outline

1. Boundaries
2. Muscles
3. Arteries
4. Veins
5. Nerves

The infratemporal region comprises the infratemporal fossa and its contents. The space is located behind the maxilla, below and deep to the zygomatic arch between the mandible and pharynx. It contains the parotid gland, muscles of mastication, branches of the mandibular nerve, maxillary artery and the pterygoid venous plexus.

1. Boundaries (Walls)

Lateral: Ramus and coronoid process of the mandible.

Medial: Lateral pterygoid plate (of pterygoid process of sphenoid), the small pyramidal process of the palatine bone.

Anterior: Posterior surface of the maxilla.

Posterior: Anterior surface of the condylar process of the mandible.

Roof: Infratemporal surface of the greater wing of the sphenoid and small part of squamous temporal.

2. Muscles

Lateral Pterygoid

Lateral pterygoid is a conical shaped muscle that has two heads:

- **Origin: Upper head** from the infratemporal surface and crest of the greater wing of sphenoid and adjacent squamous temporal (roof of the infratemporal fossa).
- **Inferior head:** Lateral surface of lateral pterygoid plate.
- **Insertion:** Fibers run backward and laterally to the front of the neck of the mandible (pterygoid fovea),

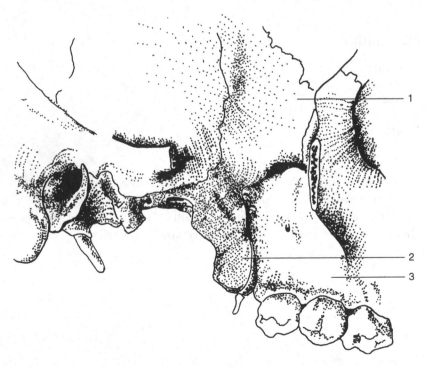

FIG. 10-1 Boundries of infratemporal fossa. (1) Greater wing of sphenoid; (2) Lateral pterygoid plate; (3) Poserior surface of maxilla.

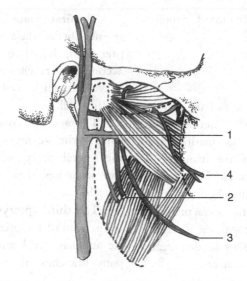

FIG. 10-2 Contents of infratemporal fossa. (1) Maxillary a; (2) Inferior alveolar n; (3) Lingual n; (4) Buccal branch of V3.

FIG. 10-3 Pterygoid muscles. (1) Lateral pterygoid m; (2) Medical pterygoid m.

capsule and articular disc of the temporomandibular joint.

■ **Action:** Protracts the mandible and disc in synchrony and depresses the mandible (by pulling the mandibular condyle down the

articular eminence) bilaterally or unilaterally.

■ **Nerve supply:** Branch of the anterior trunk of mandibular nerve (V3).

Medial Pterygoid

This muscle also has two heads of origin.

■ **Origins: Superficial head:** Smaller; arises from the maxillary tuberosity and pyramidal process of the palatine bone.

■ **Deep head:** Larger; arises from the medial surface of the lateral pterygoid plate.

■ **Insertion:** Medial surface of the angle and ramus of the mandible behind the mylohyoid groove.

■ **Action:** Elevates mandible (with masseter) and assists in protraction.

■ **Nerve supply:** Branch of mandibular nerve (V3).

3. Arteries

Maxillary artery is one of the two terminal branches of the external carotid artery, it begins in the parotid gland behind the neck of the mandible. It passes forward between the mandible and sphenomandibular ligament along the lower border of the lateral pterygoid muscle to the pterygopalatine fossa.

Lateral pterygoid muscle demarcates three segments of the artery.

The **first (mandibular) part**, is the segment before the artery reaches lateral pterygoid, has five branches which accompany branches of the mandibular nerve through bony canals in the base of the skull.

The **second (pterygoid) part** is usually the segment located behind the lateral pterygoid muscle. It has five branches that supply five local muscles.

The **third (pterygopalatine) part** lies in front of the lateral pterygoid muscle and has five branches that accompany branches of the sphenopalatine

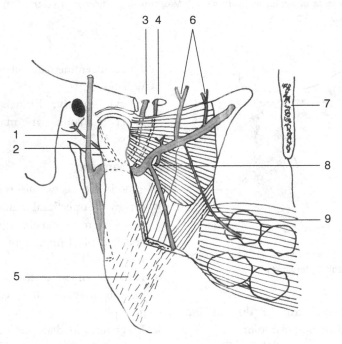

FIG. 10-4 Maxillary artery. (1) Deep auricular branch; (2) Anterior tympanic branch; (3) Middle meningeal branch; (4) Accessory meningeal branch; (5) Inferior alveolar branch; (6) Deep temporal branch; (7) Pterygoid branches; (8) Masseteric branch; (9) Buccal branch.

ganglion (see details under ptery-gopalatne fossa).

Branches of the first part:
Anterior tympanic artery passes to the middle ear tympanum through the petrotympanic fissure (with the chorda tympani nerve). It supplies the tympanum and tympanic membrane.

Deep auricular artery (may be a branch of anterior tympanic artery) enters the external acoustic meatus between the tympanic temporal bone and cartilage with a branch of the auriculotemporal nerve. It supplies the external acoustic meatus.

Middle meningeal artery passes between two roots of auriculotemporal nerve then enters foramen spinosum with the meningeal branch of the mandibular nerve (V3). It supplies the trigeminal ganglion and dura.

Accessory meningeal artery enters foramen ovale with the mandibular nerve (V3). It supplies the trigeminal ganglion and dura.

Inferior alveolar artery accompanies the inferior alveolar nerve into the mandibular foramen and canal. It supplies the mandibular teeth and some of the buccal and labial mucosa.

Branches of the second part:
Masseteric branch crosses the mandibular notch to enter masseter muscle.

Anterior and posterior temporal branches cross the infratemporal crest to supply temporalis muscle.

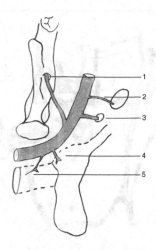

FIG. 10-5 Horizontal section through infratemporal fossa showing branches of the first part of maxillary artery. (1) Inferior alveolar a; (2) Accessory meningeal a; (3) Middle meningeal a; (4) Tympanic branch; (5) Deep auricular branch.

Pterygoid branches enter lateral and medial pterygoid muscles directly.

Buccal branch descends with the buccal nerve (branch of V3) to supply the buccinator muscle (an accessory muscle of mastication).

Branches of the third (pterygopalatine) part Run with branches of V3 through foraminae see Chapter 11.

4. Veins

Pterygoid venous plexus is a venous network overlying the lateral and medial pterygoid muscles. It is formed by vessels that correspond to the branches of the maxillary artery. The principal drainage pathway is the short **maxillary vein** (at the lower border of lateral

FIG. 10-6 Pterygoid venous plexus.

pterygoid) that unites with the superficial temporal vein to form the retromandibular vein.

5. Nerves

MANDIBULAR DIVISION OF THE TRIGEMINAL NERVE

The trigeminal nerve provides the principal cutaneous innervation to the scalp and face, motor innervation to the muscles of mastication, and sensory innervation to the mucous membrane of the facial cavities. The sensory and motor roots leave the pons and meet at the trigeminal ganglion near the apex of the petrous temporal bone.

Mandibular nerve (V3) is the largest division of the trigeminal nerve arises from the trigeminal ganglion and emerges from the cranial cavity into the infratemporal fossa via the foramen ovale. It divides behind the lateral pterygoid muscle into two unequal divisions after giving off the meningeal and the medial pterygoid branches.

Meningeal branch is a recurrent branch containing sensory fibers passes back through the foramen spinosum to supply the dura mater.

The **nerve to medial pterygoid** innervates the medial pterygoid muscle, gives a branch to the otic ganglion and also supplies tensor veli palatini and tensor tympani muscles.

Anterior Division

The anterior division is small and has one sensory branch and three motor branches.

Masseteric branch ascends through the mandibular notch. It innervates its

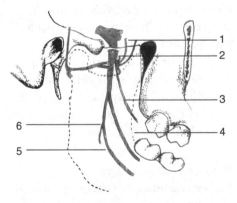

FIG. 10-7 Mandibular division of trigeminal nerve (V3). (1) Temporal n; (2) Nerve to masseter; (3) Buccal n; (4) Lingual n; (5) Inferior alveolar n; (6) Mylohyoid n.

deep surface masseter and provides a branch to the temporomandibular joint.

Deep temporal branches into two or three nerves leave the infratemporal fossa by passing over its crest and enter the deep surface of temporalis.

Lateral pterygoid branch enters the muscle on its deep surface.

Buccal branch is a sensory nerve emerges between the heads of lateral pterygoid toward buccinator, where it supplies the skin of the cheek and mucous membrane of the mouth.

Posterior Division

The posterior division is large and has three sensory branches and one motor branch.

Auriculotemporal nerve arises by two roots which encircle the middle meningeal artery just below the foramen spinosum and forms a single trunk. It passes backward medial to the lateral pterygoid muscle then between the sphenomandibular ligament and capsule of the temporomandibular joint. It crosses the root of the zygoma and enters the temporal region with the superficial temporal artery.

Commuicating postganglionic parasympathetic fibers from the otic ganglion join the auriculotemporal nerve to reach the parotid gland.

The facial nerve receives sensory fibers from the auriculotemporal nerve in the substance of the parotid gland.

The auriculotemporal nerve gives branches to the skin of the upper half of the outer surface of the auricle, upper external acoustic meatus and tympanic membrane, scalp above the meatus and posterior part of temporomandibular joint.

Lingual nerve descends medial to the lateral pterygoid muscle and emerges at its lower border where it receives the chorda tympani nerve (a branch of the facial nerve mediating taste sensation and parasympathetic secretomotor supply to the submandibular and sublingual glands). The nerve passes between medial pterygoid and the ramus of the mandible, over the superior constrictor and styloglossus muscles, to reach the submandibular region. It continues between the hyoglossus muscle and the submandibular gland, over the submandibular duct, and on the undersurface, to the tip of the tongue.

Chorda tympani nerve is a branch of the facial nerve carrying preganglionic parasympathetic fibers to the submandibular and sublingual ganglia as well as seromucous glands in the oral mucosa. These (preganglionic) fibers leave the lingual nerve to synapse in the ganglion and then postganglionic fibers continue to the buccal, labial, lingual, submandibular and sublingual glands.

Inferior alveolar nerve descends deep to and emerges at the lower border of lateral pterygoid. It continues between the sphenomandibular ligament and ramus of the mandible to pass through the mandibular foramen behind the lingula into the mandibular canal. It supplies the pulp of the molar and premolar teeth.

Branches before entering the mandibular foramen: **Mylohyoid branch** is given off immediately above mandibular foramen. It grooves the medial side of the ramus supplying mylohyoid muscle and anterior belly of the digastric muscle.

Branches after entering the mandibular canal: **Inferior dental branches** in the mandibular canal form the inferior dental plexus to supply the mandibular teeth.

Mental branch is the continuation of the stem of the inferior alveolar nerve. It supplies the skin of the chin and lower lip.

Incisive branch is the terminal branch after giving off the mental nerve. It supplies the canine and incisor

teeth (area of supply may cross the midline).

Otic ganglion is a parasympathetic ganglion that lies in the infratemporal fossa on the medial side of the mandibular nerve close to foramen ovale. Fibers connected to the ganglion (roots) are:

Parasympathetic innervation is from the preganglionic cell bodies that are located in the **inferior salivatory nucleus** in the brainstem, pass in the glossopharyngeal (IX) nerve, re-enter the skull in the tympanic branch of IX then lesser petrosal nerve to relay in the otic ganglion.

Sympathetic preganglionic cell bodies are located in the thoracic spinal cord and postganglionic cell bodies in the **superior cervical ganglion**. Axons form a plexus around the middle meningeal artery to reach and pass through the ganglion.

Sensory roots have their cell bodies in the **trigeminal ganglion** and

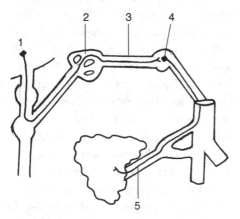

FIG. 10-8 Autonomic innervation of parotid gland (parasympathetic). (1) LX; (2) Tympanic plexus; (3) Lesser petrosal n; (4) Otic ganglion; (5) Auriculotemporal n.

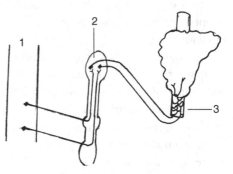

FIG. 10-9 Autonomic innervation of parotid gland (sympathetic). (1) Spinal cord; (2) Superior cervical ganglion; (3) Plexus on middle meningeal a.

pass through the ganglion without synapsing.

The otic ganglion, uniquely among cranial parasympathetic ganglia, has a motor root. Neurons in the **trigeminal motor nucleus** in the brainstem have axons that pass through the otic ganglion. They reach the tensor veli palatini and tensor tympani muscles.

Chapter 11

Pterygopalatine Fossa

Chapter Outline

1. Boundaries
2. Contents
3. Arteries

The pterygopalatine fossa is a deeply placed, triangular interval between the maxilla in front and the anterior surface of the root of the pterygoid process of the sphenoid behind. It communicates laterally with the infratemporal fossa, medially with the nasal cavity, anteriorly, with the orbit and posteriorly, with the middle cranial fossa and pharynx.

1. Boundaries (walls)

Anterior: The posterior surface of max-illa. Near the roof anteriorly, the infe-rior orbital fissure leads into the orbit. Below, the pterygoid process behind meets the anterior wall enclosing a ver-tical canal, the greater palatine canal. From this canal, the lesser palatine canals pass down through the pyrami-dal process of the palatine bone.

Posterior: Pterygoid process of the sphe-noid. Near the roof posteriorly, fora-men rotundum leads back to the middle cranial fossa. More medially, two paral-lel canals lead back toward the nasal part of the pharynx. The *pterygoid canal* passes through the root of the ptery-goid process of the sphenoid to reach the front of foramen lacerum. More medially, the palatinovaginal canal leads to the root of the pharynx.

Medial: Formed by the perpendicular plate of palatine bone that gives off two processes above. The orbital process anteriorly and sphenoidal process poste-riorly. With the body of the sphenoid above, they enclose the sphenopalatine foramen which opens into the nasal cav-ity (closed *in vivo* by mucoperiosteum).

Lateral: Open (pterygomaxillary fis-sure).

Roof: Body of sphenoid and orbital sur-face of palatine bone.

2. Contents

The contents of the pterygopalatine fossa are the maxillary nerve, third or terminal part of the maxillary vessels and the pterygopalatine ganglion.

Maxillary nerve (V2) is entirely sensory. It arises from the middle of the trigeminal ganglion and leaves the middle cranial fossa through the foramen rotun-dum. It crosses the pterygopalatine fossa to enter the infraorbital canal where it becomes known as the infraorbital nerve.

In the infraorbital canal, the infra-orbital nerve provides the middle supe-rior alveolar nerve which descends in the lateral wall of the maxillary sinus to innervate the upper bicuspid teeth; and the anterior superior alveolar nerves that descend in the anterior wall of the sinus to innervate the upper canine and incisor teeth.

Branches in the pterygopalatine fossa: Two branches of the pterygopala-tine nerves arise in the pterygopalatine fossa and descend to the pterygopalatine ganglion.

Posterior superior alveolar nerves arise just before the maxillary nerve enters the infraorbital groove. These branches enter canals on the back of the maxilla and innervate the maxillary sinus, gingiva, molar and premolar teeth.

Zygomatic nerve enters the orbit through the inferior orbital fissure and divides into its zygomaticotemporal and zygomaticofacial branches.

Zygomaticotemporal nerve passes along the lateral wall of the orbit, gives parasympathetic nerve fibers to the lacrimal nerve for the lacrimal gland. It passes through a foramen in the zygomatic bone to supply the skin of the temple.

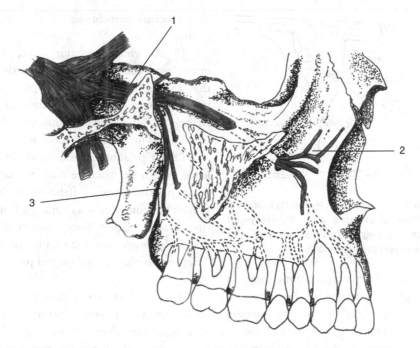

FIG. 11-1 Maxillary division of trigeminal nerve (V2). (1) Maxillary n; (2) Infraorbital n; (3) Posterior superior alveolar n.

Zygomaticofacial nerve crosses to the lower lateral side of the orbit, passes through a foramen in the zygomatic bone and supplies a region of skin on the face.

Pterygopalatine ganglion is a parasympathetic ganglion that is suspended by two pterygopalatine nerves below the maxillary nerve in the upper part of the pterygopalatine fossa opposite the sphenopalatine foramen. It is joined from behind by the nerve of the pterygoid canal (its facial root).

ROOTS

Parasympathetic (secretomotor) axons have their cell bodies in the superior salivatory nucleus in the brainstem. Preganglionic axons travel in nervus intermedius, through the geniculate ganglion as the greater petrosal nerve. In the foramen lacerum, they join the deep petrosal nerve to form the nerve of the pterygoid canal which joins the ganglion and the parasympathetic axons synapse. Postganglionic axons distribute to the lacrimal gland and all mucous glands of the nose, paranasal sinuses and palate.

Sympathetic preganglionic neurons are located in the thoracic spinal cord. Postganglionic neurons are located in the superior cervical ganglion and their axons join the carotid plexus. They leave as the deep petrosal nerve in foramen lacerum joining the deep petrosal nerve

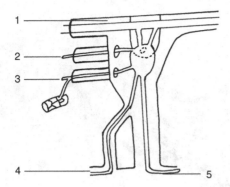

FIG. 11-2 Pterygopalatine ganglion. (1) Maxillary n; (2) Nerve of pterygoid canal; (3) Pharyngeal branch of pterygopalatine ganglion; (4) Lesser palatine n; (5) Greater palatine n.

and reach the pterygopalatine ganglion through the nerve of the pterygoid canal. They pass through the ganglion without synapsing and send vasoconstrictor fibers to the blood vessels of the nose, palate, and paranasal sinuses.

Sensory fibers come from the maxillary nerve with cell bodies in the trigeminal ganglion.

Its branches of distribution are mixed nerves containing sensory, sympathetic and parasympathetic fibers. They are branches to the nose, soft and hard palates, orbit and pharynx.

Ascending: Small branches reach the periosteum of the orbit.

Descending: Greater palatine nerves passes through the greater palatine canal to emerge from the greater palatine foramen and reach the hard palate. It divides into branches that run forward in grooves in the palate almost as far as the incisor teeth. It supplies the gingiva

and mucous membrane of the hard palate.

In the greater palatine canal, the nerve gives off inferior nasal branches that pierce the perpendicular plate of the palatine to supply the mucous membrane of the maxillary sinus and the middle and inferior nasal conchae.

Lesser palatine nerves, usually two, leave the greater palatine canal through lesser palatine canals (offshoots of the greater palatine canals) and emerge from the lesser palatine foraminae in the pyramidal process of the palatine bone. They supply the soft palate and palatine tonsil.

Medial: Posterior superior lateral nasal nerve passes through the sphenopalatine foramen to enter the nasal cavity to supply the mucosa of the posterior superior nasal region.

Posterior superior medial nasal are longer branches that enter the nasal cavity through the sphenopalatine foramen but cross the roof of the nose to reach the mucosa of the posterior superior nasal septum.

Posterior: Pharyngeal branch passes posteriorly through the palatinovaginal canal. It supplies the mucous membrane of the upper nasopharynx and sphenoidal sinus.

3. Arteries

Arteries are branches of the third (pterygopalatine) part of the **maxillary artery**. They accompany the nerves that are branches of the pterygopalatine ganglion.

Posterior superior alveolar artery descends on the infratemporal surface of the maxilla dividing into branches that enter the posterior alveolar canals. They supply the molar, premolar teeth and maxillary sinus. Other branches continue without entering the bone and supply the gingiva of the maxillary posterior teeth.

Infraorbital artery is the continuation of the trunk of the maxillary artery, it runs with the infraorbital nerve through the infraorbital canal. It supplies the orbit. In the infraorbital canal, it gives branches to the orbit, canine and incisor teeth. It appears on the face beneath levator labii superioris. On the face, it supplies the medial angle of the orbit, lacrimal sac, lower eyelid, upper lip and cheek.

Descending palatine artery enters the greater palatine canal and gives two branches:

Greater palatine artery enters the greater palatine canal (with the greater palatine nerve). It appears at the greater palatine foramen, and runs forward along a groove in the hard palate. It supplies the palatine mucous glands and the mucosa of the hard palate.

Lesser palatine artery from the greater palatine artery, passes through the lesser palatine canals and emerge through the lesser palatine foramen in the pyramidal process of the palatine. They supply the muscles and mucosa of the soft palate.

Artery of the pterygoid canal runs backward through the pterygoid canal. It supplies the roof of the pharynx.

Pharyngeal artery runs backwards through the palatinovaginal canal. It supplies the roof of the nose and pharynx.

Sphenopalatine artery is the largest branch of the third part of the maxillary artery. It passes medially traversing the sphenopalatine foramen to supply the superior nasal mucosa and paranasal sinuses.

Chapter **12**

Mandible and Temporomandibular Joint

Chapter Outline

1. Mandible
2 Temporomandibular Joint

The lower jaw, or mandible, is the largest and strongest bone in the face. It carries the mandibular teeth and the muscles of mastication. It provides attachment for the muscles of the tongue and of the floor of the mouth and articulates with the skull at the temporomandibular joint.

1. Mandible

The mandible consists of a body and a pair of rami. The body meets the ramus at the angle of the mandible and is fused with its opposite anteriorly at the symphysis menti. Each ramus divides into two processes, the condylar process posteriorly and coronoid process anteriorly with the wide mandibular notch in between.

LATERAL SURFACE OF BODY

The symphysis is a faint median ridge externally indicating the line of fusion of fetal halves of the mandible.

Mental protuberance is an anterior triangular elevation projecting forward at the lower end of the symphysis.

Mental tubercles are raised projection at the basal angles of the protuberance.

The oblique line is a faint ridge extends backward and posteriorly from the mental tubercle to the anterior border of the ramus. Buccinator is attached to its posterior part.

The mental foramen is located above the oblique line midway, between the upper and lower borders of the body and opposite the second premolar tooth. It transmits the mental nerve (branch of the inferior alveolar branch of V3) and mental vessels.

The masseter muscle is attached to the outer surface of the ramus from the mandibular notch to the angle.

MEDIAL SURFACE OF BODY

Genial tubercles are small elevations that project from the region behind the middle of the symphysis. The upper tubercles give origin to the genioglossus and lower tubercles give origin to geniohyoid.

The mylohyoid line is an oblique ridge that extends from below the tubercles above the digastric fossa to a point below the third molar tooth. It gives attachment to the mylohyoid and superior constrictor, pterygopalatine ligament and buccinator posteriorly.

Sublingual fossa is a depression above the mylohyoid line. It houses the sublingual salivary gland.

The submandibular fovea is a depression below the mylohyoid line. It houses the submandibular salivary gland.

The digastric fossa is a depression at the side of the symphysis menti. It gives attachment to the anterior belly of the digastric muscle.

MEDIAL SURFACE OF RAMUS

The mandibular foramen is located slightly above the center of the ramus. It opens into the mandibular canal that transmits the inferior alveolar nerve (branch of V3) and vessels (from the maxillary artery) continuing to the midline. They supply the mandibular teeth and give mental branches opposite the second premolar tooth.

Lingula is a triangular plate projecting upward and forming the anterior boundary of the mandibular foramen. The lingula gives attachment to the sphenomandibular ligament.

The mylohyoid groove extends down and forward from the mandibular foramen behind the lingula on the inner surface of the ramus. The mylohyoid nerve from V3 (supplying mylohyoid and anterior belly of digastric muscles) and vessels are lodged in the groove.

Medial ptergyoid muscle is attached to the roughened area below the mylohyoid groove down to the angle.

Coronoid process is a triangular, upward projection of the anterior border of the ramus. Its margin and medial surface provide insertion to the temporalis muscle.

Condyloid process is the enlarged posterior projection of the ramus and is divided into a head and a neck. The head articulates with the disc of the temporomandibular joint. The neck is the constricted part below the head to which the joint capsule and lateral ligament is attached. The lateral pterygoid muscle is attached to the pterygoid fovea anteriorly.

Mandibular notch is the thin upper border of the ramus bounded anteriorly by the coronoid process and posteriorly by the condylar process. The masseteric nerve (from V3) and vessels (from the maxillay artery) cross from the infratemporal fossa through the notch to reach the deep surface of masseter muscle.

2. Temporomandibular Joint

- **Type:** The temporomandibular joint is of the ginglymoarthrodial type; the upper part glides, and the lower part is a hinge.

- **Articulating elements:** Head of the mandible has its long axis directed backwards and medially. It is covered with fibrocartlage. Mandibular fossa and the eminence of the temporal bone are also covered with fibrocartilage.
- **Articular capsule:** Thin, loose, and attached near the margins of the articular surfaces. Above the capsule is attached posteriorly to the squamotympanic and petrosquamous sutures, and anteriorly in front of the articular eminence. Below, the capsule is attached to the articular margin of the head strengthened by the lateral ligament. Behind the capsule is a venous plexus and elastic fibers. The capsule is lined separately by synovial membrane.
- **Ligaments:** Temporomandibular ligament is a thickened lateral part of the capsule that extends from the zygoma and tubercle at its root to the lateral anterior and posterior surfaces of the neck of the mandible.

Articular disc is an oval, fibrocartilaginous disc that divides the temporomandibular joint into upper and lower compartments. The circumference of the disk is fused with the capsule and the tendon of lateral pterygoid is attached to the front edge of the discand front of capsule as well as the pterygoid fovea on the front of the neck of the mandible. The lower surface of the articular disc is

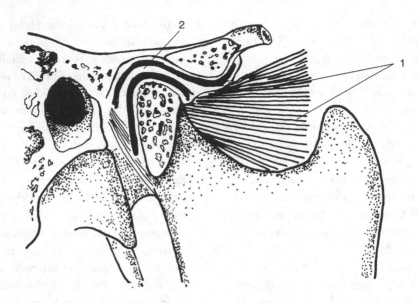

FIG. 12-1   Sagittal section through temporomandibular joint. (1) Lateral pterygoid m; (2) Articular disc.

concave and its upper surface is concavoconvex.

■ **Accessory ligaments:** Spheno-mandibular ligament is a thin and flat band descending from the spine of the sphenoid bone to the lingula of the mandible. This is part of the perichondrium of the embryonic mandibular arch completed by the anterior ligament of the malleus, malleus and incus above and Meckel's cartilage below.

Stylomandibular ligament is a band of fascia that extends from the apex of the styloid process to the posterior border of the ramus of the mandible near its angle.

■ **Nerve supply:** Auriculotemporal nerve (from V3) and a branch of the masseteric nerve (V3).

■ **Movements:** Movements of the mandible are controlled by the balance of muscles and the occlusion of teeth. They are highly individual. Movements are depression, elevation (elevation), protrusion (protraction), retraction (retrusion), and grinding (Bennett movement).

Depression is caused by lateral pterygoids (pulls the mandible downward and forward on the articular eminence), digastric and gravity. Against a resistance, digastric, geniohyoid and mylohyoid muscles contribute.

Temporalis (mostly anterior fibers), masseter and medial pterygoid elevate the mandible. Temporalis maintains a position of rest.

Lateral and medial pterygoids protrude the mandibles

Retraction of the mandible is brought by the posterior fibers of temporalis, with contribution of the middle and deep parts of masseters, digastric and geniohyoid.

Lateral movement comes from temporalis and masseter ipsilaterally, and medial and lateral pterygoids contralaterally.

Bennett movement is brought about by lateral bodily swing of the mandible in which both condylar processes are displaced.

Chapter 13

Mouth, Teeth and Tongue

Chapter Outline

1. Mouth
2. Lips
3. Cheeks
4. Gingiva
5. Teeth
6. Tongue
7. Palate
8. Deglutition

The oral cavity is the beginning of the alimentary canal and comprises the mouth cavity proper and vestibule of the mouth. It is lined with a mucous membrane whose stratified squamous epithelium keratinizes completely only on the dorsum of the tongue, hard palate and gingiva. Gingiva is the specialized region of oral mucosa surrounding the necks of the teeth. It is pale pink in color and stippled where underlying collagen fibers connect the epithelium to underlying bone.

1. Mouth

The mouth cavity is proper is located within the arches of the maxilla and mandible.

BOUNDARIES

Anterior and lateral: Alveolar arches, teeth and gingiva.
Roof: Hard and soft palates.
Floor: Tongue and soft tissues of the sublingual region including the mylohyoid muscles that form the diaphragma oris.
Posterior: Communicates with the vestibule and oropharynx through the pharyngeal (faucial) isthmus marked on either side by the palatoglossal arches.

Vestibule of the mouth is the space external to the teeth that separates the lips and cheeks from the gums and teeth. The vestibule receives openings of labial and buccal glands while the parotid duct opens into the vestibule opposite the second upper molar tooth.

There are three types of oral mucosa: masticatory mucosa bears the major forces of mastication; lining mucosa is not exposed to major forces of mastication e.g. covering lips, cheeks, soft palate and inferior surface of the tongue; and specialized mucosa lining the dorsum of the tongue.

2. Lips

Lips are mobile, musculofibrous folds bounding the opening of the mouth. In section, the lips comprise five layers.

Skin contains hair follicles, sweat and sebaceous glands.

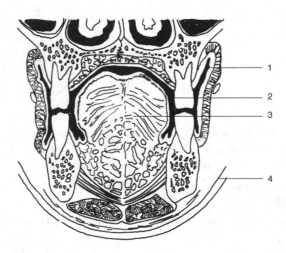

FIG. 13-1 Coronal section through oral cavity. (1) Vestibule; (2) Oral cavity proper; (3) Tongue; (4) Mylohyoid m.

Lips also comprise loose connective tissue, orbicularis oris muscle, and submucosa containing seromucous labial glands blood vessels and lymphatics.

Mucous membrane on the inner surface of the lip is covered by stratified epithelium and lacking hair follicles.

Philtrum is a shallow median groove in the upper lip.

Frenulum of the lip is an internal median fold of mucous membrane by which each lip is connected with adjoining gingiva.

3. Cheeks

Cheeks resemble the lips in structure but contain the buccinator muscle and seromucous buccal glands.

Buccal pad of fat overlies buccinator and passes deep to the masseter muscle. It assists in suckling by preventing the cheek from being drawn into the mouth.

Parotid duct is the main excretory duct of the parotid gland. It opens into the vestibule opposite the maxillary second molar tooth.

4. Gingiva

Gingiva is part of the oral mucosa. It consists of vascular fibrous connective tissue and covering the alveolar margins of the maxilla and mandible. The mucous membrane varies in degree of keratinization and is continuous with that of the lips and cheeks at the mucogingival junction.

Masticatory mucosa bears the major forces of mastication and covers the gingiva and hard palate.

The lining mucosa is not exposed to major forces and lines the vetibule, lips and cheeks.

The blood supply to the periodontium can be divided into three zones: that to the periodontal ligament; that to the gingiva facing the oral cavity; and that to the gingiva facing the tooth. Anastomoses between the three areas allows considerable collateral circulation in the supporting tissues of the teeth.

5. Teeth

Each tooth consists of specialized connective tissue (pulp) covered by three calcified tissues: dentine, cementum, and enamel. The crown varies in shape; it is chisel-shaped in incisor teeth; conical in canine teeth; has two cusps in bicuspid (premolar) teeth; and has three to four cusps in molar teeth.

Anatomical crown is the part of the tooth covered by enamel.

Clinical crown is the part of the tooth exposed in the oral cavity (it varies with the degree of gingival recession).

Root is the part of the tooth covered by cementum and embedded in an alveolar socket. Through its attachment to the periodontal ligament, the root is

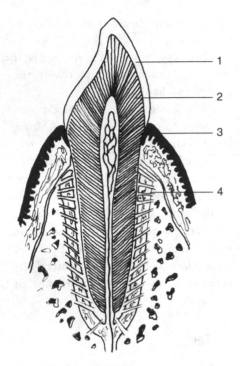

FIG. 13-2 Longitudinal section of a tooth. (1) Enamel; (2) Dentine; (3) Pulp; (4) Periodontal ligament.

responsible for anchoring the tooth to alveolar bone.

Neck (or cervix of the tooth) is the constriction at the cementoenamel junction. It is normally not seen as it is covered by gingiva.

Pulp cavity is an internal cavity of the tooth comprising a pulp chamber and one or more root canals inside the tooth. It contains the pulp, a specialized loose connective tissue, a ground substance, vessels and nerves that have entered the tooth via the apical foramen.

DENTAL MORPHOLOGY

Incisors have a long, thin cutting edge. There are two central incisors and two lateral incisors in each jaw. In the maxilla, they are carried in the premaxillary segment of the maxilla. Incisors are primarily used to cut or incise food.

Canines (cuspids) have a single prominent cusp. They assist in seizing and cutting food.

Premolars (bicuspids) usually have two cusps separated by a sagittal groove. There are two in each buccal segment. They assist in cutting food. During development, they replace deciduous molars.

Molars have 3–5 cusps. Upper (maxillary) molars have three roots and lower (mandibular) molars two. They are used to grind food.

DENTAL TERMINOLOGY

Mesial is medial in anterior teeth and anterior in posterior teeth.

Distal is lateral in anterior teeth and posterior in posterior teeth.

Buccal and labial are the surfaces of the posterior and anterior teeth respectively that face the vestibule.

Lingual is the surface facing the tongue.

Occlusal and incisal are the surfaces of posterior and anterior teeth respectively. They are the surfaces that come in contact with its opposite when the jaws are closed.

Primary (deciduous) teeth appear in the mouth from six months to two and a half years. They are 20 in number

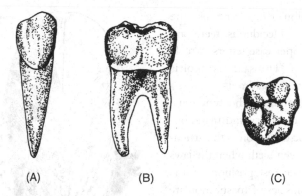

(A)   (B)   (C)

FIG. 13-3 Permanent teeth (A) right mandibular cuspid (B) right second mandibular molar (C) occlusal surface of maxillary right first permanent molar.

comprising two incisors, one canine and two molars in each quadrant.

The permanent dentition begins to replace the deciduous dentition at six years, and the last permanent tooth to erupt is the third molar (variably from 17 years). There are 32 permanent teeth comprising two incisors, one canine, two bicuspids and three molars in each quadrant.

TOOTH NUMBERING

Numbering systems have been developed to have a standard way of referring to particular teeth. Three are in common use:

The **Universal Numbering System** adopted by the ADA (American Dental Association) identifies permanent teeth are numbered 1–32 from the upper third molar to the lower right third molar. Thus the upper left canine is identified as 12. Deciduous teeth are numbered as A–T from upper right second molar to

lower right second molar. Thus the lower left second molar for example is identified as K.

The **Federation Dentaire Internationale** or **Two Digit Notation** (FDI) system identifies permanent teeth by quadrant (upper right one, upper left two, lower left three and lower right four) and numbered 1–8 with the central incisor numbered one. Thus the upper left canine is identified as 23. Deciduous teeth are numbered one to five with each quadrant numbered five upper right, six upper left, seven lower left and eight lower right. Teeth of the upper left quadrant for example become 61, 62, 63, 64, 65.

The **Palmer Notation** divides the mouth into quadrants with permanent teeth numbered 1–8 (upper right central incisor is one and the third molar eight. The numbers sit inside an L-shaped symbol used to identify the quadrant (L is the right side and backward L is the left side. Mandibular quadrants use the L upside down. Deciduous teeth are numbered by upper case letters A to E in each quadrant. The same symbol is used to identify the quadrants.)

Occlusion refers to the functional relation of the maxillary and mandibular teeth. Centric occlusion is the articular relation between teeth when the jaws are closed. In the rest position of the mandible, the muscles of mastication are at rest and the teeth are not in contact.

6. Tongue

The tongue is a mobile, mucosa-covered muscular organ situated in the floor of the mouth. It is attached by muscles to the hyoid bone, mandible, styloid process and pharynx and is important in the appreciation of taste, in mastication, swallowing and speech.

PARTS OF THE TONGUE

Tip (Apex) lies free in the oral cavity and rests against the anterior teeth. The margins of the tongue are also free and rest against the gingiva and posterior teeth.

Dorsum is convex partly in the oral cavity and partly in the oropharynx.

Sulcus terminalis is a V-shaped groove on the dorsum marking the boundary between the anterior two-thirds or oral part of the tongue and the posterior third, or pharyngeal part of the dorsum of the tongue.

Foramen cecum is a blind pit at the apex of the sulcus terminalis. It is

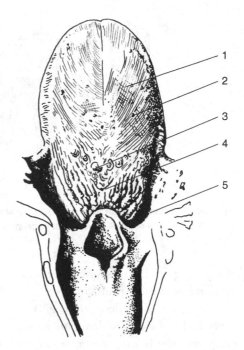

FIG. 13-4 Dorsum of tongue. (1) Filiform papillae; (2) Fungiform papillae; (3) Vallate papillae; (4) Foramen caecum; (5) Lymphoid tissue.

the remnant of the primordium of the thyroid gland and thyroglossal duct.

Median glosso-epiglottic fold is a fold of mucous membrane connecting the tongue to the epiglottis in the midline. Contained within it is the hyoepiglottic ligament.

Lateral glossoepiglottic folds are folds of mucous membrane joining the lateral edge of the epiglottis to the lateral wall of the pharynx.

Epiglottic valleculae are depressions on either side of the median glossoepiglottic fold. They are a common site of lodgement of foreign bodies such as fish bones.

Lingual papillae are vertical projections of the mucosa on the dorsal surface of the tongue. There are four types of papillae:

Filiform papillae are the predominant form of papilla forms the plush of the tongue. Each papilla is a primary elevation of the lamina propria (primary papilla) subdivided into one or more secondary papillae. Their thinly keratinized epithelial cover ends in tapered points which point towards the pharynx.

Fungiform papillae are club-like projections with a narrow stalk scattered singly among filiform papillae and projecting above neighboring filiform papillae. A thin, translucent epithelial cover and central vessels give these papillae a red colour. Some bear taste buds.

Vallate (cricumvallate) papillae 7–11 in number and much larger than the other papillae, make a "v" shaped row demarcating the body from the root of the tongue. They have the shape of an inverted cone (but do not extend above the dorsal surface) and are surrounded by a trench. The walls of the trench contain taste buds and serous glands open into the trench presumably to remove old taste stimuli.

Foliate papillae are parallel mucosal folds on the lateral margins of the tongue located at the junction of the body and root. They are folds of lamina propria covered with stratified squamous epithelium. They bear taste buds and serous glands open into the base of the trenches between folds. Foliate papillae are marked in infants but regress in adults.

Pharyngeal part of the dorsum of the tongue contributes to the anterior wall of the oropharynx. The covering mucous membrane is devoid of papillae but the surface is uneven because of lymphoid tissue (the lingual tonsils) in the underlying submucosa.

Lingual tonsils comprise lymphoid follicles and diffuse lymphoid tissue around crypts lined with stratified non-keratinized epithelium. Mucous lingual glands open into the base of crypts.

Inferior surface or undersurface of the free part of the tongue is smooth, devoid of papillae and covered with mucous membrane.

Frenulum of the tongue is a median fold of mucous membrane connecting the tongue to the floor of the mouth.

Fimbriated fold (plica fimbriata) is a fringed ridge of mucous membrane ridge that passes on the lateral sides of the inferior surface of the tongue from the floor almost to the tip.

Sublingual fold is a fold of mucosa raised by the sublingual gland. 15–20 sublingual ducts open separately along the sublingual fold.

Sublingual caruncle is a (paired) elevation at the side of the frenulum on the top of which the duct of the submandibular gland opens into the oral cavity.

Root is the fixed part of the tongue resting on the floor of the mouth and connected by muscles to the mandible in front, and to the hyoid bone, below and behind.

MUSCLES OF THE TONGUE

Muscles of the tongue are divided into two groups: the extrinsic and intrinsic.

Extrinsic Muscles

Extrinsic muscles are attached to bone and alter the position and shape of the tongue. Intrinsic muscles lie wholly within the tongue and can only alter the shape of the tongue.

Genio-glossus

Genio-glossus is a fan shaped muscle with apex at the origin. Each is vertically placed, contacts its fellow medially, and makes up most of the bulk of the posterior part of the tongue.

- **Origin:** Superior genial tubercle on inner surface of mandible near the symphysis.
- **Insertion:** Body of hyoid bone (posterior fibers), inferior aspect of tongue from root to tip (anterior fibers). They are separated from one another by a midline fibrous septum.
- **Nerve supply:** Hypoglossal nerve (XII).
- **Action:** Mainly a depressor of the tongue. Inferior fibers raise tongue and hyoid bone, draw tongue forwards and protrude it to opposite side. Anterior fibers withdraw the tip of the protruded tongue.

Flaccidity of genioglossus during anaesthesia or REM stage of sleep contributes to airway occlusion and sleep apnea. Fractures of the anterior part of the mandible including the origin of genioglossus in traumatic accidents may result in airway obstruction particularly if the patient is supine.

Hyoglossus

Hyoglossus is a flat, quadrangular muscle located mostly under cover of mylohyoid muscle.

- **Origin:** From side of the body, all of the greater horn and the lesser horn of the hyoid.
- **Insertion:** Margins of the posterior half of the tongue, interdigitating with stylo-glossus.
- **Nerve supply:** Hypoglossal nerve (XII).
- **Action:** Depresses the sides of the tongue making the surface convex transversely. With genioglossus and styloglossus, it retracts the tongue.

Hyoglossus is the key to the suprahyoid region. Its deep relations are the middle constrictor, sylohyoid ligament, glossopharyngeal nerve, second part of the lingual artery, genioglossus and part of the insertion of geniohyoid.

Styloglossus

- **Origin:** Front of the apex of the styloid process and stylohyoid ligament.
- **Insertion:** Side of dorsum of tongue, interdigitating with hyo-glossus.
- **Nerve supply:** Hypoglossal nerve (XII).

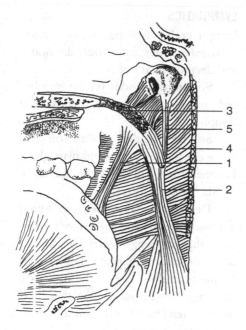

FIG. 13-5 Muscles of the oral pharynx. (1) Palatoglossus m; (2) Palatopharyngeus m; (3) Musculus uvulae; (4) Superior constrictor m; (5) Salpingopharyngeus m.

- **Action:** Draws tongue upwards and backwards and makes superior surface concave transversely.

Palatoglossus

- **Origin:** Inferior and lateral surface of soft palate and palatine aponeurosis.
- **Insertion:** Side and dorsum of tongue.
- **Nerve supply:** Pharyngeal plexus by fibers from the internal branch of the accessory nerve (XI).
- **Action:** Approximates palatoglossal folds shutting off the oral from

Intrinsic Muscles

Intrinsic muscles are entirely contained within the substance of the tongue, and are all supplied by hyoglossal nerve.

Superior longitudinal muscles are found on the right and left side of the tongue. Longitudinal fibers lie under the mucous membrane. It arises from the median glossoepiglottic fold, and from the septum along the mid line. The fibers pass obliquely outwards, the anterior fibers being longitudinal, to the side of the tongue.

Inferior longitudinal muscle of the tongue is a band of muscle fibers running along the under surface of the tongue from base to tip on each side. It lies between the genioglossus and hyoglossus muscles. It arises from the septum at the base of the tongue, is joined anteriorly with some fibers of the styloglossus, and passes to the tip.

Transverse muscle of the tongue forms a horizontal layer of muscular fibers between the superior and inferior longitudinal muscles. The fibers arise from the septum and pass laterally to the submucosa of the sides of the tongue.

Vertical muscle of tongue are located at the borders of the apex, and pass from the superior to the inferior surface.

The septum of the tongue is a vertical fibrous partition extending, in the muscular portion, from the hyoid bone to the tip but not reaching the dorsum.

ARTERIES

Lingual, and tonsillar branch of facial artery supplies the tongue.

NERVES

Lingual branch of the mandibular nerve (including the chorda tympani) supplies to the anterior two-thirds of the tongue; glossopharyngeal branch supplies to posterior third of the tongue; and the hypoglossal branch supplies to muscles of the tongue.

Lingual nerve is a part of the post-trematic nerve of the first arch and supplies the sensation for anterior two thirds of the tongue.

Chorda tympani nerve is the pretrematic nerve to the first pharyngeal arch. It carries taste and secretomotor fibers for the submandibular gland, sublingual gland and intrinsic glands of the anterior two thirds of the tongue.

Glossopharyngeal nerve supplies sensation and taste for the posterior third of the tongue.

Internal laryngeal branch of the vagus receives general sensation and taste from a region near the epiglottis.

Hypoglossal nerve supplies somatic motor innervation of the intrinsic and extrinsic (except palatoglossus) muscles of the tongue.

LYMPHATICS

Lymph nodes draining the tongue are submental nodes, submandibular nodes and deep cervical nodes

Submental nodes on the mylohyoid muscle below the mandibular symphysis.

Submandibular nodes on the surface of the submandibular gland between the gland and the mandible beside the facial artery.

Deep cervical nodes are found alongside the internal jugular vein especially the jugulodigastric node (situated where the internal jugular vein is crossed by the posterior belly of the digastric), and juguloomohyoid (where the intermediate tendon of omohyoid crosses the internal jugular vein).

Lymphatics may drain to one or both sides of the tongue. Those from the oral part of the tongue either pierce mylohyoid muscle or pass laterally on the surface of hyoglossus ending in submental, submandibular or deep cervical nodes (especially juguloomohyoid node. Lymphatics from the pharyngeal part of the tongue pierce the superior constrictor to end in upper deep cervical nodes especially the jugulodigastric node.

7. Palate

The palate forms the roof of the mouth and the floor of the nasal cavity. It consists of two parts: an anterior two thirds, the hard palate, and a posterior third, the soft palate.

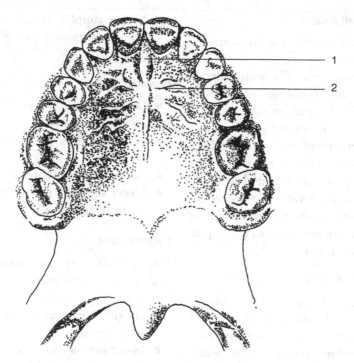

FIG. 13-6 Hard and soft palates. (1) Incisive papilla; (2) Rugae.

Hard palate is formed by the palatine processes of the maxillae in front and the horizontal plates of the palatine behind. It is covered above by the mucous membrane of the nasal cavity and below by mucoperiosteum which contains blood vessels, nerves, and a large number of mucous palatine glands posteriorly. A median ridge in the mucoperiosteum, the median raphe, ends anteriorly in the incisive papilla. Extending sideways from the raphe are a number of transverse palatine folds (rugae).

Soft palate (velum palatinum) is a movble fibromuscular fold suspended from the posterior edge of the hard palate. It consists of an aponeurosis, muscles, blood vessels, and nerves covered by mucous membrane. It is continuous with the palatoglossal and palatopharyngeal folds laterally. The uvula is an inferomedian projection from the free posterior border.

Palatine aponeurosis attached to the posterior border of the hard palate. It arises from the expanded tendon of tensor veli palatini after it hooks around the pterygoid hamulus. It supports the muscles of the soft palate all of which participate in swallowing.

Levator Veli Palatini

Levator veli palatini is located lateral to the choanae.

- **Origin:** Apex of petrous temporal bone and medial side of the cartilage of the auditory tube.
- **Insertion:** Palatine aponeurosis and muscle of the opposite side.
- **Action:** Principal elevator of the soft palate drawing it backwards. Also a key elevator of the pharynx.
- **Nerve supply:** Internal branch of accessory nerve through the pharyngeal plexus.

Tensor Veli Palatini

- **Origin:** Scaphoid fossa at base of pterygoid plate, spine of sphenoid, and cartilage of the auditory tube.
- **Insertion:** Tendon winds around pterygoid hamulus, passes above buccinator to form the broad palatine aponeurosis.
- **Action:** Tenses soft palate thereby flattening and slightly depressing it.

- **Nerve supply:** Mandibular nerve from a twig passing through the otic ganglion.

Musculus Uvulae

- **Origin:** Posterior nasal spine and palatine aponeurosis.
- **Insertion:** Mucous membrane of the uvula.
- **Action:** Raises the uvula.
- **Nerve supply:** Vagus nerve through the pharyngeal plexus.

Palatoglossus

Palatoglossus occupies the palatoglossal fold and forms the anterior arch of the tonsillar fossa.

- **Origin:** Undersurface of palatine aponeurosis.
- **Insertion:** Side of tongue.
- **Action:** Approximates palatoglossal folds shutting off the oral cavity from the oropharynx.
- **Nerve supply:** Vagus through the pharyngeal plexus.

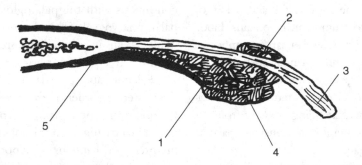

FIG. 13-7 Sagittal section of soft palate. (1) Levator veli palatini m; (2) Palatopharyngeus m; (3) Musculus uvulae; (4) Palatoglossus m; (5) Tensor aponeurosis.

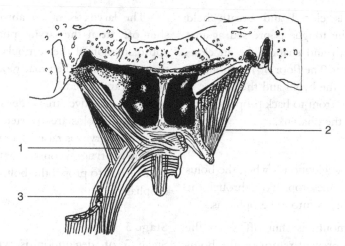

FIG. 13-8 Muscles of the soft palate. (1) Levator veli palatine m; (2) Tensor veli palatine m; (3) Palatoglossus m.

Palatopharyngeus

Palatopharyngeus forms the posterior arch of the tonsillar fossa occupying the palatopharyngeal fold.

- **Origin:** Posterior border of hard palate and palatine aponeurosis. It is arranged in two bands, medial and lateral separated by levator veli palatini.
- **Insertion:** Posterior border of thyroid cartilage.
- **Action:** Approximates palatopharyngeal folds separating oropharynx from nasopharynx.
- **Nerve supply:** Vagus through the pharyngeal plexus.

NEUROVASCULAR SYSTEM OF PALATE

Arteries: Ascending palatine branch of the maxillary artery, the ascending palatine branch of the facial artery, and the palatine branch of the ascending pharyngeal artery.

Veins: Drain to the pterygoid and tonsillar plexuses.

Nerves: The sensory innervation is provided by branches of the palatine and nasopalatine nerves.

The motor innervation of palatal muscles is provided by the internal branch of the accessory nerve through the pharyngeal plexus and the nerve to the medial pterygoid muscle (tensor veli palatini).

8. Deglutition

Three stages of shallowing are described.

Stage 1

Stage 1 of deglutition is when the passage of bolus from mouth to oropharynx.

The jaw is closed and breath held reflexly. The tongue is pressed against incisor teeth and shaped like an elongated trough. The floor of the mouth is raised with the hyoid and the tongue is raised from front to back propelling the bolus into the pharynx.

Stage 2

Stage 2 of deglutition is when the bolus passes from the oropharynx through the laryngopharynx into the esophagus.

The mouth is shut off from the phaynx to prevent return of the bolus into the mouth. The nasopharynx is shut off from the oropharynx to prevent the bolus from ascending into the nose.

The larynx is raised abruptly and shut off from the laryngopharynx by approximation of the vestibular folds to the epiglottis and the rimae glottidis and vestibuli are closed.

The sensitive free edges of the aryepiglottic folds are inverted.

The pharynx is raised over the bolus and the pharyngeal constrictors contract in sequence to propel the bolus into the esophagus.

Stage 3

Stage 3 of deglutition is when The bolus passes from the esophagus to the stomach.

The bolus is propelled in the esophagus to the stomach.

Chapter 14

Pharynx

Chapter Outline

1. Divisions of the Pharyngeal Cavity
2. Structure of the Pharynx

The pharynx is a fibromuscular tube dependent from base of skull attached to and lying behind the nose, mouth, and larynx and communicating with them. It extends from the skull base to the level of the sixth cervical vertebra behind or lower border of the cricoid cartilage in front. Above, where it is a part of both respiratory and alimentary tracts, its walls are rigid and always patent. Below the level of laryngeal aperture, it is normally a transverse slit continuous with the esophagus which opens upon deglutition. Loose areolar tissue and fascia separate it from the prevertebral muscles.

1. Divisions of the Pharyngeal Cavity

In front, the incomplete anterior wall of the pharynx opens directly into the nasal, oral, and laryngeal cavities. It includes the soft palate, dorsum of the tongue and arytenoid and cricoid cartilages.

Nasopharynx is the part of the pharynx behind the nose and in front of the atlas and axis.

Oropharynx is the part in front of the axis and CIII behind the mouth.

Laryngopharynx is the part in front of CIII to CVI (the "typical" cervical vertebrae) below and behind the opening of the larynx.

Nasopharynx extends from the base of the skull to the level of the soft palate.

Choanae are the two posterior nasal apertures placed in the upper part of the anterior wall.

Pharyngeal ostium of auditory tube is the triangular shaped opening of the auditory tube on the lateral wall. It has rounded margins containing lymploid tissue (tubal tonsil).

Infection may track along the auditory tube producing inflammation of the middle ear and mastoid antra (otitis media).

Torus tubarius (tubal elevation) is the elevation behind the pharyngeal ostium. It is produced by underlying

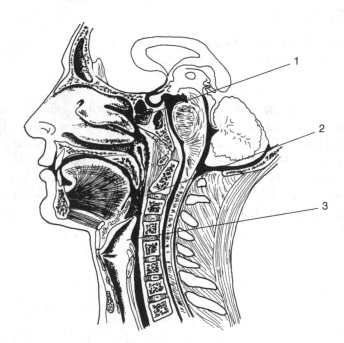

FIG. 14-1 Sagittal section of head and neck showing extent of pharynx. (1) Nasopharynx; (2) Oropharynx; (3) Laryngopharynx.

cartilage of the medial end of the auditory tube.

Salpingopharyngeal fol is produced by the underlying salpingopharyngeus muscle. It extends from the lower part of the torus to the lateral wall of the pharynx.

Torus levatorius is a rounded bulge running downward below the ostium to the soft palate raised by the underlying levator veli palatin muscle.

Pharyngeal recess is the space behind the torus tubarius.

Pharyngeal tonsil is the mass of lymphoid tissue embedded in the posterior wall of the nasopharynx.

Hypertrophy of the pharyngeal tonsil ("adenoids") may interfere with nasal respiration and phonation.

Pharyngeal isthmus is the interval between the soft palate and the back of the pharynx, marking the junction of nasopharynx and the oropharynx. It is bounded by the soft palate, palatopharyngeus muscle and pharyngeal fold.

RECESSES OF THE PHARYNX

Pharyngeal recess is a narrow slit in the nasopharynx between the auditory tube and prevertebral muscles.

In the oropharynx lies the valleculae, between the base of tongue and epiglottis separated by the median glossoepiglottic fold. Each vallecula is limited behind by lateral glossoepiglottic (pharyngoepiglottic) folds.

Piriform fossae are situated below, in the laryngopharynx between the laryngeal aperture and the side wall of the pharynx.

Valleculae are all common sites of impaction of foreign bodies (e.g. fish bones) and in danger of perforation in endoscopy.

Oropharynx extends from the soft palate to the hyoid bone.

Median glossoepiglottic fold is a median fold of mucous membrane running from the epiglottis to the base of the tongue. Embedded in it is the hyoepiglottic ligament.

Lateral glossoepiglottic folds are lateral folds of mucous membrane running from the epiglottis to the lateral wall of the pharynx.

Epiglottic valleculae are spaces in each side of the median glossoepiglottic folds (between the median and lateral glossoepiglottic folds).

Fauces is the term applied to the space surrounded by the palate, tonsils, and uvula. It includes the oropharyngeal isthmus, the arches, the tonsillar fossa and the tonsil.

Faucial (oropharyngeal) isthmus is the aperture through which the mouth communicates with the oropharynx. It is bounded by the tongue, palatoglossal arches and soft palate.

Palatoglossal arch is a fold of mucous membrane produced by the underlying palatoglossus muscle. It extends from the side of the root of the tongue to the soft palate in front of the tonsil.

Palatopharyngeal arch is a fold of mucous membrane produced by the underlying palatopharyngeus muscle. It extends from the end of the soft palate to the pharyngeal wall.

The two diverging palatoglossal and palatopharyngeal arches are known as the anterior and posterior pillars of the fauces and enclose the tonsillar fossa.

PALATINE TONSIL

Palatine tonsil is a mass of lymphoid tissue located between the palatoglossal and palatopharyngeal folds. The medial surface of the tonsil presents an intra-tonsillar cleft (known as the supratonsillar fossa). The front of the tonsil is covered with mucous membrane which bounds the supratonsillar fossa superiorly as the semilunar fold. The fold continues down in front of the tonsil as the triangular fold, which is separated from the tonsil by the tonsillar sinus. The lateral surface of the tonsil is covered by a fibrous capsule.

The arrangement of lymphoid tissue around the fauces is known as the tonsillar ring. It consists of the lingual tonsil, anteriorly; the palatine tonsils, laterally; and the pharyngeal tonsil, posteriorly.

The bed of the tonsil lateral to the loose areolar tissue surrounding the fibrous capsule of the tonsil is made of four thin sheets, two areolar and two fleshy. They are (a) pharyngobasilar fascia; (b) palatopharyngeus; (c) superior constrictor; and (d) buccopharyngeal fascia. Structures passing to the tongue in the lower part of the bed include

(a) styloglossus and (b) the glossopharyngeal nerve. The paratonsillar vein from the soft palate pierces the wall of the tonsillar bed.

Neurovascular supply of the tonsil is supplied by the **external carotid artery** (chiefly by the tonsillar branch of the facial artery); external palatine (paratonsillar) vein; and the lymphatics of the tonsil drain to the upper deep cervical, especially the jugulodigastric node which is the so called sentinel or tonsillar node.

Laryngopharynx extends from the upper border of the epiglottis to the lower border of the cricoid cartilage where it is continuous with the esophagus.

Aditus laryngis is the anterior opening (inlet) of the larynx. It faces superiorly and posteriorly.

Piriform recess (fossa) is part of the cavity of the laryngopharynx on each side of the aditus. It is bounded by the thyrohyoid membrane and thyroid cartilage laterally and arytenoid and cricoid cartilages medially.

Foreign bodies may lodge in the piriform recesses where the internal laryngeal nerve and superior laryngeal vessels lie immediately beneath the mucous membrane.

2. Structure of the Pharynx

The pharyngeal wall comprises four coats or tunics (from within outward).

The mucous membrane covering epithelium is pseudostratified ciliated cells in the nasopharynx, and stratified squamous in the oropharynx and laryngopharynx. Mixed serous and mucous glands are located in the mucous membrane.

Fibrous coat, pharyngobasilar fascia (pharyngeal aponeurosis) is the fibrous submucosal layer forming the basis of the walls of the nasopharynx. It is thin below, but strong above, where it fills in the space above the upper margin of the superior constrictor, over which the auditory tube passes. It limits the deformation of the nasopharynx and keeps it patent. The fascia is attached above to the body of the occiput and petrous portion of the temporal and strengthened in the midline by a process of fascia attached to the pharyngeal tubercle (the pharyngeal raphe). Inferiorly, the fascia becomes lost between the muscular and mucous membrane coats.

The muscular coat comprises five paired voluntary muscles arranged in two layers. The superior, middle and inferior constrictors are (constrictors) in the circular coat and stylopharyngeus and palatopharyngeus are (levators) in the longitudinal coat.

Circular coat consists of the three constrictor muscles. These are fan-shaped and incomplete anteriorly. Each overlaps its neighbor from below upwards like three flower pots placed one inside the other.

Superior Constrictor

- **Origin:** (1) Posterior border of medial pterygoid lamina; (2) pterygomandibular raphe; (3) mylohyoid line of mandible; (4) side of root of the tongue.
- **Insertion:** Pharyngeal tubercle at the base of skull and the median pharyngeal raphe, a fibrous cord extending in the mid-line from pharyngeal tubercle to blend below with the back of the esophagus.

Each constrictor blends with its fellow of the opposite side down to level of vocal folds. The muscle surrounds the lower part of the pharyngobasilar fascia.

Middle Constrictor

- **Origin:** Upper border of great cornu of hyoid bone; lesser cornu and lower part of stylohyoid ligament.
- **Insertion:** Median raphe overlapping superior constrictor down to level of the vocal folds.

Inferior Constrictor

- **Origin:** (1) Outer surface of thyroid cartilage behind oblique line; (2) cricoid cartilage.
- **Insertion:** Thyroid part into median raphe below level of the vocal folds.

The muscle wall is thin near the insertion of the inferior constrictor and may be the site of a diverticulum.

Cricopharyngeus

Fibers arising from the cricoid are sphincteric (they do not have a raphe) and are named cricopharyngeus muscle.

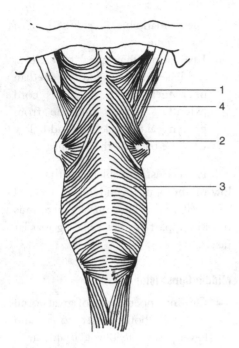

FIG. 14-2 Pharyngeal constrictor muscles from behind. (1) Superior constrictor; (2) Middle constrictor; (3) Inferior constrictor; (4) Stylopharyngeus.

- **Action:** Constrict the wall of the pharynx and participate in act of swallowing.
- **Nerve supply:** From pharyngeal plexus. The inferior constrictor muscle is also innervated by branches of the external and recurrent laryngeal nerves.

The constrictors are supplied by the accessory nerve (cell bodies in nucleus ambiguus) through the pharyngeal plexus. Cricopharyngeus (the part of inferior constrictor arising from the arch of the cricoid and providing a sphincter at the origin of the esophagus) is supplied by recurrent laryngeal nerve.

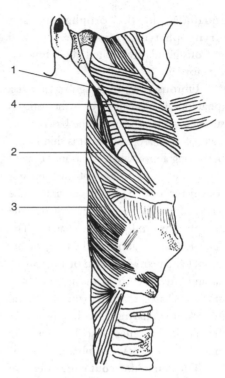

FIG. 14-3 Pharyngeal constrictors from the side. (1) Superior constrictor; (2) Middle constrictor; (3) Inferior constrictor; (4) Stylopharyngeus.

Longitudinal coat also consists of three muscles that originate separately and blend at their insertions.

Stylopharyngeus

- **Origin:** Medial side of styloid process near its root. It descends between superior and middle constrictors.
- **Insertion:** Posterior border of thyroid cartilage.
- **Action:** Elevates larynx and pharynx.
- **Nerve supply:** Glossopharyngeal nerve, IX.

Palatopharyngeus

Palatopharyngeus lies in the palatopharyngeal fold.

- **Origin:** Posterior border of hard palate and palatine aponeurosis as two strands separated by levator veli palatini.
- **Insertion:** Joins fibers of stylopharyngeus muscle to insert on the posterior border of the thyroid cartilage.
- **Action:** Approximates the palatopharyngeal folds.
- **Nerve supply:** Pharyngeal plexus.

Salpingopharyngeus

Salpingopharyngeus lies in the salpingopharyngeal fold, in the lateral wall of the pharynx.

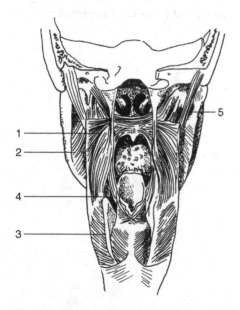

FIG. 14-4 Muscles of the pharynx from within. (1) Superior constrictor; (2) Middle constrictor; (3) Inferior constrictor; (4) Palatopharyngeus; (5) Stylopharyngeus.

- **Origin:** Lower part of the auditory tube.
- **Insertion:** Joins fibers of palatopharyngeus and reaches to posterior border of the thyroid cartlage.
- **Action:** Assists in raising the walls of the pharynx in swallowing.
- **Nerve supply:** Pharyngeal branch of the vagus through the pharyngeal plexus comprising fibers from the cranial part of the accessory nerve.

FASCIAL COAT

Buccopharyngeal fascia is the areolar coat covers the pharyngeal muscles and buccinator and contains pharyngeal venous and nerve plexuses. It blends above with the pharyngobasilar fascia.

NEUROVASCULAR SUPPLY OF THE PHARYNX

Arteries: The ascending pharyngeal and inferior thyroid (of the external carotid) arteries, ascending palatine branches of the facial artery and the greater palatine branch of the maxillary artery.

Lymphatics: Drains directly to upper deep cervical nodes or indirectly through retropharyngeal nodes. Nasopharynx and pharyngeal tonsils drain to retropharyngeal then upper deep cervical nodes along the internal jugular vein. The laryngopharynx drains through the thyrohyoid membrane to upper deep cervical nodes.

Veins: An extensive pharyngeal venous plexus is formed on the posterior and lateral walls of the pharynx. It drains into the internal jugular vein and communicates with the pterygoid plexus.

Nerves: The pharyngeal plexus, situated on the middle constrictor, comprises branches from IX, pharyngeal branches of X and the superior cervical sympathetic ganglion. Sensory fibers travel in IX. Motor fibers in the plexus are from the cranial accessory nerve via the vagus to all muscles of the pharynx and soft palate except tensor veli palatini (trigeminal nerve).

PHARYNGEAL GAPS

Above and below the narrow origins of the three pharyngeal constrictors, the overlap is deficient leaving four lateral gaps through which structures enter of leave the pharynx. The gaps are normally closed by two layers of fascia (pharyngobasilar and buccopharyngeal).

The uppermost gap between superior constrictor and base of the skull is traversed by (1) auditory tube; (2) leator veli palatini; and (3) ascending palatine artery.

The second gap between superior and middle constrictors is traversed by (1) stylopharyngeus; and (2) the glossopharyngeal nerve.

The third gap between middle and inferior constrictor extends forwards to include the gap between hyoid and thyroid (closed by thyrohyoid membrane) is traversed by (1) internal laryngeal nerve; and (2) superior laryngeal artery; and (3) superior laryngeal vein.

The fourth gap is below the inferior constrictor and traversed by (1) inferior laryngeal nerve; (2) inferior laryngeal artery; and (3) inferior laryngeal vein.

Chapter 15

Larynx

Chapter Outline

1. Cartilages
2. Joints
3. Ligaments and Membranes of the Larynx
4. Cavity of the Larynx
5. Muscles

The larynx lies in the front and upper part of the neck, being placed below the tongue and hyoid bone and between the large vessels of the neck. In the midline, it is covered only by skin and cervical fascia; but laterally, it is overlaid by the sternohyoid and sternothyroid, by the thyrohyoid and origin of the inferior constrictor muscles.

The larynx is involved in the production of speech sounds and phonation and as a valve it controls the flow of air and the distribution of air in the lungs during respiration and act as a sphincter to guard the air passages during swallowing.

1. Cartilages

The larynx is a specialized valve connecting the lower pharynx with the trachea. It is composed of four major cartilages joined by synovial joints. Ligaments and membranes fill the gaps between, muscles control the relationship between the cartilages and the whole is lined by mucous membrane.

There are nine laryngeal cartilages: three unpaired: thyroid, cricoid, and epiglottic; three paired: arytenoid, cuneiform, and corniculate.

THYROID CARTILAGE

Thyroid cartilage is the largest of the laryngeal cartilages.

Laminae are flat quadrilateral plates of hyaline cartilage fused in front at an acute angle (thyroid angle). The angle forms a median projection known as the laryngeal prominence (Adam's apple) separated above by a V-shaped notch, the superior thyroid notch. Posteriorly the laminae diverge widely. Each lamina has a free border which projects upward and downward as the superior and inferior horns. The inferior horns have articular facets at their lower ends for articulation with the cricoid cartilage.

Outer Surface

The oblique line passes upwards and backwards from the inferior thyroid tubercle in the middle of the inferior border to the superior thyroid tubercle in front of the root of the superior horn. It gives attachment to three muscles: sternothyroid, thyrohyoid, and below the oblique line to the inferior constrictor.

Inner Surface

The thyroid angle is in the midline at the junction of right and left alae. Attached to it are the petiolus of the epiglottis, the ventricular and vocal folds, the thyroarytenoid and thyroepiglottic muscles.

Superior Border

Thyrohyoid membrane connects the superior border to the hyoid bone.

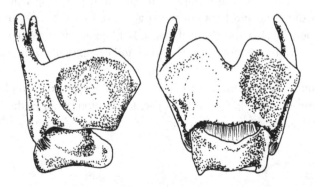

FIG. 15-1 Thyroid cartilage articulating with the cricoid cartilage.

Inferior Border

Cricothyroid ligament connects the thyroid with the upper border of the cricoid near the midline.

Cricothyroid muscle connects the thyroid with the upper border of the cricoid laterally.

Posterior Border

The posterior border is thick and rounded and receives the insertions of stylo-, salpingo- and palatopharyngeus muscles.

Thyrohyoid ligament is attached to the tips of the superior horns.

Inferior horns have a small facet on their medial surface for articulation with the cricoid at a synovial joint.

Cricoid Cartilage

Cricoid cartilage is located below the thyroid cartilage and has the shape of a signet ring. It has a wider lamina posteriorly and an arch, the narrower anterior part of the ring.

Outer Surface

The anterior half gives attachment to cricothyroid muscles, and behind this to the cricopharyngeus part of inferior constrictor. The posterior half is broad and thick, presenting a ridge in the midline for attachment of some longitudinal fibers of the esophagus. On each side of the ridge is a depression that gives attachment to the posterior cricoarytenoid. At the junction of the arch and lamina is a small facet for

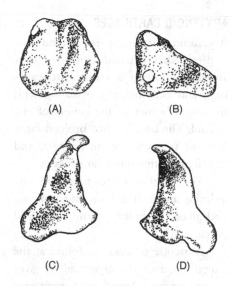

(A)   (B)

(C)   (D)

FIG. 15-2 (A) Cricoid cartilage (posterior aspect); (B) Cricoid cartilage (lateral aspect); (C) Right arytenoid (frontal aspect); (D) Right arytenoid (medial aspect).

articulation with the inferior horn of the thyroid cartilage.

Inner Surface

It is smooth and lined with mucous membrane.

Superior Border

Superior border gives attachment anteriorly to the cricothyroid ligament, and laterally to the cricovocal membrane and lateral cricoarytenoid. The upper border of each lamina has facets for articulation with the arytenoid cartilages.

Inferior Border

Inferior border is horizontal and connected with the first ring of trachea by the cricotracheal ligament.

ARYTENOID CARTILAGES

Arytenoid cartilages are shaped like a three-sided pyramid; with an apex above supporting a corniculate cartilage; and base below articulating with the upper border of the lamina of the cricoid. The base has two projections, a forward projecting vocal process, and lateral projecting muscular process.

Vocal process are found in the anterior angle. It is long and pointed, and gives attachment to the vocal fold (vocal cord).

Muscular process are found at the lateral angle of the arytenoid. It gives attachment to lateral and posterior cricoarytenoid muscles.

Antero-laterally, it contains a deep depression, the triangular fovea above, and a more shallow depression, the oblong fovea below. The two depressions are separated from each other by part of the arcuate crest.

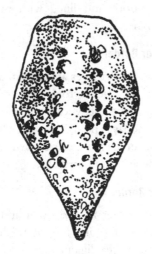

FIG. 15-3 Laryngeal aspect of epiglottis.

Thyroa-rytenoid muscle is attached to the oblong fovea and the posterior edge of the vestibular fold is attached to the triangular fovea.

Posterior surface is hollowed for the attachment of transverse arytenoid muscle.

Medial surface is narrow and smooth, covered with mucous membrane.

Base consists of concave facet which articulates with the cricoid cartilage, forming a gliding synovial joint.

Apex supports the corniculate cartilage.

EPIGLOTTIC CARTILAGE

Epiglottic cartilage covers the superior aperture of the larynx. It is a leaf-shaped lamina of yellow elastic cartilage serving as the skeleton of the epiglottis. It is located behind the base of the tongue and in front of the upper aperture of the larynx. Mucous membrane covers the upper third of the anterior surface and the entire posterior surface.

Apex: The stem (petiolus) is attached to the thyroid angle by the thyroepiglottic ligament a little below the thyroid notch.

Base: It is broad, rounded, and free, directed upwards.

On the anterior surface the epiglottic cartilage is covered in the upper part by mucous membrane, which passes forwards as the median glossoepiglottic fold on to the tongue. Beneath the fold is the hyoepiglottic ligament which attaches the epiglottis to the hyoid bone.

Laterally the lateral glossoepiglottic (pharyngoepiglottic) folds pass on to the pharyngeal wall. Between the lateral and medial folds are the shallow fossae, the epiglottic valleculae.

On the posterior surface, the cartilage is covered by pits made by glands and foramina for nerves and blood vessels. It is covered by mucous membrane. The lower convex part is known as the tubercle of the epiglottis.

CORNICULATE CARTILAGES

Corniculate cartilages are small, conical shaped yellow elastic cartilage attached to the apex of each arytenoid cartilage, and giving attachment to the aryepiglottic fold. It produces the corniculate tubercle in the aryepiglottic fold.

CUNEIFORM CARTILAGES

Cuneiform cartilages are small, rod-shaped yellow elastic cartilages located in front of the corniculate cartilages within the aryepiglottic folds. They may be absent.

2. Joints

There are two joints on each side. In both the participating parts are the hyaline cartilages.

Cricothyroid Joint

- **Type:** Synovial.
- **Articulating elements:** Inferior cornu of the thyroid cartilage and lower posterior facet on the side of the cricoid arch.

- **Articular capsule.**
- **Movement:** Rotation of the thyroid cartilage around a horizontal axis through the joints of both sides. Allows a visor type movement. Some gliding may occur.

Cricoarytenoid Joint

- **Type:** Synovial.
- **Articulating elements:** Concave on the base of arytenoid cartilage and the convex sloping upper facet of the cricoid cartilage.
- **Articular Capsule:** Reinforced by the posterior cricoarytenoid ligament.
- **Movements:** There are two types of movement by the arytenoid, rotation around a vertical axis, and

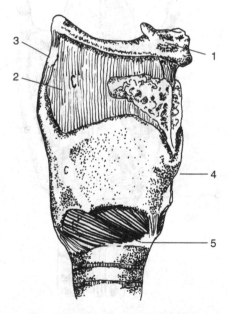

FIG. 15-4   Lateral aspect of larynx. (1) Hyoid bone; (2) Thyrohyoid membrane; (3) Triticial cartilage; (4) Laryngeal prominence; (5) Cricoid cartilage.

gliding. The vocal processes are carried laterally or medially (rima glottidis is widened or narrowed).

3. Ligaments and Membranes of the Larynx

Thyrohyoid membrane is attached above to the upper border of the posterior surface of the body and greater horns of the hyoid bone and extends down to the upper margin of the thyroid cartilage and to the front of the superior horns. A bursa (the retrohyoid bursa) separates it from the posterior surface of the hyoid bone and a fat pad

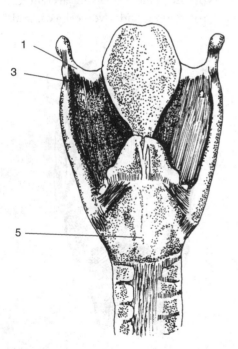

FIG. 15-5 Posterior view of skeleton of larynx. (1) Hyoid bone; (3) Triticial cartilage; (5) Cricoid cartilage.

from the anterior surface of the epiglottis.

The thyrohyoid membrane forms the base of the piriform fossa and is pierced by the superior laryngeal vessels and internal laryngeal nerve of each side. The median part of the membrane is thickened — the median thyrohyoid ligament, which extends from the body of the hyoid bone to the thyroid notch. Fibers thicken the lateral margins of the membrane and convert them to cords, the lateral thyrohyoid ligaments. These connect the tips of the greater horns of the hyoid bone with the superior horns of the thyroid cartilage. In the posterior margin of this ligament is a fibrocartilaginous nodule, the triticeal cartilage

The fibroelastic membrane within the cartilaginous skeleton of the larynx is divided by a horizontal cleft into an upper quadrangular membrane and a lower cricothyroid (cricovocal) membrane.

Cricovocal membrane (conus elasticus, lateral parts of the cricothyroid ligament) (Fig 15-9) is attached below to the upper border of the cricoid arch and above its free border is the vocal ligament. The latter is attached anteriorly to the deep surface of the thyroid laminae and posteriorly, to the vocal process of the arytenoid. This band, is composed of yellow elastic fibers.

Cricothyroid ligament connects the middle of the inferior border of the thyroid cartilage above to the arch of the cricoid cartilage below. This ligament

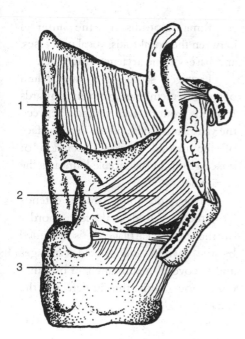

FIG. 15-6 Lateral aspect of larynx (lamina of thyroid cartilage removed). (1) Thyrohyoid membrane; (2) Vestibular ligament; (3) Conus elasticus.

may also be referred to as the anterior part of the conuselasticus.

Quadrangular membrane extends from the lateral edge of the epiglottic cartilage to the medial margin of the triangular upper fovea of the arytenoid cartilage. This ligament has two free margins and two surfaces. The medial surface is covered by mucous membrane and the lateral surface by the aryepiglottic muscle. The inferior margin of this membrane is free, forming the vestibular ligament ("false cord"). The long upper free margin is thickened and forms the aryepiglottic ligament.

In acute respiratory obstruction, cricothyrostomy may be performed when endotracheal intubation is not possible. With the patient supine and head and neck fully extended, the groove between the cricoid and thyroid is located and a short transverse incision made through the skin and cricothyroid membrane and an endotracheal or tracheostomy tube inserted.

4. Cavity of the Larynx

Aditus or inlet of the larynx is the oval, nearly vertical communication with the pharynx. It is bounded in front by the

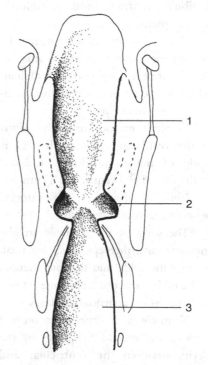

FIG. 15-7 Coronal section of larynx; (1) Vestibule; (2) Ventricle; (3) Infraglottic cavity.

epiglottis, behind are the summits of the arytenoid cartilages and corniculate cartilages, and, laterally, the aryepiglottic folds. A cuneiform cartilage located within each fold produces an elevation, the cuneiform tubercle, near its posterior end. The posterior boundary of the inlet is formed by the apices of the corniculate and arytenoid cartilages and the interarytenoid notch.

Cavity of the larynx extends from the inlet to the lower border of the cricoid cartilage. It may be divided into three areas by two projecting folds on either side: the ventricular and vocal folds (false and true vocal cords) into vestibule, ventricle, and infraglottic compartment.

The whole cavity is covered by mucous membrane; the true cords are lined by stratified squamous epithelium cells, but below this it is lined by columnar ciliated epithelium.

Vestibule extends from the aditus to the vestibular folds. Each fold is a ridge of mucous membrane produced by the underlying ventricular ligaments.

Rima vestibuli is the interval between the vestibular folds.

The walls of the vestibule are the epiglottic (and thyroepiglottic) ligament, aryepiglottic folds, and the interartenoid fold, a fold of mucous membrane extending between the arytenoid cartilages.

Ventricle is a depression on the sides of the wall of the cavity of the larynx between the ventricular and vocal folds.

Rima glottidis is the interval between the vocal folds, vocal processes, and bases of the arytenoid cartilages.

Laryngeal saccule is a diverticulum of mucous membrane extending upwards from the front of each ventricle between the ventricular folds and the inner surface of the thyroid cartilage. Secretion of mixed glands in the saccule lubricates the vocal folds.

Infraglottic compartment extends from the vocal folds to the lower border of the cricoid cartilage. It is bounded by the thyroid and cricoid cartilages and cricovocal membrane and becomes continuous with the lumen of the trachea.

5. Muscles

Muscles of the larynx are divided into two groups: extrinsic muscles and intrinsic muscles

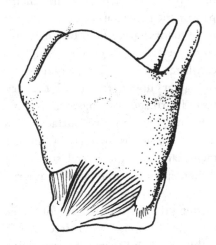

FIG. 15-8 Cricothyroid muscle.

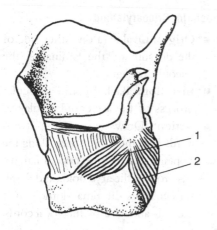

FIG. 15-9 (1) Lateral; (2) and posterior cricoary-tenoid muscles.

Extrinsic muscles suspend the larynx from the surrounding bony structures. They are elevators or depressors and move the larynx as a whole.

Intrinsic muscles are limited to the larynx and act upon it as sphincters or adjusters of the tension of the vocal cords. There are nine intrinsic muscles, eight paired and one unpaired. Six of the paired and the unpaired muscle work directly on the arytenoid.

Cricothyroid

It is triangular in shape.

- **Origin:** Anterior and lateral surfaces of the cricoid cartilage.
- **Insertion:** Lower border of the lamina of the thyroid and front of the inferior cornu.
- **Action:** Swings the cricoid arch upwards when working bilaterally with a fixed thyroid cartilage. This

produces increased tension in the vocal cords, lengthening of the cords. It also results in adduction of the cords.

Lateral Cricoarytenoid

It lies lateral and parallel to the vocal cord. It supports the wall of the ventricle.

- **Origin:** Upper border of the arch of the cricoid.
- **Insertion:** Muscular process of arytenoid.
- **Action:** Adduction of cords by rotation forwards of muscular process.

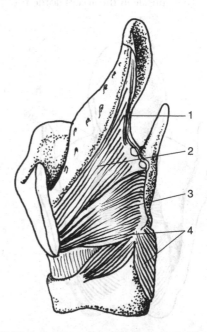

FIG. 15-10 Lateral aspect of larynx showing intrinsic muscles. (1) Aryepiglotticus m; (2) Thyroepiglotticus m; (3) Thyroarytenoid m; (4) Cricoarytenoid m.

Transverse Arytenoid

It is the only unpaired laryngeal muscle.

- **Origin:** Back of one arytenoid.
- **Insertion:** Back of other arytenoid.
- **Action:** Approximates arytenoids, adducting the cords (closing the rima).

Oblique Arytenoid

It consists of muscle bands on the transverse arytenoid.

- **Origin:** Muscular process of the arytenoid cartilages.
- **Insertion:** Apex of the other arytenoid muscle, continuing to the edge of the epiglottis as the aryepiglotticus muscle in the aryepiglottic fold.

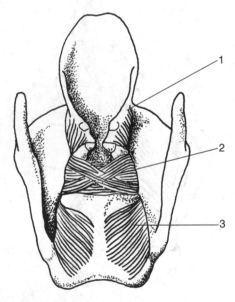

FIG. 15-11 Intrinsic muscles of the larynx — posterior view. (1) Aryepiglotticus m; (2) Arytenoid m; (3) Posterior cricoarytenoid m.

Posterior Cricoarytenoid

- **Origin:** Oval area on either side of the dorsum of the lamina of the cricoid cartilage.
- **Insertion:** Back of the muscular process of the arytenoid cartilage.
- **Action:** Draws arytenoids downwards (and therefore apart) along the sloping facets on the cricoid lamina. Also rotates the arytenoids and therefore enlarges the rima glottidis. It is the only abductor of the vocal cords.

THYROARYTENOID

- **Origin:** Medial surface of the thyroid lamina from junction of alae just lateral to attachment of vocal cord. Lies in the vocal fold.
- **Insertion:** Lateral surface of the arytenoid cartilage.

A compact bundle of muscle fibres deep to the lower fibres of thyroarytenoid muscle is the vocalis muscle. It is applied to the inferolateral aspect of the vocal ligament.

THYROEPIGLOTTICUS

- **Origin:** Back of thyroid just lateral to the angle. It lies in the sagittal plane and skirts the lateral side of the saccule in the aryepiglottic fold.
- **Insertion:** Side of epiglottis
- **Action:** A sphincter closing the aperture of the larynx by approximating arytenoids to the tubercle of the epiglottis.

SUMMARY OF ACTION OF LARYNGEAL MUSCLES

1. Tension in the vocal cords is mostly the result of action of cricothyroid and thyroarytenoid. The cords are lengthened by cricothyroid (tilting the thyroid cartilage) and shortened by thyroarytenoid (tilting the arytenoid cartilages forward).

2. Narrowing and widening of the glottis is mostly the function of posterior cricoarytenoid, lateral cricoarytenoid and arytenoids. The vocal cords are opened by posterior cricoarytenoid, whose upper horizontal fibers rotate arytenoids and whose vertical fibres separate arytenoids bodily from each other. The vocal cords are closed by lateral cricoarytenoids (opponent of rotary fibers of posterior cricoarytenoid — they rotate the arytenoids medially) and transverse arytenoid (opponent of vertical fibres of posterior cricoarytenoid — they approximate the arytenoids adducting the cords).

3. The aditus of the larynx is closed by aryepiglottic muscles. Fibers of this muscle pass lateral to the saccule and help to empty it. In addition, they act to invert the sensitive free edge of the aryepiglottic fold initiating the cough reflex if food is directed to the larynx.

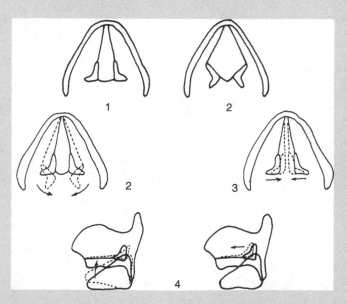

FIG. 15-12 Shape and size of the laryngeal aperture. (1) Normal respiration; (2) Forced inspiration; (3) Phonation; (4) Cricothyroid action in tensing the vocal cords.

4. The neurovascular system of the larynx is supplied by arteries: laryngeal branches of the superior and inferior thyroid arteries; veins: superior and inferior laryngeal veins; nerves: internal laryngeal nerve innervates the mucous membrane on the back of the larynx and on its interior as far as the vocal folds, external laryngeal nerve innervates the cricothyroid muscle, and recurrent laryngeal nerve innervates all the other laryngeal muscles.

5. All of these nerves have their cell bodies in the nucleus ambiguus in the brainstem. Cricothyroid is innervated by the external laryngeal nerve, and the other muscles are innervated by the recurrent laryngeal nerve.

Neck

Chapter Outline

1. Bony-Cartilaginous Framework
2. Fascia

The neck extends from a line along the lower border of the mandible to its angle, to the mastoid process and superior nuchal line of the occipital bone above, to the sternal notch, the clavicles, and a line from the acromioclavicular joint to the spinous process of CV 7 below. It provides a strong but mobile support for the head by means of its cervical vertebrae and muscles.

1. Bony-Cartilaginous Framework

The upper parts of the respiratory and digestive passageways both pass through the neck. Arterial blood courses upwards from the thorax to the head and neck while veins return blood to the heart. Several cranial nerves leave the brainstem through the foramina in the base of the skull and descend through the neck. Shorter structures are either component parts of the longer structures or are separate organs.

Hyoid bone is a U-shaped bone. It is located in the front of the neck between the mandible and the larynx at the level of the third cervical vertebra. It does not articulate with any other bone but is suspended from the styloid processes by the stylohyoid ligament.

Body is the median part of the U-shaped bone that is flattened anteroposteriorly. Posteriorly, it is separated from the thyrohyoid membrane by a bursa. The upper anterior surface gives

attachment to geniohyoid, mylohyoid and genioglossus while the lower anterior surface gives attachment to the sternohyoid and omohyoid.

Greater horns (cornua) form the limbs of the U-shaped bone projecting backwards from the body. Their medial surface gives attachment to the thyrohyoid membrane and lateral surface to the middle constrictor, hyoglossus and at their junction with the fascial sling of digastric and insertion of stylohyoid.

Lesser horns project backwards and upwards from the junction of the body and the greater horns. They give attachment to the stylohyoid ligaments and middle constrictors.

2. Fascia

Cervical fascia enables structures to move over one another, allows slippage during rotation of the neck and easy passage of vessels and nerves.

SUPERFICIAL CERVICAL FASCIA

The superficial fascia is a layer of loose connective tissue between the dermis and deep fascia. It is a thin layer that may contain considerable amounts of fat. It is continuous with the superficial fascia over the deltoid and pectoral regions and backward onto the dorsum of the neck. The platysma is embedded in this layer.

DEEP CERVICAL FASCIA

The deep cervical fascia lies deep to platysma and invests muscles of the

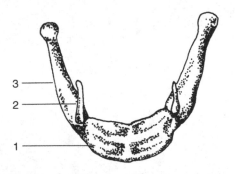

FIG. 16-1 Hyoid bone. (1) Body; (2) Lesser horn; (3) Greater horn.

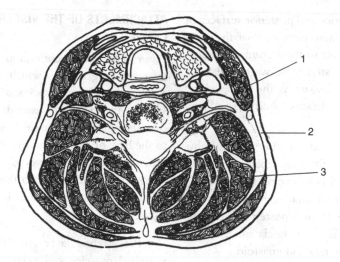

FIG. 16-2 Chief layers of cervical fascia below the level of the hyoid bone. (1) Sternomostoid m; (2) Superficial lamina; (3) Prevertebral layer.

neck from the base of the skull to the root of the neck. It consists of fibroareolar tissue situated between muscles, viscera, vessels and nerves. In some situations, it is arranged in well-defined sheets described in four layers, investing, muscular, visceral, and prevertebral, that are separated by fascial clefts.

Superficial (investing) layer forms a complete collar around the neck. It splits to form an investment of trapezius, and from its anterior border covers the posterior triangle and splits again to envelop the sternocleidomastoid.

Between the angle of the mandible and mastoid process, the superficial layer is known as the parotid fascia covering the gland and attaching to the zygomatic process. From here, it continues upwards to the superior temporal line as the temporal fascia. The deeper layer, known as the submandibular ligament passes deep to the parotid gland and attaches to the styloid process and tympanic bone.

From attachment to the greater cornu of the hyoid bone, the layer splits above, a deep layer passes upwards beneath mylohyoid attaching to the mylohyoid line and a superficial layer attaches to the border of the mandible, thereby enclosing the submandibular gland.

ATTACHMENTS OF THE SUPERFICIAL (INVESTING) LAYER

Anterior: Body and greater horn of the hyoid bone.
Posterior: External occipital protuberance, nuchal ligament and spine of CV7.
Above: Lower border of the mandible, zygomatic arch, cartilage of the ear, mastoid process, superior nuchal line.

Below: Anterior and posterior surfaces of the clavicle, acromion, crest of the spine of the scapula to the acromion laterally, manubrium sterni. Below the clavicle, this layer is known as the clavipectoral fascia that ensheaths subclavius and pectoralis minor.

Muscular layer encloses the infrahyoid muscles, forming separate lamellae. A superficial layer encloses sternohyoid and omohyoid attaching above to the hyoid bone. It fuses posteriorly with the superficial layer of fascia on the deep aspect of sternocleidomastoid.

ATTACHMENTS OF THE MUSCULAR LAYER

Above: Lateral border of omohyoid.
Below: Follows inferior attachment of sternohyoid to back of sternum.
Lateral: Covers omohyoid. More laterally, it invests the inferior belly of omohyoid. The deeper lamina invests sternothyroid and thyrohyoid, ending superiorly at the hyoid.

ATTACHMENTS OF THE DEEPER LAMINA

Lateral: Extends to the carotid sheath.
Below: Follows sternohyoid to its attachment to the back of the sternum.

Visceral layer is a cylindrical fascial covering enclosing the cervical viscera, the pharynx, larynx, trachea and esophagus. It is called the pretracheal fascia at a higher level, where it encloses the pharynx, larynx and splits to form a sheath of the thyroid gland.

ATTACHMENTS OF THE VISCERAL LAYER

Above: The layer is continuous with the buccopharyngeal fascia which covers the posterior and lateral aspects of the constrictor muscles and extends forward over the buccinator muscle. It is attached to the base of the skull at the pharyngeal tubercle and to the pterygoid hamulus.
Below: It is continued into the superior mediastinum with the inferior thyroid veins.

Prevertebral layer covers the prevertebral muscles and extends laterally over the scalene and levator scapulae muscles forming a fascial floor for the posterior triangle of the neck. The subclavian artery and brachial nerves emerge from behind the subclavian vein from in front of scalenus anterior and carry the prevertebral fascia downwards and laterally behind the clavicle into the axilla as the axillary sheath.

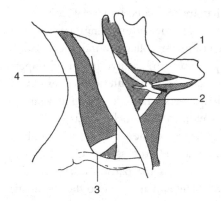

FIG. 16-3 Triangles of the neck. (1) Submandibular triangle; (2) Carotid triangle; (3) Supraclavicular or subclavian triangle; (4) Occipital triangle.

ATTACHMENTS OF THE PREVERTEBRAL LAYER

Above: Base of the skull (basilar part of the occipital bone and styloid process).
Below: Continues downward in front of the vertebral column into the superior mediastinum fusing with the anterior longitudinal ligament.
Lateral (fuses with that part of the superficial layer deep to trapezius): The tips of the transverse processes.
Anterior: The layer is separated from the pharynx and it is covered with buccopharyngeal fascia by an interval, the retropharyngeal space.

Carotid sheath is a tubular layer of deep cervical fascia enclosing the common and internal carotid arteries, internal jugular vein, and the vagus nerve. The sympathetic trunk is embedded in its posterior wall. The sheath is a condensation of fascia that blends in front with the pretracheal and superficial layer and behind with the prevertebral layer of deep cervical fascia.

ATTACHMENTS OF THE CAROTID SHEATH

Above: Base of the skull.

Below: Continues downwards into the superior mediastinum.

Fascial spaces are potential spaces that exist between the layers of fascia or by splitting of a layer.

Premuscular space lies between the investing layer and the middle layer. Previsceral space is located between the visceral layer and the muscular layer anterolaterally, and the carotid sheath laterally. Lateral pharyngeal space extends from the base of the skull to the hyoid bone. It contains styloglossus and stylopharyngeus and opens medially into the retropharyngeal space. The retropharyngeal space lies between the visceral layer and the prevertebral layer. It communicates with the pretracheal space and ends at the bifurcation of the trachea.

The prevertebral layer of cervical fascia encloses the vertebral muscles and is attached to the transverse processes of the cervical vertebrae. It is considered to split over front of the vertebrae into alar and prevertebral layers enclosing a so called "danger space". Infections can track in this space from the base of the skull to the bifurcation of the trachea in the thorax.

Chapter **16**

Anterior Cervical Triangle

Chapter Outline

1. Boundaries

Sternocleidomastoid muscle serves to divide the side of the neck into the anterior and posterior cervical triangles.

BOUNDARIES

Anterior: Midline of the neck from the chin to the jugular notch.
Above: Lower margin of the body of the mandible and a line extended from the angle of the mandible to the mastoid process.
Posterior: Anterior border sternocleidomastoid.

The anterior triangle is crossed by the digastric and stylohyoid and superior belly of omohyoid and with the hyoid can subdivided into four triangles: submental, digastric, carotid, and muscular. The ability to identify these boundaries and the knowledge of the contents of these triangles are important in clinical examination.

2. Submental (Suprahyoid) Triangle

BOUNDARIES

Lateral: Anterior belly of digastric.
Below: Body of the hyoid.
Floor: Mylohyoid.
The submental triangle extends across the midline.
Contents: Submental lymph nodes.

3. Digastric (Submandibular) Triangle

BOUNDARIES

Above: Inferior border of the body of mandible extended to the mastoid process.
Medial: Anterior and posterior bellies of digastric.
Floor: Mylohyoid muscle in front (over-lapping) hyoglossus muscle behind.
Contents: Submandibular salivary gland and related vessels, nerves and lymph nodes.

Submandibular gland almost fills the triangle. The greater part of the gland is superficial (cervical) but it is folded around the posterior border of mylohyoid and projects into the floor of the oral cavity (as the deep process). The deep process lies between mylohyoid and hyoglossus, the lingual nerve above and hypoglossal nerve below.

Facial vein descends diagonally behind tie facial artery on the face, pierces the deep fascia at the lower border of the mandible and crosses behind sternohyoid, posterior belly of digastric and the submandibular gland to join the retromandibular vein.

Submandibular lymph nodes is about five nodes that lie in the sulcus between the submandibular gland, and the body of the mandible. Lymphatics from the medial corner of the eye, greater part of the external nose and most of the lower part of the face drain into these nodes.

Facial artery arises from the external carotid artery above the level of the tip of the greater cornu of the hyoid (sometimes in common with the lingual artery). It passes forward and upward (on superior constrictor) grooving the deep surface of the submandibular gland to the inferior border of the mandible. It appears in front of the facial vein and provides submental branches then passes up onto the face.

Mylohyoid nerve descends in the mylohyoid sulcus on the medial side of the mandible onto mylohyoid muscle beneath the submandibular gland. It passes forwards to innervate mylohyoid and anterior belly of the digastric muscle.

Tendon of digastric muscle lies on hyoglossus just above the hyoid bone. Each belly of the digastric is inserted into the tendon that is held in a loop of fascia that holds it down to the side of the body and adjacent part of the greater cornu of the hyoid bone.

The digastric is a landmark for the facial nerve (above it) and hypoglossal nerve (below it) while the occipital artery runs beneath its posterior belly.

Hypoglossal nerve (XII) from the hypoglossal canal, the hypoglossal nerve passes downwards and forwards crossing the external carotid and lingual arteries beneath the submandibular gland then between mylohyoid and hyoglossus. It ends by dividing into branches on genioglossus.

EXTRINSIC MUSCLES OF THE TONGUE

Bilateral extrinsic muscles arise from bone and insert into the tongue alter the position and shape of the tongue. They lie deep to the digastric triangle. All except palatoglossus are innervated by the hypoglossal nerve (XII).

Intrinsic muscles lie wholly within the tongue and change the shape of the tongue only.

Genioglossus

Genioglossus is a fan-shaped muscle, with its apex at the origin placed vertically in contact medially with its fellow.

- **Origin:** Superior mental spine on the inner surface of the mandible near the symphysis.
- **Insertion:** Radiates into the tongue from the root to its apex. The lowest fibers have a membranous attachment to the front of the body of the hyoid.
- **Action:** Posterior fibers protrude the tongue and anterior fibers retract the protruded apex. Acting bilaterally, the muscles depress the median part of the tongue forming an elongated gutter for the passage of food.

In accidents where the mandible is fractured anteriorly, the tongue may occlude the upper airway when the protrusive action of genioglossus is lost.

Styloglossus

Styloglossus is the shortest of the three styloid muscles.

- **Origin:** Tip of the styloid process and adjoining stylohoid ligament, it fans out as it approaches the side of the tongue. Posterior fibers decussate with hyoglossus.
- **Insertion:** Side of the tongue along its entire length.
- **Action:** Draws the margin of the tongue up and back helping to form a median gutter during swallowing.

Hyoglossus

Hyoglossus is quadrate in shape.
- **Origin:** Part of body and entire greater horn of the hyoid bone (external to middle constrictor).
- **Insertion:** Posterior half of the margin of the tongue.
- **Action:** Depresses the tongue to produce a trough, as in sucking.

Hyoglossus is a key structure in the region. Note the deep relations of hyoglossus — middle constrictor, stylohyoid ligament, glossopharyngeal nerve, second part of the lingual artery (and dorsal lingual branches), genioglossus and part of the insertion of geniohyoid.

Palatoglossus, see soft palate.

4. Muscular Triangle

BOUNDARIES

Medial: Midline from hyoid to the jugular notch.
Lateral: Superior belly of omohyoid muscle and anterior margin of lower part of sternocleidomastoid muscle.

Floor: Sternohyoid, sternothyroid, and thyrohyoid muscles.

INFRAHYOID MUSCLES

The infrahyoid muscles develop from a sheet of muscle in the front of the neck analogous to rectus abdominis in the anterior abdominal wall. The sheet splits into superficial and deep layers and again into two, the superficial layer longitudinally and the deep layer transversely.

Sternohyoid

Sternohyoid is a thin, narrow, strap-like muscle converging upwards in the front of the neck.
- **Origin:** Behind the sternoclavicular joint which is the back of manubrium and sternal end of the clavicle.
- **Insertion:** Medial half of lower border of body of hyoid.
- **Action:** Draws hyoid bone down.
- **Nerve supply:** Ansa cervicalis (C1–3).

The ventral rami of cervical nerves C1 to C3 supply geniohyoid and the infrahyoid muscles.

C1 and 2 form a loop that gives branch to the hypoglossal nerve. Some of these fibers leave the hypoglossal nerve to supply geniohyoid and thyrohyoid. Others continue as the superior root of ansa cervicalis.

C2 and 3 unite to form the inferior root of ansa cervicalis. The roots join at the level of the cricoid to form the ansa cervicalis. From the ansa, branches are

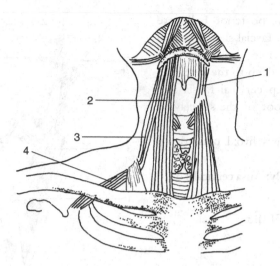

FIG. 16-4 Infrahyoid muscles. (1) Thyrohyoid m; (2) Sternohyoid m; (3) Superior belly of omohyoid; (4) Inferior belly of omohyoid.

given to the three long infrahyoid muscles (sternohyoid, sternothyroid and omohyoid).

Sternothyroid

Sternothyroid is a wide muscle that lies beneath the sternohyoid muscle.

- **Origin:** Back of manubrium and first costal cartilage.
- **Insertion:** Oblique line of thyroid lamina.
- **Action:** Draws thyroid cartilage down (depressing the larynx).
- **Nerve supply:** Ansa cervicalis (C1–3).

Thyrohyoid

Thyrohyoid is small and quadrilateral in shape.

- **Origin:** Oblique line of thyroid cartilage.

- **Insertion:** Lower border of body and greater cornu of the hyoid.
- **Action:** Depresses the hyoid or elevates the larynx depending on which attachment is fixed.
- **Nerve supply:** Branch of hypoglossal (fibers from C1).

Omohyoid

Omohyoid is a two bellied muscle united by an intermediate tendon extending from the hyoid bone to the scapula.

- **Origin:** The superior belly is lower border of body of hyoid lateral to sternohyoid.
- The inferior belly is suprascapular ligament and adjacent upper border of scapula.
- **Insertion:** Intermediate tendon over the carotid sheath beneath sternocleidomastoid. The tendon

and adjacent posterior belly are covered by a fascial sling attached to the clavicle and first rib. This sling is deep to the investing layer of deep cervical fascia and forms the roof of the subclavian triangle.

- **Action:** Draws hyoid bone down and back.
- **Nerve supply:** Ansa cervicalis (C1).

5. Carotid Triangle

BOUNDARIES

Above and in front: Posterior belly of digastric.
Below and in front: Superior belly of omohyoid.
Posterior: Anterior margin of upper part of sternocleidomastoid.

ARTERIES

Common carotid artery lies in the lower part of the triangle along the medial side of the internal jugular vein. At the level of the upper border of the thyroid lamina, it bifurcates into the external and internal carotid arteries.

External carotid artery begins at the division of the common carotid and ascends from the upper border of the thyroid cartilage to the neck of the mandible where it divides into superficial temporal and maxillary arteries. It is relatively superficial in the carotid triangle covered by superficial layer of cervical fascia and overlapped by the anterior border of sternocleidomastoid.

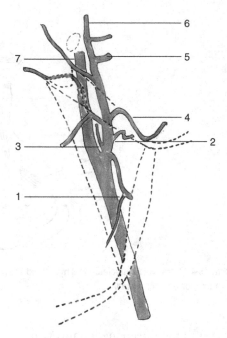

FIG. 16-5 Contents of carotid triangle. (1) Superior thyroid a; (2) Lingual a; (3) Ascending pharyngeal a; (4) Facial a;(5) Maxillary a; (6) Superficial temporal a; (7) Internal carotid a.

The external carotid artery has eight branches of which four branches arise in the carotid triangle.

Superior thyroid artery is the first branch arising opposite the thyrohyoid membrane (but may arise from the carotid bifurcation). It arches upwards then descends beside the pharynx and under omohyoid to the thyroid gland.

Lingual artery arises opposite the greater cornu of the hyoid. It ascends to disappear beneath the posterior border of hyoglossus and is crossed by the hypoglossal nerve on its way to the tongue.

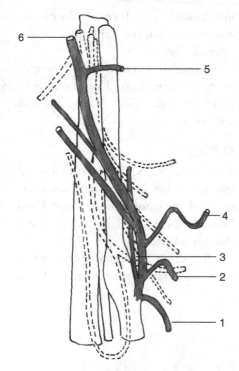

FIG. 16-6 Branches of the external carotid artery. (1) Superior thyroid a; (2) Lingual a; (3) Ascending pharyngeal a; (4) Facial a; (5) Maxillary a; (6) Superficial temporal a.

Facial artery arises immediately above the lingual artery. It passes obliquely upward to arch over deep to the posterior belly of digastric and stylohyoid where it grooves the submandibular gland. It curves over the lower margin of the mandible at the anterior margin of masseter to enter the face.

Ascending pharyngeal artery arises from the medial aspect of the external carotid in the carotid triangle and ascends on the pharynx. It gives pharyngeal branches to the superior and middle constrictors and a palatine branch that arches over the superior constrictor to reach the soft palate and palatine tonsil.

Occipital artery branches from the posterior aspect of the external carotid near the lower margin of the posterior belly of the digastric. It courses upward and backward to occupy a groove on the mastoid process. It pierces the superficial layer of cervical fascia between sternocleidomastoid and trapezius along the superior nuchal line to enter the scalp. Branches are given to sternocleidomastoid, meninges (through the jugular foramen), and the auricle.

Internal carotid artery ascends beside the pharynx to the base of the skull where it enters the carotid canal to become the main artery of the brain and orbit. It has no branches in the neck.

Carotid sinus is a dilated part of the terminal part of the common carotid and the root of the internal carotid. The walls of the sinus are modified and contain nerve endings of the glossopharyngeal nerve. The sinus is responsive to changes in blood pressure.

Carotid body is an oval specialized cell mass containing many blood capillaries embedded at the bifurcation angle of the common carotid artery. It responds to hypoxia mediating appropriate cardiovascular responses.

Both the carotid body and sinus are innervated by carotid branches of the glossopharyngeal and vagus nerves.

Sympathetic external and internal carotid plexuses, derived from the superior cervical ganglion, envelop the respective arteries.

VEINS

Internal jugular vein is the largest vein of the head and neck collecting blood from the brain, superficial face and the neck. It extends from the jugular foramen and ends below behind the sternoclavicular joint in the brachiocephalic vein. It is the largest structure in the carotid triangle descending vertically posterolateral to the internal and common carotid arteries and anterolateral to the vagus nerve within the carotid sheath. The right lymphatic duct (or thoracic duct on the left side) opens into the internal jugular vein or junction of the internal jugular vein and subclavian vein. The internal jugular vein receives four tributaries in the carotid triangle:

Pharyngeal vein begins from a venous plexus on the outer surface of the pharynx, it descends to join the internal jugular vein.

Common facial vein begins in the carotid triangle below the angle of the mandible by the union of the facial and the anterior division of the retromandibular veins. It descends to join the internal jugular vein at the level of the hyoid.

Lingual vein is formed by the convergence of four veins from the dorsum, sides and under surface of the tongue at the posterior border of the

hyoglossus muscle. It crosses the carotid arteries to form the upper part of the internal jugular vein.

Superior thyroid vein emerges from the thyroid gland and ascends to join the upper part of the internal jugular vein.

NERVES

Vagus nerve (X) lies in the carotid triangle, it lies in a separate compartment of the carotid sheath behind and between the internal jugular vein and the carotid arteries.

Superior laryngeal nerve arises from the vagus near the base of the

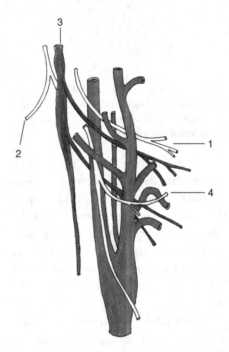

FIG. 16-7 Relationship of the last four cranial nerves to the carotid arteries. (1) IX; (2) XI; (3) X; (4) XII.

skull and descends medial to the internal and external carotid arteries. At a variable point, it divides into the internal and external laryngeal nerves. The internal branch swings forwards and pierces the thyrohyoid membrane and supplies the mucous membrane of the epiglottis, base of the tongue, aryepiglottic fold and upper larynx. The external branch descends on superior constrictor to the lower border of the thyroid where it innervates cricothyroid muscle.

External branch of the accessory nerve (XI) enters the carotid triangle crossing the upper part of the internal jugular vein superficially. It is accompanied by lymph glands and a sternocleidomastoid branch of the occipital artery piercing the deep aspect of sternocleidomastoid.

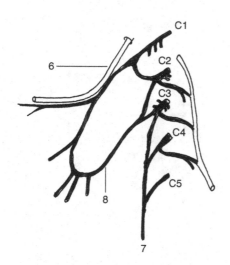

FIG. 16-8 Cervical plexus. (1) C1; (2) C2; (3) C3; (4) C4; (5) C5; (6) XII; (7) Phrenic n; (8) Ansa cervicalis.

Hypoglossal nerve (XII) descends between the vagus and accessory nerves, between the internal carotid artery and internal jugular vein to appear in the carotid triangle at the lower border of the posterior belly of the digastric. It curves forward around the occipital artery, crosses the external carotid artery and lingual artery then between mylohyoid and hyoglossus to end in genioglossus. In the carotid triangle it gives off two branches:

Nerve to thyrohyoid passes around the greater cornu of the hyoid to supply thyrohyoid. It appears to be a branch of the hypoglossal but in fact consists of fibers from C1.

Superior root of ansa cervicalis consists of fibers from C1 given off the hypoglossal nerve as it hooks around the occipital artery. It passes down over the carotid sheath joining in a loop (ansa cervicalis) with a branch from CN2 and 3 (inferior root of ansa cervicalis).

Inferior root of ansa cervicalis consists of fibers from the ventral rami of C2 and 3. The nerve fibers unite on the lateral side of the internal jugular vein on which it descends to join the ansa cervicalis. It provides no branches.

Ansa cervicalis is a loop formed by the union of superior and inferior roots; thus it contains fibers from C1, 2, and 3. The loop (ansa) lies on the common carotid artery at the level of the cricoid cartilage. It provides branches that innervate the sternothyroid and sternohyyoid muscles and both bellies of the omohyoid muscle.

Chapter 17

Submandibular Region

Chapter Outline

1. Muscles
2. Salivary Glands
3. Arteries
4. Veins
5. Nerves

The submandibular region is the suprahyoid part of the anterior cervical triangle. It lies under cover of the body of the mandible and anterior and posterior bellies of diagstric form its sides. Superficially, it includes the submental and digastric triangles and structures related to the mylohyoid and hyoglossus muscles — submandibular and sublingual glands, lingual and hypoglossal nerves, lingual artery and vein, suprahyoid muscles and extrinsic muscles of the tongue.

1. Muscles

Digastric Muscle

The digastric muscle consists of two bellies united by an intermediate tendon.

Anterior belly

- **Origin:** Digastric fossa of the inner side of the lower border of mandible near the symphysis.
- **Insertion:** Intermediate tendon which is connected to the body of the hyoid by a loop of fibrous connective tissue.
- **Nerve supply:** Mylohyoid nerve (from inferior alveolar branch of the mandibular nerve).

Posterior belly

- **Origin:** Mastoid notch of the temporal bone.
- **Insertion:** Intermediate tendon.
- **Nerve supply:** Facial nerve (VII).

- **Action:** Both bellies usually work together and depress the mandible in resisted opening of the jaw or raise the hyoid (and floor of the mouth) during swallowing.

Anterior bellies help to extend the head against a resistance and the posterior bellies open the jaw against a resistance.

The digastric bounds the digastric triangle and the muscle is a landmark for the facial nerve (above it) and hypoglossal nerve (below it).

Stylohyoid

Stylohyoid courses along the posterior belly of the digastric muscle, the tendon of which perforates the muscle.

- **Origin:** Posterior edge and lateral surface ot the styloid process in its middle third.

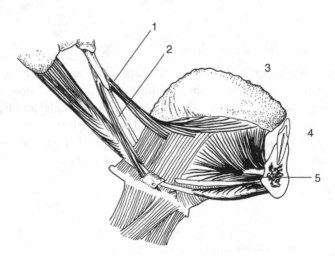

FIG. 17-1 Extrinsic muscles of the tongue. (1) Styloglossus; (2) Stylohyoid; (3) Hyoglossus; (4) Genioglossus; (5) Geniohyoid.

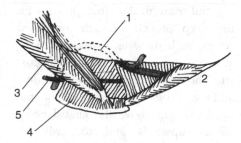

FIG. 17-2 Right submandibular triangle. (1) Facial a; (2) Anterior belly of digastric m; (3) Posterior belly of digastric m; (4) Stylohyoid m; (5) XII.

- **Insertion:** It is perforated below by the intermediate tendon of the digastric and inserts into the hyoid at the junction of body with greater cornu.
- **Action:** Draws the hyoid up and back during swallowing.
- **Nerve supply:** Facial nerve (VII).

The attachments of the stylohyoid and posterior belly of digastric are almost continuous, that they have the same innervation (digastric branch of the facial nerve) and have the same action (elevation of the hyoid).

Mylohyoid

Mylohyoid forms the muscular floor (diaphragm) across the floor of the mouth.

- **Origin:** Mylohyoid line of mandible.
- **Insertion:** Body of hyoid bone and mylohyoid raphe.
- **Action:** Raises hyoid bone during swallowing.
- **Nerve supply:** Mylohyoid nerve.

Mylohyoid and anterior belly of digastric are derived from the same muscle mass and share the same innervation, contiguous attachment and common action.

Geniohyoid

Geniohyoid is a narrow accessory muscle of the tongue located above the medial border of the mylohyoid.

- **Origin:** Lower mental spine.
- **Insertion:** Front to the body of the hyoid in contact with its fellow.
- **Action:** Draws hyoid bone forwards and upwards and indirectly the tongue.
- **Nerve supply:** Hypoglossal nerve (XII) carrying C1 fibers.

2. Salivary Glands

SUBMANDIBULAR GLAND

Submandibular gland is a mixed mucous and serous secreting salivary gland lying partly under cover of the mandible between the mandible and the hyoid bone. It overlaps both bellies of the digastric muscle. The greater part of the gland is superficial but it has a deep part which is prolonged forward from its medial surface between the mylohyoid and hyoglossus muscles. The deep part lies on hyoglossus, genioglossus, lingual nerve, submandibular duct, submandibular ganglion and third part of the lingual artery.

The **neurovascular system** of the submandibular gland is supplied by: *Arteries*: branches of the facial and lingual arteries;

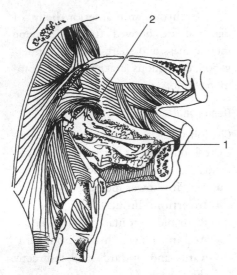

FIG. 17-3 Medial view of dissection of floor of the mouth. (1) Submandibular duct; (2) Lingual n.

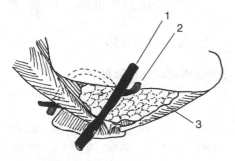

FIG. 17-4 Contents of the submandibular triangle. (1) Facial v; (2) Facial a; (3) Submandibular gland.

Veins: drain into the external jugular vein; nerves: sympathetic nerves are derived from the plexus around the facial artery. Parasympathetic (secreto-motor) nerves pass via the chorda tympani (of the facial nerve) to the lingual nerve and then to the submandibular ganglion where they synapse. Postganglionic fibers then supply the gland.

Submandibular duct passes with the deep process beneath mylohyoid resting on hyoglossus. It passes forward at first on hyoglossus then on genioglossus. The lingual nerve lies above the duct and the hypoglossal nerve below but the duct crossing above the lingual nerve passes upwards and forwards on genioglossus and opens into the mouth on the sublingual caruncle, at the side of the frenulum of the tongue. It has thin walls and a narrow orifice.

SUBLINGUAL GLAND

Sublingual gland is a small, flattened predominantly mucus secreting salivary gland lies on the anterior oral surface of mylohyoid. It raises the sublingual fold in the mucous membrane of the oral cavity. About 12 small sublingual ducts from the upper surface of the gland open separately along the sublingual fold.

The gland presents two surfaces. Lateral: mandible; medial: genioglossus and geniohyoid, submandibular duct and branches of the lingual nerve.

The **neurovascular system** of the sublingual gland is supplied by:

Arteries: Sublingual branch of the lingual artery and submental branch of the facial artery;

Nerves: Parasympathetic preganglionic fibers originate in the superior salivatory nucleus in the brainstem and travel in the chorda tympani branch of the facial nerve and lingual branch of the mandibular nerve to reach the submandibular ganglion where they synapse.

Parasympathetic postganglionic fibers leave the submandibular ganglion and reach the gland with peripheral branches of the lingual nerve.

3. Arteries

FACIAL ARTERY

Facial artery arises from the external carotid artery near the angle of the mandible. It ascends on the superior constrictor beneath posterior belly of the digastric and stylohyoid muscles to groove the deep surface of the submandibular gland. It then descends forward to the lower border of the mandible at the anterior border of the masseter muscle to continue as the facial part of the artery.

Cervical branches:

Ascending palatine artery ascends between styloglossus and superior constrictor and turns down over the upper border of superior constrictor. It pierces pharyngobasilar fascia and supplies the soft palate and pharyngeal tonsil.

Tonsillar artery ascends to pierce the superior constrictor and supplies the pharyngeal tonsil.

Glandular branches two or three branches supplying the submandibular gland.

Submental artery arises near the lower border of the mandible on the superficial surface of mylohyoid. It supplies adjacent muscles and the submandibular and sublingual glands.

LINGUAL ARTERY

Lingual artery arises from the external carotid artery at the level of the tip of the greater cornu of the hyoid. It loops up on the middle constrictor then runs forward above and parallel with the greater cornu. In its course, it is divided into three parts (posterior, deep and anterior to hyoglossus).

The first part lies in the carotid triangle, it is S-shaped and ends at the posterior border of the hyoglossus muscle.

The second part lies beneath hyoglossus, it courses on the upper border of the hyoid bone on the middle constrictor.

The third part courses on the genioglossus in the tongue.

The branches of the lingual artery are:

Suprahyoid artery from the first part of the artery, it runs along the

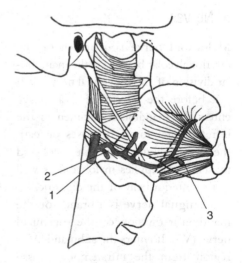

FIG. 17-5 Lingual artery (1) Dorsal lingual branches; (2) Suprahyoid branch; (3) Sublingual artery.

upper border of the hyoid to supply adjacent muscles.

Dorsal lingual branches from the second part of the artery ascends medial to hyoglossus to supply the dorsum of the tongue, supplying the tonsil, epiglottis and soft palate.

Sublingual artery from the third part supplies the sublingual gland.

Deep lingual artery runs with the lingual nerve. It is a terminal branch to the tip of the tongue.

4. Veins

Lingual vein is formed at the posterior border of the hyoglossus muscle from veins accompanying the lingual artery and the hypoglossal nerve. It drains into the internal jugular vein.

5. Nerves

Mylohyoid nerve is a branch of the inferior alveolar branch of the mandibular division of the trigeminal nerve (V3) just before the inferior alveolar nerve enters the mandibular foramen in the infratemporal fossa. It passes beneath the submandibular gland on mylohyoid muscle and supplies mylohyoid as well as the anterior belly of the digastric.

Lingual nerve is a branch of the mandibular division of the trigeminal nerve (V3). It enters the submandibular region from the infratemporal fossa where it is joined by the chorda tympani. It lies in front of the inferior alveolar

nerve and passes between medial pterygoid and the ramus of the mandible. It then passes over the styloglossus and hyoglossus and the deep part of the submandibular gland. It passes across the submandibular duct then curves upwards into the tongue, spirals under the duct and passes medially, providing general sensation to the anterior two thirds of the tongue. It gives branches to the pharyngeal isthmus, submandibular ganglion, mucous membrane of the mouth and gingiva of the mandibular bicuspid and molar teeth.

Submandibular ganglion containing the cell bodies of postganglionic parasympathetic neurons. It is located between the hyoglossus and the submandibular gland below the lingual nerve.

The roots are:
Sensory from the lingual nerve. Their cell bodies are in the trigeminal ganglion.

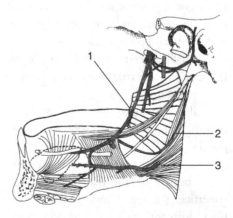

FIG. 17-6 Submandibular and infratemporal regions. (1) Lingual n; (2) IX; (3) XII.

They do not synapse in the submandibular ganglion but pass through it to reach the mucous membrane of the mouth.

Sympathetic from a plexus around the facial artery. Cell bodies are in the superior cervical ganglion and axons pass through the ganglion without synapsing to reach the submandibular and sublingual glands.

Parasympathetic neurons giving rise to preganglionic fibers are situated in the superior salivatory ganglion in the brainstem. They enter the chorda tympani nerve then join the lingual nerve and reach the ganglion where they synapse.

Postganglionic fibers supply to both submandibular and sublingual glands.

The branches are:
Several small branches supply the salivary glands, oral mucous membrane and submandibular duct.

Hypoglossal nerve (XII) enters the submandibular region from the carotid triangle. It passes along the lower border of the posterior belly of digastric curving around the occipital artery, it crosses the external carotid and lingual arteries then between mylohyoid and on hyoglossus. It continues forward on hyoglossus beneath the submandibular gland and mylohyoid to terminate in genioglossus. The hypoglossal nerve contains only somatic motor nerve cells and supplies all of the extrinsic (except palatoglossus) and intrinsic muscles of the tongue.

Glossopharyngeal nerve (IX) is a mixed nerve with visceral sensory and motor components as well as special visceral (taste) and branchial motor (to stylopharyngeus) fibers originating from the medulla oblongata. It presents superior and inferior ganglia as it enters the neck through the jugular foramen. It descends between the interior jugular vein and internal carotid artery behind the muscles attached to the styloid process. The nerve curves forward along the lower border of stylopharyngeus to pass between the internal and external carotid arteries and passes deep to the hyoglossus muscle, where it divides to supply the pharyngeal tonsil and mucous membrane of the posterior third of the tongue.

Its branches are:
Tympanic nerve carries preganglionic parasympathetic fibers arising from the inferior salivatory nucleus in the brainstem. It enters the tympanic canaliculus to reach the tympanic cavity, where it ramifies on the promontory to contribute to formation of tympanic plexus. Some fibers of the tympanic plexus reunite to form the lesser petrosal nerve that passes forward through the petrous temporal bone and emerges in the middle cranial fossa just lateral to the hiatus of VII. It leaves the cranial cavity usually through the foramen ovale to reach the otic ganglion. Postganglionic fibers travel on the auriculotemporal nerve to reach the parotid gland.

Nerve to stylopharyngeus descends along the posterior border of stylopharyngeus then curves over its lateral border.

It innervates stylopharyngeus, the only muscular distribution of the nerve.

Pharyngeal branches (3 or 4) join pharyngeal branches of the vagus and cervical sympathetic trunk on the middle constrictor to form the pharyngeal plexus.

Nerve to carotid sinus from the glossopharyngeal just below the jugular foramen. It descends in front of the internal carotid (receiving components from the vagus and sympathetic trunk) ending in the carotid sinus. It carries afferent information from blood pressure receptors in the carotid sinus.

Tonsillar branches are given off deep to hyoglossus and form a network around the palatine tonsil and soft palate.

Lingual branches provide sensation for taste to the vallate papillae and general sensation from the mucous membrane of the posterior third of the tongue and in front of the epiglottis.

Chapter 18

Posterior Cervical Triangle

Chapter Outline

1. Boundaries
2. Fascia
3. Contents
4. Muscles of the Floor of the Posterior Triangle

Sternocleidomastoid bisects the neck diagonally from the sternum and clavicle below to the mastoid process and occipital bone above. The posterior cervical triangle is the second of the two large triangles of the side of the neck situated behind sternocleidomastoid muscle.

1. Boundaries

Anterior: Posterior border of sternocleidomastoid.
Posterior: Anterior border of trapezius.
Base: Middle third of the clavicle.
Roof: Superficial fascia.
Floor: Deep fascia, muscles (see below).

The inferior belly of the omohyoid muscle divides the posterior triangle into the larger occipital triangle above and smaller omoclavicular triangle below.

2. Fascia

Superficial deep to the skin and superficial to the investing layer of the neck. It is continuous with the fascia of the face and upper extremity. Platysma, coursing through the layer, is attached

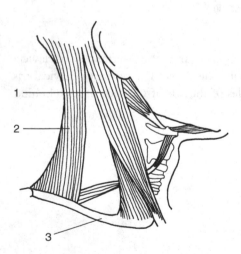

FIG. 18-1 Boundaries of the posterior triangle. (1) Sternodleidomastoid m; (2) Trapezius; (3) Clavicle.

to the skin over the pectoral and deltoid regions and above to the body of the mandible.

Investing layer of deep cervical fascia extends between the sternocleidomastoid and trapezius muscles. It is attached:
Above: To the external occipital protuberance, superior nuchal line, mastoid process and inferior border of the mandible. It is continuous between the mastoid process and mandible with the parotid fascia.
Below: To the manubrium, clavicle, spine of the scapula and spinous process of the seventh cervical vertebra. It splits above the manubrium into anterior and posterior arches enclosing the suprasternal space occupied by the jugular arch.
Behind: It splits to enclose trapezius.

Prevertebral layer covers the muscles of the floor of the triangle. It is attached to the base of the skull and transverse processes of the cervical vertebrae. It covers the prevertebral muscles, the scalene muscles and phrenic nerve and deep muscles of the back. It is prolonged laterally as the axillary sheath investing the brachial plexus and brachial artery.

3. Contents

SUPERFICIAL STRUCTURES

Occipital artery from the posterior surface of the external carotid artery, runs beneath digastric and mastoid process then in the occipital groove before crossing the apex of the occipital triangle in company with the greater occipital nerve (C2).

External jugular vein drains most of the tissue of the posterior triangle. It begins by the union of two veins, the posterior auricular vein and a branch of the retromandibular vein after it leaves the parotid gland. It enters the anterior lower corner of the triangle before it pierces the investing layer of cervical fascia and then joins the subclavian vein.

Transverse cervical vein passes a little above the clavicle to join the external jugular vein.

Suprascapular vein courses with the suprascapular artery to empty into the external jugular vein.

CERVICAL PLEXUS

The cervical plexus is formed by the union of ventral rami of the first four cervical nerves and lies on scalenus medius under cover of the upper half of sternocleidomastoid. Each nerve in the plexus divides into two, an ascending and a descending branch that connect the nerve above and the nerve below.

The first loop lies in front of the transverse process of the atlas and the second and third loops are directed backwards.

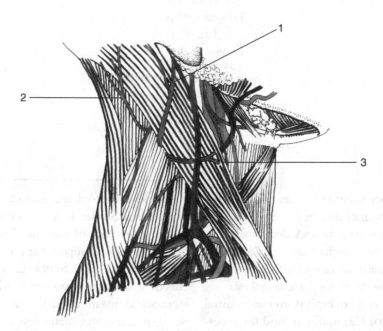

FIG. 18-2 Peripheral branches of cervical plexus. (1) Greater auricular n; (2) Lesser occipital n; (3) Transverse cervical n.

The loop between C2 and C3 gives off cutaneous branches to the head and neck.

The loop between C3 and C4 gives off superficial cutaneous branches to the shoulder and chest with muscular and communicating branches.

The lower branch of C4 joins C5 in the formation of the brachial plexus.

Branches of the cervical plexus are disposed in two groups, superficial (cutaneous) and deep (muscular) and the deep branches are arranged in anterior and posterior groups.

Superficial	*Deep* (*Anterior*)	*Deep* (*Posterior*)
Transverse cervical (C2, 3)	Phrenic (C3, 4, 5)	Sternocleidomastoid (C2)
Lesser occipital (C2)	Muscular branches to:	Levator scapulae (C3, 4)
Great auricular (C2, 3)	Thyrohyoid (C1)	Trapezius (C3, 4)
Supraclavicular (C3, 4)	Geniohyoid (C1)	Scalenus medius (C3 to 7)
	Rectus capitis	
	lateralis (C1)	
	Longus capitis	
	(C1 to 4)	
	Longus colli	
	(C3 to 8)	
	Scalenus anterior	
	(C3 to 8)	
	Intertransversales	
	(C1 to 8)	

Four superficial cutaneous branches can be identified at the middle of the posterior border of sternocleidomastoid. They pierce the superficial layer of deep fascia near their destinations which is an important consideration in local anaesthesia.

Lesser occipital nerve (ventral ramus of C2) hooks around the accessory nerve at the posterior border of sternocleidomastoid and ascends along its posterior border. It supplies the skin of the upper third of the cranial surface of the auricle and adjacent scalp.

Great auricular nerve (C2 and 3) curves around the posterior border of sternocleidomastoid just below the point of emergence of the lesser occipital nerve to ascend on the superficial surface of sternocleidomastoid parallel with the external jugular vein. It divides near the auricle into anterior branch supplies the skin over the parotid and ramus of the mandible; and posterior branch that supplies the lower part of the front and back of the auricle.

Transverse cervical nerve (C2 and 3) also hooks around the posterior border of sternocleidomastoid just below the great auricular nerve, crosses its superficial surface, pierces the deep fascia, and divides into an ascending branch that supplies the skin of the upper anterior part of the neck; and a descending branch that supplies the skin of the side and front of the neck.

Supraclavicular nerves (C3 and 4) appear just below the transverse cervical nerve from a common trunk beneath sternocleidomastoid. The trunk divides into three branches, medial, intermediate and lateral supraclavicular nerves which cross the clavicle superficially.

Medial supraclavicular nerves cross sternocleidomastoid and the sternal third of the clavicle and supply skin and the sternoclavicular joint.

Intermediate supraclavicular nerves cross the middle of the clavicle and supply the skin over pectoralis major and deltoid.

Lateral supraclavicular nerves pass onto the trapezius muscle and across the lateral third of the clavicle and supply the skin over the shoulder.

External branch of accessory nerve (XI) descends from the jugular foramen along the internal jugular vein to below the mastoid process, where it pierces (and supplies) sternocleidomastoid. It emerges into the posterior triangle at the junction of the upper and middle thirds of the muscle and crosses the triangle on the middle of levator scapulae, passes deep to trapezius (supplying it) a little above the clavicle, and ramifies.

Branches of C2 and 3 to sternocleidomastoid and C3 and 4 to trapezius are sensory (proprioceptive).

DEEP STRUCTURES

Transverse cervical artery is a branch of the thyrocervical trunk (a branch of the first part of the subclavian artery), the transverse cervical artery crosses the lower third of the triangle over the brachial plexus to the anterior border of levator scapulae muscle where it divides into superficial and deep branch.

Superficial branch crosses levator scapulae and ascends deep to trapezius to supply cervical muscles.

Deep branch passes deep to levator scapulae to the scapula. It may arise separately from the subclavian as the descending or dorsal scapular artery.

Suprascapular artery is also a branch of the thyrocervical trunk. It crosses scalenus anterior and the phrenic nerve then laterally behind and parallel to the clavicle to the upper border of the scapula. It passes over the suprascapular ligament supplying supraspinatus then winds around the neck of the scapula to reach dorsal infraspinous muscles.

Subclavian artery (third part) the third part of the subclavian is that part from lateral edge of the scalenus anterior to lateral border of the first rib. It enters the posterior triangle at the lateral border of scalenus anterior a little above the clavicle, descends laterally and leaves the triangle at the outer margin of the first rib (apex of the axilla) and becomes the axillary artery. It has no branches in the triangle.

The subclavian artery can be felt (from behind) pulsating and can be compressed against the first rib.

Subclavian vein is the continuation of the axillary vein enters the triangle behind the middle of the clavicle. It ascends medially on the front of scalenus anterior (which separates it from the subclavian artery) and joins the internal jugular vein to form the brachiocephalic vein behind sternocleidomastoid.

Brachial plexus emerges into the triangle at the lateral border of scalenus anterior (between scalenus medius and anterior) above the third part of the subclavian artery. It passes behind the transverse cervical and suprascapular arteries and leaves the triangle behind the middle third of the clavicle to enter the axilla. In the posterior triangle the branches of the brachial plexus are those from its roots:

Dorsal scapular nerve (C5 ventral ramus) pierces scalenus medius, runs across it then enters levator scapulae which it supplies.

Upper roots of the long thoracic nerve (C5 and 6) pierce scalenus medius and descend to be joined by a root from C7. The nerve descends over the first rib to supply serratus anterior.

Ventral rami of C5 and 6 unite to form the upper trunk of the brachial plexus near the lateral edge of scalenus medius. The upper trunk gives two branches:

Suprascapular nerve runs parallel with the upper trunk and supplies muscles on the dorsum of the scapula.

Subclavian nerve descends superficially to supply subclavius.

Below, the middle and lower trunks cross scalenus medius, the lower trunk lying in the sulcus for the subclavian artery on the first rib.

4. Muscles of the Floor of the Posterior Triangle

The muscular floor of the posterior triangle is formed mainly by three muscles whose fibers run posteroinferiorly, splenius capitis, levator scapulae and scalenus medius. They lie deep to the prevertebral layer of cervical fascia.

Semispinalis capitis whose fibers run vertically lies in the triangle only at its apex.

Splenius Capitis

Splenius capitis is a thick flat muscle running obliquely across the back of the neck.

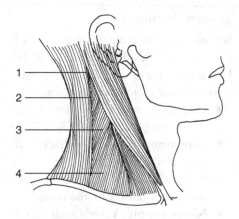

FIG. 18-3 Muscles of the floor of the posterior triangle. (1) Semispinalis capitis; (2) Splenius capitis; (3) Levator scapulae; (4) Splenius medius.

- **Origin:** Lower part of ligamentum nuchae and spines of the seventh cervical and upper thoracic vertebrae.
- **Insertion:** Lateral third of the superior nuchal line extending to the mastoid part of the temporal bone.
- **Action:** Acting bilaterally they extend the neck. Acting singly, they turn the head toward the ipsilateral shoulder (synergists to the contralateral sternocleidomastoid).
- **Nerve supply:** C4 to 8 (posterior primary rami).

Levator Scapulae

- **Origin:** Posterior tubercles of transverse processes of upper four cervical vertebrae.
- **Insertion:** Vertebral border of the scapula from the superior angle to the root of the spine.

- **Action:** Raises and maintains the level of the shoulder (assisting upper fibers of trapezius).
- **Nerve supply:** C3 and 4 (ventral primary rami) assisted by the dorsal scapular nerve (to the rhomboids — C5).

Lateral Vertebral Muscles

Scalenus anterior lies under sternocleidomastoid.

- **Origin:** Anterior tubercles of the transverse processes of C3 to 6 (the "typical" cervical vertebrae).
- **Insertion:** Scalene tubercle of first rib (between the subclavian artery and vein).
- **Action:** Elevates first rib and bends the neck laterally.
- **Nerve supply:** Ventral rami C4 to 7.
- **Relations of scalenus anterior:**
(1) The first branches of the subclavian artery pass superiorly on its anterior surface.
(2) Roots of the cervical plexus and brachial plexus emerge at its posterolateral edge.
(3) The phrenic nerve lies on its anterior surface *en route* to the diaphragm.

Ventral rami innervate the prevertebral muscles.

Scalenus Medius

Scalenus medius is the longest and largest muscle of this group. It lies posterior to roots of the brachial plexus (the brachial plexus emerges between scalenus anterior and scalenus medius).

- **Origin:** Posterior tubercles of the transverse processes of C2 to 6.
- **Insertion:** First rib behind its subclavian groove.
- **Action:** Elevates first rib and bends the neck laterally.
- **Nerve supply:** Ventral rami of C3 to 7.

Scalenus Posterior

Scalenus posterior is the smallest muscle of the group. It lies behind and may be blended with the medius muscle but passes across the first rib to reach the second rib.

- **Origin:** Posterior tubercles of the transverse processes of C5 to 7.
- **Insertion:** Outer surface of second rib.
- **Action:** Elevates the second rib and lateral flexion of the neck.
- **Nerve supply:** Ventral rami of C5 to 7.

Chapter 19

Root of the Neck

Chapter Outline

1. Thymus
2. Thyroid and Parathyroid Glands
3. Trachea and Esophagus (Cervical Part)
4. Deep Arteries
5. Deep Veins
6. Deep Nerves: Vagus Nerve and Sympathetic Trunk
7. Prevertebral Muscles

The root of the neck is the region occupied by the structures that enter or leave the thoracic cavity. In the neck, these structures lie largely behind the sternocleidomastoid muscle that must be displaced when this area is examined.

1. Thymus

The thymus is a **lymphoid organ** that lies on the trachea, left brachiocephalic vein, (and inferior thyroid tributaries), roots of the brachiocephalic trunk, left common carotid artery, aortic arch and the pericardium and behind the infrahyoid muscles. Within the thymus, T lymphocyte production occurs after stem cells reach it from the bone marrow. In addition, the thymus produces thymopoietin which induces the formation of specific receptors on T lymphocytes, thymosin that stimulates differentiation and maturation of T lymphocytes and thymic humoral factor.

The thymus has two halves or lobes connected in the midline by connective tissue. A capsule encloses each lobe from which interlobular septa extend inward to separate lobes into lobules and trabeculae project from both capsule and septa to further divide the cortex. Lobules have a cortex densely populated with lymphocytes sharply demarcated from an axial less dense medulla.

The thymus is normally large at birth and in childhood but begins to involute after puberty. In the adult, it consists of fibrous and fatty tissue.

The thymus extends from the neck into the thorax over the great vessels and pericardium.

The **neurovascular system** of the thymus is supplied by *arteries*: Twigs from the internal thoracic, inferior thyroid and anterior intercostal arteries;

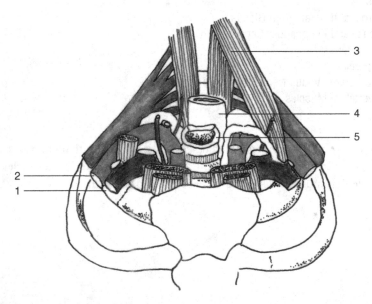

FIG. 19-1 Structures at the root of the neck. (1) Subclavian v; (2) Subclavian a; (3) Scalenus anterior m; (4) Esophagus; (5) Trachea.

veins: Tributaries drain mostly into the left brachiocephalic vein; *nerves*: Parasympathetic branches from the vagi and cardiac sympathetic fibers from plexuses around blood vessels; *lymphatics*: Drain mainly into sternal, anterior mediastinal and tracheobronchial lymph nodes.

2. Thyroid and Parathyroid Glands

The thyroid gland is an **endocrine gland** that controls the rate of metabolism by secreting into the blood the hormones thyroxine (T4) and tri-iodothyronine (T3). These hormones are produced from tyrosine and inorganic iodine within epithelial cells of thyroid follicles. A second hormone, thyrocalcitonin, that lowers blood calcium levels, is formed by parafollicular cells or C cells also located in the walls of follicles. Formation and secretion of thyroid hormones are primarily under the influence of hypophyseal hormone TSH.

The thyroid is a soft, highly vascular gland that lies in the front of the neck at the level of C5 to C7. It has a thin, fibrous, true capsule and an outer sheath of pretracheal fascia or false capsule. The fascial sheath is attached to the thyroid and cricoid cartilages so that the gland rises and falls in swallowing.

The gland usually comprises two pear shaped lobes joined near their lower poles by an isthmus over tracheal cartilages two to four. Each lobe

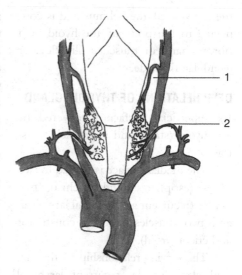

FIG. 19-2 Blood supply of the thyroid gland from behind. (1) Superior thyroid a; (2) Inferior thyroid a.

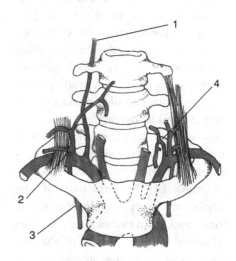

FIG. 19-3 Subclavian artery. (1) Vertebral a; (2) Thyrocervical trunk; (3) Internal thoracic a; (4) Costocervical trunk.

extends laterally to the oblique line of the thyroid laminae. Occasionally, a pyramidal lobe projects upward from

the left side of the isthmus and is connected to the body of the hyoid by a
fibrous band or muscular slip (levator
glandulae thyroideae).

DEEP RELATIONS OF THYROID GLAND

Its superficial surface is covered by
infrahyoid (strap) and sternocleidomastoid muscles.

The medial surface is related to two
tubes (esophagus and trachea), two
nerves (recurrent and external laryngeal)
and two muscles (inferior constrictor
and cricothyroid).

The varying relationship of the thyroid gland to the recurrent laryngeal
nerve that can be damaged in surgical
approaches to the gland. The nerve may
be well posterior to the gland, pass
through a condensation of connective
tissue in the attaching the gland to the larynx (the suspensory or lateral ligament)
or traverse the gland for a short distance.

The **neurovascular system** of the
thyroid gland is supplied by:
Arteries: The superior thyroid artery
from the external carotid artery, the inferior thyroid artery from the thyrocervical trunk of the first part of the
subclavian artery supply the gland.
Sometimes, a **thyroidea ima** artery, an
occasional unpaired branch of the brachiocephalic artery contributes;
Veins: Arise from a plexus between the
sheath and capsule on the surface of the
gland. The superior thyroid veins empty
into the upper part of the internal jugular
vein. The middle thyroid veins empty into
the lower part of the internal jugular vein.

And the inferior thyroid veins empty into
the brachiocephalic veins;
Nerves: Mostly sympathetic from the
cervical ganglia and vagal fibers form
vascular plexuses. Stimulation or abla-
tion of these nerves does not affect the
function of the gland. Vasoconstrictor
nerves originate from the cervical sympathetic ganglia.

PARATHYROID GLANDS

The parathyroid glands are two pairs of
small ovoid-shaped **endocrine glands**
within their own capsule situated in the
back of the thyroid between capsule
and the gland. There are two superior
parathyroid glands (parathyroids 4) at
the middle of the posterior of the lobe
of the thyroid and two inferior parathyroid glands (parathyroids 3) at various
positions, in the fascial sheath near
the inferior pole of the thyroid, behind
the thyoid outside the fascial sheat
of the thyroid or in the substance of the
thyroid near the inferior border of
the thyroid. Chief or principal cells
of the parathyroids produce parathyroid
hormone (PTH) that raises blood calcium by multiple actions on several
organ systems including bone, kidney
and the gut.

The **neurovascular system** of the
parathyroid glands is supplied by:
Arteries: Superior thyroid artery supplies
the superior parathyroids and inferior
thyroid artery supplies the inferior
parathyroids. Both share anastomoses
between thyroid, laryngeal, pharyngeal
and esophageal arteries;

Nerves: Vasomotor sympathetic arise from the superior or middle cervical ganglia.

3. Trachea and Esophagus (Cervical Part)

The trachea is continuous with the larynx beginning opposite CV 6 below the cricoid cartilage, to which it is attached by the cricotracheal ligament. It is about 10–12 cm long and ends opposite the sternal angle (TV 5) there dividing into the right and left bronchi. The tube is extensile and comprises about 15–20 U-shaped hyaline cartilaginous bars open posteriorly. It is lined with respiratory mucous membrane (cilia beat toward the laryngopharynx). The posterior wall of the trachea is flat and closed by the trachealis muscle (smooth muscle). The cervical part of the trachea lies behind the infrahyoid muscles and in front of the esophagus.

Posterior to the trachealies the esophagus, **recurrent laryngeal nerves** (in the tracheoesophageal grooves).

The cervical esophagus is a short, flattened muscular tube continuous with the pharynx (also at the level of C6) ending at the upper edge of T1. It lies behind the trachea and is attached to a median ridge on the cricoid lamina by the cricoesophageal tendon. The inner circular coat is continuous with the cricopharyngeal part of inferior constrictor muscle. The lower part of the esophagus lies in the thorax and abdomen.

RELATIONS

Anterior: Trachea, recurrent laryngeal nerves.

Posterior: Vertebral column, longus colli, prevertebral layer of cervical fascia.

Lateral: Common carotid arteries, posterior part of thyroid, thoracic duct on right side.

The **neurovascular system** of the esophagus is supplied regionally along its length by:

Arteries: Inferior thryroid artery;

Veins: Inferior thyroid veins;

Nerves: Autonomic nerve supply to glands and striated muscle in the cervical and upper thoracic parts of the esophagus. Recurrent laryngeal nerves and esophageal plexus provide sensory supply to the mucosa. Parasympathetic motor fibers supply smooth muscle and parasympathetic secretomotor fibers to glands in the wall, Sympathetic innervation is provided from the cervical sympathetic trunk;

Lymphatics: Lymphatics in the cervical esophagus drain to paratracheal and inferior deep cervical lymph nodes.

4. Deep Arteries

The **brachiocephalic artery** arises from the beginning of the arch of the aorta behind the middle of the manubrium of the sternum. It ascends obliquely behind the right sternoclavicular joint where it divides into the **right common carotid** and the **right subclavian arteries**.

CERVICAL PART OF COMMON CAROTID ARTERIES

The **right common carotid artery** arises with the right subclavian artery from the brachiocephalic artery behind the sternoclavicular joint. The **left common carotid artery** arises from the aortic arch. Neither have branches until they enter the carotid triangle, where they divide into the external and internal carotid arteries at the level of the upper border of the thyroid cartilage.

SUBCLAVIAN ARTERIES

The subclavian arteries are the principal arteries of the upper limbs.

The **right subclavian artery**, the other terminal branch of the brachiocephalic artery arches from behind the sternoclavicular joint to the outer border of the first rib, where it becomes the **axillary artery**. On the first rib, it lies between scalenus anterior and medius.

The **left subclavian artery** arises directly from the arch of the aorta on the left side of the trachea in the thorax. It takes the same course in the neck as the right subclavian artery.

The subclavian artery is divided into three parts by the scalenus anterior muscle.

PRESCALENE PART

The first, or prescalene part extends from the beginning of the vessel (behind the sternoclavicular joint) to the

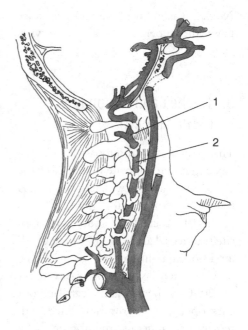

FIG. 19-4 Vertebral and internal carotid arteries. (1) Vertebral a; (2) Internal carotid a.

medial border of scalenus anterior. Its branches are:

Vertebral artery arises from the subclavian artery and ascends in the neck behind the carotid sheath to reach the foramen transversarium of C6. The entire vertebral artery is divided into four parts.

The first (cervical) part ascends behind the carotid sheath then between scalenus anterior and longus cervicis to enter the foramen transversarium of the sixth cervical vertebra.

The second (vertebral) part ascends through the foramina transversaria of the upper cervical vertebrae.

The third (suboccipital) part winds backward around the lateral mass of the atlas in the suboccipital triangle where the first cervical nerve lies below it.

The fourth (intracranial) part pierces the posterior atlanto-occipital membrane and enters the cranial cavity through foramen magnum and unites to form the unpaired basilar artery.

Branches of the first part or cervical part of the vertebral artery gives the spinal branches which enter intervertebral foramina to reach dorsal root ganglia and nerve roots; and the muscular branches which supply deep cervical muscles.

The triangle of the vertebral artery has the following boundaries:

Base: First part of the subclavian artery.
Lateral: Scalenus anterior.
Medial: Longus colli.
Apex: Anterior tubercle of the sixth cervical transverse process.
Floor: Transverse process of C7, ventral ramus of C8, neck of the first rib, cupola of pleura.
Contents: Vertebral artery and vein, sympathetic trunk. In front of the triangle are carotid sheath and contents, phrenic nerve, inferior thyroid and thoracic duct (on the left arch in front of the triangle).

Internal thoracic artery from the first part of the subclavian artery descends below the clavicle to the posterior surface of the first costal cartilage. It passes through the thorax behind the upper six costal cartilages.

Thyrocervical trunk is a short trunk that also arises from the subclavian artery, closer to scalenus anterior than does the vertebral artery. It ascends for a short distance, then divides into:

(1) Transverse cervical artery passes laterally over the scalene muscles, phrenic nerve and brachial plexus under omohyoid to the lateral border of levator scapulae where it divides. A superficial branch ascends beneath trapezius to supply surrounding muscles and a deep branch passes backward to the superior angle of the scapula then along the medial border of the scapula with the dorsal scapular nerve.

(2) Suprascapular artery runs laterally across scalenus anterior and the phrenic nerve to enter the posterior triangle. It passes over the suprascapular ligament to enter the supraspinous fossa supplying supraspinatus. It winds around the neck of the scapula to reach and supply the infraspinous fossa and infraspinatus.

(3) Inferior thyroid artery ascends along the medial border of scalenus anterior to the level of C6, where it curves medially behind the sheath of the carotid and sympathetic trunk to the lower posterior thyroid gland.

Its branches are:

Ascending cervical artery ascends medial to the phrenic nerve providing muscular branches to scalenus anterior and longus colli.

Inferior laryngeal artery passes upwards on the trachea to the larynx with the recurrent laryngeal branch of X.

Tracheal and esophageal branches supply the trachea and cervical part of the esophagus.

Glandular branches which are several small branches to the lower pole of the gland.

RETROSCALENE PART

Second or (retroscalene) part of the subclavian artery is also the highest part of the artery. Its sole branch is:

Costocervical trunk arises from the back of the subclavian artery, it bends over the cupola of the pleura to the neck of the first rib, where it divides into:

Deep cervical artery (cf posterior branch of an aortic intercostal artery) courses backward between C7 and the neck of the first rib. It ascends beneath semi-spinalis capitis and supplies the muscles of this region.

Superior (supreme) intercostal artery descends behind the pleura and divides to supply the first two posterior inter-costal arteries.

POSTSCALENE PART

Third or postscalene part is located in the posterior triangle (supraclavicular triangle of). It extends from the lateral border of scalenus anterior to the outer border of the first rib.

5. Deep Veins

Vertebral vein descends from veins draining deep muscles of the back of the neck that enter the foramina trans-versaria of the atlas then remaining upper five cervical vertebrae terminating in the brachiocephali vein. The vertebral vein receives the anterior vertebral vein and deep cervical vein that correspond to the ascending cervical artery and deep cervical arteries respectively.

Internal jugular vein is the largest vein in the neck. It begins as a continua-tion of the sigmoid sinus as the dilated superior bulb at the jugular foramen. It descends on the lateral side of the inter-nal and common carotid arteries near the medial border of the scalenus anterior and joins the subclavian vein to form the brachiocephalic vein. It is crossed by the XI nerve and inferior root of the ansa cervicalis. Near its lower end it is dilated as the inferior bulb.

Tributaries are inferior petrosal sinus at the jugular foramen, lingual veins, pharyngeal veins, common facial vein, superior thyroid vein in the carotid triangle and middle thyroid vein in the root of the neck.

Inferior thyroid vein arises from a venous plexus on the isthmus and medial parts of the lobes of the gland. There may be three or four veins terminating in the brachiocephalic veins.

Subclavian vein is a short vein that begins as a continuation of the axillary vein at the outer border of the first rib. It terminates by joining the internal jugular vein to form the brachiocephalic vein close to the sternoclavicular joint. It usually receives the external jugular vein. At the level of scalenus anterior, it passes in front of the muscle by which it is separated from the subclavian artery.

Brachiocephalic veins begin behind the clavicle by the union of the subclavian and internal jugular veins. They terminate by joining each other to form the superior vena cava behind the sternal angle. The left vein is longer than the right, and both lack valves. Tributaries of the left vein are the left vertebral, left internal thoracic, left inferior thyroid, thymic, mediastinal, pericardiacophrenic and left superior intercostal veins.

6. Deep Nerves

Vagus nerve (X) is a mixed nerve containing visceral motor and secretomotor fibers (from the dorsal motor nucleus of the vagus), branchial motor fibers (from nucleus ambiguus) and sensory fibers including taste (reaching nucleus tractus solitarius) from neurons in the superior and inferior ganglia).

The nerve emerges from the side of the medulla and from the skull at the jugular foramen, (where it exhibits superior and inferior ganglia). The cranial root of the accessory nerve (XI) joins the nerve at the inferior ganglion. The composite nerve descends in the carotid sheath, between the internal jugular vein and carotid arteries to the root of the neck.

The *right vagus* passes between the internal jugular vein and the first part of the subclavian artery.

The *left vagus* passes between the left common carotid artery and the first part of the left subclavian artery. Both nerves continue into the thorax.

Branches of the vagus nerve in the head and neck are found in the head and neck regions.

IN THE HEAD

Two branches arise in the jugular fossa meningeal and auricular branch.

Meningeal: A recurrent branch to the dura mater of the posterior cranial fossa.

Auricular: From the superior ganglion, it passes into the mastoid canaliculus starting in the lateral wall of the jugular fossa. It emerges through the tympanomastoid fissure on the side of the skull just behind the external acoustic meatus. It supplies skin of the posterior meatus and auricle and part of the tympanic membrane.

IN THE NECK

Pharyngeal branches. (from the cranial root of the accessory nerve) descend between the external and internal carotid arteries, to the side of the pharynx where they ramify and, together with branches of cranial nerve IX and the sympathetic

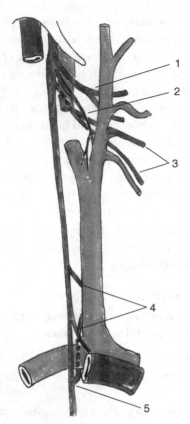

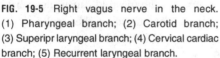

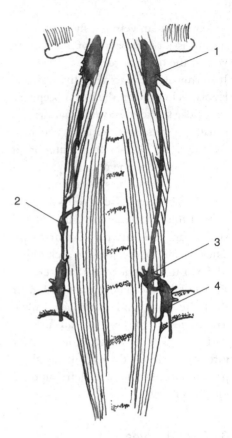

FIG. 19-5 Right vagus nerve in the neck. (1) Pharyngeal branch; (2) Carotid branch; (3) Superipr laryngeal branch; (4) Cervical cardiac branch; (5) Recurrent laryngeal branch.

FIG. 19-6 Sympathetic chain in the neck. (1) Superior cervical ganglion; (2) Middle ganglion; (3) Vertebral ganglion; (4) Stellate ganglion.

nerves form the pharyngeal plexus. It contributes motor branches to the constrictors of the pharynx and soft palate.

Superior laryngeal nerve descends deep to the internal carotid artery where it divides into the following branches:

- *Internal larygeal branch* (sensory) pierces the thyrohyoid membrane and supplies the mucous membrane of the piriform fossa of the larynx above the vocal folds.

- *External laryngeal branch* (notor) courses deep to the sternothyroid muscle and thyroid gland to reach and supply cricothyroid.

- *Cervical cardiac branches* arise from the trunk of X at different levels. They join the cardiac branches of the cervical sympathetic ganglia.

- *Superior cardiac branch* descends behind the carotid plexus and major vessels to reach the deep cardiac plexus.

- *Inferior cardiac branch* arises at the root of the neck and on the right side it passes into the thorax at the side of the trachea to join the deep cardiac plexus. On the left side, this branch descends over the arch of the aorta and curves medially to end in the superficial cardiac plexus.
- *Recurrent laryngeal nerve* on the **right** side from the vagus nerve, it crosses the subclavian artery, hooks around the artery, and ascends on the side of the esophagus and trachea.
- On the left side, it arises as the vagus nerve crosses the aortic arch, and hooks around it.

In the neck, each nerve ascends on the side of the trachea, passes under the lower border of the inferior constrictor muscle, and reaches the larynx supplying all of the muscles of the larynx except cricothyroid.

Branches of the **recurrent laryngeal nerve** are:
- *Cardiac branches*: To the deep cardiac plexus, tracheal and esophageal branches. To the muscles and mucosa of the trachea and esophagus.
- *Pharyngeal branches*: To the inferior constrictor muscle.
- *Laryngeal branches*: To the intrinsic muscles of the larynx (except the cricothyroid)

SYMPATHETIC TRUNK

In the neck, it lies behind the carotid sheath and consists of three interconnected ganglia and their branches. It has no white rami communicates but consists of ascending preganglionic axons that have entered the trunk from Tn 2 and 3.

Superior cervical ganglion is a spindle-shaped ganglion is the largest of the three (about 2.5 cm long) comprising the cervical sympathetic trunk. It is located on the pre-vertebral fascia in front of the transverse process of the second and third cervical vertebrae.

Its branches are:
It communicates with ventral rami of C1–4. Pharyngeal branches descend from the front of the superior ganglion to join branches of the vagus and glossopharyngeal nerves to form the pharyngeal plexus on the middle constrictor. Internal carotid nerve accompanies the internal carotid artery and forms the internal carotid plexus. External carotid nerve forms a plexus on the external carotid artery and its branches. Superior cervical cardiac nerves descend behind the carotid sheath in front of the recurrent laryngeal nerve to the subclavian artery along the brachiocephalic trunk to the deep cardiac plexus. The left nerve descends along the left common carotid artery crossing the aorta to join the superficial cardiac plexus.

Middle cervical ganglion is the smallest of the ganglia located at the level of C6. Its branches are Communicating branches to the ventral rami of C5 and 6; the thyroid branch runs along the inferior thyroid artery supplying the larynx below the vocal folds, pharynx, trachea and esophagus; **vertebral ganglion** is commonly found at the level of C7. Its branches are communicating with ventral rami of C7 and 8 inferior cardiac nerve descends behind the subclavian artery joining the recurrent laryngeal nerve and ends in the deep cardiac plexus vascular branches form a plexus around the vertebral artery.

Cervicothoracic (stellate) ganglion is formed by fusion of the inferior cervical and first thoracic ganglia. It may be absent. The ganglion gives branches to C6 to T2 and, with the vertebral ganglion, forms a vertebral plexus.

7. Prevertebral Muscles

Prevertebral muscles form a group of four muscles lying in front of the vertebral column.

Longus Capitis

Longus capitis is the upper half of the prevertebral band of muscle interrupted by attachment to the anterior tubercles of the "typical" cervical vertebrae (the lower half represents scalenus anterior). Longus capitis is broad and thick above and narrow below.

- **Origin:** Anterior tubercles of transverse processes of the lower (3 to 6) cervical vertebrae.
- **Insertion:** Lower surface of the basilar part of occipital bone in front of rectus capitis anterior.

- **Action:** Flexes head and upper part of the neck.
- **Nerve supply:** Branches from ventral rami of C1 to 3.

Longus Colli

Longus colli clothes the anterior tubercles of the upper 10 vertebrae (cf scalenus anterior). It consists of three parts:

Superior oblique part
- **Origin:** Anterior tubercles (transverse processes) of C3 to 5.

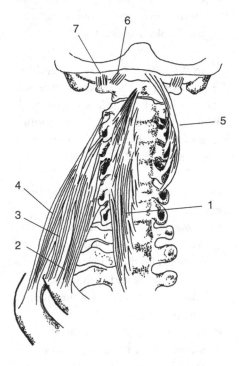

FIG. 19-7 Prevertebral muscles. (1) Longus colli; (2) Scalenus anterior; (3) Scalenus medius; (4) Scalenus posterior; (5) Longus capitis; (6) Rectus capitis anterior; (7) Rectus capitis lateralis.

- **Insertion:** Anterior tubercle on arch of atlas.
- **Action:** Flexion and lateral flexion of the neck.
- **Nerve supply:** Anterior rami of C2 to 4.

Vertical Part

- **Origin:** Anterior surface of bodies of 1 to 4 thoracic vertebrae.
- **Insertion:** Anterior surface of bodies of 2 to 4 cervical vertebrae.
- **Action:** Flexes the neck forward.
- **Nerve supply:** Anterior rami of C2–6.

Inferior Oblique Part

- **Origin:** Anterior surface of bodies of one to two thoracic vertebrae.
- **Insertion:** Anterior tubercles of transverse processes of five to six.cervical vertebrae.
- **Action:** Flexes neck and slightly rotates the cervical part of the vertebral column.
- **Nerve supply:** Branches from ventral rami of C2 to 7.

Rectus Capitis Anterior

Rectus capitis anterior is a short and flat muscle beneath longus capitis.

- **Origin:** Front of lateral mass of atlas.
- **Insertion:** Inferior surface of basilar part of occipital bone close to the condyle beneath longus capitis.
- **Action:** Flexes head at atlanto-occipital joint.

- **Nerve supply:** Branches of ventral rami of C1 and 2 (first loop of the cervical plexus).

Rectus Capitis Lateralis

Rectus capitis lateralis is a short and flat muscle.

- **Origin:** Transverse process of atlas.
- **Insertion:** Inferior surface of jugular process of occipital bone.
- **Action:** Lateral flexion of the head to the same side.
- **Nerve supply:** Branch of ventral rami (the loop between Cn 1 and 2).

Sternocleidomastoid

Sternocleidomastoid extends obliquely up the neck from the sternoclavicular joint to the mastoid process, dividing the side of the neck into anterior and posterior triangles.

- **Origin:** By two heads.
- **Sternal head:** From the upper lateral part of the front of the manubrium. This part is rounded and tendinous.
- **Clavicular head:** Is fleshy and broad arising from the medial third of the clavicle.
- **Insertion:** Side of the base of the mastoid process and lateral third of the superior nuchal line on the occipital bone.
- **Action:** Acting together, they flex the neck primarily at the articulation of lower cervical and uppermost thoracic vertebrae. They extend the head primarily at the upper crainio-cervical articulations. Acting alone, each muscle flexes the head and

bends it to the same side and rotates the head so that the face looks upward to the opposite side.

- **Nerve supply:** External branch of the accessory nerve (XI) and ventral rami (probably sensory) of C2 and 3.

Important relations of sternocleido-mastoid:

(1) Sterncleidomastoid divides the side of the neck into anterior and posterior triangles.

(2) The carotid sheath is covered by the lower part of the muscle and the anterior border above.

(3) The cervical plexus is covered by its upper part.

(4) Nerves of the cervical and brachial plexuses appear at its anterior border.

(5) The spinal part of the accessory nerve (XI) runs downward and posteriorly through its deep fibers.

Upper Limb

The upper limb is connected to the trunk by the pectoral or shoulder girdle comprising the scapula and clavicle. It is completed in front by the manubrium of the sternum with which the clavicle articulates.

The three segments of the free upper limb are the arm, forearm, and hand.

There are analogous structures in both upper and lower limbs e.g. the thumb and big toe, radius and tibia, ulna and fibula and humerus and femur.

During development, the upper and lower limbs appear as limb buds — longitudinal ridges on the body wall. The rostral edge of each ridge is the preaxial border and the caudal edge is the postaxial border. The buds elongate and rotate in their long axis but in opposite directions. Thus, the flexor aspect of the upper limb faces forwards while the flexor aspect of the lower limb faces posteriorly. In the anatomical position, the big toe is located medially while the thumb is located laterally.

The upper limb is adapted for working and compared with the lower limb its stability is sacrificed for mobility. This may be attested to by the attachment of the limb including the pectoral girdle being attached to the trunk only at one point, the small sternoclavicular joint and the resulting great range of movement that is possible by the upper limb. The upper limbs are moored to the ribs and vertebrae by muscles (mooring muscles) that contribute to their support and stability.

Chapter **20**

Superficial Structures of the Back

Chapter Outline

1. Superficial Fascia

The superficial fascia covering superficial muscles of the back and neck contains fat and varies in thickness.

CUTANEOUS NERVES

Dorsal rami of the spinal nerves supply (i) deep dorsal muscles of the sacrospinalis group that lie deep to the thoracolumbar fascia and (ii) sensation from skin outward to the posterior axillary line.

Each nerve divides into medial and lateral branches that innervate the deep muscles of the back but only one branch appears on the surface as a cutaneous nerve. C1 has no cutaneous branch and the lower three cervical nerves (and lower two lumbar nerves) reach the skin.

Above the midthorax, the terminal cutaneous nerves arise from the medial branches; below this level, they arise from the lateral branches.

Cutaneous blood vessels accompany the cutaneous nerves. They arise from the ascending branch of the transverse cervical artery in the neck, the intercostal arteries in the thorax, and the lumbar arteries in the loin.

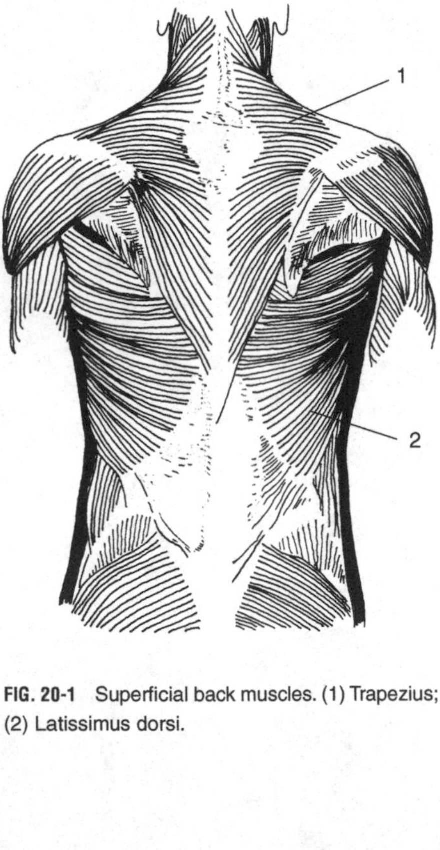

2. Superficial Muscles of the Back

These are grouped as posterior muscles connecting the limb to the vertebral column.

Five superficial muscles of the back, latissimus dorsi, trapezius, levator scapulae and the rhomboids (major and minor), connect the upper limb through the pectoral girdle to the vertebral column. Although topographically related to the back, they developed from a ventrolateral sheet of trunk musculature that migrated backward. As a result, they are innervated by ventral rami of the spinal nerves. They are arranged in two layers, trapezius and latissimus dorsi are superficial, and the levator scapulae and the two rhomboids lie deep to the trapezius with a continuous insertion on the vertebral border of the scapula.

FIG. 20-1 Superficial back muscles. (1) Trapezius; (2) Latissimus dorsi.

Trapezius

Trapezius is a large, triangular sheet of muscle covering the back of the neck and upper part of the back.

- **Origin:** Medial half of superior nuchal line (the occipital part), nuchal ligament (the cervical part) and spines and supraspinous ligaments of cervical vertebra 7 to thoracic vertebra 12 (the thoracic part).
- **Insertion:** Posterior border of the lateral third of the clavicle (occipital part) medial border of the acromion and crest of the spine of the scapula (cervical part) and tubercle of the crest of the scapular spine (thoracic part).
- **Action:** The entire muscle rotates the scapula (with serratus anterior) and elevates its lateral angle in raising the arm. Both muscles acting together brace the shoulders back.
- **Occipital part:** Elevates the scapula sharing the actions of levator scapulae (as in shrugging).

Cervical part draws the scapula toward vertebral column. Thoracic part depresses the medial end of the scapula rotating the scapula in abduction of the arm. Trapezius is an important postural muscle keeping the shoulder in its normal elevated position.

- **Nerve supply:** Upper part by the external branch of the accessory nerve. (XI), lower part by branches of ventral rami of C3 and 4.

Latissimus Dorsi

Latissimus dorsi is a large, thin, triangular sheet of muscle, covering the lower part of the back.

- **Origin:** By fleshy fibers from the spines and supraspinous ligaments of thoracic vertebrae 7–12 and via the thoracolumbar fascia from the spines of lumbar vertebrae 1–5. It also has a fleshy origin from the posterior iliac crest.

The iliac fibers of latissimus dorsi form the posterior boundary of the lumbar triangle. This triangle is the site of the rare lumbar hernia and is a surgical approach to the kidney.

The upper edge of latissimus dorsi (with trapezius and the scapula) helps to form the triangle of auscultation. The cardiac orifice of the stomach lies deep to this triangle on the left side and fluid can be heard to enter the stomach by stethoscope.

- **Insertion:** The muscle narrows to a flat tendon and inserts in the floor of the intertubercular groove.
- **Action:** With the trunk fixed, it adducts, extends, and medially rotates the humerus against a resistance (as in breast-stroke swimming). With the humerus flexed, it elevates the trunk (as in climbing or chinning a horizontal bar).

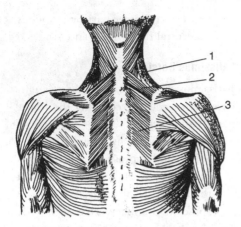

FIG. 20-2 Second layer of superficial back muscles. (1) Levator scapulae; (2) Rhombiod minor; (3) Rhombiod major.

- **Nerve supply:** Thoracodorsal nerve from the posterior cord of the brachial plexus (C6, 7, 8).

Levator Scapulae

Levator scapulae comprises four muscular slips that connect some vertebral transverse processes and the scapula.

- **Origin:** Posterior tubercles of the transverse processes of cervical vertebrae 1 to 4.
- **Insertion:** Vertebral (medial) border of the scapula from the superior angle to the root of the spine.

The muscle fibers twist posteriorly reversing the order of attachment that allows the muscle to rotate the superior angle of the scapula strongly forward.

- **Action:** Elevates the scapula (supports the weight of the limb) and tends to rotate it lowering its

lateral angle (rotates the glenoid downward). With the scapula fixed, it rotates the neck to the same side.

- **Nerve supply:** Ventral rami of C3 and 4 directly from the cervical plexus and C5 via the dorsal scapular nerve.

Rhomboid Major

Rhomboid muscles form a thin, flat sheet beneath trapezius from the vertebral spines to the vertebral border of the scapula. They may be fused.

- **Origin:** Spines and supraspinous ligaments of thoracic vertebrae 2 to 5.
- **Insertion:** Vertebral border of the scapula from the root of the spine of the scapula to its inferior angle.
- **Action:** Together they brace the shoulders backward (or draw the scapula towards the vertebral column) or elevate and maintain the levelof the scapula.
- **Nerve supply:** Dorsal scapular nerve from the ventral rami forming the brachial plexus (chiefly C5).

Rhomboid Minor

Rhomboid minor is a slender muscular band that parallels rhomboid major and may be poorly separated from it.

- **Origin:** Extends from the lower part of ligamentum nuchae and the spines and supraspinal ligaments of cervical vertebra 7 and thoracic vertebra 1.

- **Insertion:** Vertebral border of the scapula opposite the root of its spine.
- **Action:** Braces the shoulders back and elevates and draws the scapula towards the vertebral column (maintains the level of the scapula).
- **Nerve supply:** Dorsal scapular nerve from the ventral rami forming the brachial plexus (chiefly C5).

Chapter 21

Deltoid and Scapular Regions

Chapter Outline

The deltoid and scapular muscles cover the shoulder joint and give the shoulder its characteristic outline. They arise from the shoulder girdle and insert into the humerus and are supplied by ventral rami of C5 and 6 through branches of the brachial plexus.

1. Fascia

SUPERFICIAL

The fascia of the deltoid and scapular regions contains a variable amount of fat and the lateral part of the platysma which inserts into the skin.

Cutaneous nerves

Cutaneous branches of dorsal rami runs from the upper thoracic nerves especially T2 reach the back of the shoulder. Lateral supraclavicular nerves (C3 and 4) cross the clavicle and acromion deep to platysma to supply the upper half of the deltoid region.

Lateral brachial cutaneous nerve is a branch of the axillary nerve that appears at the lower edge of deltoid and spreads forward over the lower half of deltoid.

Cutaneous vessels

Branches of the thoracoacromial artery (from the first part of the axillary artery) and circumflex humeral artery (from the third part of the axillary artery).

DEEP

Deltoid muscle is invested by fascia that passes between its fasciculi and is continuous with the fascia covering the pectoralis major in front and with that covering the infraspinatus muscle behind. It is attached above to the clavicle, acromion, and scapular spine and is continuous below with the deep brachial fascia. It is thin over the supraspinatus muscle, but dense over infraspinatus and teres major and minor.

2. Muscles

DELTOID

Deltoid is a thick, triangular, coarse-textured, muscle covering the shoulder joint. Its acromial fibres are multipennate.

- **Origin:** Clavicular part: Lateral third of the front and lateral side of the clavicle; Acromial part: Lateral border of the acromion; Spinous part: Lower lip of the spine of the scapula.
- **Insertion:** Deltoid tuberosity of the mid-shaft of the humerus.
- **Action:** Clavicular (anterior) part: Flexes and medial rotates the humerus against a resistance; Acromial part: Abducts the arm

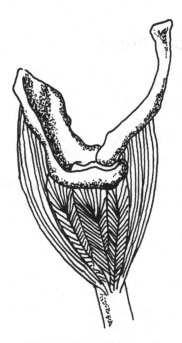

FIG. 21-1   Right deltoid muscle.

to the horizontal position; Spinous (posterior) part: Extends and laterally rotates the arm against a resistance.

- **Nerve supply:** Axillary nerve from the posterior cord of the brachial plexus (C5 and 6) winds around the neck of the humerus under deltoid.

The deltopectoral triangle between the anterior border of deltoid and upper border of pectoralis major allowing passage of the cephalic vein, the deltoid artery and lymph vessels.

Supraspinatus

Supraspinatus is a triangular muscle filling the supraspinatus fossa.

- **Origin:** Medial three fourths of the supraspinous fossa and from

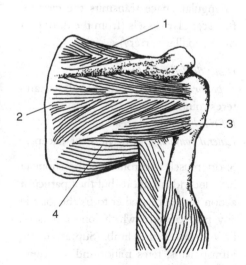

FIG. 21-2 Scapular muscles. (1) Supraspinatus; (2) Infraspinatus; (3) Teres minor; (4) Teres major.

the fascia that covers the muscle. A bursa separates supraspinatus from the neck of the scapula.

- **Insertion:** Superior facet on the greater tuberosity of the humerus and its tendon blends with the capsule of the shoulder joint.
- **Action:** Abductor of the shoulder but particularly a fixator of the shoulder joint.
- **Nerve supply:** Suprascapular nerve (C5 and 6 from the upper trunk of the brachial plexus).

Infraspinatus

Infraspinatus is also a triangular muscle that fills most of the infraspinous fossa. A bursa separates the muscle from the neck of the scapula.

- **Origin:** Infraspinous fossa and fascia that covers the muscle.
- **Insertion:** Middle facet on the greater tuberosity of the humerus and capsule of the shoulder joint.
- **Action:** Lateral rotator of the humerus and a fixator of the shoulder joint.
- **Nerve supply:** Suprascapular nerve (C5 and 6 from the upper trunk of the brachial plexus).

Teres Major

- **Origin:** Back of the inferior angle and axillary border of the scapula below teres minor.
- **Insertion:** Medial lip of intertubercular groove of the humerus.
- **Action:** Adductor (with latissimus dorsi) and medial rotator of the arm.

- **Nerve supply:** Lower subscapular nerve (C6 and 7 from the posterior cord of the brachial plexus).

Subscapularis

Subscapularis is a large, triangular, multipennate muscle filling the subscapular fossa. It covers the front of the shoulder joint and forms the posterior wall of the axilla (completed below by teres major and latissimus dorsi).

- **Origin:** Subscapular fossa except at the neck of the scapula where a bursa separates the muscle from the underlying bone.
- **Insertion:** Lesser tuberosity of the humerus.
- **Action:** Adductor, fixator and medial rotator of the arm.
- **Nerve supply:** Upper and lower subscapular nerves (5, 6 and 7 from the posterior cord of the brachial plexus).

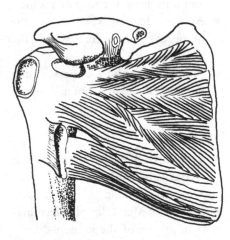

FIG. 21-3 Right subscapularis.

The interval between teres minor and subscapularis above and teres major below and the surgical neck of the humerus laterally is divided by the long head of triceps into a quadrangular space laterally and triangular space medially.

QUADRANGULAR SPACE

Quadrangular space transmits the axillary nerve and posterior humeral circumflex vessels.

Its boundaries are:
Superior boundary: Subscapularis and teres minor.
Inferior boundary: Teres major.
Lateral boundary: Surgical neck of the humerus.
Medial boundary: Long head of triceps.

TRIANGULAR SPACE

Triangular space transmits the circumflex scapular vessels (from the third part of the axillary artery).

It has the following boundaries:
Superior boundary: Subscapularis and teres minor.
Inferior boundary: Teres major.
Lateral boundary: Long head of triceps.

Short muscles surrounding the shoulder have individual actions but their particular action is to act together to fix the joint in any position into which longer muscles have moved the limb. Supraspinatus, infraspinatus, teres minor and subscapularis are called "rotator cuff" muscles because together they form a cuff around the shoulder joint. All, except supraspinatus, are short lateral rotators (cf short lateral rotators of the thigh).

3. Deep Arteries

Anterior humeral circumflex artery is a branch of the third part of the axillary artery. It winds around the surgical neck of the humerus and anastomoses with the posterior humeral circumflex artery.

Muscular branches: Supply coracobrachialis, long head of biceps, and a descending branch to pectoralis major.

Ascending branch: Courses in the intertubercular sulcus to the shoulder joint.

Posterior humeral circumflex artery arises from the third part of the axillary artery. It accompanies the axillary nerve through the quadrangular space and anastomoses with the acromiothoracic, profunda brachii and anterior humeral circumflex arteries.

Muscular branches: Supply deltoid, teres minor and long head of triceps.

Articular branch: Supplies the shoulder joint.

Nutrient branch: Supplies the shoulder joint and head of the humerus.

Circumflex scapular artery is a branch of the subscapular artery from the third part of the axillary artery. It passes backward through the triangular space to reach the infraspinous fossa.

Anterior branch: Reaches subscapularis and supplies it.

Descending branch: Runs down between the teres muscles (supplying them) to reach the angle of the scapula.

Suprascapular artery is a branch of the thyrocervical trunk from the first part of the subclavian artery. It passes laterally crossing scalenus anterior and the phrenic nerve then the subclavian artery and cords of the brachial plexus. It crosses over the suprascapular ligament to reach the supraspinous fossa

Muscular branches: Supply neighboring muscles such as supraspinatus and infraspinatus.

Articular branches: Supply the shoulder and acromioclavicular joints.

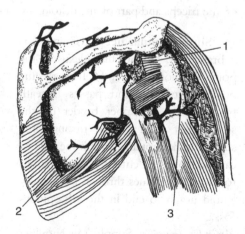

FIG. 21-4 Deep arteries of deltoid — scapular region. (1) Suprascapular a; (2) Circumflex scapular a; (3) Posterior humeral circumflex a.

Nutrient branches: Supply the clavicle and scapula.

ARTERIAL ANASTOMOSIS AROUND THE SCAPULA (COLLATERAL CIRCULATION)

This anastomosis establishes an important link between the first part of the subclavian artery and third part of the axillary artery. The vessels involved form an extensive network on both surfaces of the scapula and since they have no valves, they can interconnect.

The arterial anastomosis is derived from three main sources:

(1) Subscapular and circumflex scapular arteries (lateral border of scapula).
(2) Dorsal scapular artery (medial border of scapula).
(3) Suprascapular artery (superior border of scapula).

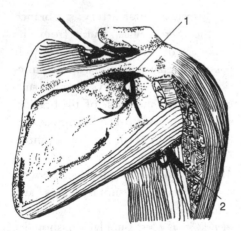

FIG. 21-5 Deep netves of the deltoid-scapular region. (1) Suprascapular n; (2) Axillary n.

4. Veins

Veins accompany the arteries superficially and end in the external jugular vein.

5. Deep Nerves

Axillary nerve (C5 and 6) is a branch of the posterior cord of the brachial plexus. It passes backwards with the posterior circumflex artery and leaves the axilla at the lower border of subscapularis through the quadrangular space. It gives an articular branch to the shoulder joint and then divides into two branches:

Anterior branch: Winds around the neck of the humerus supplying deltoid and overlying skin.

Posterior branch: Provides branches to teres minor and deltoid. It pierces the deep fascia and as the lateral brachial cutaneous nerve, supplies the skin over the long head of the triceps and part of the deltoid.

Suprascapular nerve (C5 and 6) is a branch of the brachial plexus at the point of union of C5 and 6. It crosses the posterior triangle and passes beneath trapezius to the upper border of the scapula. It then passes through the suprascapular notch (beneath the suprascapular notch) to enter the supraspinous fossa and continues through the spinoglenoid notch to end in the infraspinous fossa.

Articular branches: Supply the shoulder joint.

Muscular branches: Supply the supraspinatus and infraspinatus.

Chapter 22

Pectoral Region

Chapter Outline

1. Superficial Fascia
2. Breast
3. Deep Fascia
4. Muscles

The breast (mammary gland) is situated on the front of the thorax between the superficial and deep layers of superficial fascia, and is intimately related to the pectoral region.

1. Superficial Fascia

Superficial fascia of the pectoral region contains a variable amount of fat and splits to enclose the breast. Fat is particularly abundant in the region of the mammary gland in the female. The superficial fascia contains the cutaneous nerves.

CUTANEOUS NERVES

Supraclavicular nerves (medial, intermediate and lateral from C3 and 4) that pass over the clavicle onto the pectoral region supplying the skin from the midline to that over the shoulder.

The upper six terminal branches of the intercostal nerves emerge through the intercostal spaces at the border of the sternum to pierce the pectoralis major muscle and deep fascia. Each nerve divides into a short medial branch, that passes to the midline, and a longer lateral branch that passes sideways over the pectoral muscles to innervate the skin and fascia. In the female, it provides medial mammary branches to the breast.

The third to sixth intercostal nerves give branches that pierce the musculature covering the ribs at the midaxillary line and then divide into *anterior* and *posterior* branches. The anterior branches curve forward around the lateral margin of pectoralis major and supply the lateral part of the anterior thoracic wall.

Cutaneous arteries are small perforating branches of the internal thoracic artery (branch of the first par to the subclavian artery) or a branch of a posterior intercostal artery (from the aorta). The upper two posterior intercostal arteries are branches of the supreme intercostal artery (in turn a branch of the costocervical artery given off the costocervical trunk) and accompany the anterior cutaneous nerves. In the female, the branches in the second to fourth intercostal spaces are large because they supply the mammary gland. The lower anterior intercostal arteries are branches of the musculophrenic artery (a branch of the internal thoracic artery) and supply part of the anterior abdominal wall.

2. Breast

The breast or mammary gland is a modified apocrine sweat gland rudimentary in the male, in the aged female and in children. Each breast overlies pectoralis major, serratus anterior and the external oblique muscles and extends from the second to the sixth rib and from the sternum to the midaxillary line. Glandular tissue may extend into the axilla (the axillary tail), over the clavicle, median plane and/or the epigastrium.

MAMMARY GLAND

Parenchyma consists of 15 to 20 compound tubuloalveolar glands (lobes) with each opening onto the nipple by a separate lactiferous duct. Ducts have dilatations (lactiferous sinuses) near their termination. The ducts then narrow

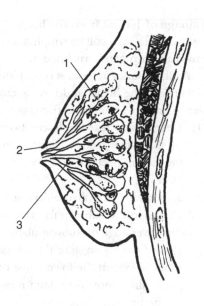

FIG. 22-1 Sagittal section of breast. (1) Lobules; (2) Nipple; (3) Lactiferous duct and sinus.

and directly traverse the nipple to its summit, where they open in very constricted individual orifices.

STROMA

Fibrous connective tissue loosely envelops the entire gland and extends into its substance to enclose the lobes, lobules, and alveoli. In addition, bundles of collagenous fibers, the suspensory ligaments, define and interconnect the lobes and extend from the skin to the deep layer of superficial fascia.

SKIN

At the level of the fourth intercostal space, a papilla-like projection, the nipple, contains the openings of the lactiferous ducts. There is no hair on the

nipple. It contains free nerve endings and Meissner's corpuscles and helicoidally arranged smooth muscle fibers while the overlying ridged epidermis contains sebaceous glands.

The areola is the circular area of skin that surrounds the nipple. It is pink in the nullipara but during pregnancy it darkens and increases in diameter. Small elevations of its surface are produced by the numerous, large, underlying tubuloalveolar areolar glands as well as sebaceous glands. These glands also enlarge during pregnancy and produce a lipoid material which lubricates and protects the nipple during nursing. In the deeper parts of the dermis, smooth-muscle fibers, (mammillary muscle) generally arranged helicoidally, accounts for the erectile capacity of the nipple.

At puberty, the female breast and areola enlarge. Ducts bud and proliferate as the menstrual cycle becomes established. Lactiferous ducts elongate. These changes are mostly in response to estrogen. Stroma rich in adipose tissue inreases considerably.

In pregnancy, there is a considerable multiplication and branching of tubuloalveoli in response to estrogen, progesterone, human placental lactogen and corticoids.

Secretion (lactation) begins toward the end of pregnancy but flow occurs only after parturition and absence of the placenta with decrease of estrogen and progesterone renders the gland more sensitive to prolactin.

In senile involution, there is a reduction of tubuloalveoli and lobules almost totally disappear.

The neurovascular system of the breast is supplied by:

Arteries: Medial mammary branches of the second to fourth perforating branches of the internal thoracic artery and the lateral mammary branches of the third to fifth lateral cutaneous branches of the posterior intercostal arteries and thoracic branches of the axillary artery.

Veins: Superficial veins drain into the perforating branches of the internal thoracic veins; the deep veins drain into the perforating branches of the internal thoracic, axillary, and intercostal veins.

Nerves: From the anterior and lateral branches of the intercostal nerves (particularly T4-6).

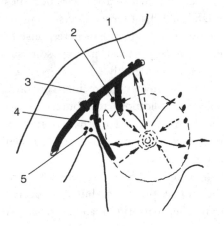

FIG. 22-2 Lymphatic drainage of the right breast. (1) Apical nodes; (2) Pactoral nodes; (3) Lateral nodes; (4) Subscapular nodes; (5) Central nodes.

Drainage of lymph from the breast is mostly laterally into axillary lymph nodes or medially into parasternal nodes.

Axillary nodes act as a series of filters between the breast and venous circulation and are arranged in five groups.

(1) *Central nodes* near the fascial floor of the axilla (receive lymph from the lateral, pectoral and posterior nodes).

(2) *Subscapular* (or subscapular) *nodes* alongside the subscapular vein at the lower edge of subscapularis.

(3) *Pectoral nodes* alongside the lateral thoracic vein at the lower edge of pectoralis minor (they drain most of the breast).

(4) *Lateral nodes* alongside the axillary vein (drain the upper limb).

(5) *Apical nodes* at the apex of the axilla where the lymph drains into the subclavian lymph trunk (receive sslymph from all other nodes in the region).

Two main lymphatic channels can be recognized leading from the breast, one from the skin envelope extending beyond the areola and the other from the parenchyma of the gland itself including the nipple and areola.

The skin envelope contains a dense cutaneous plexus that drains into the subareolar plexus beneath the skin. It drains laterally into pectoral and apical nodes, and medially into parasternal nodes alongside the internal thoracic vessels.

Lymph from the substance of the gland drains to the subareolar plexus into which lymph from the nipple and areola also drain. From the subareolar plexus, two main collecting trunks, medial and lateral drain to the pectoral group of axillary nodes.

Some drainage from the may occur directly through pectoralis major to reach the apical nodes. Some drainage may come from the contralateral side. If blockages are present in normal routes of lymph flow, drainage may be diverted to reach lower deep cervical nodes or through lymphatics that pierce linea alba to reach upper abdominal nodes.

The lymphatic drainage of the breast is particularly important because of involvement of the lymphatic system in carcinoma of the breast.

3. Deep Fascia

The pectoral fascia is thin and encloses the pectoralis major muscle. It is attached above to the clavicle below, it is continuous with the deep fascia over the abdominal muscles; laterally, with fascia over the deltoid muscle and medially to the sternum.

The deeper clavipectoral fascia envelops subclavius and pectoralis minor muscles. It is attached to anterior and posterior borders of subclavius enclosing it. The part of the fascia between subclavius and pectoralis minor is known as the costocoracoid ligament. It descends from the lower border of pectoralis minor to the axillary fascia as the suspensory ligament of the axilla.

4. Muscles

Pectoralis Major

The pectoralis major is a flat, large, multilaminar superflcial muscle of the chest.

- **Origin:** *Clavicular head*: Anterior surface of the medial half of the clavicle; *Sternocostal head:* Anterior surface of the sternum and upper six costal cartilages; *Abdominal head*: Rectus sheath (aponeurosis of the external abdominal oblique).
- **Insertion:** Whole length of the lateral lip of the intertubercular groove.

The tendon of insertion is folded to become bilaminar — clavicular fibers join the anterior lamina and sternocostal fibers form the posterior lamina.

- **Action:** As a whole, the muscle adducts and medially rotates the

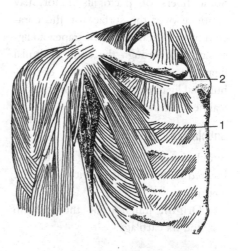

FIG. 22-3 Muscles of the pectoral region. (1) Pectoralis minor m; (2) Subclavius m.

arm. The clavicular part flexes the arm and the sternocostal part extends the fully flexed arm.

- **Nerve supply:** Lateral (C6, 7) and medial pectoral (C8, T1) nerves from the lateral and medial cords.

Pectoralis Minor

The pectoralis minor is a flat, elongated, triangular muscle that lies beneath pectoralis major in the same plane as subclavius.

- **Origin:** Sternal ends of the third to fifth ribs near their cartilages.
- **Insertion:** By a tendon to the coracoid process of the scapula.
- **Action:** Draws the shoulder downwards and forwards and raises the ribs when the shoulder is fixed.
- **Nerve supply:** Medial pectoral nerve (C8, T1) from the medial cord of the brachial plexus.

Some fibers of pectoralis minor may continue over the surface of the coracoid process in front of the trapezoid ligament and through the coracoacromial ligament to blend with the coracohumeral ligament and reach the anatomical neck of the humerus.

Subclavius

Subclavius is a small, elongated muscle located between the clavicle and the first rib.

- **Origin:** Junction of the first rib and its cartilage.

- **Insertion:** Groove on the lower surface of the clavicle.
- **Action:** Draws the clavicle downwards and backwards. It braces the sternal end of the clavicle against the articular disc and resists distraction.
- **Nerve supply:** Subclavian nerve (C5 and 6 as a branch of the upper trunk of the brachial plexus).

Serratus Anterior

Serratus anterior is a broad, flat muscle that forms the medial wall of the axilla.

- **Origin:** Digitations from the lateral surfaces of the upper eight ribs.
- **Insertion:** First digitation inserts into the superior angle of the scapula, the second to fourth digitations insert into the medial border of the scapula, the fifth to the eighth digitation inserts into the inferior angle of the scapula.
- **Action:** Acting as a whole the muscle draws the scapula forward as in pushing or thrusting. The lower fibers rotate its inferior angle of the scapula around the chest as in elevation of the upper limb. Its main action is as a fixator of the scapula in all movements of the upper limb.
- **Nerve supply:** Long thoracic nerve (C5, 6 and 7) branches supply digitations serially as the nerve descends on the outer surface of the muscle.

Chapter **23**

Axilla

Chapter Outline

1. Boundaries
2. Axillary Artery
3. Axillary Vein
4. Brachial Plexus

The axilla is the hollow between the trunk and the arm. It is a pyramidal space with three main walls and an almost linear lateral wall. The main vessels and nerves of the upper limb enter the apex from the posterior triangle of the neck so that the apex is sometimes known as the cervicoaxillary canal.

1. Boundaries

Anterior wall: Pectoralis major and minor.
Posterior wall: Subscapularis, teres major and latissimus dorsi.
Medial wall: Serratus anterior muscle, first to fifth ribs and intercostal muscles.
Lateral wall: Intertubercular groove of humerus (biceps and coracobrachialis descend laterally between the anterior and posterior walls).
Apex (truncated).
Anterior: Posterior border of clavicle.
Medial: Lateral border of first rib.
Posterior: Superior border of scapula.
Base: axillary fascia.

The axillary artery, vein and brachial plexus descend through the axilla to the arm. They are enclosed within the axillary sheath, a fascial extension of the prevertebral layer of cervical fascia covering the scalene muscles. The axilla also contains the tendon of long

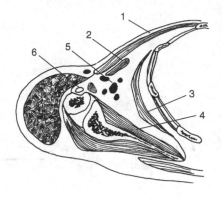

FIG. 23-2 Horizontal section through the axilla. (1) Pectoralis minor m; (2) Pectoralis major m; (3) Serratus anterior m; (4) Subscapularis m; (5) Axillary a; and n; (6) Biceps long head.

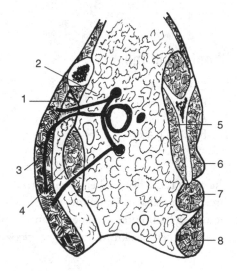

FIG. 23-3 Axilla sagittall section. (1) Costocoracoid membrane; (2) Thoracoacromial a; (3) Pectoralis major m; (4) Pectoralis minor m; (5) Subscapularis m; (6) Teres minor m; (7) Teres major m; (8) Latissimus dorsi m.

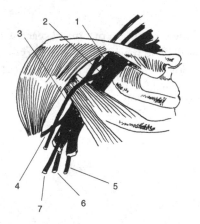

FIG. 23-1 Contents of the axilla. (1) Axillary a; (2) Axillary v; (3) Cephalic v. (4) Musculocutaneous n; (5) Ulnar n; (6) Radial n; (7) Median n.

head of biceps, the short head of biceps, and coracobrachialis muscles. Axillary lymph nodes are located within the fat and loose areolar tissue of the axilla.

2. Axillary Artery

The axillary artery, the continuation of the subclavian artery is defined by the limits of the axilla. It extends from the lateral border of the first rib to the lower border of the teres major and continues down the arm as the brachial artery. The artery is divided into three parts located (1) above; (2) behind; and (3) below pectoralis minor, respectively. Each part has the same number of branches as its name indicates.

FIRST PART

The first part of the axillary artery extends from the outer border of the

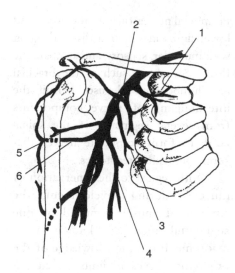

FIG. 23-4 Braches of axillary artery. (1) Superior thoracic a; (2) Thoracoacromial a; (3) Lateral thoracic a; (4) Subscapular a; (5) Anterior circumflex humeral a; (6) Posterior circumflex humeral a.

first rib to the upper border of pectoralis minor and gives off a single branch.

Superior thoracic artery arises at the lower border of subclavius, courses inferomedially, behind the axillary vein to reach the first intercostal space. It supplies the first intercostal muscle and the upper part of serratus anterior.

SECOND PART

The second part of the axillary artery lies behind the tendon of pectoralis minor and gives off two branches.

Thoracoacromial artery is a short trunk arising beneath the upper border of pectoralis minor from the anterior surface of the axillary artery. It curves around the muscle to pierce the clavipectoral fascia then divides into four branches.

The acromial branch runs over the coracoid process to reach the acromion on which it ramifies.

The deltoid branch descends beside the cephalic vein in the deltopectoral triangle and gives branches to both deltoid and pectoralis minor and major muscles.

The pectoral branch is a large branch that descends between and supplies the pectoral muscles. In the female, the pectoral branch may be large as it also supplies the deep aspect of the mammary gland.

The clavicular branch is a small branch that supplies subclavius muscle and the sternoclavicular joint.

Lateral thoracic artery is variable, descends along the lateral border of pectoralis minor to the thoracic wall. It supplies the pectoral muscles and serratus anterior and also the breast in the female (lateral mammary branches).

THIRD PART

The third part of the axillary artery extends from the lower border of pectoralis minor to the lower border of teres major. It gives off three branches.

Subscapular artery is the largest branch of the axillary artery. It arises opposite the lower border of subscapularis and courses with the thoracodorsal nerve toward the inferior angle of the scapula.

Circumflex scapular branch is given off near the origin of the scapular artery, traverses the triangular space and winds around the axillary border ending in infraspinatus. The parent trunk continues as the thoracodorsal artery accompanying the thoracodorsal nerve to supply the wall of the thorax.

Anterior humeral circumflex artery is an inconstant branch that winds around the surgical neck of the humerus. One branch ascends in the intertubercular sulcus to reach the shoulder.

Posterior humeral circumflex artery is a large branch passing backward through the quadrangular space with the axillary nerve. It supplies the deltoid, head of the humerus, shoulder joint, teres minor and long head of triceps. A descending branch anastomoses with the profunda brachii artery.

3. Axillary Vein

A continuation of the brachial and basilic veins begins at the lower border of subscapularis, ascends medial to the axillary artery and becomes the subclavian vein at the lateral border of the first rib. Tributaries of the axillary vein correspond to the branches of the axillary artery except for the thoracoacromial artery whose venae comitantes end in the cephalic vein. It also receives thoracoepigastric veins from the subcutaneous parts of the lower thoracic and upper abdominal regions providing a collateral circulation for venous return if the inferior vena cava is obstructed.

4. Brachial Plexus

Peripheral nerves of both the upper and lower limbs innervate a distinct region with muscular, sensory, and sympathetic fibers. Structures such as the brachial and lumbosacral plexuses permit the intermingling of nerve components from several segments of the spinal cord, to form composite nerves that supply individual territories.

With regard to the innervation of muscles, during development, the amount of muscle supplied by one segmental nerve is known as a myotome. From the knowledge of the origin of nerves to limb muscles, it has become established that most muscles are supplied equally from two spinal cord segments. Muscles with a

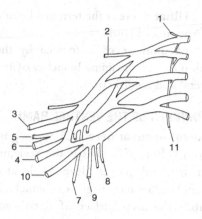

FIG. 23-5 Plan of the brachial plexus. (1) Dorsal scapular n; (2) Suprascapular n; (3) Musculocutaneous n; (4) Median n; (5) Axillary n; (6) Radial n; (7) Subscapular n; (8) Medial brachial cutaneous n; (9) Medial antebtachial cutaneous n; (10) Ulnar n; (11) Long thoracic n.

common primary action on a joint are supplied by the same segments and their opponents are supplied by the same segments. Joints more distal in the limbs are supplied by nerves removed one segment lower in the spinal cord.

With regard to the innervation of the skin, the area of skin supplied by a single spinal nerve is known as a dermatome. Compared with the innervation of the trunk where segmental nerves supply concentric bands of skin, the bands in the limbs are drawn out and follow a sequence from the upper surface of the shoulder, down the outer aspect of the limb then up along the medial aspect.

Segmental nerves come together in plexuses at the area of constriction where the limb is attached to the trunk

then disperse, however, there are some principles of their organization:

1. The limb plexuses are formed by anterior primary rami (never by posterior primary rami that supply dorsal muscles).
2. Nerves making up the plexuses divide into anterior and posterior divisions, anterior divisions supply flexor muscles and posterior divisions supply extensor muscles.
3. The flexor compartment is more sensitive and has muscles concerned with precise voluntary control and as a result, the most caudal nerve in a limb plexus is distributed to the flexor compartment only.

The **brachial plexus** consists of roots, trunks, divisions, cords, and terminal branches. It lies partly in the neck (roots lie between scalenus anterior and medius) then trunks cross the posterior triangle. Divisions are placed behind the clavicle and cords enter the axilla surrounding the axillary artery. Branches from the cords lie around the third part of the axillary artery.

ROOTS

The roots of the brachial plexus are formed by the ventral rami of C5 to 8 and T1 that are sometimes joined by contributions from C4 and T2. The roots emerge between scalenus anterior and medius in line with similar roots constituting the cervical plexus.

TRUNKS

The roots unite to form three trunks.

Upper trunk: The ventral rami of C5 and 6 unite at the lateral border of scalenus anterior to form the upper trunk.

Middle trunk: The ventral ramus of C7 continues directly to form the middle trunk.

Lower trunk: The ventral rami of C8 and T1 unite to form the lower trunk.

DIVISIONS

Each trunk divides into an anterior and a posterior division.

CORDS

Divisions join to form three cords, named with respect to the second part of the axillary artery behind pectoralis minor.

Posterior cord: Posterior divisions of all trunks combine to form the posterior cord.

Lateral cord: Anterior divisions of the upper and middle trunks form the lateral cord.

Medial cord: Anterior division of the lower trunk forms the medial cord.

TERMINAL BRANCHES

Cords divide into four terminal branches at the inferolateral border of pectoralis minor and lie around the third part of the axillary artery. The lateral and medial cords bifurcate.

Radial nerve is the terminal branch of the posterior cord.

Musculocutaneous nerve is the terminal branch of the lateral cord.

Ulnar nerve is the terminal branch of the medial cord.

Median nerve is formed by the union of the remaining branches of the lateral and medial cords.

BRANCHES OF THE VENTRAL RAMI

Dorsal scapular nerve (C5) arises posteriorly from C5, pierces the scalenus medius and passes deep to levator scapulae to which it gives a branch. It enters the deep surface of the rhomboid muscles.

Long thoracic nerve (C5, 6 and 7) arises posteriorly by three roots in and in front of scalenus medius descending behind the brachial plexus and axillary artery to reach the surface of serratus anterior which it supplies segmentally.

Injury to the long thoracic nerve may result from penetrating wounds, thoracic surgery or removal of axillary lymph nodes following mastectomy. Resultant paralysis of serratus anterior results in the medial border of the scapula becoming prominent ("winging") paticularly noticable when pushing against a resistance.

BRANCHES OF THE TRUNKS

Subclavian nerve (mostly C5) from the front of the roots, the nerve passes downward in front of the third part of the subclavian artery to supply subclavius.

Continuation of the nerve, the accessory phrenic nerve may continue into the thorax in front of the subclavian vein to supply the diaphragm.

Suprascapular nerve (C5 and 6) passes beneath trapezius to the upper border of the scapula and enters the suprascapular fossa through the suprascapular notch. It supplies supraspinatus then winds through the spinoglenoid notch, supplies the shoulder joint and ends in infraspinatus.

BRANCHES OF THE CORDS
From the Lateral Cord

Lateral pectoral nerve (C5, 6 and 7) gives a communicating branch to the medial pectoral nerve and passes in front of the axillary vessels, pierces the clavipectoral fascia to innervate the clavicular and upper sternocostal portions of pectoralis major and through the communicating branch reaches pectoralis minor.

Musculocutaneous nerve (C5 to 7) pierces coracobrachialis and is ultimately distributed to the flexor muscles on the front of the arm, to the skin on the lateral side of the forearm, and to the elbow joint.

Lateral head of the median nerve (C(5), 6)

From the Medial Cord

Medial head of median nerve (C7 and 8, T1) is the combined median nerve containing fibers from C5, 6, 7, 8 and T1 descends along the anterolateral side of the axillary artery to innervate most of the flexors of the forearm, most of the short muscles of the thumb, the skin on the front and lateral

sides of the hand, and the joints of the elbow and hand.

Medial pectoral nerve (C8 to T1) passes forward between the axillary artery and vein. It gives a branch that communicates with the lateral pectoral nerve, forming a loop around the artery. It then enters the deep surface of pectoralis minor which it pierces and supplies then sends a terminal branch to the pectoralis major.

Medial brachial cutaneous nerve (T1) passes behind the axillary vein then descend along the medial side of the vein. It pierces the deep fascia at the middle of the arm and supplies the skin of the medial and posterior sides of the arm.

Medial antebrachial cutaneous nerve (C8, T1) descends on the medial side of the axillary and brachial arteries to pierce the deep fascia about the middle of the arm. It divides into two branches — *anterior* which passes behind the median basilic vein which supply the skin on the medial forearm to the wrist and *posterior* that winds around the medial epicondyle to supply the back of the medial side of the forearm.

Ulnar nerve (C7, 8 and T1) is a terminal branch of the medial cord. It descends along the medial side of the axillary and brachial arteries. It is distributed to some of the flexors of the forearm, to the short muscles of the hand, to the skin on the front and back of the medial side of the hand, and to the joints of the elbow and the hand.

From the Posterior Cord

Subscapular nerves (C5, 6). Three nerves supply the muscles of the posterior wall of the axilla.

Upper subscapular nerve (C5), after a short course downwards and backwards, it enters the upper part of subscapularis.

Middle subscapular (Thoracodorsal) nerve (C7 and 8) passes behind the axillary artery and descends across subscapularis and teres major to the inferior angle of the scapula. It terminates in latissimus dorsi.

Lower subscapular nerve (C5 and 6) after descending laterally across subscapularis, the lower part of which it supplies, it terminates in teres major.

Axillary nerve (C5 and 6, posterior cord of the brachial plexus) is a branch of the posterior cord that lies behind the axillary artery and in front of subscapularis lateral to the radial nerve. At the lower border of subscapularis muscle, it curves posteriorly to leave the axilla by passing below the capsule of the shoulder joint medial to the surgical neck of the humerus and through the quadrangular space. It innervates the shoulder joint, deltoid and teres minor as well as the skin of the back of the shoulder (through the upper lateral brachial cutaneous nerve).

Radial nerve (C5 to 8 and T1 terminal branch of the posterior cord) descends behind the axillary artery anterior to subscapularis, teres major, and latissimus dorsi. It innervates the extensors of the arm and forearm, skin on the back of the arm, forearm, and hand.

Chapter **24**

Front of Arm, Back of Arm and Cubital Fossa

Chapter Outline

1. Fascia
2. Muscles
3. Deep Artery
4. Deep Veins
5. Deep Nerves

The front of the arm extends from the shoulder joint to the elbow.

1. Fascia

SUPERFICIAL FASCIA

Superficial fascia of the front of the arm contains a variable amount of fat, parts of the cephalic and basilic veins, their tributaries, and terminal ramifications of cutaneous nerves.

Cephalic vein ascends in the subcutaneous tissue from a dorsal network on the radial side of the forearm. It ascends alongside the lateral border of biceps and pierces the brachial fascia and continues upwards in the groove between deltoid and pectoralis major. At the deltopectoral triangle, it pierces the

clavipectoral fascia and ends in the axillary vein.

Basilic vein ascends from a dorsal network on the ulnar side of the forearm and winds to the front of the medial epicondyle in the medial bicipital groove. It pierces the brachial fascia in the middle of the arm, and ascends to the axilla, where it joins the brachial vein to form the axillary vein.

Median cubital vein is a communicating vein that runs upward and medially from the cephalic vein to join the basilic vein. It lies above the bicipital aponeurosis separating the vein from the artery.

The pattern of cutaneous veins in the region of the cubital fossa may vary. A common variation is for a median antebrachial vein to ascend from the forearm between the cephalic and basilic veins dividing equally in the cubital fossa into median cephalic and median basilic veins draining into the cephalic and basilic veins respectively.

CUTANEOUS NERVES

Upper lateral brachial cutaneous nerve arises from the posterior branch of the axillary nerve (C5, 6), turns around the posterior border of deltoid and supplies an area of skin on the back of the arm over deltoid and long head of triceps.

Posterior brachial cutaneous nerve is a branch of the radial nerve (C5–T1) that arises in the axilla. It passes to the medial side of the arm to supply the skin on the back of the arm below deltoid nearly to the olecranon.

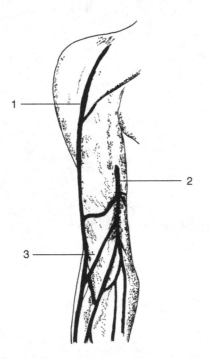

FIG. 24-1 Superfical veins of the arm. (1) Cephalic v; (2) Basilic v; (3) Median cunital v.

Intercostobrachial nerve (T2) is the lateral cutaneous branch of the second intercostal nerve. It emerges from the second intercostal space, pierces the intercostal and serratus anterior muscles, crosses the axilla to the medial side of the arm, and pierces the brachial fascia to become superficial near the posterior axillary fold. It innervates the skin of the medial and posterior surfaces of the arm from the axilla down to the elbow.

Medial brachial cutaneous nerve arises in the axilla from the medial cord of the brachial plexus (T1). It descends on the medial side of the brachial artery to the middle of the arm, where it pierces the deep fascia. It innervates the skin of the medial and posterior aspect of the lower third of the arm.

DEEP FASCIA

The brachial fascia or deep fascia of the arm, completely envelops the arm. It is continuous above with the fascia over deltoid, the pectoral muscles and axillary fascia. Below, it is attached to the olecranon and epicondyles of the humerus and is continuous with the fascia of the forearm. It gives off intermuscular septa which attach to the humerus and separate flexor from the extensor groups of muscles.

Medial intermuscular septum is attached above to the insertion of coracobrachialis and below to the medial epicondyle supracondylar ridge. It lies between the coracobrachialis and brachialis anteriorly and triceps posteriorly. It is pierced by the ulnar nerve as it approaches the back of the medial epicondyle.

Lateral intermuscular septum is attached below to the lateral epicondyle, supracondylar ridge and humerus and it extends above to the deltoid tuberosity.

2. Muscles

Coracobrachialis

Coracobrachialis is a short, band-like muscle that lies in the upper, medial part of the arm.

- **Origin:** Tip of the coracoid process (with the short head of biceps) by a fleshy origin (the origin of biceps is tendinous).
- **Insertion:** Middle of the medial border of shaft of the humerus by a flat tendon.
- **Action:** Assists in flexion and adducts the arm.
- **Nerve supply:** Musculocutaneous nerve (C6 and 7) pierces the muscle and supplies it.

Biceps Brachii

Biceps brachii is long, large and spindle-shaped and the most superficial muscle of the front of the arm. It extends from the scapula to the radius and thus crosses two joints (the shoulder and elbow).

- **Origin:** Is by two heads.

Long head: Arises by a tendon from the supraglenoid tubercle (and glenoid labrum) of the scapula within the

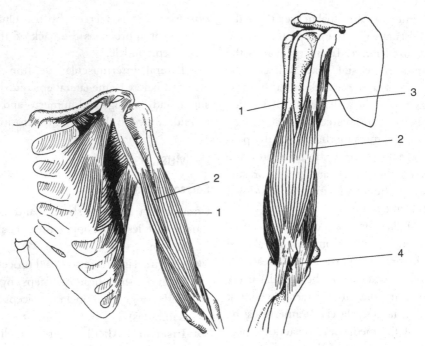

FIG. 24-2 Muscles of the front of the arm. (1) Biceps (long head) m; (2) Biceps (short head) m; (3) Coracobrachialis m; (4) Bicipital aponeurosis.

fibrous capsule of the shoulder joint where it is covered by reflected synovial membrane. It descends in the intertubercular groove between the greater and lesser tubercles where it is held in place by the transverse humeral ligament.

Short head: Tip of the coracoid process lateral to coracobrachialis.

The bellies of the two heads unite immediately below the middle of the arm.

- **Insertion:** Back of the tuberosity of the radius and by the bicipital aponeurosis, a fascial expansion that crosses the brachial artery, median nerve and origin of flexors of the forearm and into the fascia of the forearm and thereby to the ulna.

- **Action:** Flexor of the shoulder and elbow and a strong supinator of the forearm. The long head enhances the stability of the shoulder joint.

- **Nerve supply:** Musculocutaneous nerve (C5 and 6).

The median nerve and brachial artery lie in a groove along the medial border of biceps then pass in the cubital fossa beneath the bicipital aponeurosis. The median basilic vein, commonly used in venepuncture lies superficial to the aponeurosis.

Brachialis

Brachialis lies deep to the lower half of biceps.

- **Origin:** Anterior surface of the lower half of the shaft of the humerus and medial and lateral intermuscular septa.
- **Insertion:** The muscle converges to a thick tendon that adheres to the capsule of the elbow joint and inserts on the tuberosity and an impression on the anterior surface of the coronoid process of the ulna.
- **Action:** A flexor of the forearm in any position of the limb.

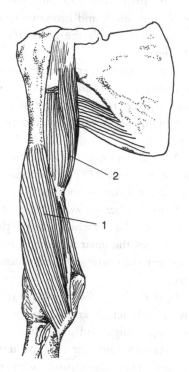

FIG. 24-3 Brachialis. (1) Brachialis m; (2) Coracobrachialis m.

- **Nerve supply:** Musculocutaneous nerve (C5 and 6). The radial nerve also sends a small branch to the lateral head.

Flexion of the forearm when the bicipital aponeurosis is tapped is known as the biceps jerk. It tests the integrity of the musculocutaneous nerve or C5 and 6 segments of the spinal cord.

3. Deep Artery

BRACHIAL ARTERY

The continuation of the axillary artery extending from the lower border of teres major to the cubital fossa opposite the neck of the radius where it divides into the radial and ulnar arteries. The upper half of the artery lies on the medial side of the arm and the lower half lies on the front of the arm (it can be compressed against the shaft of the humerus). It starts its course in the arm on the long head of triceps then on to medial head of triceps and coracobrachialis before coursing on brachialis and entering the cubital fossa deep to the bicipital aponeurosis.

The brachial artery is not crossed by any muscle and can be surgically exposed without cutting a muscle. It is located along a line connecting the middle of the clavicle to a point midway between the epicondyles of the humerus.

BRANCHES

Profunda brachii artery arises from the medial and posterior aspect of the

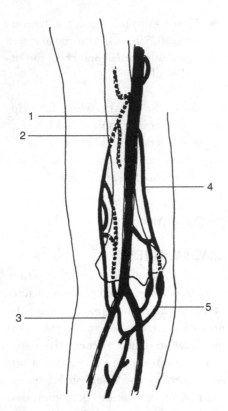

FIG. 24-4 Arteries of the front of the arm.
(1) Profunda brachii a; (2) Radial collateral a;
(3) Radial a; (4) Superior ulnar collateral a;
(5) Inferior ulnar collateral a.

brachial artery early opposite the lower border of the teres major. It courses backwards with the radial nerve in the sulcus for the radial nerve of the humerus. It supplies neighboring muscles and ends as a posterior branch behind the lateral epicondyle. The profunda brachii has branches that contribute to the arterial anastomoses around the shoulder and elbow joints.

Ascending (deltoid) branch ascends between lateral and long heads of triceps and anastomoses with the posterior humeral circumflex humeral artery.

Radial collateral artery accompanies the radial nerve into the forearm. It anastomoses with the radial recurrent artery.

Middle collateral artery descends posteriorly through the medial head of triceps to the lateral epicondyle. It anastomoses with the interosseous recurrent artery.

Anterior branch is a terminal branch of the profunda brachii, passes between brachialis and pronator teres. It anastomoses with the radial recurrent artery.

Posterior branch a terminal branch of the profunda brachii passes across humerus deep to triceps and anastomoses with the interosseous recurrent artery.

Nutrient (humeral) branch enters a nutrient foramen of the humerus to supply the marrow of the shaft.

Superior ulnar collateral artery is a branch of the brachial artery that accompanies the ulnar nerve and anastomes with the posterior ulnar recurrent artery.

Inferior ulnar collateral artery arises from the brachial artery above the medial epicondyle and then divides on brachialis into anterior and posterior branches. They anastomose with the anterior and posterior branches of the ulnar recurrent artery.

4. Deep Vein

BRACHIAL VEINS

Paired veins accompany the brachial artery. They arise at the elbow by the union of the venae comitantes of the radial and ulnar arteries. Tributaries of these veins run with branches of the brachial artery. They are located in the neurovascular compartment on the medial side of the arm. The lateral brachial vein crosses the axillary artery near the lower border of subscapularis or teres major to join the medial brachial vein. Near the point of union, the basilic vein joins the medial brachial vein forming a single axillary vein.

5. Deep Nerves

MEDIAN NERVE (C5,6,7,8 AND T1)

Median nerve is a terminal branch of the brachial plexus arising from two roots, one from the lateral cord and the other from the medial cord. It descends on the lateral side of the axillary artery to the middle of the arm, where it gradually crosses to the medial side of the artery.

In the cubital fossa, the median nerve lies behind the medial cubital vein and bicipital aponeurosis.

The median nerve gives no branches in the arm.

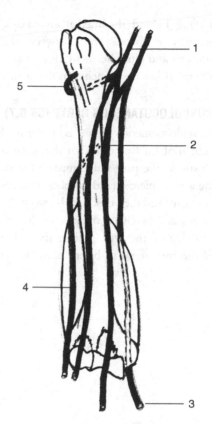

FIG. 24-5 Nerves of front of arm. (1) Musculocutaneous n; (2) Median n; (3) Ulnar n; (4) Radial n; (5) Axillary n.

ULNAR NERVE (C8 AND T1)

Ulnar nerve is the terminal branch of the medial cord of the brachial plexus. It descends on the medial side of the axillary and brachial arteries. At the middle of the arm, it pierces the medial intermuscular septum, descends with the ulnar collateral artery on the medial head of triceps. It enters the forearm by passing behind the medial epicondyle beneath a fibrous arch between the humeral and ulnar heads of flexor carpi ulnaris. It has no branches in the arm.

MUSCULOCUTANEOUS NERVE (C5,6,7)

Musculocutaneous nerve is a terminal branch of the lateral cord of the brachial plexus to the pre-axial compartment of the arm. It pierces (and supplies) coraco-brachialis and then descends between the biceps and brachialis muscles to the lateral side of the arm. A little above the elbow and lateral to biceps tendon, the nerve pierces the brachial fascia and becomes the lateral antebrachial cutaneous nerve that supplies the skin of the lateral aspect of the forearm via anterior and posterior branches (the former reaching the ball of the thumb).

Branches in the arm are:

- *Muscular branches* to coracobrachialis, brachialis, and biceps brachii.
- *Articular branch* to the elbow joint.
- *Bumeral branch* coursing with the nutrient artery to the shaft of the humerus.

Back of Arm

Chapter Outline

257

1. Fascia

SUPERFICIAL FASCIA

Superficial fascia of the back of the arm contains a variable amount of fat and branches of cutaneous nerves from the front of the arm. The medial brachial cutaneous nerve (C8, T1) arises from the medial cord of the brachial plexus and supplies the skin of the medial and posterior arm. The upper lateral cutaneous nerve arises from the axillary nerve and supplies the skin over the posterior border of deltoid and long head of triceps. Branches of the posterior brachial cutaneous nerve which arises in the axilla from the radial nerve. These pass posteromedially to innervate the skin of the middle of the back of the arm.

A bursa over the olecranon (the subcutaneous olecranon bursa) is enlarged in "student's elbow."

DEEP FASCIA

The deep fascia, the brachial fascia, completely envelopes the arm and is thicker over triceps than over biceps. Both lateral and medial intermuscular septa add to the origin of triceps in the back of the arm.

2. Muscles

Triceps

Triceps forms the muscle mass of the back of the arm. It has three heads of origin arranged in two planes, superficial and deep.

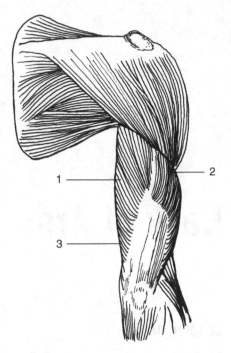

FIG. 24-6 Muscles of the arm — posterior aspect. (1) Triceps m. (long head); (2) Triceps m. (lateral head); (3) Triceps m. (medial head)

- **Origin:**
  Long head: By a strong tendon from the infraglenoid tubercle of the scapula. The belly of the long head descends between teres major and teres minor and joins the other heads to form a common tendon.
  Lateral head: Arises from the posterior and lateral surface of the humerus above the sulcus for the radial nerve and from the lateral intermuscular septum.
  Medial head: Arises by fleshy fibers from the medial surface of the humerus below the sulcus for

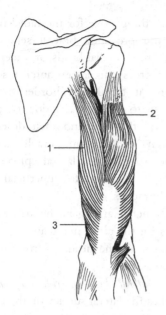

FIG. 24-7 Triceps from behind with deltoid removed. (1) Triceps m (long head); (2) Triceps m (lateral head); (3) Triceps m (medial head).

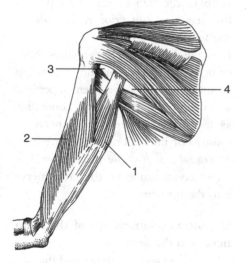

FIG. 24-8 Triceps from behind with deltoid removed. (1) Triceps m (long head); (2) Triceps m (lateral head); (3) Quadrangular space; (4) Triangular space.

the radial nerve and from the entire medial and lateral intermuscular septa below the sulcus.

- **Insertion:** By a flat tendon (covering the distal two fifths of the muscle) to the posterior surface of the upper part of the olecranon.
- **Action:** Extensor of the forearm. The long head is also an adductor of the forearm and supports the shoulder joint in abduction.
- **Nerve supply:** Radial (C7–C8). A distinct branch innervates each head.

Extension of the elbow or twitch of triceps that occurs following tapping of the triceps tendon is known as the "triceps jerk." It is elicited to test the integrity of the C7 and C8 segments of the spinal cord.

There is a subcutaneous bursa (the olecranon bursa) between the triceps tendon and olecranon just proximal to the insertion of the tendon.

Anconeus

Anconeus represents a small, triangular continuation of triceps extends from the lateral epicondyle to the humerus.

- **Origin:** Lateral epicondyle of the humerus.
- **Insertion:** Fibers diverge and insert into the lateral surface of the olecranon and posterior surface of the ulna.
- **Action:** Assists triceps as an extensor of the elbow and stabilizes the elbow joint during pronation and supination.

■ **Nerve supply:** Radial nerve (via branch to medial head of triceps).

3. Deep Nerves: Radial

RADIAL NERVE

Radial nerve (C5–C8, T1) is a terminal branch of the posterior cord of the brachial plexus. It descends behind the third part of the axillary artery and the proximal part of the brachial artery. It winds obliquely across the back of the humerus under the lateral head of tri-

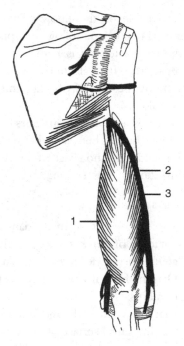

FIG. 24-9 Back of shoulder and arm. (1) Medial head of triceps; (2) Radial n; (3) Profunda brachii a.

ceps in the groove for the radial nerve with the profunda brachii separating lateral from medial heads of triceps. It then pierces the lateral intermuscular septum at the lateral border of the humerus to enter the anterior compartment of the arm and then descends between brachialis and brachioradialis to the front of the lateral epicondyle, where it divides into superficial and deep branches.

Branches arising in the arm in the sulcus for the radial nerve are:

Muscular: To the heads of triceps and anconeus.

Lower lateral brachial cutaneous nerve: Supplies the lateral surface of the lower part of the arm in front.

Posterior antebrachial cutaneous nerve: Pierces the lateral head of triceps and supplies the skin of the lower part of the arm and posterior radial side of the forearm to the wrist.

Deep branch of the radial nerve: Arises above or below the lateral epicondyle. It winds around the radius between superficial and deep parts of supinator continuing as the posterior interosseous nerve to muscles of the back of the forearm.

Superficial branch of the radial nerve: The direct continuation of the radial nerve into the forearm.

Important relationships of the main nerves in the arm

1. *Axillary nerve* winds around the surgical neck of the humerus (may be

damaged in fractures of the neck of the humerus).

2. *Radial nerve* winds around the back of the shaft of the humerus (may be damaged in fractures of the shaft of the humerus).

3. *Musculocutaneous nerve* passes across the front of the shaft of the humerus.

4. *Ulnar nerve* passes behind the medial epicondyle (may be damaged in fractures of the epicondyle).

5. *Median nerve* lies lateral to the upper half of the brachial artery and medial to its lower half.

Features related to the surgical neck of the humerus

1. *Axillary nerve.*
2. *Posterior humeral circumflex artery.*

Features related to the middle of the shaft of the humerus

1. Distal extent of attachments of deltoid and coracobrachialis.

2. The ulnar nerve pierces the medial intermuscular septum and enters the posterior compartment.

3. The radial nerve pierces the lateral intermuscular septum and enters the anterior compartment.

4. The *basilic vein* and *medial brachial cutaneous* (from 1, medial cord) and *antebrachial cutaneous nerves* (from C8 and T1, medial cord) pierce the deep fascia medially.

5. The *inferior lateral brachial cutaneous* and *posterior antebrachial cutaneous nerves* (branches of the radial nerve) pierce the deep fascia laterally.

Features related to the distal end of the shaft of the humerus

1. The ulnar nerve passes behind the *medial epicondyle.*

These features may be damaged by fractures of the proximal humerus, shaft or distal shaft.

Cubital Fossa

Chapter Outline

1. Boundaries
2. Contents (Lateral to Medical)

The cubital fossa is the triangular hollow lying in front of the elbow joint.

1. Boundaries

Inferior and lateral: Medial border of bra-chioradialis.

Inferior and medial: Lateral border of pronator teres.

Apex: Intersection of the medial and lateral boundaries.

Base: Imaginary line joining the medial and lateral epicondyles

Roof: Skin, superficial and deep fascia

Floor: Brachialis (of the arm) medially and supinator (of the forearm) inferolaterally.

2. Contents (Lateral to Medial)

Lateral antebrachial cutaneous nerve: This nerve is the superficial continuation of the musculocutaneous nerve between biceps and brachialis becoming superficial on brachioradialis.

Tendon of biceps: The central structure in the fossa, the tendon of the biceps descends to its insertion on the tuberosity of the radius. The bicipital aponeurosis spans the cubital fossa medially.

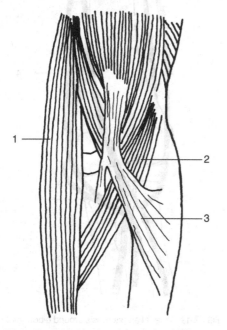

FIG. 24-10   Cubital fossa — boundaries. (1) Brachioradialis m; (2) Pronator teres m; (3) Bicipital aponeurosis.

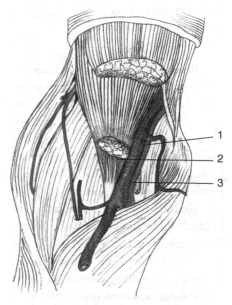

FIG. 24-11   Cubital fossa — contents. (1) Brachial a; (2) Tendon of biceps; (3) Median n.

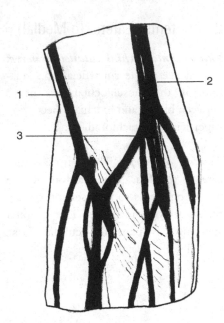

FIG. 24-12 Cunbital fossa — veins. (1) Cephalic v; (2) Basilic v; (3) Median cubital v.

Brachial artery lies beneath the bicipital aponeurosis and bifurcates opposite the neck of the radius at the apex of the fossa forming the radial and ulnar arteries medial to the tendon. The ulnar artery leaves the fossa under pronator teres while the radial artery leaves the fossa at its apex.

Median nerve lies medial to the brachial artery about half way between the tendon of biceps and the medial epicondyle. It leaves the fossa passing between the two heads of pronator teres.

Medial antebrachial cutaneous nerve lies in front of the lateral border of pronator teres.

The following structures are not within the fossa but lie beneath its boundaries:

Anterior ulnar recurrent artery ascends over brachialis beneath pronator teres. It anastomoses with the inferior ulnar collateral artery.

The *radial nerve* lies close but not in the fossa, it lies in the groove between brachialis and brachialis. It divides in front of the lateral epicondyle

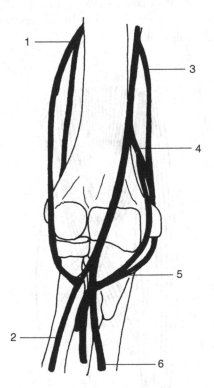

FIG. 24-13 Arterial anastomosis around elbow joint. (1) Radial recurrent a; (2) Radial a; (3) Superior ulnar collateral a; (4) Inferior ulnar collateral a; (5) Ulnar recurrent a; (6) Ulnar a.

into superficial and deep branches. The superficial branch descends into the forearm on supinator and beneath brachioradialis. The deep branch pierces supinator toward the back of the forearm.

COLLATERAL CIRCULATION AROUND THE ELBOW

An arterial anastomosis around the elbow provides functional contact between the brachial artery (proximal to the superior ulnar collateral artery). It ensures adequate blood supply to the distal segment around the joints with considerable angulation.

Major vessels in the anastomosis are:

In front of the lateral epicondyle: Radial collateral artery (of profunda brachii) with the radial recurrent (of the radial artery).

Behind the lateral epicondyle: Middle collateral artery (of profunda brachii) to the interosseous recurrent (of the ulnar artery).

In front of the medial epicondyle: Anterior branch of the inferior ulnar collateral with the anterior ulnar recurrent (of the ulnar artery).

Behind the medial epicondyle: Superior ulnar collateral (of the brachial) with the posterior branch of the inferior ulnar collateral (of the ulnar artery).

25

The Front and Back of the Forearm

Chapter Outline

1. Fascia
2. Muscles
3. Deep Arteries
4. Deep Nerves

The front of the forearm extends from the elbow to the distal crease of the wrist. It contains two bones, the radius and ulna while attached muscles mostly cross into the hand.

1. Fascia

SUPERFICIAL FASCIA

The superficial fascia of the front of the forearm contains vessels, nerves, and a variable amount of fat.

Superficial Veins

The cephalic vein arises from the lateral end of the dorsal venous arch of the hand. It passes obliquely to the front of the forearm and ascends in front of the elbow to the arm.

The basilic vein arises from the medial end of the dorsal venous arch of the hand. It ascends along the medial side of the forearm and into the arm in front of the medial epicondyle.

The median antebrachial vein begins on the front of the hand and ascends on the middle of the front of the forearm. It joins the basilic or the median cubital vein.

CUTANEOUS NERVES

The medial antebrachial cutaneous nerve (C8, T1, medial cord of the brachial plexus) descends alongside the axillary and brachial arteries to the middle of the arm where it pierces the brachial fascia and divides. The anterior branch, descends to innervate the skin of the anteromedial surface of the forearm, and the posterior branch, innervates the posteromedial surface of the forearm.

The lateral antebrachial cutaneous nerve is the continuation of the musculocutaneous nerve in the forearm. It pierces the brachial fascia lateral to the tendon of biceps then passes behind the cephalic vein to divide into an anterior branch that innervates the skin on the anterolateral side of the forearm, and a posterior branch, that passes backward to innervate the skin on the posterolateral side of the forearm.

DEEP FASCIA

The antebrachial fascia is continuous with the brachial fascia and encloses both flexor and extensor muscles. It gives partial origin to the muscles that arise from the epicondyles. Septa extend from the fascia between the fleshy bellies of the muscles. In the lower forearm, the fascia splits into two layers, a superficial layer encloses flexor carpi ulnaris and flexor carpi radialis and a deep layer separates these muscles from the other flexors. Transverse fibers of the deep layer give rise to a strong flexor retinaculum (transverse carpal ligament) that stretches across the front of the wrist and joins the scaphoid and trapezium laterally and pisiform and hamate medially.

2. Muscles

The muscles of the flexor-pronator group are arranged on the front of the forearm in three layers: superficial, intermediate, and deep. They are innervated by two nerves, the median and ulnar.

SUPERFICIAL LAYER

The four superficial muscles share a common tendinous origin from the front of the medial epicondyle of the humerus (and fascia over the muscles) — the common flexor origin. They are flexors or pronators of the forearm. From lateral to medial, they are:

Pronator Teres

Pronator teres is confined to the upper forearm and has two heads of origin.

FIG. 25-2   Brachioradialis.

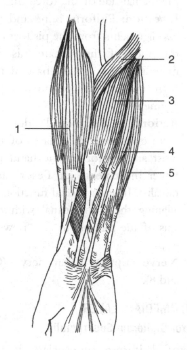

FIG. 25-1   Muscles of the front of the forearm – superfical layer. (1) Brachioradialis m; (2) Pronator teres m; (3) Flexor carpi radialis m; (4) Palmaris longus m; (5) Flexor carpi ulnaris m.

- **Origin:**
  Humeral head: Medial supracondylar ridge and medial epicondyle.
  Ulnar (deep) head: Distal to tubercle on the medial side of the coronoid process.
- **Insertion:** Rough oval impression on the lateral surface of the radius at its greatest convexity.
- **Action:** Flexor of the forearm only against a resistance. Primarily a pronator of the forearm.

■ **Nerve supply:** Median nerve (C6). The nerve to pronator arises in the cubital fossa as the median nerve passes between the two heads of pronator.

Flexor Carpi Radialis

Flexor carpi radialis is located between the pronator teres and the palmaris longus.

■ **Origin:** Medial epicondyle by the common tendon, intermuscular septum and deep fascia of the forearm.

■ **Insertion:** Base of the second metacarpal (compare with radial extensors).

■ **Action:** Flexor and abductor of the hand at the radiocarpal and mid-carpal joints. It fixes the wrist (with flexor carpi ulnaris and the carpal extensors) in movements of the fingers as in writing.

■ **Nerve supply:** Median nerve (C6 and 7).

Palmaris Longus

Palmaris longus parallels flexor carpi radialis but may be absent in about 10% of cases. It has a short belly and long tendon.

■ **Origin:** Medial epicondyle by the common tendon.

■ **Insertion:** Palmar aponeurosis and flexor retinaculum.

■ **Action:** An aid in flexing the wrist and tenses the palmar fascia.

■ **Nerve supply:** Median nerve (C6 and 7).

Flexor Carpi Ulnaris

Flexor carpi ulnaris is the most medial of the superficial flexors. It has two heads of orign.

■ **Origin:**

Humeral head: Medial epicondyle by the common tendon and a fibrous arch between the medial epicondyle and the olecranon (beneath which the ulnar nerve enters the forearm):

Ulnar head: By a strong aponeurosis to the upper three fourths of the posterior border of the ulna.

The muscle belly extends down the medial side of the forearm.

■ **Insertion:** Pisiform bone and via two ligaments from the pisiform to the hook of the hamate (pisoha-mate ligament) and the base of the fifth metacarpal (pisometacarpal ligament).

■ **Action:** A flexor and adductor (with extensor carpi ulnaris) of the wrist and fixator of the hand for finger movements (cf flexor carpi radialis). It has a special function of aligning the carpal canal with the axis of the forearm in the "power" grip.

■ **Nerve supply:** Ulnar nerve (C7 and 8).

INTERMEDIATE LAYER
Flexor Digitorum Superficialis

Flexor digitorum superficialis is the superficial flexor of the fingers and the largest of the superficial group. It may be considered a superficial lamina of the deep flexors.

- **Origin:**

Humeroulnar head: Medial epicondyle by a common tendon, the ulnar collateral ligament of the elbow joint, and the medial surface of the coronoid process.

Radial head: Oblique line on the upper two-thirds of the anterior border of the radius.

 The two heads are connected by a fibrous bridge which forms the superficialis arch. In the wrist, the tendons for the third and fourth digits are superficial to the tendons for the second and fifth digits. These tendons diverge in the palm. Each tendon enters a fibrous flexor sheath of a finger and is split to transmit the tendons of flexor digigorum profundus.

- **Insertion:** Margins of the shafts of the middle phalanges of the second to fifth digits (not the thumb).
- **Action:** Flexes the middle and proximal phalanges and the wrist.
- **Nerve supply:** Median nerve (C7 and 8).

There is a neurovascular plane between the intermediate layer and flexor digitorum profundus. It is exposed by dividing the septum between the two covering muscles, flexor digitorum superficialis and flexor carpi ulnaris. As these two muscles are innervated by different nerves (the median and ulnar nerves respectively), their surgical separation does not endanger their nerve supply.

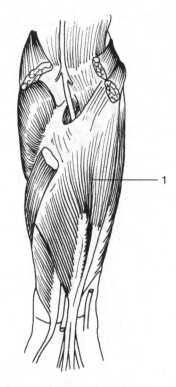

FIG. 25-3 Muscles of the front of the forearm – intermediate layer. (1) Flexor digitorum superficialis m.

DEEP LAYER
Flexor Digitorum Profundus

Flexor digitorum profundus clothes the anterior and medial surface of the ulna.

- **Origin:** Upper two-thirds of the anterior and medial surface of the ulna, the medial side of the coronoid process, and the interosseous membrane. Four tendons descend side by side over pronator quadratus (next to the tendon of flexor pollicis longus) and traverse the carpal canal. They are retained

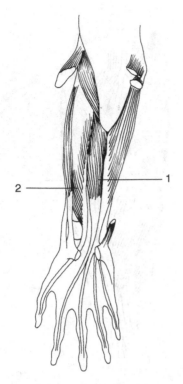

FIG. 25-4 Front of forearm – deep layer. (1) Flexor digitorum profiunds m; (2) Flexor pollicis longus m.

against the plamar surfaces of the proximal and middle phalanges by fibrous flexor sheaths, and perforate a tendon of flexor digitorum superficialis.

- **Insertion:** By tendons into the palmar surface of the base of the distal phalanx of the second to fifth digits.
- **Action:** Prime mover in flexion of the terminal phalanges of the second to fifth digits but aids in the flexion of the fingers as a whole and the wrist.

- **Nerve supply:** Anterior interosseous nerve (of the median C7 and 8) to the lateral half (index and middle fingers) and ulnar nerve to the medial half (ring and little fingers). Compare the innervation pattern of the lumbrical muscles in the hand.

Flexor Pollicis Longus

Flexor pollicis longus clothes the radius behind the flexor digitorum superficialis.

- **Origin:** Upper three-fourths of the anterior surface of the radius and the adjacent interosseous membrane. Its tendon passes through the carpal tunnel, enters the palm and enters a fibrous flexor sheath.
- **Insertion:** Base of the distal phalanx of the thumb.
- **Action:** Flexor of the distal phalanx of the thumb and secondarily the other joints which it crosses.
- **Nerve supply:** Anterior interosseous nerve (of the median nerve C7, 8 and T1).

Pronator Quadratus

Pronator quadratus is located at the distal end of the forearm just above the wrist deep to the flexor tendons and sheaths, median nerve and radial vessels.

- **Origin:** Pronator ridge on the distal quarter of the anterior surface of the ulna.

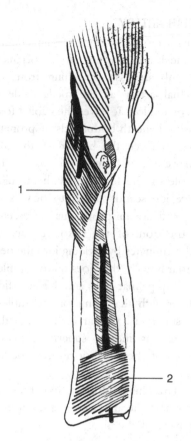

FIG. 25-5 Muscles of the front of the forearm – deep layer. (1) Supinator m; (2) Pronator quadratus m.

- **Insertion:** Distal quarter of the anterior surface of the radius and a triangular area on the medial surface of the radius.
- **Action:** Pronator of the forearm (and by pulling the radius around the ulna, it pronates the hand).
- **Nerve supply:** Anterior interosseous nerve (of the median nerve, C7, 8 and T1).

3. Deep Arteries

RADIAL ARTERY

Radial artery is a terminal branch of the brachial artery. Beginning from the bifurcation of the brachial artery in the cubital fossa opposite the neck of the radius, it descends to the styloid process of the radius. In the upper third of the forearm, it courses between brachioradialis and pronator teres. In the lower two-thirds, it lies medially to the tendon of brachioradialis under the deep fascia only. At the distal end of the radius, it curves around the lateral border of the wrist (where its pulse can be felt) under abductor pollicis longus and extensor pollicis brevis, crossing the "anatomical snuff box". Under extensor pollicis longus, it reaches the proximal end of the first inte metacarpal space to complete the deep palmar arch.

BRANCHES IN THE FOREARM

Radial recurrent artery arising in the cubital fossa, it ascends between brachioradialis and brachialis to the front of the lateral epicondyle. It supplies the adjacent muscles and the elbow joint and anastomoses with the radial collateral branch of the profunda brachii artery.

Muscular branches supply the muscles on the radial side of the forearm.

Palmar carpal branch is a small vessel arising at the distal border of

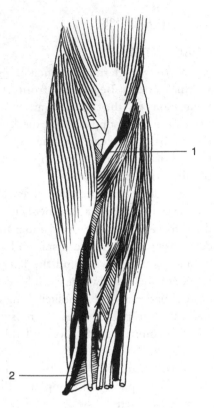

FIG. 25-6 Neurovascular plane of front of forearm. (1) Radial artery; (2) Superficial branch of radial n.

pronator quadratus. It crosses the wrist to unite with the corresponding branch of the ulnar artery forming the palmar carpal arch.

Superficial palmar branch arises near the wrist joint just before the radial artery curves around the wrist. It descends through the thenar muscles, supplying them, and joins the superficial palmar arch by anastomosing with the terminal part of the ulnar artery.

ULNAR ARTERY

Ulnar artery is the larger terminal branch of the brachial artery. It also begins in the cubital fossa branching from the brachial artery at the medial side of biceps tendon. It leaves the cubital fossa passing beneath both heads of pronator teres to reach the border of the ulna under cover of flexor carpi ulnaris at the middle of the forearm. With the ulnar nerve, it descends straight to the wrist on flexor digitorum profundus. It descends on the front of the flexor retinaculum and terminates by dividing into the deep palmar branch (that goes on to complete the deep palmar arch) and the superficial palmar arch that commonly completes the superficial palmar arch with either the radialis indicis, superficial palmar branch of the radial or princeps pollicis arteries).

The ulnar pulse can sometimes be felt where the artery crosses anterior to the head of the ulna.

BRANCHES IN THE FOREARM

Ulnar recurrent artery ascends between brachioradialis and pronator teres or gives off anterior and posterior branches which pass in front and behind the medial epicondyle respectively. The anterior ulnar recurrent artery supplies pronator teres and brachialis and anastomoses with the inferior ulnar collateral artery. The posterior ulnar recurrent artery ascends between flexor digitorum superficialis and profundus and ascends behind the medial epicondyle to anastomose with the superior ulnar collateral artery.

Common interosseous artery arises shortly after the ulnar artery begins. It ascends to the upper border of interosseous membrane and divides into anterior and posterior interosseous branches.

Anterior interosseous artery descends on the front of the interosseous membrane accompanied by the interosseous nerve (note it is part of the pimitive axial artery and the median artery its main trunk). It pierces the interosseous membrane and descends to join the dorsal carpal network (anastomose with the posterior interosseous artery).

Posterior interosseous artery passes above the interosseous membrane to supply the back of the forearm.

Muscular branches supply adjacent muscles.

Nutrient branches enter nutrient foraminae of the radius and ulna.

Palmar carpal branch curves around the medial side of flexor digitorum profundus then over the distal end of the radius to form the anterior carpal arch with the anterior carpal branch of the radial artery.

Dorsal carpal branch arises just above the pisiform, winds back beneath flexor carpi ulnaris reaching the back of the carpus contributing to the posterior carpal arch.

4. Deep nerves

MEDIAN NERVE

Median nerve comprise of C5, 6, 7, 8 and T1 by two roots, one from the lateral cord and the other from the medial cord of the brachial plexus. Median nerve leaves the cubital fossa and enters the forearm by passing between the two heads of the pronator teres muscle. It crosses the ulnar artery and descends through the middle of the forearm on the deep surface of flexor digitorum superficialis. At the wrist, it lies between the tendons of flexor digitorum superficialis and flexor carpi radialis.

Branches in the cubital fossa and forearm are:

Anterior interosseous nerve arises from the median nerve in the cubital fossa. It descends on the front of the interosseous membrane accompanied by the anterior interosseous artery. It supplies the three adjacent "clothing" muscles of the deep layer then passes deep to pronator quadratus to supply the joints of the wrist.

Articular branches to the elbow joint.

Muscular branches to all of the superficial flexor muscles except flexor carpi ulnaris.

Palmar cutaneous branch arises just above the wrist running into the hand to supply the skin of the thenar eminence and central palm.

ULNAR NERVE (C8 AND T1 FROM THE MEDIAL CORD OF THE BRACHIAL PLEXUS)

Ulnar nerve passes behind the medial epicondyle of the humerus entering the forearm between the two heads of flexor carpi ulnaris. It descends under cover of flexor carpi ulnaris on flexor digitorum profundus to the wrist at the lateral side of the pisiform, where it terminates by dividing into a superficial and a deep branch. Above the middle of the forearm, the nerve meets the ulnar artery becoming superficial on the lateral side of flexor carpi ulnaris.

BRANCHES IN THE FOREARM

Articular branch: To the elbow joint.

Muscular branch: To flexor carpi ulnaris and the medial half of flexor digitorum profundus.

The **palmar cutaneous** branch arise in the middle of the forearm and descends into the palm to supply the medial side of the palm.

Dorsal cutaneous arises immediately above the wrist, it passes deep to flexor carpi ulnaris to supply the dorsum of the hand.

RADIAL NERVE (C5,6,7,8 FROM THE POSTERIOR CORD OF THE BRACHIAL PLEXUS)

Radial nerve enters the forearm between brachioradialis and brachialis. It divides immediately into superficial and deep branch.

The superficial branch, the smaller of the two terminal branches of the radial nerve, is purely sensory. It arises in front of the lateral epicondyle and descends along the lateral border of the forearm under brachioradialis. It accompanies the radial artery in the upper forearm but the two diverge distally. The nerve provides articular branches to the elbow joint. It curves beneath the tendon of brachioradialis to lateral side of the dorsum of the hand.

The deep branch is the larger of the terminal branches of the radial nerve. It is articular and muscular in distribution. It begins beneath the origin of brachioradialis and winds laterally around the radius between fibers of supinator to lie between superficial and deep extensor muscles. It reaches the interosseous membrane passing deep to extensor pollicis longus then in the groove for extensor digitorum on the back of the radius ending in an enlargement on the back of the carpus.

Each of the three motor nerves to the forearm enters the forearm by passing between the humeral and ulnar heads of a muscle. The median nerve passes between the heads of pronator teres, the ulnar nerve passes between the heads of flexor carpi ulnaris and the deep branch of the radial passes between the heads of supinator.

Back of Forearm

Chapter Outline

1. Fascia
2. Muscles
3. Deep Arteries
4. Deep Nerves

The back of the forearm extends from the back of the elbow to the back of the wrist. Its muscles arise from the supracondylar ridge and lateral epicondyle of the humerus (common extensor tendon) or the posterior surfaces of the radius and ulna.

277

1. Fascia

SUPERFICIAL FASCIA

The superficial fascia of the back of the forearm contains a variable amount of fat, unnamed arteries derived from the deep arteries, lymphatics and cutaneous nerves.

Superficial Veins

Accessory cephalic vein arises from a venous plexus on the dorsum of the hand and forearm. It ascends diagonally to join the cephalic vein near the elbow.

Cutaneous Nerves

Posterior antebrachial cutaneous nerve arises from the radial nerve as it passes through the sulcus for the radial nerve. It pierces the lateral head of triceps below the deltoid tuberosity and descends to innervate the skin on the lower lateral third of the arm and middle of the back of the forearm as far as the wrist.

Ulnar branch of the medial antebrachial cutaneous nerve is derived from the medial antebrachial cutaneous nerve (C8, T1, from the medial cord of the brachial plexus). It innervates the skin on the posteromedial surface of the forearm.

Posterior branch of the lateral antebrachial cutaneous nerve is the other terminal branch of the medial antebrachial cutaneous nerve. It innervates the skin on the posterolateral surface of the forearm.

DEEP FASCIA

Deep fascia is dense and strong on the back of the forearm where it gives origin to muscles. It sends septa between muscles. It is firmly attached to the posterior border of the ulna. At the elbow, it is reinforced by the tendon of triceps and at the junction with the wrist it is thickened to form the extensor retinaculum.

2. Muscles

Muscles of the back of the forearm are extensors of the wrist and fingers or supinators of the forearm (except brachioradialis). They are divided into superficial and deep groups and are supplied by the radial nerve or by its deep branch (that becomes the posterior interosseous nerve).

SUPERFICIAL GROUP

These muscles arise from the lateral epicondyle (the common extensor origin). From lateral to medial, the muscles of this group are:

Brachioradialis

Brachioradialis lies on the radial side of the forearm.

- **Origin:** Upper part of the lateral supracondylar ridge of the humerus and the lateral intermuscular septum.
- **Insertion:** Lateral surface of the distal end of the radius just above the styloid process.

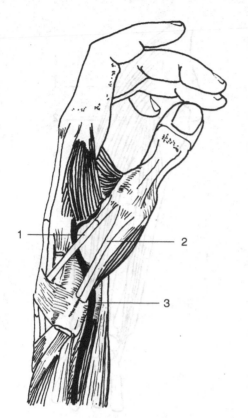

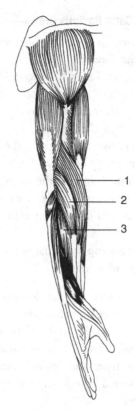

FIG. 25-7 Back of the forearm — anatomical snuff box. (1) Extensor pollicis longus m; (2) Abductor pollicis longus m; (3) Radial a.

FIG. 25-8 Lateral view of radial group of forearm muscles. (1) Brachioradialis m; (2) Extensor carpi radialis longus m; (3) Extensor carpi radialis brevis m.

- **Action:** Flexes the elbow.

  Although it belongs to the lateral group of superficial extensors, brachioradialis has lost its extensor function.

- **Nerve supply:** Radial nerve (C5 and 6).

Extensor Carpi Radialis Longus

This muscle lies behind brachioradialis.

- **Origin:** Lower third of the lateral supracondylar ridge and the lateral intermuscular septum.
- **Insertion:** Base of the second metacarpal.
- **Action:** Extends and abducts the wrist.

  Extensor carpi radialis longus fixes the wrist when long flexors are acting on the fingers.

- **Nerve supply:** Radial nerve.

Extensor Carpi Radialis Brevis

This muscle lies on the posterolateral side of the forearm.

- **Origin:** Lateral epicondyle by means of the common tendon. The origin is commonly injured in "tennis elbow".
- **Insertion:** Base of the third metacarpal bone.
- **Action:** Extends the hand (with extensor carpi ulnaris which is associated with finger flexion in making a fist).
- **Nerve supply:** Radial nerve or the deep radial branch.

Extensor carpi radialis longus arises from the lower supracondylar ridge and is inserted into the base of the second metacarpal. Extensor carpi radiais brevis arises from the lateral epicondyle and is inserted into the base of the third metacarpal.

The three lateral superficial extensors lack attachment to the underlying radius and are often referred to as the "mobile wad of three".

The medial group of superficial extensors are:

Extensor Digitorum

Extensor digitorum lies on the back of the forearm.

- **Origin:** Lateral epicondyle by means of the common tendon. Above the wrist, the muscle divides into four tendons which diverge on

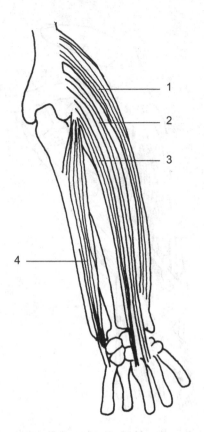

FIG. 25-9 Superfical muscles of back of forearm. (1) Brachioradialis m; (2) Extensor carpi radialis longus m; (3) Extensor carpi radialis brevis longus m; (4) Extensor carpi ulnaris m.

the back of the hand but remain connected by bands.

- **Insertion:** By four flat tendons that spread on the back of the hand, one going to the second to fifth digits. The tendon on the back of each finger is the extensor expansion.

- **Action:** Extends the phalanges and secondarily the wrist.
- **Nerve supply:** Deep branch of the radial nerve (C7 and 8).

Extensor Digiti Minimi

Extensor digiti minimi is often fused with the proximal part of extensor digitorum.

- **Origin:** By the common tendon from the lateral epicondyle of the humerus. The muscle becomes tendinous and passes beneath the extensor retinaculum.
- **Insertion:** The tendon splits, one part joins the tendon of extensor digitorum and both parts insert into the extensor aponeurosis of the fifth digit.
- **Action:** Extends the proximal phalanx of the little finger.
- **Nerve supply:** Deep branch of the radial nerve.

Extensor Carpi Ulnaris

Extensor carpi ulnaris lies on the posterolateral side of the ulna.

- **Origin:** By the common tendon from the lateral epicondyle of the humerus and an aponeurosis from the posterior margin of the ulna.
- **Insertion:** Tubercle on medial side of base of the fifth metacarpal.
- **Action:** Extends and adducts the hand (with flexor carpi ulnaris). It

acts as a fixator of the wrist when the fingers are in action.
- **Nerve supply:** Deep branch of the radial nerve.

ANCONEUS

Anconeus appears to be continuous with the lateral fibers of the medial head of triceps.

- **Origin:** Medial part of the back of the lateral epicondyle. It covers supinator and the interosseous artery.
- **Insertion:** Triangular area above the oblique line running from the radial notch to the posterior margin of the ulna.
- **Action:** Assists triceps in extending the elbow. It also stabilizes the ulna and slightly abducts and rotates the ulna.
- **Nerve supply:** Radial nerve.

DEEP GROUP

Deep group comprise supinator muscle and a series of long extensors of the thumb and index finger.

Supinator

Supinator surrounds the upper third of the radius. It is arranged in two layers separated by the deep branch of the radial nerve that supplies it.

- **Origin:** Deep (transverse) fibers arise from the supinator crest and fossa of the ulna. Superficial (oblique) fibers arise from the lateral

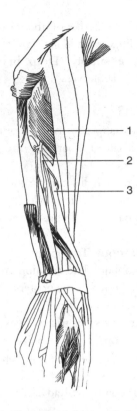

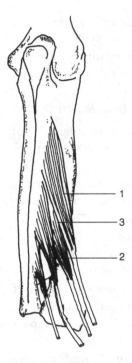

FIG. 25-10 Deep dorsal structures in the forearm. (1) Supinator m; (2) Posterior interosseous a; (3) Deep branch of radial n.

FIG. 25-11 Deep layer of dorsal forearm muscles. (1) Abductor pollicis longus m; (2) Extensor pollicis longus m; (3) Extensor pollicis brevis m.

epicondyle of the humerus, radial collateral and annular ligaments.

- **Insertion:** The deep fibers encircle the radius and insert into the upper part of the shaft (from the radial tuberosity above to the insertion of pronator teres below). Superficial fibers insert into an oblique line on the radius extending between the radial tuberosity and insertion of pronator teres.

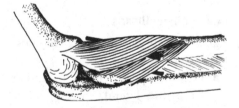

FIG. 25-12 Right supinator muscle. Deep radial n dividing supinator into two laminae.

- **Action:** Supinates the pronated forearm and hand.
- **Nerve supply:** Deep radial nerve (divides the supinator muscle into two lamellae).

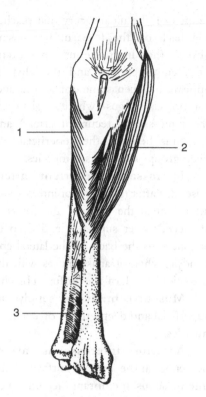

FIG. 25-13 Supinator and pronator muscles. (1) Supinator m; (2) Pronator teres m; (3) Pronator quadratus m.

Abductor Pollicis Longus

Abductor pollicis longus lies deep in the middle third of the forearm and superficial on the back and radial side of the lower third of the forearm.

- **Origin:** Lateral part of the posterior surface of the ulna, adjacent interosseous membrane, middle third of the posterior surface of the radius.

  The tendon runs obliquely across those of extensor carpi radialis longus and brevis.

- **Insertion:** Lateral side of the base of the first metacarpal.
- **Action:** Abducts the thumb (first metacarpal at the carpometacarpal joint) and stabilizes the metacarpal during movement of the fingers.
- **Nerve supply:** Posterior interosseous nerve (C7 and 8).

Extensor Pollicis Brevis

Extensor pollicis brevis is the deepest extensor muscle arising from the back of the radius and interosseous membrane.

- **Origin:** Distal part of posterior surface of the radius and adjacent interosseous membrane. Its tendon courses with that of the extensor pollicis longus muscle.
- **Insertion:** Dorsum of the base of the proximal phalanx of the thumb.
- **Action:** Extends the thumb (first phalanx) and abducts the wrist.
- **Nerve supply:** Posterior interosseous nerve (C7 and 8).

Extensor Pollicis Longus

Extensor pollicis longus lies deep in the back of the middle third of the forearm and superficial on the back of the wrist and thumb.

- **Origin:** Middle of the posterior surface of the ulna and adjacent interosseous membrane. Its tendon changes direction around the dorsal tubercle of the radius. Its tendon also runs obliquely across those of both radial extensors of the wrist.

More distally, with the tendons of extensor pollicis brevis and abductro pollicis longus.

Extensor pollicis longus contributes to the boundaries of the "anatomical snuff box" in the base of which, the base of the first metacarpal, trapezium, scaphoid, radial styloid and the radial artery can be palpated.

- **Insertion:** Base of dorsum of the distal phalanx.
- **Action:** Extends the distal metacarpophalangeal joint (and secondarily abducts the wrist).
- **Nerve supply:** Posterior interosseous nerve (C7 and 8).

Extensor Indicis

Extensor indicis lies deep in the lower third of the forearm and superficial on the dorsum of the wrist and hand.

- **Origin:** Small area on the distal part of the posterior surface of the ulna and the adjacent interosseous membrane.
- **Insertion:** Ulnar side of the tendon of the extensor expansion of the index finger.
- **Action:** Extends the metacarpophalangeal joint of the index finger.
- **Nerve supply:** Posterior interosseous nerve (C7 and 8).

3. Deep Arteries

Posterior interosseous artery is a branch of the common interosseous branch of the ulnar artery that reaches the back of the forearm by passing above the interosseous membrane between the radius and the ulna. It appears between supinator and abductor pollicis longus (where it gives off the interosseous recurrent artery) and descends between the superficial and deep groups of extensor muscles.

Interosseous recurrent artery arises just after the posterior interosseous artery enters the back of the forearm. It ascends over supinator and deep to anconeus to the back of the lateral epicondyle, where it anastomoses with the posterior branch of the profunda brachii.

Muscular branches supply the superficial and deep groups of extensor muscles.

Anterior interosseous artery descends on the anterior surface of the inte-rosseous membrane accompanied by the anterior interosseous branch of the median nerve. It pierces the interosseous membrane above the wrist and emerges on the dorsum of the forearm and terminates in the dorsal carpal arch.

4. Deep Nerves

Deep branch of the radial nerve arises beneath brachioradialis and winds laterally around the radius, between the superficial and deep layers of supinator to reach the back of the forearm. It descends between the superficial and deep groups of muscles accompanied by the posterior interosseous artery. From here, it is known as the posterior interosseous nerve. At about the middle of the forearm, it passes deep to the extensor pollicis longus muscle and descends on the interosseous membrane to the wrist. It terminates as an enlargement which gives off twigs to the intercarpal joints.

Muscular branches supply extensor carpi radialis brevis and supinator and the other extensors except brachioradialis and extensor carpi radialis longus.

Chapter 26

Dorsum and Palm of the Hand

Chapter Outline

1. Fascia
2. Deep Arteries
3. Fascial Spaces

The dorsum of the hand extends from the back of the wrist to the tips of the digits.

1. Fascia

SUPERFICIAL FASCIA

Superficial fascia of the dorsum of the hand is arranged in two layers, superficial (thin and fatty) and deep (membranous containing superficial vessels and nerves). It is not anchored to deep underlying fascia therefore deep infection of the hand may be indicated by swelling of the dorsum.

Cutaneous Arteries

A dorsal carpal network is formed on the back of the distal row of carpal bones by the end of the posterior interosseous artery from above and the dorsal carpal branches of the radial and ulnar arteries from the sides.

Cutaneous Veins

The dorsal digital veins, usually two longitudinal channels, are located along the sides of the back of each digit. They anastomose freely across the dorsum and receive veins from the palmar surface of the hand. They join near the level of the metacarpals to form usually three dorsal metacarpal veins that in turn drain into a dorsal venous arch. On the radial side of the hand, the arch and several smaller channels drain into the cephalic vein while the ulnar end of the arch drains into the basilic vein.

The dorsal venous arch (cf dorsal venous arch of the foot) lies above and proximal to the heads of the metacarpals. It receives the dorsal metacarpal veins. The radial and ulnar ends join the beginning of the cephalic and basilic veins respectively. Blood is forced into the arch from the palm by the pressure of gripping.

Cutaneous Nerves

The superficial branch of the radial nerve passes distally leaving the distal third of the forearm beneath the tendon of brachioradialis. It winds around the back of the hand supplying the overlying skin, and finally divides into five digital nerves. They supply the skin on the back of the proximal phalanges of the first three and a half digits.

The dorsal branch of ulnar nerve also arises in the distal half of the front of the forearm. It descends, winding medially around the ulnar side of the forearm. It pierces the deep fascia a little above the wrist and passes down the back of the hand supplying the overlying skin and finally divides into three digital nerves. They supply the skin on the back of the proximal phalanges of the last one and a half digits.

The palmar digital branches of the median and ulnar nerves provide branches that supply the skin on the back of the middle and distal phalanges.

The dorsal nerves do not supply the nail beds or skin over the distal phalanx. The nail beds are supplied by dorsal branches of palmar digital nerves (of the median and ulnar nerves).

The territory supplied by the ulnar nerve may include the entire ring finger and half to the middle finger.

DEEP FASCIA

Deep fascia is very thin on the dorsum of the hand. It is attached medially the fifth metacarpal bone and laterally to the dorsum of the second metacarpal bone. It envelops the extensor tendons and their expansions. The deep fascia of the dorsum may be attached to the dorsal surfaces of the second to fifth metacarpal bones then extending around the borders of the hand to be continuous with fascia over the thenar and hypothenar muscles.

EXTENSOR RETINACULUM

The extensor retinaculum is the thickened, oblique band of deep fascia at the junction of the wrist and forearm. It extends from the anterolateral border of the distal end of the radius to the triquetrum and styloid process of the ulna crossing the wrist joint. It retains the extensor tendons and so prevents them from bowstringing. Septae extend from the deep aspect of the retinaculum to ridges on the back of the radius and to the back of the head of the ulna. Six compartments or osseofibrous tunnels are formed which contain the nine extensor tendons.

From lateral to medial, their location and contents of the compartments are:
1. Lateral surface of radius: Abductor pollicis longus and extensor pollicis brevis.

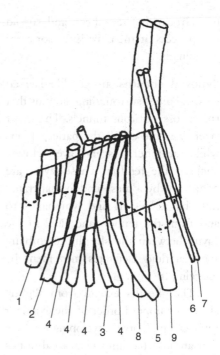

FIG. 26-1 Arrangements of tendons at the right wrist. (1) Extensor carpi ulnaris; (2) Extensor digiti minimi; (3) Extensor indicis; (4) Extensor digitorum; (5) Extensor pollicis longus; (6) Extensor pollicis brevis; (7) Abductor pollicis longus; (8) Extensor carpi radialis brevis; (9) Extensor carpi radialis longus.

2. Lateral wide groove on back of radius: Extensor carpi radialis longus and extensor carpi radialis brevis.
3. Narrow, oblique groove medial to dorsal tubercle of radius: Extensor pollicis longus.
4. Medial, wide groove on back of radius): Extensor digitorum and extensor indicis.
5. Between radius and ulna: Extensor digiti minimi.

6. Groove between head and styloid process of the ulna: Extensor carpi ulnaris.

Synovial sheaths are double layered synovial tubes surrounding tendons that run in osseofibrous tunnels. The outer layer lines the canal (the parietal layer) while the inner (visceral layer) closely encloses the tendon. These layers are connected by a mesotendon that transmits blood vessels and lymphatics to and from the tendon and fluid in the cavity between the sheaths resembling synovial fluid facilitates movement by reducing friction.

The sheaths extend from slightly above the upper border of the extensor retinaculum. Those enveloping tendons that are inserted into metacarpal bones end just short of their insertions. The sheaths investing the tendons that are inserted into phalanges reach only to the middle of the metacarpals.

There are six synovial sheaths for the nine tendons that pass under the extensor retinaculum.

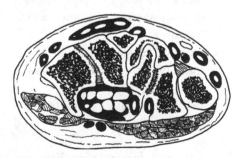

FIG. 26-2 Cross section of wrist at the flexor retinaculum and carpal tunnel.

2. Deep Arteries

These deep arteries course beneath the extensor tendons.

RADIAL ARTERY

At the wrist, the radial artery winds dorsally around the lateral side of the scaphoid and trapezium and passes deep to abductor pollicis longus, extensor pollicis longus and brevis, across the floor of the "anatomical snuff box" (where it can be felt pulsating). At the proximal end of the first intermetacarpal space, it passes between the heads of the first dorsal interosseous muscle to enter the palm.

Dorsal carpal branch arises where the radial artery is crossing the "snuff box". It anastomoses with the corresponding branch from the dorsal carpal branch of the ulnar artery and the terminal part of the anterior and posterior interosseous arteries to form the dorsal carpal arch. From the arch, three dorsal metacarpal arteries descend on the dorsal interossei to the level of the heads of the metacarpal bones where they bifurcate, forming the proper dorsal digital arteries. These arteries descend on the contiguous sides of the digits to supply the proximal phalangeal area. The proper palmar digital arteries supply the greater part of the digits.

The dorsal metacarpal arteries are linked with the palmar arterial arches by two sets of anastomoses. The proximal anastomosis occurs between the superior perforating branches of the dorsal metacarpal arteries and the perforating branches of the deep palmar arch. The distal anastomosis occurs between the inferior perforating branches of the dorsal metacarpal arteries and the perforating branches of the common palmar digital arteries of the superficial palmar arch.

3. Fascial Spaces

Two clefts present on the dorsum of the hand between the layers of fascia.

Dorsal subcutaneous (supratendinous) space extends between the superficial and deep fascia.

Dorsal subaponeurotic space extends between the deep fascia of the dorsum of the hand (and its extensions between the extensor tendons) and the fascia covering the interossei. Its apex is at the wrist and base at the knuckles. The membranes connecting the dorsal expansions to each other and metacarpals two and five limit the space medially and laterally.

Palm of Hand

Chapter Outline

1. Fascia
2. Compartments of the Hand

The front of the hand or palm extends from the distal crease at the wrist to the tips of the digits. The palm transmits long tendons, nerves and vessels of the digits and has special fascial layers and intrinsic muscles important in controlling the fingers and the thumb.

1. Fascia

Infections confined to the fascial spaces of the hand are extremely common.

SUPERFICIAL FASCIA

Superficial fascia is a tough layer containing a moderate amount of fat divided into fine lobules by septa that connect the skin to the deep fascia. These fibro-fatty pads aid in grasping. The layer contains vessels, nerves, and a muscle.

Cutaneous arteries arise from deep arteries.

Cutaneous veins drain into the dorsal venous arch reaching the cephalic and basilic veins.

Cutaneous Nerves

Two nerves supply the skin of the palm both arising in the distal third of the forearm and both cross in front of the flexor retinaculum.

The palmar branch of ulnar nerve descends in the forearm beneath the deep fascia. Near the wrist, it pierces the deep fascia on the lateral side of flexor carpi ulnaris and crosses superficial to the flexor retinaculum to innervate the skin of the ulnar half of the palm.

Just above the flexor retinaculum, the median nerve provides a branch the palmar branch of median nerve that pierces the deep fascia. It passes superficial to the flexor retinaculum to innervate the skin of the radial half of the palm.

Palmaris Brevis

Palmaris brevis comprises a few transverse skeletal muscle fibres in the superficial fascia of the hypothenar eminence.

- **Origin:** Front of flexor retinaculum and ulnar margin of the palmar aponeurosis.
- **Insertion:** Skin on the ulnar border of the hand.
- **Action:** Draws skin on the ulnar side inwards to provide a firmer grasp and deepens the hollow of the palm.
- **Nerve supply:** Superficial branch of the ulnar nerve.

DEEP FASCIA

The antebrachial fascia of the front of the forearm is continuous with the fascia of the palm of the hand. It is thickened at the wrist to form the palmar carpal ligament. The deep fascia is thin over the thenar and hypothenar eminences (enclosing the thenar and hypothenar spaces respectively) and attached along the margins of the first and fifth metacarpal respectively. It is thickened at the wrist to form the flexor retinaculum, in the central part of the palm to form the palmar aponeurosis, and over the fingers to form the fibrous flexor sheaths.

FLEXOR RETINACULUM

Flexor retinaculum (transverse carpal ligament) is a thick band forming a bridge across the wrist. It is attached to the pisiform and hook of the hamate on the ulnar side and to the tubercle of the scaphoid bone and lips of the groove of the trapezium on the radial side. It converts the arch of the carpus into a fibroosseous tunnel, the carpal tunnel, for the passage of the flexor tendons (enclosed in their synovial sheaths) and the median nerve into the palm. The tendon of flexor carpi radialis with its synovial sheath lies in a groove on the front of the trapezium between the attachments of the retinaculum to the trapezium.

Relationships of the Flexor Retinaculum

1. Tendon of flexor carpi radialis within the retinaculum medial to the tubercle of the scaphoid but not within the carpal tunnel.
2. Tendon of palmaris longus passes in front of the proximal part of the retinaculum but is attached to the distal part.
3. Ulnar nerve and vessels pass in front of the main retinaculum but are protected by a superficial band.
4. Median nerve and nine flexor tendons pass deep to the retinaculum through the carpal tunnel.

Palmar aponeurosis is a triangular thickening of the deep fascia between the thenar and hypothenar muscles. Its apex is continuous with the distal edge of the flexor retinaculum and tendon of palmaris longus. There are two layers of fibers in the aponeurosis.

A superficial longitudinal stratum begins at the apex of the aponeurosis as

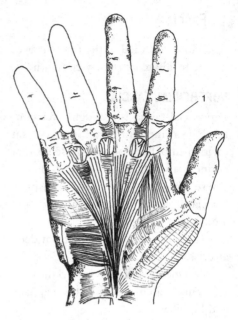

FIG. 26-3   Palmar aponeurosis. (1) Digital a and n.

a continuation of the tendon of palmaris longus and divides near the base into four digital slips superficial to the flexor tendons. The divisions pass deeply on either side of the finger to fuse with the fibrous flexor sheath, capsule of the metacarpophalangeal joint and the proximal phalanx.

A deeper transverse stratum of the palmar aponeurosis is continuous with the thenar and hypothenar fascia at its sides. The transverse fibers that interconnect the diverging digital slips are thickened near the heads of the metacarpals to form the superficial transverse metacarpal ligaments.

Distally, at the root of the fingers, the webs of the fingers are reinforced by transverse fibers known as the transverse fasciculi that are attached to the fibrous flexor sheaths at the bases of the first phalanges. In the short intervals between the superficial transverse metacarpal ligament and the transverse fasciculi, the digital vessels and nerves are visible.

The digital slips provide superficial fibers that become attached to the skin creases of the palm at the base of the digit. The deeper part of each slip is divided into three parts by the flexor tendons that pierce them. The central part contributes to the formation of a fibrous flexor sheath; the marginal processes arch deeply between the heads of the metacarpus to become attached to the deep transverse metacarpal ligament. There is no digital slip to the thumb, but longitudinal fibers extend from the palmar aponeuroses onto the thenar fascia.

Fibrous flexor sheaths are thickenings of the deep fascia of the front of the digit arranged as an elongated arch. Each sheath is attached to the margins of the palmar ligaments of the joints, to the margins of the proximal and middle phalanges, and to the palmar surface of the distal phalanges beyond the insertion of the flexor tendon. As a result, osseoaponeurotic canals for these tendons are formed by the fascia of the fibrous flexor sheaths. Opposite the bodies of the phalanges the sheath is strong and made up of transverse fibers known as annular ligaments to retain the tendons. Opposite

the interphalangeal joints, the sheath is thin, has oblique fibers that permit flexion known as the cruciform ligaments. The outer layer of the synovial sheaths which envelop the tendons serve as the linings of the fibrous sheaths.

SHORT MUSCLES OF THE THUMB

Short muscles of the thumb are arranged in a two superficial and one deep pattern and all the muscles of this group are innervated by the recurrent branch of the median nerve.

Abductor Pollicis Brevis

Abductor pollicis brevis forms the greater part of the lateral side of the thenar eminence.

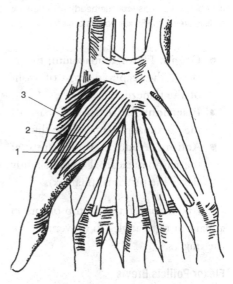

FIG. 26-4 The thenar muscles. (1) Flexor pollicis brevis; (2) Abductor pollicis brevis; (3) Opponens pollicis.

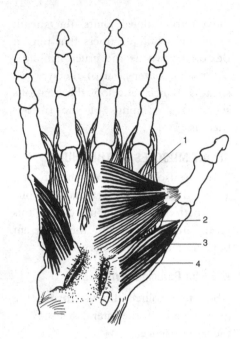

FIG. 26-5 Short muscles of the thumb. (1) Abductor pollicis (transverse head); (2) Abductor pollicis (oblique head); (3) Abductor pollicis brevis; (4) Flexor pollicis brevis.

- **Origin:** Flexor retinaculum, tubercle of the scaphoid, crest of front of the trapezium.
- **Insertion:** Radial side of base of the proximal phalanx of the thumb.
- **Action:** Abducts the thumb at the carpometacarpal joint (slightly flexes the proximal phalanx).

Abduction of the thumb occurs in a plane at right angles to movements of the fingers.

Flexor Pollicis Brevis

Flexor pollicis brevis forms the medial part of the thenar eminence.

- **Origin:** Distal border of the flexor retinaculum and crest of the trapezium.
- **Insertion:** Radial side of base of proximal phalanx of thumb by a tendon containing sesamoid bones.
- **Action:** Flexes the proximal phalanx in its own long axis.

Opponens Pollicis

Opponens pollicis is a deep muscle that extends beneath the preceding two muscles.

- **Origin:** Flexor retinaculum and tubercle of the trapezium.
- **Insertion:** Entire radial border of the first metacarpal bone.
- **Action:** Opposition of thumb i.e. flexion, adduction, and medial rotation so that the tip of the thumb can meet each of the tips of the four fingers.

Opponens plays an important role in the "precision" grip that involves the radial two fingers and thumb.

SHORT MUSCLES OF THE LITTLE FINGER

Short muscles of the little finger are also arranged in a two superficial and one deep pattern. All are innervated by the deep branch of the ulnar nerve.

Abductor Digiti Minimi

Abductor digiti minimi is the most lateral muscle of the group.

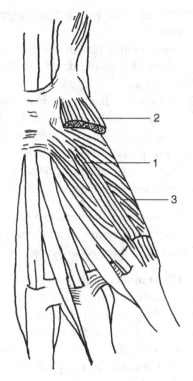

FIG. 26-6   The hypothenar muscles. (1) Flexor digiti minimi brevis; (2) Abductor digiti minimi brevis; (3) Opponens digiti minimi brevis.

- **Origin:** Pisiform bone (may be continuous with flexor carpi ulnaris).
- **Insertion:** Ulnar (medial) side of base of the proximal phalanx of the little finger.
- **Action:** Abducts the little finger at the metacarpophalangeal joint.

Flexor Digiti Minimi Brevis

Flexor digiti minimi brevis extends on the radial side of the hypothenar eminence.

- **Origin:** Flexor retinaculum and hook of the hamate.
- **Insertion:** Tendon fuses and inserts with the abductor muscle.
- **Action:** Flexes the proximal phalanx.

Opponens Digiti Minimi

- **Origin:** Flexor retinaculum and hook of the hamate.
- **Insertion:** Entire ulnar margin of the fifth metacarpal.
- **Action:** Opposition; flexion; adduction; and rotation of the fifth metacarpal.

All of the hypothenar muscles are supplied by the ulnar nerve (T1).

2. Compartments of the Hand

Septae extend from the margins of the palmar aponeurosis.

The thenar (lateral fascial) septum extends from the lateral side of the palmar aponeurosis and attaches to the palmar aspect of the first metacarpal bone and with the intermediate septum, bounds the thenar space.

The hypothenar (medial fascial) septum extends down to attach to the radial side of the fifth metacarpal and with the intermediate septum, bounds the midpalmar space.

An intermediate fascial septum extends from the palmar aponeurosis to the third metacarpal subdividing the central compartment into a midpalmar space medially and thenar space laterally.

These spaces are commonly sites of rapid spread of infection from penetrating wounds. Infections can spread proximally alongside flexor tendons, through the carpal tunnel and into the forearm. Distally, infections can spread alongside the lumbrical tendons into the web spaces into which the compartments open.

Superficial palmar branch of radial artery supplies the thenar muscles. It arises from the radial artery in the forearm, continues in the line of the radial artery across the retinaculum and into the thenar eminence. It curves around to the back of the hand, courses deep to the abductor pollicis brevis muscle, and often helps complete the superficial palmar arch.

Recurrent branch of median nerve arises as the median nerve emerges from under the flexor retinaculum. It provides a branch that courses directly under the deep fascia and curves radially to supply the muscles of the thenar compartment (except the adductor pollicis).

The motor nerve enters the muscles of both eminences between the two superficial muscles (but subsequent courses differ greatly).

Ulnar nerve (C8, T1 medial cord) crosses the wrist superficial to the flexor retinaculum and passes to the base of the hypothenar eminence, where it divides into superficial and deep branches.

The superficial branch passes beneath (innervates) palmaris brevis and divides into two palmar digital branches:

(1) proper digital branch that descends along the ulnar side of the little finger and,

(2) common digital branch that descends to the web of the fourth interdigital space and divides into two proper digital nerves that supply the contiguous sides of the ring and little fingers.

The superficial branch also sends a communicating branch to a palmar digital branch of the median nerve.

Ulnar artery passes superficial to the flexor retinaculum on the radial side of the pisiform (where it may be covered by a slip of the retinaculum). At the base of the hypothenar eminence, it divides into superficial and deep branches.

The superficial branch gives a proper palmar digital branch that runs along the ulnar side of the little finger. The ulnar artery curves laterally (a finger's breadth more distal than the deep palmar arch) to form the superficial palmar arch.

The superficial arch is variable in form and may be absent.

Branches of the superficial arch are:

Common palmar digital arteries (3) pass toward the medial three clefts. Each is joined by a palmar metacarpal artery from the deep arch and a distal perforating branch of a dorsal metacarpal artery. They bifurcate into proper palmar digital arteries supplying adjacent sides of the fingers.

The deep branch of the ulnar artery sinks with the ulnar nerve between the origins of abductor minimi and flexor digiti minimi brevis and perforates the opponens digiti minimi. It supplies the muscles then crosses into the central compartment lying on interosseous muscles. It joins the radial artery to form the deep palmar arch.

Proper plamar digital artery to the ulnar side of the little finger.

THENAR AND MIDPALMAR SPACES (THE CENTRAL SPACE)

The central space is bounded by:
Anteriorly: The deep surface of the palmar aponeurosis.
Laterally: Fascia covering the thenar and hypothenar muscles.
Posteriorly: Deep pad of fat lying on interosseous fascia and fascia over adductor pollicis muscle.

Superficial palmar arch is formed by the distal curved terminal portion of the ulnar artery and the superficial palmar branch of the radial artery (or occasionally the radialis indicis or princeps pollicis branches of the radial artery). This arch lies deep to the palmar aponeurosis between the two transverse skin creases and is accompanied by venae comitantes. It provides branches to the medial three and a half digits.

Common palmar digital arteries arise from the arch and descend on the lumbrical muscles, between the flexor tendons, to the webs of the fingers. Each artery anastomoses with a palmar metacarpal artery of the deep palmar arch and a distal perforating branch from a dorsal metacarpal artery. The short trunk thus formed divide into two proper palmar digital arteries that supply the contiguous sides of the fingers. These arteries supply the skin, tendons and joints of the digits, and the dorsum of the middle and distal phalanges.

The **median nerve** (C5–C8) passes through the carpal tunnel to enter the palm lateral to the tendon of palmaris longus. It branches under cover of the palmar aponeuosis into lateral and medial divisions.

The lateral division gives the important muscular (recurrent) branch (that supplies the thenar muscles) and then three palmar digital nerves for both sides of the thumb and lateral aspect of the index finger.

The medial division breaks up into two common palmar digital nerves that descend in front of the second and third lumbricals then divide into proper palmar digital nerves to supply the adjacent sides of the second and third clefts.

The five muscles supplied by the median nerve are usually the three thenar muscles and two lateral lumbricals.

The five palmar digital branches of the median nerve supply the dorsal side of the second and third digits and a variable portion of the fourth digit distal to the proximal interphalangeal joints.

The palmar cutaneous branch supplies the radial side of the palm and the palmar aspect of the thumb.

FLEXOR TENDONS, THEIR SHEATHS, AND THE LUMBRICAL MUSCLES
Tendons of Flexor Digitorum Superficialis

Four tendons pass deep to the flexor retinaculum and into the palm. In the proximal carpal tunnel, they lie in anterior and posterior pairs but distally they lie anterior to the corresponding deep tendons. There, each tendon runs towards its finger, where it enters a fibrous flexor sheath. Each tendon divides above the base of the proximal phalanx into medial and lateral slips which pass deep to the

subjacent deep tendon, twist through 180° and reunite behind the tendon opposite the base of the middle phalanx to allow passage for its corresponding profundus tendon.

Tendons of the Flexor Digitorum Profundus

In the wrist, they lie between the flexor superficialis tendons and the carpus. In the palm, the diverging tendons give origin to the lumbrical muscles and each runs towards its finger, where it enters its fibrous flexor sheath. Above the proximal phalanx, the profundus tendon

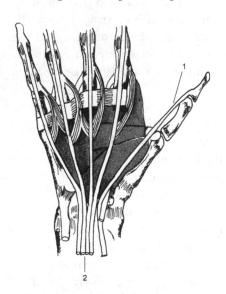

FIG. 26-7 Long tendons to the hand and lumbrical muscles. (1) Flexor pollicis longus; (2) Tendons of flexor digitorum profundus.

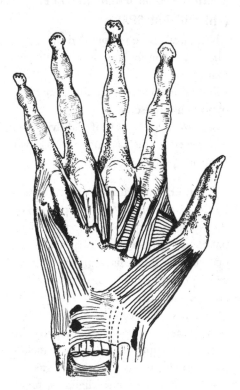

FIG. 26-8 Synovial tendon sheaths of the palm.

passes through its corresponding superficialis tendon to insert in the base of the distal phalanx.

The surgical significance of the arrangement of flexor digitorum profundus traversing the tubular canal formed by flexor digitorum superficialis is that the profundus lacks protection of superficialis throughout the length of the finger. A deep cut on the palmar surface of a finger may primarily effect profundus.

Synovial sheath of the flexor pollicis longus (the radial bursa) encloses the tendon and extends from just proximal to the flexor retinaculum to the insertion into the base of the terminal phalanx of the thumb. There is commonly a small connection between the radial with the common synovial sheath of the flexor of superficialis and profundus (ulnar bursa) permitting spread of infection to or from the thumb to little finger.

Synovial sheath of the flexor carpi radialis encloses the tendon of flexor carpi radialis en route to the base on the second (and third) metacarpal splits the lateral attachment of the flexor retinaculum as it lies in the groove on the trapezium. The split layers are attached to the lips of the groove. A synovial sheath extends from proximal to the flexor retinaculum, invests the tendon and encloses the tendon to its insertion.

Common flexor sheath (ulnar bursa) extends from above the flexor retinaculum to the middle of the palm, envelopes the flexor superficialis and profundus tendons arranged in two rows. Its most

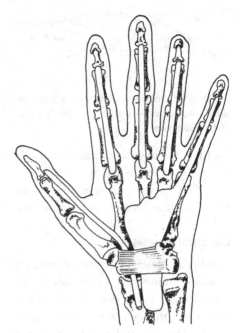

FIG. 26-9   Synovial tendon sheaths of the palm.

medial part of the sheath is prolonged to the insertion of the profundus tendon of the fifth digit at its distal phalanx. The main part of the sheath envelops both the flexor tendons and the proximal part of the lumbrical muscles.

Where tendons are retained in osseofibrous tunnels, friction is minimized by two layered synovial sheaths containing synovial fluid. Individual digital sheaths extend from the bases of the metacarpals, and terminate at the insertion of the tendons in the bases of the distal phalanges.

Within the fibrous flexor sheaths of each digit, vinculae, or vascular folds (mesotendons) represent duplications between the outer and inner synovial

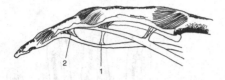

FIG. 26-10 Arrangement of superficial and deep flexor tendons within a digit. (1) Vincula longa; (2) Vincula brevia.

sheaths. A triangular vinculum breve is located near the insertion of each tendon. The vincula longa extend between the tendons and the proximal phalanges.

Lumbrical Muscles

Lumbrical muscles are four cylindrical muscles numbered from lateral to medial and join the deep flexors to the extensors (through the dorsal expansions). The first and second arise by a single head innervated by the digital branches of the median nerve and the third and fourth arise from two heads innervated by the deep branch of the ulnar nerve.

- **Origin:** Tendons of the flexor digitorum profundus muscle.

   First and second: lateral side of the tendons to the index and middle fingers.

   Third and fourth: contiguous sides of the tendons for the middle ring, and little fingers.

   Each muscle passes in front of the deep transverse metacarpal ligament and then fans out.

- **Insertion:** Lateral border of the extensor expansions at the level of the metacarpophalangeal joint.

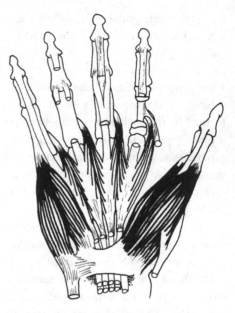

FIG. 26-11 Thenar, hypothenar and lumbrical muscles.

- **Action:** Flex the proximal phalanges and extension of the middle and distal phalanges i.e. flexion of the extended finger.
- **Nerve supply:**
   Two lateral lumbricals are innervated by digital branches of the median nerve. Two medial lumbrical muscles by the deep branch of the ulnar nerve.

   This innervation reflects the innervation of the muscle to whose tendon the lumbrical is attached.

DEEP PALMAR SPACES

Deep palmar spaces are the fascial clefts, or potential spaces in the palm, should not be confused with the fascial compartments.

Thenar Space

Thenar space extends from the thenar eminence to the third metacarpal and from the flexor retinaculum to 2.5 cm proximal to the webs. Its boundaries are:

- *Anterior*: First lumbrical and flexor tendons of index finger.
- *Posterior*: Transverse head of adductor pollicis muscle.
- *Lateral*: Flexor pollicis longus tendon and sheath.
- *Medial*: Posterior half of fibrous septum extending from palmar aponeurosis to third metacarpal.

Midpalmar Space

Midpalmar space extends from the third metacarpal to the hypothenar eminence and from the flexor retinaculum to 2.5 cm proximal to the webs. Its boundaries are:

- *Anterior*: Second, third, and fourth lumbrical muscles and flexor tendons of medial three digits.
- *Posterior*: Medial two and a half metacarpals and enclosed interosseous muscles.
- *Lateral*: Posterior half of fibrous septum extending from the palmar aponeurosis to third metacarpal.
- *Medial*: Hypothenar septum.

Radial artery enters the palm from the back of the hand passing across the "anatomical snuff box" then between the heads of origin of the first dorsal interosseous muscle at the base of the first intermetacarpal space. It curves medially between the heads of adductor pollicis and gives off two branches.

Princeps pollicis artery descends along the medial side of the first metacarpal then divides into two palmar digital arteries that descend on both sides of the thumb.

Radialis indicis artery between the transverse head of the adductor pollicis muscle and the first dorsal interosseous muscle, this artery runs along the lateral side of the index finger (a palmar digital artery).

Deep palmar arch is formed between the terminal part of the radial artery and the deep branch of the ulnar artery. It lies on the interossei deep to the flexor tendons curving medially between the transverse and oblique heads of adductor pollicis then across the third and fourth metacarpals and interossei, to the base of the fifth metacarpal, where it joins the deep palmar branch of the ulnar artery.

Branches

Palmar metacarpal arteries (3) arise from the convexity of the arch and descend on the palmar interosseous fascia and supply the interossei and lumbricals. Near the webs of the fingers, they join corresponding common palmar digital arteries from the superficial palmar arch.

Recurrent carpal branches are small branches that ascend to the wrist where they supply the intercarpal joints and terminate in the palmar carpal network.

Perforating branches pass backwards between the heads of the metacarpals between the heads of origin of the dorsal interossei to join the dorsal metacarpal arteries (from the arterial arch on the dorsum of the hand).

Deep palmar branch of the ulnar artery descends on the medial side of the hook of the hamate and descends between abductor and flexor digiti minimi. It may course through opponens, superficial to it or deep to it. It supplies adjacent muscles, the third and fourth lumbricals, and interossei.

Deep branch of the ulnar nerve arises at the root of the hypothenar eminence. The ulnar nerve branches supply the hypothenar muscles and then runs across the palm with the deep palmar arch, to supply the third and fourth lumbricals and all the interossei. It terminates by supplying the adductor pollicis muscle.

DEEP MUSCLES

Adductor pollicis and the interossei are the deepest structures in the palm. The interossei and metacarpals are enclosed between the dorsal and palmar interosseous fasciae.

Adductor Pollicis

Adductor pollicis arises by two heads.
- **Origin:**
  Oblique head: Capitate and trapezoid and front of the bases of the second and third metacarpals.

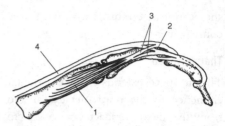

FIG. 26-12 Extensor aponeurosis. (1) Dorsal interosseous m; (2) Lateral extensor band; (3) Interosseous bands; (4) Extensor digitorum tendon.

Transverse head: Palmar surface of the third metacarpal.
- **Insertion:** medial aspect of the base of the proximal phalanx of the thumb by a common tendon (with the first palmar interosseous) containing a (medial) sesamoid bone.
- **Action:** Adducts the thumb (and aids in opposition).
- **Nerve supply:** Deep branch of the ulnar nerve.

Interossei

There are four dorsal and four palmar interossei in the hand located mostly between the metacarpals. Except for the first dorsal and palmar interossei, their tendons pass behind the deep transverse ligament and insert mainly into the extensor expansion.

All of the interossei are innervated by the deep branch of the ulnar nerve.

Dorsal Interossei

There are four bipennate muscles.
- **Origin:** By two heads from adjacent sides of two metacarpals.

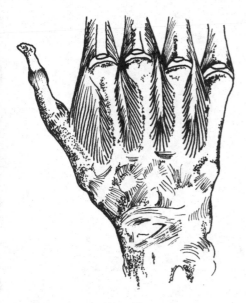

FIG. 26-13 Dorsal interosseous muscles.

■ **Insertion:** Lateral (abduction) sides of second and third digits (first and second) and medial sides of third and fourth digits (third and fourth) and the extensor expansions.

■ **Action:** Abducts extended fingers from the longitudinal axis of the middle finger and flexion of the metacarpophalangeal joints. (mnemonic-DAB-dorsal abduct).

Palmar Interossei

These are four unipennate muscles.

■ **Origin:** Palmar surfaces of first, second, third, and fifth metacarpal bones.

■ **Insertion:** Ulnar (adduction) sides of the proximal phalanges of the first and second digits and radial side of the proximal phalanges of the fourth and fifth digits and the extensor expansions.

■ **Action:** Adduct fingers toward the longitudinal axis of the third finger. In addition, they flex metacarpophalangeal joints and extend the interphalangeal joints (mnemonic-PAD-palmar adduct).

The **layers of the palm** in order are:

(1) Palmar fascia
(2) Superficial arterial arch and median nerve
(3) Superficial and deep flexor tendons and lumbricals
(4) Adductor pollicis, deep arterial arch and ulnar nerve
(5) Interossei

Chapter 27

Bones of Upper Limb

Chapter Outline

1. Shoulder Girdle
2. Arm
3. Forearm
4. Hand

The upper limb is adapted for prehension and working. Stability is sacrificed for mobility and the skeleton of the limb is loosely attached to the trunk at a single small joint, the sternoclavicular joint. The pectoral or shoulder girdle comprises the clavicle and scapula. By moving together, the girdle essentially enhances the mobility of the shoulder joint on the thorax.

1. Shoulder Girdle

CLAVICLE

The clavicle or collar bone, forms the anterior part of the pectoral girdle. It articulates medially with the manubrium of the sternum and first costal cartilage (an articular disc intervening) laterally with the acromion of the scapula. It is unique in lacking a medullary cavity.

The shaft is S-shaped with its medial two thirds convex anteriorly and its lateral third convex posteriorly.

The sternal (medial) end is thick and rounded. It has a saddle-shaped facet that continues inferiorly to provide an articular surface for the cartilage of the first rib.

The acromial (lateral) end is broad and flattened with a sloping acromial articular facet facing inferolaterally.

The impression for the costoclavicular ligament is an oval impression on the inferior surface near the sternal end

FIG. 27-1 Clavicle — superior surface. (1) Sternal end; (2) Acromial end; (3) Shaft.

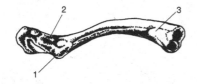

FIG. 27-2 Clavicle — inferior surface. (1) Conoid tubercle; (2) Trapezoid ridge; (3) Coracoclavicular impression.

of the clavicle. The costoclavicular ligament runs from the costal cartilage and adjacent body of the first rib to this impression resisting lateral displacement of the clavicle.

Conoid tubercle is a projection at the posterior edge of the lateral third of the underside of the clavicle. It attaches the conoid ligament that runs to the coracoid process and prevents forward displacement of the scapula at the acromioclavicular joint.

Trapezoid line is a ridge on the inferior surface running anterolaterally from the conoid tubercle. This gives attachment to the two parts of the coracoclavicular ligament (the conoid and trapezoid ligaments).

SCAPULA

The scapula or shoulder blade forms the posterior part of the shoulder girdle. It has a thin, flat, triangular body, coracoid process and spine. The glenoid cavity placed at the lateral angle articulates with the head of the humerus.

The body of the scapula has two surfaces: costal surface and dorsal surface. Costal surface forms a broad, slightly concave, subscapular fossa that contains several ridges and is occupied by subscapularis muscle.

Dorsal surface is unequally divided by the spine into the smaller supraspinous fossa above the spine to which supraspinatus muscle is attached and the larger infraspinous fossa below the spine which gives attachment to the medial three quarters of infarspinatus muscle.

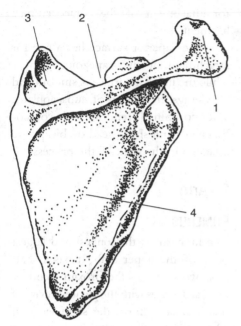

FIG. 27-3 Scapula — dorsal surface. (1) Acromion; (2) Scapular notch; (3) Supraspinous fossa; (4) Infraspinous fossa.

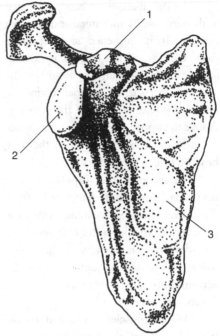

FIG. 27-4 Scapula — costal surface. (1) Coracoid process; (2) Glenoid fossa; (3) Subscapular fossa.

Borders of the Scapula

Superior border from the superior angle to the root of the coracoid process. The scapular notch is on the lateral end of the border medial to the root of the coracoid process. It is bridged in vivo by the superior transverse scapular ligament and traversed by the suprascapular nerve.

Medial border extends from the superior to the inferior angle. It parallels the vertebral column and is divided into three parts by the root of the spine. Levator scapulae is attached above the root of the spine, rhomboid minor opposite the root of the spine and rhomboid major below the root of the spine.

Lateral border the thickest border, extends from the lowest part of the glenoid cavity to the inferior angle. It gives attachment to the teres minor and major. The infraglenoid tubercle is the rough impression at its upper end for attachment of the long head of triceps muscle.

Angles of the Scapula

The superior angle is at the junction of the superior and medial borders.

The inferior angle is at the junction of the medial and lateral borders. It gives attachment to a small part of latissimus dorsi.

The lateral angle (or head) is located at the junction of the lateral and superior borders. It is thickened and separated from the rest of the scapula by a constricted neck. Laterally, the head has the shallow, pear-shaped glenoid cavity for articulation with the head of the humerus. Above the upper apex of the cavity is the supraglenoid tubercle to which the long head of biceps is attached.

The spine of the scapula is a narrow ridge projecting backwards from the dorsal surface. It extends from the neck to the medial border. It separates the supraspinous from the infraspinous fossa and extends laterally into the acromion. Its posterior surface, or crest, is subcutaneous.

Deltoid muscle is attached to the lateral border of the acromion and the entire lower lip of the spine and trapezius is attached to the medial border of the acromion and upper lip of the spine.

Acromion is the bony ledge that projects laterally from the crest of the spine to overhang the shoulder joint. Its medial border has an ovoid articular facet for articulation with the clavicle.

The coracoid process extends upwards, forwards and laterally from the upper aspect of the head. It has two borders:

The medial border is continous with the superior border of the body (beginning at the suprascapular notch) which gives attachment to the pectoralis minor.

The lateral border, beginning at the upper edge of the glenoid cavity, is rough for attachment of the coracoacromial ligament.

The superior surface has a line for attachment of the trapezoid ligament while the inferior surface is smooth and overhangs the tendon of subscapularis. The combined tendons of coracobrachialis and short head of biceps are attached to the apex of the process.

2. Arm

HUMERUS

The humerus is the longest and largest bone of the upper limb. It has a shaft and two ends — the proximal end or head articulates with the glenoid cavity of the scapula to form the shoulder joint while the distal end articulates through the trochlea with the trochlear notch of the ulna and the capitulum with the radial fossa of the radius to form the elbow joint.

PROXIMAL END

The head is large, convex, smooth. It articulates with the glenoid cavity of the scapula.

The **anatomical neck** is marked by the ridge around the margin of the head. It is the site of the epiphyseal plate. The capsule of the shoulder joint is attached to the anatomical neck except medially where it is attached about two centimeters below the articular edge.

The constriction of the humerus below the head and tuberosities is the surgical neck of humerus. It is a common site of fractures.

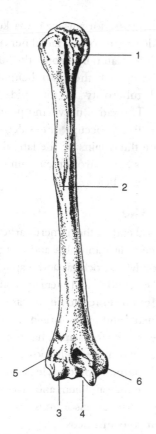

FIG. 27-5 Right humerus — anterior aspect. (1) Interubercular groove; (2) Deltoid tubersity; (3) Capitulum; (4) Trochlea; (5) Radial fossa; (6) Coronoid fossa.

The greater tubercle is a prominent blunt projection on the lateral side of the head. On its upper surface, it presents three facets (for muscular attachments — from front to back, supraspinus, infraspinus and teres minor).

The lesser tubercle projects anteriorly on the medial side from the head of the humerus. It gives attachment to the subscapularis tendon.

The intertubercular groove is located between the greater and lesser tubercles and is extended down onto the shaft. The greater tubercle continues downward on the shaft and forms the lateral lip of the groove. It gives attachment to pectoralis major. The lesser tubercle continues downward on the shaft and forms the medial lip of the groove. It gives attachment to teres major. The groove contains the tendon of long head of biceps and the attachment of latissimus dorsi.

SHAFT

In cross section, the shaft is cylindrical above and triangular below.

Its medial border extends from the lesser tubercle to the medial epicondyle. Near the middle of the shaft it has a rough area for attachment of coracobrachialis. Distally, it forms part of the medial supracondylar ridge which is part of the common origin for long flexors of the hand.

Its lateral border extends from the back of the greater tubercle to the lateral epicondyle. Proximally, it is indistinct, in the middle it is interrupted by the sulcus for the radial nerve. Distally, it continues into the lateral supracondylar ridge that is part of the common extensor origin for long extensors of the wrist and hand.

Its anterior border extends from the greater tubercle to the lateral lip of the trochlea. Proximally it is part of the crest of the greater tubercle. In the middle it forms the anterior margin of the deltoid tuberosity. Below the middle, it is

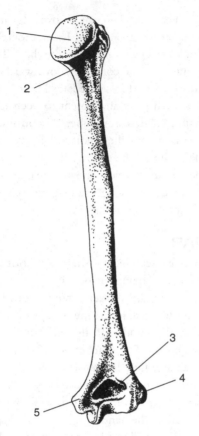

FIG. 27-6 Right humerus — posterior aspect. (1) Head; (2) Neck; (3) Olecranon fossa; (4) Lateral epicondyle; (5) Medical epiondyle.

The posterior surface is marked by the wide sulcus for the radial nerve that runs inferolaterally from the upper medial third of the shaft behind the deltoid tuberosity to the mid-lateral border. The radial nerve and profunda brachii artery occupy the sulcus and separate the origins of the lateral head of triceps above from the origin of the medial head below.

DISTAL END

The distal end of the humerus articulates with the radius and ulna and comprises the condyle (trochlea and capitulum), medial epicondyle and lateral epicondyles.

Medial epicondyle is large and prominent and gives origin to flexor muscles of the forearm. It has the sulcus for the ulnar nerve on its posterior side.

The **lateral epicondyle** is small and lies above the capitulum and gives origin to supinator, extensor muscles of the forearm and anconeus.

The **capitulum** is a round eminence on the anterior part of the articular surface. It articulates with the head of the radius. Above it lies the radial fossa that receives the head of the radius in flexion of the elbow.

The **trochlea** is the spool-shaped medial part of the articular surface articulates with the ulna. Its medial lip is more prominent, but the lateral lip separates the trochlea from the capitulum.

The coronoid fossa lies above the trochlea in the front and receives the coronoid process of the ulna in flexion of the elbow.

ill-defined, and distally it is part of the ridge between the radial and coronoid fossae.

The anteromedial surface forms the floor of the intertubercular sulcus proximally. In the middle it is rough; distally it is smooth.

The anterolateral surface is smooth proximally. The large, triangular, rough deltoid tuberosity for insertion of the deltoid muscle is located near its middle.

The olecranon fossa triangular in shape, lies above the trochlea posteriorly and receives the olecranon process of the ulna in extension of the elbow.

3. Forearm

ULNA

The ulna is the medial and longer bone of the forearm. It articulates with the radius and humerus as follows:

1. The radial notch of the ulna articulates with the head of the radius (to form the superior radio-ulnar joint).
2. The head of the ulna articulates with the ulnar notch of the radius (to form part of the inferior radio-ulnar joint) and
3. The trochlear notch of the ulna articulates with the trochlea of the humerus (to form part of the elbow joint).

The head of the ulna is covered with an articular disc and therefore does not form part of the elbow joint.

Proximal End

The **olecranon process** is the projection from the shaft of the ulna on the back of the elbow. Its anterior surface forms part of the trochlear notch. Its posterior (subcutaneous) surface is smooth and triangular and is covered by a bursa. Its superior surface is rough behind for insertion of triceps.

The **coronoid process** is the triangular projection from the front of the ulna has a smooth superior surface and forms the lower part of the trochlear notch. The anterior surface is rough and presents the tuberosity of the ulna (where it meets the shaft) for insertion of brachialis muscle.

The radial notch is a depression on the lateral surface of the coronoid process for articulation with the head of the radius.

The trochlear notch is the wide concavity divided by a longitudinal ridge that is formed partly by the olecranon process and partly by the coronoid process. It articulates with the trochlea of the humerus.

Shaft

The shaft tapers gradually from above downward. It has three borders and three surfaces.

Anterior border extends from the coronoid process above to the styloid process below. Its lower fourth is part of the pronator ridge.

The interosseous border is prominent in the middle and indistinct distally. From both margins of the radial notch extend ridges, that enclose the triangular supinator fossa. The interosseous border commences from the apex of this fossa.

The posterior (subcutaneous) border extends from the back of the olecranon to the styloid process. It separates the flexor group of muscles from the extensor muscles of the forearm.

The anterior surface presents upper three-fourths which is concave for the attachment of flexor digitorum profundus. A nutrient foramen is present on its proximal part directed toward the elbow.

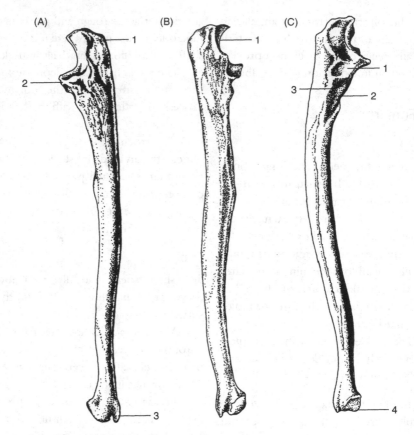

FIG. 27-7 Ulna (A) Medical aspect. (1) Olecranon; (2) Coronoid process; (3) Styloid process. (B) Posterior aspect. (1) Trochlear notch. (C) Lateral aspect. (1) Radial notch; (2) Supinator fossa; (3) Supinator crest; (4) Head.

The medial surface is smooth and subcutaneous distally. It gives attachment to flexor digitorum profundus proximally.

The posterior surface has an impression for anconeus on its upper third. Below it is divided by a vertical ridge into a narrow medial area and a lateral area. Two oblique ridges divide the lateral surface into impressions for attachements of abductor pollicis longus above, extensor pollicis longus in the middle and extensor indicis below.

Distal End

Also called the head, is small and rounded and articulates with the ulnar notch of the radius.

The head has a semilunar articular surface which is in contact with an articular disc that separates the ulna from the carpal bones.

Styloid process is short and blunt. It projects from the posteromedial end of the shaft. It has a pit at its base for attachment of the disc and the medial ligament of the wrist joint is attached to its apex.

RADIUS

The radius is the lateral of the two bones of the forearm. It has a small upper end, a shaft and an expanded lower end.

Proximal End

The proxmal end articulates with the capitulum of the humerus and the radial notch of the ulna.

Head: A disk the upper surface of which has a concave fovea.

Neck: The constricted portion of the radius immediately below the head is related to the annular ligament.

Tuberosity of the radius. On the medial side below the neck provides attachment for biceps brachii.

Shaft

The shaft has three borders and three surfaces.

The anterior border (anterior oblique line) begins below the radial tuberosity. It curves downward and laterally giving attachment to the radial head of flexor digitorum superficialis. It terminates in a tubercle at the anterior part of the base of the styloid process.

The interosseous borde starts below the radial tuberosity and divides distally into two parts to enclose the ulnar notch. The interosseous membrane is attached to this border.

The posterior border (posterior oblique line) starts from the back of the neck and extends to the posterior part of the base of the styloid process. It is indistinct except in its middle third.

The anterior surface in its proximal part, has a concavity for attachment of flexor pollicis longus and in the lower third for pronator quadratus.

The lateral surface is characterized by the rough oval impression in the middle for attachment of pronator teres.

The posterior surface is convex and below the oblique line, has a concavity for the attachment of abductor pollicis longus. Below is a narrow impression for extensor pollicis brevis.

Distal End

The distal end of the radius is quadrilateral in shape. It has an artucular surface articulating with the lunate.

The medial surface presents a narrow concave articular surface, the ulnar notch that articulates with the head of the ulna.

The lateral surface is narrow and prolonged downward as the styloid process.

The anterior surface is concave. It ends in a rough forward-projecting ridge for attachment of the anterior ligament of the wrist joint.

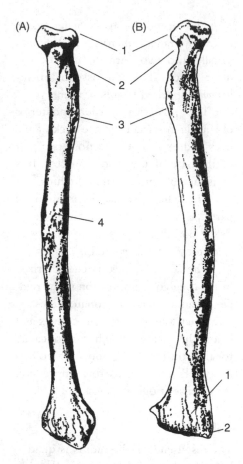

FIG. 27-8 Radius. (A) Lateral aspect. (1) Head; (2) Neck; (3) Tuberosity; (B) Dorsal aspect. (1) Dorsal tubercle; (2) Styloid process.

The posterior surface is convex. It is characterized by a prominent, subcutaneous, dorsal tubercle near the middle and by four shallow grooves for tendons of extensor digitorum and extensor indicis, extensor pollicis longus (just medial to the dorsal tubercle) and two poorly delineated grooves for extensor carpi radialis brevis and longus lateral to the tubercle.

The dorsal tubercle acts as a pulley for extensor pollicis longus.

The carpal articular surface is the concave articular surface that is divided by a slight anteroposterior ridge into a medial quadrilateral part (for articulation with the lunate) and a triangular lateral part (for articulation with the scaphoid).

4. Hand

The skeleton of the hand comprises carpal bones (or bones of the wrist or carpus), metacarpal bones (or bones of the palm) and phalanges (or bones of the digits). The digits are numbered I to V with the first digit also known as the thumb or pollex; the second digit is the index finger; third digit, the middle finger; the fourth finger, the ring finger and the fifth finger, the little finger.

CARPUS

Carpus consists of **eight** bones, arranged in two rows and named from lateral to medial and in relation to their appearance:

Proximal Row

Scaphoid (Gk) scooped out, lunate (L) moon shaped, triquetral (L) three cornered and pisiform (L) pea shaped.

Distal Row

Trapezium (Gk) quadrilateral, trapezoid (Gk) quadrilateral shaped (two sides parallel), capitate (L) head shaped or having a head, and hamate (L) possesing a hook.

All of the carpal bones (except the pisiform) are cuboid, having six surfaces, of which only the dorsal and palmar are nonarticular and give attachment to ligaments. The proximal surfaces are generally convex and the distal surfaces concave.

Features of the Carpus

1. The pisiform articulates only with one bone, the triquetral.
2. The combined articular surfaces of the remaining proximal carpal bones form an ovoid joint capable of particularly free movement.
3. The capitate is the largest centrally placed bone. Its head extends between the proximal row of bones and its distal end articulaes with three metacarpal bones.
4. The mid carpal joint between proximal and distal rows of carpal bones is sinuous allowing flexion and extension but limiting abduction and adduction.
5. The trapezium has a saddle shaped joint with the first metacarpal permitting great mobility of the thumb.
6. The second metacarpal is wedged between the trapezium, trapezoid and capitate. Adduction and abduction occur relative to a relatively fixed axis through the middle finger.
7. The scaphoid and trapezium can be palpated in the "anatomical snuff box" — the hollow between the styloid process of the radius and metacarpal of the thumb and bounded laterally by the tendons of extensor pollicis brevis and abductor

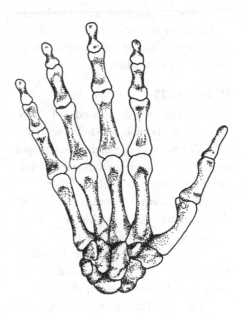

FIG. 27-9 Bones of the hand — anterior aspect.

pollicis and medially by the tendon of extensor pollicis longus.
8. The hamate articulates with the base of metacarpals four and five permitting some flexion.

The anterior projections of three carpal bones and a role of a fourth:

The tubercle of the scaphoid projects anterolaterally from the lateral part of the distal end of the scaphoid.

The tubercle of the trapezium is a vertical ridge about a centimeter lateral to its medial edge.

The pisiform is a sesamoid bone articulating with the front of the triquetral.

The hook of the hamate projects anteromedially from the distal end of the hamate.

These projections give attachment to the flexor retinaculum with which the anterior surface of the carpal bones forms the carpal tunnel.

METACARPUS

Metacarpus is composed of **five** metacarpal bones set between four carpal bones and the digits. Metacarpal bones are numbered from one to five, from the thumb to the little finger. The fourth metacarpal articulates with both the capitate and the hamate. Each metacarpal has a body and two ends.

The base is cuboid, with an articular facets for a carpal bone and on the sides for adjacent metacarpals.

The dorsal and palmar surfaces are non articular and rough for the attachment of ligaments.

The shaft is narrow, convex above, and concave below.

The head, or distal end, articulates with the base of the proximal phalanx. Its sides are flattened with tubercles for the attachment of collateral ligaments of the metacarpophalangeal joint. The palmar surface is notched for the long flexor tendons.

The metacarpal of the thumb is short, stout and rounded. It has an anterolateral surface which gives attachment to opponens pollicis and an anteromedial surface giving attachment to the first dorsal interosseous muscle. Its base is saddle (sellar) shaped articulating with the trapezium and it does not articulate with the adjacent metacarpal, factors contributing to the mobility of the thumb.

PHALANGES

Each digit has **three** phalanges, proximal or first, middle or second and distal or third. The first digit has only two phalanges. Each phalanx has a base that has a cuplike facet, a short shaft, forming the floor of osseofibrous tunnels for the long digital flexors and a convex head (except for the distal phalanx which bears the phalangeal tuberosity).

Chapter **28**

Joints of Upper Limb

Chapter Outline

Joints of the upper extremity are classified into six groups — clavicular joints (sternoclavicular and acromioclavicular), the shoulder joint, elbow joint, radioulnar joints, the wrist joint and joints of the hand.

1. Clavicular Joints

STERNOCLAVICULAR JOINT

Type

A synovial saddle, di-arthrodial joint with fibrocartilaginous surfaces.

Articulating elements

Sternal end of the clavicle articulates with the clavicular notch of the manubrium and the adjacent first costal cartilage.

Ligaments

Articular capsule is attached to the margins of the articular surfaces.

Anterior sternoclavicular ligament is a broad band, attached in front to the articular margins and margins of the articular disc. It strengthens the capsule.

Posterior sternoclavicular ligament is a broad band attached behind to the articular margins and margins of the articular disc. It strengthens the capsule behind.

Interclavicular ligament is a flattened band that extends from the sternal end of one clavicle to the other. It is attached to in the middle of the superior margin of the manubrium. It strengthens the capsule above.

Costoclavicular ligament is a short, strong band extending from the upper part of the cartilage of the first rib obliquely to the costoclavicular impression on the undersurface of the clavicle. It acts as a fulcrum of movement.

Articular disc is a flat, fibrocartilaginous disk dividing the joint into two cavities. It is attached by its periphery to the capsule, above to the upper border of the sternal end of the clavicle and below to the cartilage of the first rib near its junction with the manubrium.

Movements

This joint provides a limited amount of movement in every direction; thus functionally, it belongs to the ball-and-socket type. Elevation and depression occur in the compartment between the disc and the clavicle. Forward and backward movement occurs in the compartment between the disc and manubrium. Rotation of about 30° can occur in the long axis of the clavicle.

Blood supply is provided by the clavicular branch of the thoracoacromial artery.

Innervation: Medial supraclavicular nerve (C3 and 4).

ACROMIOCLAVICULAR JOINT

Type

Synovial gliding, plane arthrodial with fibrocartilaginous surfaces.

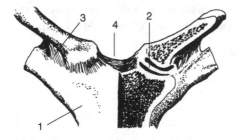

FIG. 28-1 Sternoclavicular joint. (1) Manubrium; (2) Articular disc; (3) Costoclavicular ligament; (4) Interclavicular ligament.

Articulating elements

Acromial end of the clavicle with the medial margin of the acromion of the scapula.

Ligaments

Articular capsule is attached to the articular margins and lined with a synovial membrane.

Acromioclavicular ligament extends from the superior aspect of the acromion to the superior aspect of the clavicle. This band strengthens the capsule on its superior surface.

The articular disc, when present, hangs from the articular capsule and divides the cavity incompletely.

The coracoclavicular ligament is the chief bond between clavicle and scapula. This accessory ligament has two parts:

The trapezoid ligament is flat, quadrilateral, and located anterolaterally. It extends between the upper surface of the coracoid process and the trapezoid ridge on the undersurface of the clavicle. Its lateral border is free. It is directed horizontally and outward and resists medial thrust of the scapula.

The conoid ligament is triangular and extends from the root of the coracoid process to the conoid tubercle on the undersurface of the clavicle. Its medial border is free and laterally, it is continuous with the trapezoid ligament. It helps to resist downward pull of the scapula and attached limb.

Movements

Gliding and rotation of the scapula on the clavicle.

Blood supply

Acromial branch of the thoracoacromial artery.

Innervation

Suprascapular nerves (C3 and 4).

Accessory (scapular) ligaments do not extend across a joint but connect one part of the scapula with another.

The coracoacromial ligament is triangular ligament extending from the tip of the acromion to the lateral border of the coracoid process.

The transverse scapular ligament extends from the base of the coracoid process to the medial border of the scapular notch. It converts the notch into a foramen (traversed by the suprascapular nerve).

2. Shoulder Joint

Type

Multiaxial synovial ball and socket joint with hyaline cartilaginous surfaces. The plane of the joint lies obliquely at an angle of about 45° to the sagittal plane.

Articulating elements

Head of the humerus articulates with the glenoid cavity of the scapula. The cup is deepened by the glenoid labrum, a ring of dense fibrocartilage triangular

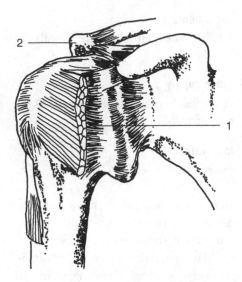

FIG. 28-2 Shoulder joint. (1) Articular capsule; (2) Coracoacromial ligament.

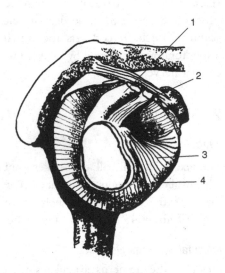

FIG. 28-3 Inside of right shoulder joint. (1) Long head of biceps; (2) Superior glenohumeral ligament; (3) Middle glenohumeral ligament; (4) Interior glenohumeral ligament.

cross section which is attached to the rim of the acetabulum.

Ligaments

The articular capsule is exceedingly loose and thin attached above to the circumference of the glenoid cavity and below to the anatomical neck of the humerus. It is strengthened by its fusion with the tendons of insertion of the supraspinatus, infraspinatus, teres minor, and subscapularis muscles.

There are two openings in the capsule: (1) at the upper end of the intertubercular groove for passage of the tendon of the long head of biceps (crossed by the transverse humeral ligament) and (2) anteriorly, below the coracoid process where the synovial cavity of the joint communicates with the subscapular bursa.

A synovial membrane lines the capsular ligament reflected over the anatomical neck of the humerus and on to the labrum. It gives a sheath to the tendon of long head of biceps as a tubular sheath and extends down the intertubercular groove. It also protrudes through the other opening in the capsule to form the subscapular bursa.

The coracohumeral ligament is a broad band on the upper part of the capsule intimately connected to it. It extends from the lateral margin of the root of the coracoid process over the articular capsule to the greater tuberosity of the humerus and is apparently a prolongation of the pectoralis minor tendon.

The glenohumeral ligaments consist of three bands between the synovial membrane and the anterior part of the articular capsule. The superior ligament extends from the labrum, anterior to the attachment of the biceps tendon, to the upper part of the lesser tuberosity. The middle ligament arises near the superior ligament and extends to the front of the lesser tuberosity. The inferior ligament extends from the anterior border of the glenoid cavity obliquely to the medial side of the humerus to the undersurface of its neck.

Movements

In all directions are possible.

Blood supply

Branches of the supra-scapular and anterior and posterior humeral circumflex arteries.

Innervation

Branches of the suprascapular, sub-scapular, and axillary nerves.

Stability of the Shoulder

(1) Bony features: Deepening of the fossa by the glenoid labrum is not significant.
(2) Ligaments: The coracoacromial ligament prevents upward dislocation.
(3) Muscular factors: Rotator cuff muscles hold the humeral head in the glenoid cavity.

Movements of the Shoulder Girdle and Shoulder Joint

Abduction (180°)

Glenohumaral — supraspinatus (initiator), and deltoid.
Glenohumeral and scapular — deltoid, serratus anterior, trapezius.
Scapular — serratus anterior, trapezius 90–180°.

Movement

Scapula rotates laterally and moves forward and upwards.

Clavicle moves in the same direction and twists (twist limited by coracoclavicular ligaments).

There is some lateral rotation of humerus in the final stage.

If the palm faces medially, movement is restricted by the greater tubercle against the coracoacromial arch.

Adduction

Gravity is the main force. Movement is controlled by progressive relaxion of abductor muscles in resistance to the pectoralis major, latissimus dorsi, and teres major muscles.

Flexion (180°)

Flexion at the glenohumeral joint: deltoid, pectoralis major and coracobrachialis.

Lateral rotation of scapula: serratus anterior and trapezius.

Medial rotation of glenohumeral joint: subscapularis, deltoid, and pectoralis major.

Extension (30°)

Teres major, latissimus dorsi and deltoid. Against resistance: pectoralis major in addition to the above.

Lateral rotation (to 50°).

Infraspinatus, teres minor, and deltoid.

Medial rotation (to 50°)

Subscapularis, pectoralis major, latissimus dorsi, teres major and deltoid.

Circumduction

Describes a cone with a moving axis.

3. Elbow Joint

Type

The elbow joint includes three articulations, humeroulnar, humeroradial and the superior radioulnar. The main joint is a synovial hinge (uniaxial) joint continuous with the superior radioulnar joint (a uniaxial pivot synovial joint).

Articulating elements

Humeroulnar: Trochlea of the humerus articulates with the trochlear notch of the ulna.

Humeroradial: The capitulum of the humerus articulates with the fovea of the head of the radius.

Superior radioulnar: The head of the radius (encircled by the annular ligament) with the radial notch of the ulna.

Ligaments

Articular capsule is thin in front and attached above to the upper margins

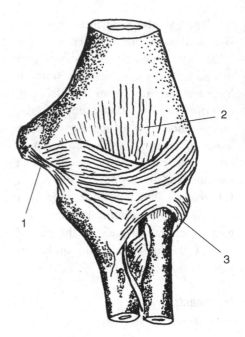

FIG. 28-4 Elbow joint — anterior aspect. (1) Ulnar collateral ligament; (2) Joint capsule; (3) Annular ligament of radius.

of the radial and coronoid fossae and to the epicondyles. Inferiorly, it is attached to the anterior border of the coronoid process of the ulna and the annular ligament of the radius. Behind, the articular capsule is also very thin. It is attached above to the margins of the olecranon fossa and to the epicondyles.

Distally, the capsule is attached to the borders of the olecranon. The capsule is thickened on its sides by the collateral ligaments. A synovial membrane lines the articular capsule and the intracapsular bony surfaces to their articular margins.

Ulnar collateral ligament is a triangular ligament, fanning out from the

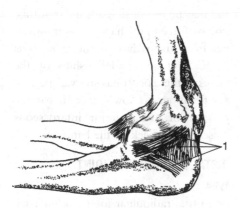

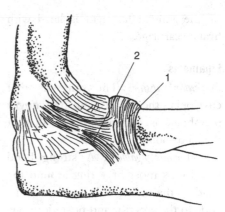

FIG. 28-5 Elbow joint — medial aspect. (1) Ulnar collateral ligament.

FIG. 28-6 Elbow joint — lateral aspect. (1) Annular ligament of radius; (2) Radial collateral ligament.

front of the medial epicondyle to the medial margin of the trochlear notch. It consists of three bands. The anterior band is thick and attached to the medial margin of the coronoid process. The oblique (posterior) band extends between the olecranon and the coronoid process. The intermediate part is thin and triangular, extending between the other two parts.

Radial collateral ligament is a triangular ligament that is a strong fan-shaped condensation of the fibrous capsule spreading out from the depression below the lateral epicondyle to the annular ligament and the margins of the radial notch of the ulna.

Movements

The elbow joint provides for flexion and extension. Flexion results from the action of brachialis, biceps brachii and brachioradialis (limited by apposition of soft parts). Extension results from the action of triceps brachii and anconeus (limited by tension in the capsule and muscle in front of the joint).

Blood supply

Derived from the anastomoses around the joint.

Innervation

By branches of the musculocutaneous, median, and radial nerves.

4. Radioulnar Joints

PROXIMAL RADIOULNAR JOINT

Type

Uniaxial synovial pivot joint

Articulating elements

The circumference of the head of the radius articulates with the radial notch of the ulna and the fibro-osseous annular ligament. The articular inner aspect of the annular ligament is lined with hyaline cartilage.

Ligaments

Articular capsule: The same as that enveloping the elbow joint. Its synovial membrane is continuous with that of the elbow joint.

 Annular ligament: A strong band that forms most of a ring around the head of the radius. It is attached by its ends to the anterior and posterior margins of the radial notch of the ulna. The ring is narrower below than above.

 Quadrate ligament: A weak horizontal band extending from the lower edge of the annular ligament to the neck of the radius.

Blood supply

From the anastomosis around the elbow joint.

Innervation

From the musculocutaneous, median, ulnar and radial nerves.

 Oblique cord is a thin band extends from the tuberosity of the ulna to the radius a little below its tuberosity.

Interosseous membrane

A broad, thin, strong sheet extending between the interosseous borders of the radius and the ulna. It begins a little below the radial tuberosity and extends to the capsule of the superior radioulnar joint. Its fibers are directed downward and medially from the radius to the ulna

and may be responsible for the transmission of force to the hand being transmitted from the radius to the ulna. It is equally taut in all relationships of the radius to the ulna. It has an oval aperture a little above its lower margin for the passage of the anterior interosseous vessels to the back of the hand.

DISTAL RADIOULNAR JOINT
Type

The distal radioulnar joint is a uniaxial, pivotal, synovial joint.

Articulating elements

The head of the ulna articulates with the ulnar notch of the radius laterally and the articular disc distally.

Ligaments

Articular capsule is attached to the margins of the disk and articular surfaces. A synovial membrane lines the L-shaped joint and may extend above the level of the articular surfaces, enclosing a space, the sacciform recess.

 Articular disc is triangular in shape. It is attached by its apex to the root of the styloid process of the ulna and by its base to the edge separating the ulnar notch from the carpal articular surface of the radius and anteriorly and posteriorlly to the ligaments of the wrist joint. The upper surface is concave and articulates with the head of the ulna; and its undersurface is concave and participates in the formation of the radiocarpal joint (articulating with the lunate and in addition the triquetral).

Movements

At the radioulnar joints both pronation and supination is possible. In pronation, the radius crosses in front of the ulna and in supination, the radius comes to lie parallel to the ulna. The axis of rotation passes through the capitulum of the humerus and the pit at the end of the ulna (to which the articular disc in attached). When the elbow is flexed at 90°, pronation of 150° results from the action of pronator quadratus and pronator teres and supination of 150° (more powerful than pronation) is caused by supinator and biceps brachii.

Blood supply

From the anterior and posterior interosseous arteries.

Intervation

From the anterior and posterior interosseous nerves.

The nerve supply of the radioulnar joints is derived from the radial, musculocutaneous and median nerves and the spinal cord segments controlling movements are supination (C6) and pronation (C7 and 8).

5. Wrist (Radiocarpal) Joint

Type

The wrist joint is an ellipsoid, synovial joint.

Articulating elements

Above, the distal end of the radius and the lower surface of the articular disc articulates with the proximal row of carpal bones (scaphoid, lunate and triquetral).

Ligaments

Articular capsule extends between the margins of the articular surfaces. Its synovial membrane lines the deep surface of the capsule.

Palmar radiocarpal ligament extends downwards and medially from the anterior margin of the distal end of the radius and its styloid process to the proximal row of carpal bones and to the capitate.

Dorsal radiocarpal ligament extends downward and medially from the posterior margin of the distal end of the radius to the back of the proximal row of carpal bones.

Radial collateral ligament extends from the tip of the styloid process of the radius to the sides of the scaphoid and trapezium.

Ulnar collateral ligament extends from the styloid process of the ulna to the pisiform and to the adjacent flexor retinaculum.

Movements (of the wrist and intercarpal joints)

The wrist joint is capable of flexion and extension; abduction and adduction; and circumduction. The movements involve those between the proximal and distal rows of carpal bones.

Flexion: Approximately 90° is the result of movement predominantly at the midcarpal joint and to a lesser extent at the wrist joint. Muscles involved are flexor carpi radialis and flexor carpi ulnaris (movement limited by tension in the extensor muscles).

Extension: Approximately 60°, is the result of movement predominantly at the wrist joint and to a lesser extent at the midcarpal joint. Muscles involved are extensor carpi radialis longus and brevis and extensor carpi ulnaris (movement limited by tension in antagonists).

Abduction: Takes place mainly at the midcarpal joint and is much more limited than adduction. Muscles involved are mainly flexor carpi radialis, extensor carpi radialis longus and brevis (movement is limited by tension in antagonistic muscles).

Adduction: Takes place predominantly at the wrist joint and the range of movement is considerably greater than abduction (limited by tension in antagonistic muscles).

Circumduction: Results from movements of flexion, adduction, extension and abduction carried out in that order in reverse.

Blood supply to the wrist joint

Provided by branches of the palmar and dorsal carpal arterial networks.

Nerve supply

By branches of the anterior and posterior interosseous nerves and the dorsal and deep branches of the ulnar nerve.

The spinal cord segments controlling movements are C6 and 7.

6. Joints of the Hand

INTERCARPAL JOINTS
JOINTS BETWEEN THE PROXIMAL ROW OF CARPAL BONES
Type

Synovial, plane (slight gliding and rotating) joints.

Articulating Elements

Scaphoid, lunate and triquetral bones.

Ligaments

Dorsal, palmar, and interosseous intercarpal ligaments.

Synovial Membrane

An extension from the mid carpal joint lining.

JOINTS BETWEEN THE DISTAL ROW OF CARPAL BONES
Type

Also synovial, plane (slight gliding and rotating) joints.

Articulating Elements

Bones of the distal row (trapezium, trapezoid, capitate and hamate bones).

Ligaments

Dorsal, palmar, and interosseous intercarpal ligaments.

Synovial membrane is an extension from the midcappal joint lining.

Joints Between the Two Rows: (Midcarpal Joint)

Type

A compound synovial sellar joint.

Articulating Elements

One joint cavity between the proximal and distal rows of carpal bones. The head of the capitate and the apex of the hamate articulate with the cup-shaped scaphoid and lunate bones. The trapezium and trapezoid articulate with the scaphoid, and the hamate articulates with the triquetrum.

Ligaments

The palmar and dorsal intercarpal and radial and ulnar collateral ligaments. The pisiform bone has its own articulation with the palmar surface of the triquetrum. The bones are united by a thin articular capsule lined with a synovial membrane.

The pisiform is connected by the pisohamate ligament to the hook of the hamate and by the pisometacarpal ligament to the base of the fifth metacarpal. These ligaments are part of the insertion of the flexor carpi ulnaris.

The synovial membrane of the intercarpal joint is very extensive. It lines the midcarpal joint and is prolonged upward and downward between the bones of the proximal and distal rows. It extends into the carpometacarpal joints (except the thumb) and also into the intermetacarpal joints. It does not communicate with the wrist joint or with the pisiform triquetrum joint, both of which have separate synovial cavities.

CARPOMETACARPAL JOINTS
CARPOMETACARPAL JOINT OF THE THUMB

Type

A synovial, sellar (saddle) joint.

Articulating Elements

The trapezium articulates with the base of the first metacarpal.

Ligaments

The articular capsule is thick but loose extending between the margins of the articular surfaces. A synovial membrane lines the capsule of this separate cavity.

Movements

This joint provides for a wide range of movements of flexion, extension, abduction, adduction, circumduction, and opposition. Movements occur at right angles to the palm.

Muscles involved are:

Flexion: Flexor pollicis longus and brevis.

Extension: Extensor pollicis longus and brevis.

Abduction: Abductors pollicis longus and brevis.

Adduction: Adductor pollicis.

Opposition: Opponens pollicis and flexor pollicis brevis. The movement includes abduction, flexion, medial rotation, and adduction.

Circumduction: Muscles acting consectively — extensors, abductors, flexors and adductors following one another in that order.

CARPOMETACARPAL JOINTS OF THE OTHER FOUR DIGITS
Type
Synovial plane joints.

Articulating Elements
The distal row of carpal bones articulates with the bases of the second to fifth metacarpal bones (trapezoid articulates with the 2nd metacarpal, capitate articulates with the 3rd metacarpal and the hamate articulates with the 4th and 5th metacarpal.

Ligaments
Dorsal, palmar ligaments and the interosseous carpometacarpal ligaments.

Nerve Supply
Median nerve (C7,8 and T1).

INTERMETACARPAL JOINTS
Type
Arthodial joints.

Articulating Elements
Contiguous sides of the bases of the second to fifth metacarpals.

Ligaments
Dorsal, palmar, and interosseous metacarpal ligaments.

METACARPOPHALANGEAL JOINTS
Type
Synovial, ellipsoid joints.

Articulating Elements
Heads of the metacarpals articulate with the bases of the proximal phalanges.

FIG. 28-7 Metacarpal and phalangeal joints — medial aspect. (1) Collateral ligament.

Ligaments

Palmar ligaments are dense fibrocartilaginous plates are located between the collateral ligaments.

Deep transverse metacarpal ligaments are three short, narrow bands that extend across the palmar surfaces of the heads of the second to fifth metacarpal bones.

Movements

Mainly of flexion and extension, with some abduction and adduction.

INTERPHALANGEAL JOINTS
Type

Uniaxial, hinge type synovial joints.

Articulating Elements

The head of the proximal phalanx articulates with the base of the distal phalanx.

Ligaments

The articular capsule and the collateral and palmar ligaments (fibrocartilaginous plate).

Movements

Flexion and extension.

FIG. 28-8 Superfical and deep flexor tendons in a digital tendon sheath — palmar aspect. (1) Flexor digitorum profundus; (2) Flexor digitorum superficialis.

The spinal cord segments controlling movements of the digits are C7 and C8.

Thorax

The thorax is the part of the trunk outlined by the extent of the rib cage and diaphragm. It contains the lungs, the heart and neurovascular connections between the neck, upper limbs and the abdomen. It communicates with the neck through the superior thoracic aperture or thoracic inlet and communicates with the abdomen through the inferior thoracic aperture or thoracic outlet that is closed in vivo by the muscular diaphragm.

The skeletal framework of the thorax comprises the sternum, ribs, costal cartilages, thoracic vertebrae and intervertebral discs. They provide protection for the heart and lungs and provide attachment for muscles of the thorax, upper limb, back and abdomen.

Chapter **29**

Thoracic Bones and Joints

Chapter Outline

1. Sternum

The sternum is a flat bone situated in the midline of the thorax. It articulates with the anterior ends of the first seven costal cartilages and medial end of the clavicle. It consists of three parts from above downward, the manubrium, body and xiphoid process. It develops from a pair of mesenchymal bands in the body wall initially with no connection with each other or the developing ribs. Following the attachment of ribs, the paired sternal bars unite in a rostro-caudal direction. Ossification centres reflect its bilateral origin and segmentation to sternebrae occur secondarily.

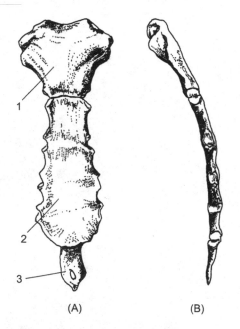

FIG. 29-1 Sternum. (A) Front view (B) Side view; (1) Body; (2) Manubrium; (3) Xiphoid process.

Manubrium is irregularly quadrilateral in shape and is the broadest part of the sternum.

Jugular notch is the concave middle part of the upper border of the manubrium. It is commonly located at the level of the third thoracic vertebra.

Clavicular notches on the lateral sides of the upper border for the sternal articular surfaces of the clavicles.

On the lateral borders of the manubrium, there are a pair of rough facets for the first costal cartilage and demifacets for the second costal cartilage. Inferiorly, the manubrium articulates with the body of the sternum at the fibrocartilaginous manubriosternal joint making the sternal angle. The joint may be ossified in the adult.

The sternal angle is an important clinical landmark. It is located at the second costal cartilage and is a reference point for counting ribs.

Body of the sternum is composed of four or less fused segments (sternebrae). Laterally above are demifacets for the second costal cartilages and below these are the costal notches for the third to sixth costal cartilages and a demifacet for the seventh cartilages. The lower margin of the body is united with the xiphoid process at the xiphisternal joint by an intervening fibrocartilage and by ligaments.

Xiphoid process is a small plate of bone surrounded by hyaline cartilage. It presents a demifacet for the seventh rib and articulates with the sternum at the xiphisternal joint (synchondrosis). It may be bifid or perforated. The rectus abdominis and linea aspera (aponeuroses of external and internal oblique muscles) are attached to the border and anterior surface cartilage and diaphragm is attached to its posterior surface.

2. Ribs

There are usually twelve ribs, or costae, on each side of the body. The first seven, or true ribs, articulate medially with the vertebrae and laterally with the sternum by means of their costal cartilages. The remaining five ribs are known as false ribs. The costal cartilages of the eighth, ninth and tenth ribs join the

costal cartilage above to form the costal margin and reach the sternum. The eleventh and twelfth ribs articulate with vertebrae only and are called floating ribs. Supernumerary ribs are common.

TYPICAL RIBS (THIRD TO NINTH)

Each typical rib can be divided into three parts.

Head is the wedge shaped medial end of the rib with two facets (separated by an interarticular crest of the head) that articulate with the numerically corresponding vertebra and with the vertebra above.

Neck is an elongated segment of the rib between the head and tubercle. It presents a crest on its upper border for attachment of the superior costotransverse ligament.

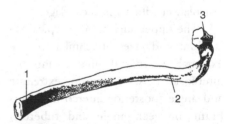

FIG. 29-3 Typical rib from within. (1) Costal groove; (2) Angle; (3) Head.

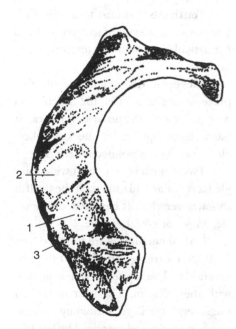

FIG. 29-2 First rib from above. (1) Scalene tubercle; (2) Groove for subclavian artery; (3) Groove for subclavian vein.

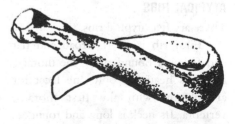

FIG. 29-4 Right rib from behind.

Tubercle is located between the neck and the shaft posteriorly on the outer aspect of the rib. It bears a smooth facet for articulation with the corresponding transverse process. Laterally and above is a nonarticular elevation for attachment of the lateral costotransverse ligament.

Shaft is thin, long, curved and flattened with internal and external surfaces and superior and inferior borders. The concave inner surface is marked by a costal groove below in which the posterior intercostal vein, artery and nerve lie (in that order from above down). The shaft presents a ridge at the angle or point of maximum curvature indicating the extent of attachment of serratus anterior and latissimus dorsi muscles. The anterior end of the shaft is oval in shape and expands into the facet for articulation with a costal cartilage.

The upper surface of a typical rib is blunt and gives attachment to the external intercostal muscle, internal intercostal between anterior extremity and angle (posterior intercostal membrane between angle and tubercle) and subcostal and intercostalis intimi muscles.

ATYPICAL RIBS

There are five atypical ribs.

First rib is short, broad and flat and forms a boundary of the thoracic inlet. Its head has only one facet for articulation with the first thoracic vertebra. Its neck is long and rounded.

A prominent tubercle is present at the junction of the neck and shaft. On the upper medial surface of the shaft are two shallow grooves, which are separated by the *scalene tubercle*. The anterior groove lodges the subclavian vein and the posterior sulcus accommodates the subclavian artery and lower trunk of the brachial plexus. Scalenus medius is attached behind the posterior groove. There is no costal groove on the inferior surface of the shaft.

Second rib is almost twice as long as the first rib but similar to it. The shaft is not twisted but has a rough eminence on its external surface for attachment of the serratus anterior. The costal groove is not very distinct.

Tenth rib is similar to a typical rib but has only a single facet on its head for articulation with the tenth thoracic vertebra.

Eleventh rib is a short rib that presents a single facet for articulation with the eleventh thoracic vertebra. It has a slight angle and no neck or tubercle. The shaft is pointed anteriorly.

Twelfth rib is very short with a single facet for articulation with the twelfth thoracic vertebra. It has no neck, tubercle, angle, or costal groove.

Costal cartilages are rounded bars of hyaline cartilage attached to the ribs anteriorly. The upper seven articulate with the sternum. The next three join the seventh cartilage combining to form a continuous costal margin. The last two costal cartilages attached to the eleventh and twelfth ribs lie between muscles in the abdominal wall.

Pectoralis major muscle is attached to the surface of the upper six cartilages, internal oblique to the surface of the seventh, eighth and ninth cartilages and rectus abdominis to the fifth, sixth, and seventh cartilages. Transversus abdominis and the diaphragm are attached to the internal surface of the lower six cartilages.

3. Joints of the Ribs and Sternum

COSTOVERTEBRAL JOINTS
Type
Plane synovial.

Articulating elements
The head of a typical rib articulates with superior and inferior costal facets of two adjacent vertebrae and intervening intervertebral disc.

Ligaments
Articular capsule is attached to the margins of articular surfaces. It is thickened anteriorly to form the radiate ligament.

Intraarticular ligament is short and horizontal and extends between the crest of the head and the intervertebral disc. It attaches to the capsule in front and behind separating the joint cavities into superior and inferior compartments.

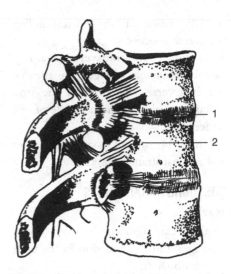

FIG. 29-5 Costovertebral joint. (1) Intervertebral disc; (2) Radiate ligament.

Innervation
Dorsal rami of intercostal nerves.

COSTOTRANSVERSE JOINTS
Type
Plane synovial.

Articulating elements
Tubercle of the rib articulates with the costal facet on the transverse process of the corresponding vertebra.

Ligaments
Articular capsule surrounds the joint and is reinforced by accessory ligaments.

Superior costotransverse ligament extends between the crest of the neck of the rib to the lower border of the transverse process above.

Costotransverse ligament extends between the back of the neck of the rib and the front of the corresponding transverse process.

Lateral costotransverse ligament extends obliquely between the apex of the transverse process and the non articular portion of the tubercle of the adjacent rib.

Ribs 11 and 12 have no tubercles and no costotransverse joints.

COSTOCHONDRAL JOINTS

Type

Hyaline cartilaginous between costal cartilages and a depression in the ends of the rib shaft.

Action

Permit gliding of the lateral end of each costal cartilage with the sternal end of its rib.

INTERCHONDRAL JOINTS

The joints between each of the fifth to ninth costal cartilages with the costal

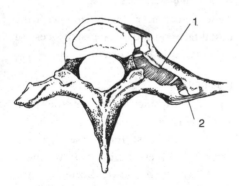

FIG. 29-6 Costotransverse joint. (1) Costotransverse ligament; (2) Lateral costotransverse ligament.

cartilage below permit gliding between the cartilages. Each joint is enclosed by an articular capsule and has a synovial cavity.

STERNOCOSTAL (STERNOCHONDRAL) JOINTS

The first rib is directly united with the sternum by a synchondrosis. The second to the seventh joints may be synovial but are commonly filled with fibrocartilaginous tissue.

Type

Commonly plane synovial.

Articulating elements

The costal cartilage of the rib articulates with the sternum.

Ligaments

Articular capsule is reinforced anteriorly by radiate ligaments.

Intra-articular ligament is horizontal and divides the joint into upper and lower cavities.

STERNAL JOINTS

Manubriosternal joint: fibrocartilage (which may ossify) unites the manubrium with the body of the sternum. In children, cartilaginous joints separate several ossification centres (sternebrae).

Xiphisternal joint: The xiphoid process and body of the sternum are united by cartilage that begins to ossify at puberty.

Chapter **30**

Walls of the Thorax

Chapter Outline

1. Intrinsic Muscles
2. Arteries
3. Veins
4. Nerves

The three muscular layers seen in the anterior abdominal wall become specialised in the thorax. External oblique muscle is represented by pectoralis major and minor, serratus anterior, rhomboids and levator scapulae. Internal oblique is represented by intersegmental bony condensations (ribs) with intervening muscle specialised into two layers, internal and external intercostals. Transversus abdominis is represented by the diaphragm and intracostal muscles. The neurovascular plane lies between the second and third layers in the abdomen and thorax.

1. Intrinsic Muscles

EXTERNAL LAYER
External Intercostals

External intercostals extend from the tubercle of a rib to the costochondral junction where it is continuous with external intercostal membrane.

- **Origin:** Outer lower border of the subcostal groove of each of the first eleven ribs.
- **Insertion:** Upper border of the rib below. Muscle fibers are directed downward and forward.
- **Action:** Elevates the ribs in inspiration.
- **Nerve supply:** Corresponding intercostal or thoracoabdominal nerves.

Levatores Costarum

Levatores costarum comprises twelve small fan shaped muscles that belong to the external layer.

- **Origin:** Transverse processes of the seventh cervical and first to eleventh thoracic vertebrae.
- **Insertion:** External surface of rib below medial to their angle.
- **Action:** Raise the ribs and thereby increase the size of the thoracic cavity in inspiration.
- **Nerve supply:** Dorsal rami of the eighth cervical and first to eleventh intercostal nerves.

MIDDLE LAYER
Internal Intercostals

Internal intercostals extend from the distal ends of the intercostal spaces at the side of the sternum to the angle of the ribs, where they are continuous with the internal intercostal membrane. The muscle fibers are directed downwards and backwards.

- **Origin:** Costal cartilages and floor of the sub-costal groove of first 11 ribs above.
- **Insertion:** Upper border of rib below.
- **Action:** Anteriorly elevates the ribs in inspiration but posteriorly depresses the ribs in expiration.
- **Nerve supply:** Corresponding intercostal or thoraco-abdominal nerves.

INTERNAL LAYER
Innermost Intercostals

Innermost intercostals are part of the innermost layer at the distal end of intercostal spaces 2–5. It is considered to be a deep layer of the internal intercostal muscles, from which it is separated by the intercostal nerves and vessels.

- **Origin:** Subcostal groove of the rib above.
- **Insertion:** Blends with internal intercostal at the upper border of a rib below. It may cross one or two intercostal spaces.

- **Action:** Fibers run in the same direction as those of the internal intercostals so they assist in raising the ribs for inspiration anteriorly but more posteriorly, they depress the ribs in expiration.
- **Nerve supply:** Corresponding intercostal or thoraco-abdominal nerves.

Subcostals

Subcostals consist of variable slips of muscle found near the angles of the ribs and often spanning two intercostal spaces.

- **Origin:** Lower margins of the ribs near their angle.
- **Insertion:** Upper margins of the second or third rib below the rib of origin.
- **Action:** Draws adjacent ribs together (elevate the ribs).
- **Nerve supply:** Intercostal or thoraco-abdominal nerves.

Transversus Thoracis (Sternocostalis)

- **Origin:** Posterior surface of xiphoid process and lower third of the body of the sternum.
- **Insertion:** Inner surface of costal cartilage of the second to seventh costal cartilages.
- **Action:** Draws anterior part of thorax downward.
- **Nerve supply:** Second to sixth intercostal nerves.

Subcostals and transversus are in the same plane as the innermost intercostals and connected to them by fascia. Innermost intercostals fuse with the diaphragm anteriorly.

2. Arteries

INTERNAL THORACIC ARTERY

Internal thoracic artery arises in the root of the neck from under the surface of the first part of the subclavian artery. It descends behind the clavicle and first costal cartilage (where it is crossed by the phrenic nerve) and then passes caudally parallel with the margin of the sternum. It ends behind the sixth intercostal space where it divides into two terminal branches, the superior epigastric and musculophrenic arteries.

BRANCHES

Pericardiacophrenic artery arises high in the thorax where it accompanies the phrenic nerve between the pleura and pericardium. It supplies the diaphragm, pericardium and pleura and anastomoses with the musculophrenic and phrenic branches of the aorta.

Anterior mediastinal artery descends to supply the thymus and pericardium.

Anterior intercostal artery passes laterally; two for each upper six intercostal spaces. They supply the intercostal muscles and intercostal branches of the aorta.

Perforating branches pass anteriorly with anterior cutaneous branches of intercostal nerves. They supply pectoralis major and the breast.

Musculophrenic artery (the lateral terminal branch of the internal thoracic) descends behind the costal attachments of the diaphragm. It pierces the diaphragm behind the eighth costal cartilage and supplies intercostal muscles, muscles of the anterior abdominal wall and the lower pericardium.

Superior epigastric artery is the medial terminal branch of the internal thoracic artery. It descends behind the seventh costal cartilage and between the costal and sternal origins of the diaphragm into the rectus sheath. It supplies the diaphragm and anterior abdominal wall.

POSTERIOR INTERCOSTAL ARTERIES

Posterior intercostal arteries have two sites of origin, the first two arise from the highest intercostal artery and the remainder arise from the back of the aorta.

The nine pairs of intercostal arteries are crossed by the sympathetic nerves and their splanchnic branches.

Highest intercostal artery arises from the costocervical trunk on the neck of the first rib. It gives a branch, the first posterior intercostal artery then continues down as the second posterior intercostal artery.

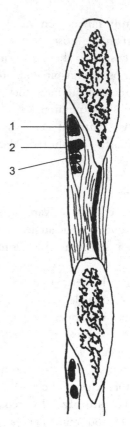

FIG. 30-1 Transverse section of intercostal space. (1) Posterior intercostal vein; (2) Posterior intercostal artery; (3) Posterior intercostal nerve.

Third to eleventh posterior intercostal arteries arise from the back of the descending aorta. They pass to the subcostal grooves, where they anastomose with the upper of the two anterior intercostal arteries.

BRANCHES

Dorsal branches pass back with the dorsal branch of intercostal nerves. They branch into muscular branches supplying the muscles of back and ending in medial and lateral cutaneous branches and the important spinal branches which supply the spinal cord.

Collateral branches arise near the angle of the rib, and run forward to the anterior end of the intercostal space. They supply intercostals, pectoralis major and minor muscles.

Lateral cutaneous branches accompany corresponding nerves through overlying muscles supplying muscles and skin.

Mammary branches arise from the third to fifth intercostal spaces supply the mammary gland.

Subcostal arteries arise from the back of the lowest part of the thoracic aorta. They pass behind the lateral arcuate ligament of the diaphragm to enter the abdomen.

In the congenital abnormality known as coarctation of the aorta, there is constriction of the aorta below the origin of the left subclavian artery. An important anastomosis can be established between intercostal arteries arising from the aorta and intercostal arteries arising from the subclavian artery around the site of constriction.

3. Veins

INTERNAL THORACIC VEINS

Two veins on each side are venae comitantes of the internal thoracic artery. They are formed by the union of the companion veins of the musculophrenic

and superior epigastric arteries and unite near the third costal cartilage. They ascend along the medial side of the internal thoracic artery to terminate in the brachiocephalic vein or superior vena cava in the right hand side.

ANTERIOR INTERCOSTAL VEINS

Two veins on each side drain the upper nine intercostal spaces and terminate in the internal thoracic vein.

POSTERIOR INTERCOSTAL VEINS

One vein in each intercostal space in the subcostal groove lies above the intercostal artery. Tributaries correspond to the arteries.

On the right side, the first intercostal vein passes upwards over the neck of the first rib, arches forward and ends in the brachiocephalic trunk. The second and third unite to form the right superior intercostal vein that joins the vena azygos. The fourth to eleventh and subcostal veins join the vena azygos directly.

On the left side, the first intercostal vein ascends to join the brachiocephalic trunk. The second and third form the left superior intercostal vein that descends over the third and fourth thoracic vertebrae then ascends to join the left brachiocephalic trunk. The fourth to eighth veins end in the superior hemiazygos vein. The ninth to eleventh and subcostal veins end in the inferior hemiazygos vein.

Blood delivered by four of the branches of the subclavian artery mostly returns to the brachiocephalic veins (that substitutes for the first part of the subclavian vein).

The vertebral and internal thoracic veins return to the brachiocephalic directly.

There are no thyrocervical or costocervical venous trunks.

The inferior thyroid vein empties into the brachiocephalic trunk.

The transverse cervical and suprascapular veins drain into the external jugular vein (that empties into the subclavian vein).

The deep cervical vein (vein from the first intercostal space) empties into the vertebral vein and thereby into the brachiocephalic vein.

4. Nerves

Typical intercostal nerves are the ventral (anterior primary) rami of the thoracic spinal nerves. Each lies below its corresponding vein and artery and pass along the posterior intercostal membrane to enter the costal groove. It continues between the internal intercostal and innermost intercostal muscles to terminate as the anterior cutaneous branch piercing the internal intercostal muscle, external intercostal membrane and pectoralis major.

BRANCHES

White ramus communicantes: One to four twigs carry sympathetic fibers to sympathetic trunk.

Muscular branches: Supply intercostal, subcostal, serratus posterior superior and transversus thoracis muscles.

Lateral cutaneous branches: Pierce intercostal and serratus anterior muscles then divide into anterior and posterior branches.

Anterior cutaneous branch: Branches medially to supply the skin over the sternum and laterally to reach the skin over the front of the thorax.

Atypical intercostal nerves:

First thoracic nerve is a short nerve that divides into a larger upper part that is incorporated into the brachial plexus. Its lower part becomes the first intercostal nerve.

Second thoracic nerve provides a lateral cutaneous branch (the intercostobrachial nerve). It innervates the floor of the axilla and the skin of the back and medial side of the arm down to the elbow.

Seventh to eleventh thoracic (thoracoabdominal) nerves course to the point where the corresponding cartilages curve upwards. They continue between the digitations of the diaphragm and transversus abdominis then down and forward between transversus and internal oblique then between rectus and the posterior layer of its sheath. They pierce the rectus and become anterior cutaneous branches.

Subcostal nerve (ventral ramus of the twelfth thoracic nerve) passes

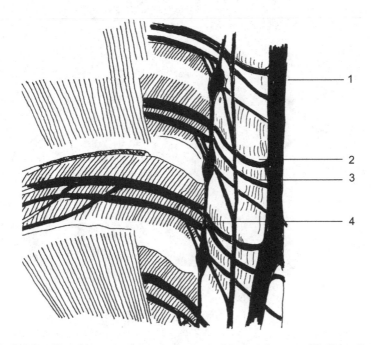

FIG. 30-2 Relationship of neurovascular structures to intercostal space. (1) Sympathetic trunk; (2) Posterior intercostal vein; (3) Posterior intercostal artery; (4) Posterior intercostal nerve.

into the abdomen obliquely behind the lateral arcuate ligament in front of quadratus lumborum. It does not lie in an intercostal space. It pierces transversus and passes forward between transversus and internal oblique muscles. It enters the sheath of rectus abdominis, pierces its anterior layer and becomes superficial half way between the umbilicus and pubic symphysis. It supplies parts of transversus and rectus and pyramidalis.

The lateral cutaneous branch pierces the internal and external oblique muscle, passes over the iliac crest and supplies an area of skin over the gluteal region as far as the greater trochanter of the femur.

Dorsal rami of the intercostal and subcostal nerves pass backward to supply muscles, bones, joints and skin of the back. Each divides into medial and lateral cutaneous branches.

Medial branches: Upper branches pierce trapezius near the spines of the vertebrae. Lower branches supply erector spinae muscles.

Lateral branches: Pursue a long downward course before becoming cutaneous. Lower branches pierce latissimus dorsi and are sensory to the skin down to the gluteal region.

Chapter 31

Superior, Anterior, Middle and Posterior Mediastinum

Chapter Outline

The mediastinum is the region of the thoracic cavity between the two pleural cavities. It is sometimes known as the interpleural space. It is divided into the superior mediastinum above the pericardium, and the inferior mediastinum below. The inferior mediastinum is again subdivided into the anterior mediastinum between the pleurae in front of the pericardium; middle mediastinum, containing the heart and pericardium, adjacent parts of great vessels main bronchi and structures at the root of the lungs and posterior mediastinum that lies behind the pericardium and diaphragm. The mediastinum is composed of loose connective tissue that suspends the thoracic organs and is covered by mediastinal pleura.

1. Superior Mediastinum

The superior mediastinum is the subdivision of the mediastinum that contains the structures passing directly to and from the neck.

BOUNDARIES

Anterior: Manubrium and origins of sternohyoid and sternothyroid muscles.
Posterior: Upper four thoracic vertebrae, intervertebral discs, anterior longitudinal ligament and lower end of longus colli.
Superior: Thoracic inlet.
Inferior: A line passing between the base of the manubrium and body of the fourth thoracic vertebra.

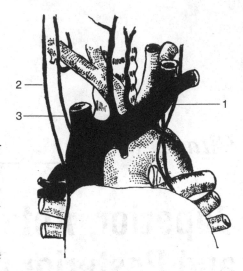

FIG. 31-2 Anterior aspect of superior mediastinum. (1) Brachiocephalic v; (2) Phrenic n; (3) Vagus n.

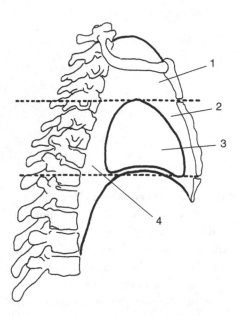

FIG. 31-1 Division of the mediastinum. (1) Superior mediastinum; (2) Anterior mediastinum; (3) Middle mediastinum; (4) Posterior mediastinum.

CONTENTS

Thymus

Thymus has two asymmetrical lobes closely connected in the midline by connective tissue. A capsule encloses each lobe and interlobular septa extend inward to separate lobes into lobules. Lobules have a cortex that is densely populated with lymphocytes (thymocytes) and a medulla that sends lateral projections into each lobule.

The thymus lies on the front of the trachea, between the carotid arteries and above the left brachiocephalic vein. It is an important source of T lymphocytes but undergoes a normal, gradual, age dependent involution beginning at puberty. No lymphatics enter the thymus.

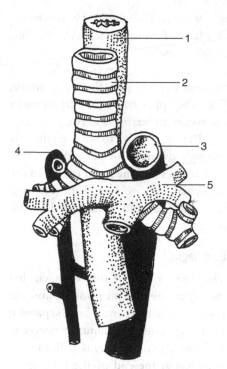

FIG. 31-3 Relationship of major structures in the superior mediastinum. (1) Esophagus; (2) Trachea; (3) Aorta; (4) Arch of azygos vein; (5) Pulmonary trunk.

Brachiocephalic Veins

Brachiocephalic veins arise in the neck from the union of the internal jugular and subclavian veins. They unite on the right side of the superior mediastinum to form the superior vena cava. In the thorax, they receive the internal thoracic and inferior thyroid veins.

The right brachiocephalic vein is short but receives the right vertebral, right internal thoracic, right inferior thyroid and right first intercostal veins. The right lymphatic duct opens into the junction between the right subclavian and internal jugular veins.

The left brachiocephalic vein is longer passing from left to right and downwards. The thoracic duct opens into the junction of the left subclavian and internal jugular veins. The left brachiocephalic vein receives the left vertebral, internal thoracic, inferior thyroid, first left intercostal, superior intercostal, thymic, mediastinal and pericardiacophrenic veins.

Superior Vena Cava

Superior vena cava begins behind the first costal cartilage near the sternum by the union of the right and left brachiocephalic veins. It enters the pericardium opposite the second costal cartilage, and descends vertically to join the right atrium opposite the third costal cartilage. The right phrenic nerve runs alongside its right side and the ascending aorta lies on its left side. It receives pericardiac and mediastinal veins and the azygos vein just as it enters the pericardium.

Arch of the Aorta

Arch of the aorta begins as the continuation of the ascending aorta behind the right sternal margin opposite the second costal cartilage. It passes in a sagittal plane behind the lower part of the manubrium giving off three major branches.

Brachiocephalic artery (trunk) arises opposite the center of the manubrium and ascends to the right into the neck taking blood to the right subclavian and right common carotid arteries.

On the left side, the left common carotid and left subclavian arteries arise from the arch and also ascend into the neck. The ligamentum arteriosum is the fetal remnant of the ductus arteriosus that connects the aortic arch to the left pulmonary artery. It was a bypass of excess blood from the pulmonary into the systemic circulation.

On the left and anterior side, the arch is crossed by left phrenic nerve and left vagus nerve. The left recurrent laryngeal nerve crosses the inferior surface of the arch to the left of the ligamentum arteriosum.

On its right and posterior side, the aortic arch lies on the trachea, esophagus (with left recurrent laryngeal nerve between), thoracic duct and vertebral bodies.

Inferiorly, the arch is related to the pulmonary trunk where it divides into right and left pulmonary arteries.

Trachea

Trachea is a flexible, fibrocartilaginous respiratory tube that lies in the median plane. It extends from the lower border of the cricoid cartilage (lower border of the larynx) opposite the lower border of the sixth cervical vertebra, and divides into two (right and left main bronchi) extrapulmonary bronchi opposite the sternal angle. The trachea has a framework of curved bars of hyaline cartilages joined by tough fibrous membranes deficient posteriorly where the wall is completed by transverse bundles of smooth muscle, the trachealis muscle.

The sensory nerves supply to the tracheal mucosa join the vagus nerve, from the upper end of the trachea join the recurrent laryngeal nerve.

Blood supply to the trachea is through branches of the inferior thyroid arteries and the lower end from the bronchial arteries. Lymphatic drainage is to tracheal and paratracheal lymph nodes.

Esophagus

The thoracic part of the esophagus lies mostly in the median plane behind the trachea then the pericardium separated from the vertebral column by prevertebral fascia and the longus colli muscle. It begins at the end of the pharynx at the level of the sixth cervical vertebra, descends through the superior and posterior mediastinum then deviates to the left below the left main bronchus and passes through the esophageal opening of the diaphragm at about the level of the eleventh or twelfth vertebra to join the stomach in the abdomen.

Phrenic Nerves

The phrenic nerves originate from the cervical plexus (C4 with branches from C3 and C5). In the thorax, they descend in front of the root of the lung between the pericardium and mediastinal pleura to supply the diaphragm.

The right phrenic nerve enters the thorax on the lateral side of the superior vena cava and the heart. It passes through the foramen for the inferior vena cava.

The left phrenic nerve enters the thorax on the lateral side of the left brachiocephalic vein and descends over the left vagus nerve and arch of the aorta.

The pleura and lung are lateral to the phrenic nerve. Both phrenic nerves are accompanied by the pericardiacophrenic vessels that provide branches to the parietal pleurae, pericardium and diaphragm.

Vagus Nerves

Vagus nerves originate in the cranial cavity (10th cranial nerve).

The right vagus nerve enters the thorax between the first part of the subclavian artery and the subclavian vein. It descends at the right side of the trachea toward the posterior part of the hilum of the lung providing branches to the esophagus, deep cardiac plexus, and anterior pulmonary plexus. It ramifies, forming the right posterior pulmonary plexus on the back of the root of the right lung, fibers then continuing into the esophageal plexus.

The left vagus nerve enters the thorax by passing between the left common carotid artery and the left brachiocephalic vein. It passes over the left subclavian artery and arch of the aorta to the root of the left lung, where it ramifies into the posterior pulmonary plexus then enters the esophageal plexus. As it crosses the arch of the aorta, it provides branches to

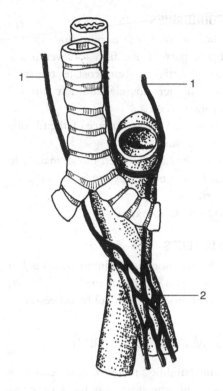

FIG. 31-4 Vagus and the esophageal plexus. (1) Vagi; (2) Esophageal plexus.

the anterior pulmonary plexus and then the recurrent laryngeal nerve.

The vagus nerves also provide cardiac branches in the thorax.

2. Anterior Mediastinum

The anterior mediastinum is the most narrow of the regions of the thoracic cavity. It is located in front of the pericardium and behind the sternum where the sternal reflexion of the left pleura fails to meet the right pleura in the median plane.

BOUNDARIES

Anterior: Body of sternum (four sternebrae), part of the fifth to seventh left costal cartilages, sternocostalis, transversus thoracis muscle and left internal thoracic vessels.

Posterior: Anterior surface of the fibrous pericardium.

Superior: Line between the sternal angle and lower border of the fourth thoracic vertebra.

Inferior: Diaphragm.

CONTENTS

The anterior mediastinum contains the lower part of the thymus, a little fat, lymph nodes, and small blood vessels.

3. Middle Mediastinum

The middle mediastinum contains not only the heart but roots of eight great vessels passing to and from the heart. Lateral to the pericardium on each side run the phrenic nerves and companion vessels (pericardiacophrenic vessels).

BOUNDARIES

Anterior: Anterior pericardium and roots of the lungs and diaphragm below.

Posterior: Vertebral column from the lower border of the fourth to twelfth thoracic vertebrae.

Superior: Line between the sternal angle and base of the body of the fourth thoracic vertebra.

Inferior: Diaphragm.

CONTENTS

The contents of the middle mediastinum are the pericardium (and its contents), phrenic nerves and accompanying vessels, arch of the vena azygos, roots of the lungs and bronchial lymph nodes.

Pericardium is an invaginated serous sac that encloses the heart and roots of the great vessels. The external surface of the parietal layer is invested by a strengthening fibrous coat.

Fibrous pericardium is the outer tough, unyielding fibrous layer of dense connective tissue. It is attached below to the central tendon of the diaphragm and is continuous with the adventitia of the great vessels.

Serous pericardium is the serous membrane lining the closed, invaginated (by the heart) sac within the fibrous pericardium and outside of the heart. The potential space between the layers of pericardium lining the heart and that lining the pericardial sac is the pericardial cavity and its mesothelium lining secretes fluid that keeps the surface layers moist and slippery. This allows frictionless movement between the heart and pericardial sac. It also provides surface tension between pericardial layers that facilitates blood return to the right atrium during inspiration.

Parietal layer of serous pericardium is the serous mesothelium that lines the fibrous pericardium.

Visceral layer of serous pericardium (epicardium) is the reflection of the mesothelium over the heart and great vessels.

SINUSES OF PERICARDIUM

In early development, the heart begins as a pair of endothelial tubes that differentiate in the cardiogenic area at the rostral end of the embryo. With head fold formation, the tubes fuse ventral to the foregut and bulge into the pericardial cavity so that the heart is suspended by a transient dorsal mesocardium (mesentery). The only region of continuity of the visceral and parietal pericardium is where veins enter the heart and arteries leave. Flexion of the tubular heart brings the ends close together but the arterial and venous ends of the cardiac tube are separated by a space, the transverse sinus of pericardium. The oblique sinus is the result of changed lines of reflection of pericardium onto pulmonary veins and vena cavae caused by absorption of these vessels into the atria.

Transverse sinus is a transverse passage, the site of breakdown of the dorsal mesocardium. It is bounded by the serous pericardium located between the ascending aorta and pulmonary trunk (in a common sleeve) and the atria.

Oblique sinus is a wide recess that lies between the left atrium and the back of the pericardium. It is bounded above by the serous pericardial reflections, at the sides by serous pericardial folds at the entrances of the

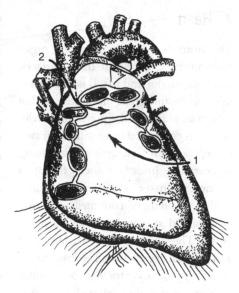

FIG. 31-5 Pericardial reflections and sinuses. (1) Oblique sinus; (2) Transverse sinus.

right and left pulmonary veins and is open below.

The neurovascular system of the pericardium is supplied by:

- *Arteries*: Pericardiacophrenic branches of the internal thoracic arteries, pericardial branches of the bronchial, esophageal and superior phrenic arteries.
- *Veins*: Correspond to the aortic branches and drain into the azygos vein.
- *Nerves*: Branches of the phrenic nerve, vagi, and sympathetic nerves via the esophageal plexus containing vasomotor and sensory nerves. The visceral layer of pericardium is insensitive.

4. Heart

The heart is a hollow muscular organ with an apex, base and three surfaces, anterior, inferior and left. It is attached at its base to the great blood vessels but otherwise is free in the pericardial cavity. Its wall is composed from without inward of epicardium (visceral pericardium), myocardium (mostly muscle fibers), a connective tissue skeleton and endocardium, a lining of endothelium. It may be regarded as a highly modified blood vessel. The heart is separated into right and left halves, each half consisting of an atrium and a ventricle whose boundaries are demarcated externally by sulci.

SURFACE MARKINGS OF THE HEART

Superior: A line from the lower border of the left second costal cartilage to the right upper border of the third costal cartilage.

Inferior: A line from the sixth right sternocostal joint to its apex.

Right side: A vertical line 4 cm from the middle of the sternum from the third to sixth costal cartilage.

Left side: From the lower border of the left second costal cartilage 2.5 cm from the sternum down and to the left to the apex.

Apex: Fifth left intercostal space 8 cm from the midline.

SULCI OF THE HEART

Coronary sulcus encircles the heart completely separating the atria from the ventricles. It contains the trunks of the circumflex branch of the left coronary artery, right coronary artery, coronary sinus and small cardiac veins embedded in fat.

Anterior interventricular sulcus extends from the left side of the root of the pulmonary trunk obliquely to the lower margin near the apex. It marks the location of the interventricular septum and contains the anterior interventricular artery (a branch of the left coronary artery) and the great cardiac vein.

Posterior interventricular sulcus is the continuation of the anterior interventricular sulcus around the apex of the heart onto the diaphragmatic surface. It marks the location of the interventricular septum and contains the posterior interventricular artery (usually a branch of the right coronary artery) and the middle cardiac vein.

LAYERS OF THE HEART

Epicardium is the visceral layer of serous pericardium. It contains a variable amount of fat and the coronary vessels before their branches enter the myocardium.

Myocardium composed mainly of specialized striated cardiac muscle. It is thin in the atria and thickest in the left ventricle. Muscle fibers of the heart originate from the annulus fibrosus and are arranged into spiralling sheets. The musculature of the artia and ventricles are however separate and do not cross the coronary sulcus.

Endocardium is a thin fibroelastic membrane covered with endothelial cells and continuous with those lining the blood vessels. At the orifices of the blood vessels, this layer is duplicated and contain a connective tissue stroma to form the cusps of the valves.

ATRIA OF THE HEART

The atria of the heart form the basal part of the heart. An ear-shaped pouch, the auricle is a small muscular pouch that extends anteriorly and whose interior is marked by parallel muscular ridges (musculi pectinatae). An interatrial septum separates right from the left atrium.

Right atrium forms the right border of the heart. It has the thinnest walls of the four chambers, is irregular in shape and is smooth walled internally. The cavity of the atrium can be divided into the sinus venarum (the area into which the vena cavae empty) and the auricle or auricular appendage.

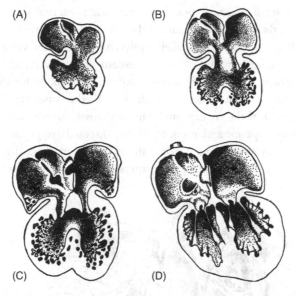

(A)   (B)

(C)   (D)

FIG. 31-6 Coronal section through heart at level of atrioventrucilar canal seen from in front. (A) 5mm; (B) 6mm; (C) 8mm; (D) Newborn. (1) Formation of a single atrial cavity by a sickle shaped septum (primum) growing from the dorsal roof; (2) Local thickening of mesenchyme (covered with endocardium — endocardial cushion) in the anterior and posterior walls of the atrioventricular canal; (3) The upper part of septum primum ruptures (C); (4) Septum secundum beginning to grow lateral to septum primum completing a one way valve allowing blood flow right atrium to left atrium (the pulmonary circulation is largely bypassed); (5) After birth, increased pulmonary resistance pushes septum secundum against septum primum obliterating septum secundum; (6) Atrioventricular valves are hollowed out from the ventricular side (C,D) but remain attached to ventricular side by chordeae tendineae.

The developing atria have a rough interior lining however in embryos of six to eight weeks the atria increase rapidly in size and the right horn of the sinus venosus is incorporated into the right atrium. This results in the superior vena cava opening directly into the cephalic wall of the atrium and inferior vena cava opening into its caudal wall. The area of absorption is smooth walled and known as the sinus venarum. The boundary between the sinus venarum and primitive atrium is the sulcus terminalis.

Auricle is a small muscular pouch projecting anteriorly from the atrium. Its interior walls are marked by parallel muscular ridges (pectinate muscles).

Sinus venarum cavarum is the smooth walled region of the atrium posterior and to the right of the crista terminalis into which the superior and inferior venae cavae, coronary sinus and anterior cardiac veins empty. It represents the partial incorporation of the embryonic sinus venosus into the right atrium.

Opening of superior vena cava enters the superior and posterior part of the atrium. It has no valves and faces downward and forward toward the right atrioventricular opening. Between the openings of the venae cavae is a muscular ridge, the intervenous tubercle that also helped to direct blood from the superior vena cava towards the right atrioventricular opening.

Opening of inferior vena cava enters the posterior and inferior wall of the atrium. A variable, rudimentary valve of the inferior vena cava forms a crescent shaped vertical flap of endocardium and subendocardial tissue in front of the opening. The valve is incompetent in the adult but in the fetus, directed oxygenated blood from the inferior vena cava through the foramen ovale to the left atrium.

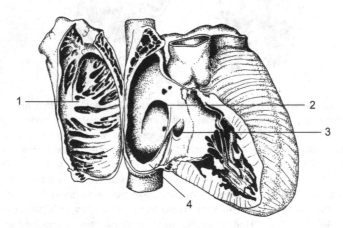

FIG. 31-7 Interior of the right atrium. (1) Auricle; (2) Fossa ovalis; (3) Valve of coronary sinus; (4) Valve of inferior vena cava.

Opening of coronary sinus is located between the atrioventricular orifice and the valve of the inferior vena cava. It is often partially guarded by a semicircular valve (valve of the coronary sinus). It does not appear to prevent reflux into the coronary sinus during atrial contraction.

Anterior cardiac veins drain the sternocostal surface of the right atrium and ventricle and open directly into the anterior wall below the auricle.

Venae cordis minimae are small vessels draining the muscular wall of the heart and open directly into pits near the septum below the superior vena cava.

Fossa ovalis is an ovoid depression on the interatrial septum above the opening of the inferior vena cava.

Fossa ovalis is the fused remnant of the incomplete septum secundum on the right side overlapping the remnant of septum primum on the left side of the posterior wall of the early common atrium. At birth, an increase in blood pressure in the left atrium causes the upper edge of septum primum to be pressed against the septum secundum obliterating the oval foramen and separating the right and left atria.

Limbus fossa ovalis is a rounded fold that forms the oval margin of fossa ovalis. Hidden under cover of the upper part of the limbus may be a slit, the remnant of the fetal foramen ovale (in one fifth of adults). The limbus is the free edge of septum secundum in the fetus.

Terminal crest is the smooth vertical muscular ridge on the right wall of the atrium which extends between the anterior margins of the openings of both vena cavae. The crest is reflected on the outside of the right border of the atrium by an indistinct line, the terminal sulcus and is the boundary between the original atrium and absorbed sinus venosus. The superior end of the crest surrounds the sinuatrial node or "pacemaker" of the heart.

Pectinate muscles are parallel muscular ridges that extend from the terminal crest spreading over the anterior and right walls of the auricle. They are confined to the auricle.

Left atrium forms part of the base and of the posterior surface of the heart. Internally, it consists of a larger smooth walled cavity and a smaller auricle whose wall is roughened by pectinate muscles.

The left atrium presents openings of four pulmonary veins (two superior and two inferior) on its posterior wall, two arising from each lung.

The relation of the pulmonary veins in the adult is the result of absorption of the posterior wall of the atrium through which the left atrium is markedly enlarged. In 6 mm embryos, a single pulmonary vein enters the caudal wall of the left atrium. This vessel bifurcates into right and left veins that in turn divide again so that two branches extend into each lung. As the atrium grows, these pulmonary vessels are drawn into the atrial wall until finally four pulmonary veins open separately into the left atrium.

From the left atrial side, the interatrial septum reflects a semilunar depression marking the fossa ovalis on the right side. The depression is bounded inferiorly by a crescent shaped ridge, the valve of foramen ovale. Behind this, there may be a communication with the right atrium.

VENTRICLES OF THE HEART

The ventricles of the heart form the anterior left (pulmonary) and lower (diaphragmatic) parts of the heart. The two chambers are completely separated by a strong, obliquely placed mostly muscular partition, the interventricular septum.

Right ventricle is triangular in cross section (in the frontal plane). It tapers above into a funnel shaped canal, the infundibulum that leads into the pulmonary trunk. The ventricle is semilunar in cross section as the thick interventricular septum bulges into its cavity. The remaining wall of the right ventricle is only one third of the thickness of the left ventricle as it is required to overcome the resistance of the vascular bed of the lungs while the left ventricle is required to overcome the resistance of the systemic circulation.

The right ventricle presents internally the following structures:

- *Trabeculae carneae* are projecting muscular bundles covered with endothelium on the internal surface of the ventricle. There are three types of trabeculae:

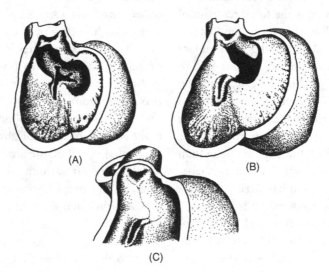

(A)

(B)

(C)

FIG. 31-8 Development of the spiral ridges and closure of the interventricular foramen. (A) 12mm; (B) 14mm; (C) 20mm. Proliferation of the right and left spiral ridges and of the posterior atrioventricular endocardial cushion closes the interventricular foramen to form the membranous part of the interventricular septum.

- *Papillary muscles* are continuous with trabeculae attached at its base to the ventricular wall. They are conical muscular projections attached at their base to the ventricular wall. Their apices are connected by slender, fibrous threads, chordae tendineae that are inserted into the underside of the flaps of the tricuspid valve.
- Some papillary muscles project inward and are attached by their bases to the ventricular wall. One of these, the septomarginal trabecula or moderator band may extend across the right ventricle from the interventricular septum to the anterior papillary muscle. The septomarginal trabecula contains a part of the impulse conducting system of the heart.
- Some papillary muscles are simple prominent ridges.
- The infundibulum is the smooth walled part of the right ventricle that leads into the pulmonary orifice. It contains the pulmonary semilunar valve.

Left ventricle receives blood from the left atrium and ejects oxygenated blood into the systemic circulation. It is longer and more conical than the right ventricle with walls three times as thick because in systole, it has to overcome the resistance of the systemic vascular bed. It forms the apex of the heart and the greater part of the diaphragmatic surface. It has a round

cavity in cross section as the shared interventricular septum bulges into the right ventricle.

The left ventricle presents internally the following structures:
- *Trabeculae carneae*: Similar to the those in the right ventricle but there is usually no septomarginal trabecula.
- *Papillary muscles*: One each arise from the anterior and posterior walls of the ventricle. They have chordae tendineae attached to the cusps of the mitral valve (more than those in the right atrioventricular orifice).
- The aortic vestibule is the region above and to the right of the mitral valve leading from the ventricle into the ascending aorta. It contains the aortic valve.

VALVES OF THE HEART

At the level of the coronary sulcus is a frame of fibrous tissue rings surrounding the atrioventricular and semilunar orifices and a membranous interventricular septum. It gives attachment to the valves and muscle bundles and is continuous with the roots of the aorta and pulmonary trunk. The annulus prevents stretching of the valve openings and forms an electrical barrier between atrial and ventricular muscle.

The *fibrous (cardiac) skeleton* is a fibrous trigone that lies between the atrioventricular orifices and the aortic semilunar valve.

There is no muscular continuity between the atria and the ventricles.

Heart valves are located at the following four orifices:

Right atrioventricular orifice is guarded by a tricuspid valve, the right atrio-ventricular valve that prevents regurgitation of blood into the atrium during ventricular systole. The valve has three triangular cusps attached to a fibrous ring. The anterior cusp is largest, the posterior (inferior) cusp is related to the diaphragmatic wall, and the medial cusp is related to the membranous part of the interventricular septum.

Left atrioventricular orifice is guarded by a (bicuspid) mitral valve that consists of two triangular cusps, a larger anterior (aortic) and smaller

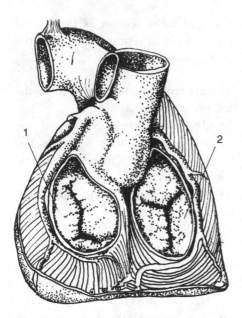

FIG. 31-9 Atrioventricular valves. (1) Left (bicuspid) valve; (2) Right (tricuspid) valve.

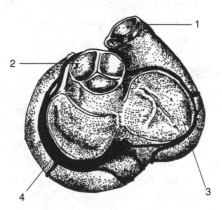

FIG. 31-10 Valves of the heart. (1) Pulmonary semilunar valve; (2) Aortic semilunar valve; (3) Right atrioventricular valve; (4) Left atrioventricular valve.

The oval membranous septum extends down from the trigone to form the upper part of the interventricular septum.

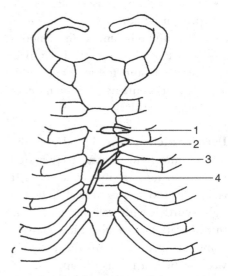

FIG. 31-11 Surface markings of hearts valves. (1) Pulmonary valve; (2) Aortic valve; (3) Left atrioventricular valve; (4) Right atrioventricular valve.

posterior cusp, attached to a fibrous ring. The valve prevents regurgitation of blood into the left atrium during ventricular systole.

Pulmonary orifice is located at the upper end of the smooth tapered infundibulum of the right ventricle. Its semilunar valve prevents regurgitation of ejected blood from the pulmonary trunk into the ventricle during ventricular diastole. The valve has three cusps; left, right and anterior that have thin sides or lunules. Their free edges

(directed toward the pulmonary trunk) are strengthened with a rim of fibrous connective tissue and a nodule in the center of the free edges completes closure of the lumen. Shallow, outward bulging of the wall of the pulmonary trunk produces sinuses which prevent the valve leaflets from sticking to the wall of the trunk when the valve is open.

Aortic orifice is also guarded by a semilunar valve with three cusps; left, right and posterior attached to the annulus fibrosus. The valve prevents regurgitation

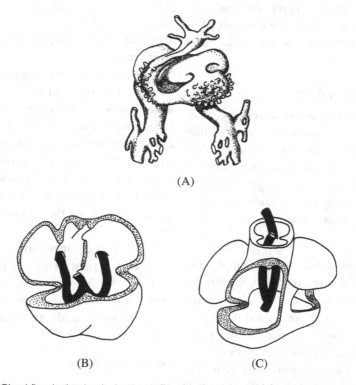

(A)

(B) (C)

FIG. 31-12   Blood flow in the developing heart. Blood in the right and left ventricular is ejected as two slightly off set streams into the conus cordis. Two spiral (trucoconal) ridges grow into the lee of the blood stream and fuse producing spiral separate channels leading to the aorta and pulmonary respectively.

of blood from the aorta into the ventricle during ventricular diastole. These valve leaflets have the same structure as leaflets of the pulmonary valve.

Three aortic sinuses are present in the wall of the aorta immediately above the aortic valve, one right (giving origin to the right coronary artery), one left (giving origin to the left coronary artery) and one (non-coronary) posterior. Blood is forced backwards into the coronary arteries during left ventricular diastole and coronary arteries arising from the sinuses are filled.

ARTERIES OF THE HEART

Coronary arteries supply the substance of the heart. The major coronary vessels lie between the epicardial and myocardial layers.

Right Coronary Artery

Right coronary artery arises from the anterior aortic sinus. It courses to the right in the atrioventricular groove to the inferior atrioventricular groove. It curves around the base of the heart and then obliquely to anastomose with the left coronary artery.

Its branches are:
Small unnamed branches that supply the aorta, pulmonary trunk, right auricle, atrium and ventricle. An important nodal branch courses toward the superior vena cava supplying the sinoatrial node and part of the right atrium.

Right marginal artery arises near the right margin of the heart passes along the inferior border of the heart toward the

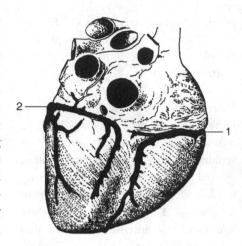

FIG. 31-13 Arteries of the heart from below. (1) Right coronary a; (2) Left coronary a.

apex of the heart. It supplies the adjacent right ventricle.

Posterior interventricular artery, the terminal part of the right coronary artery, arises where the right coronary artery enters the coronary groove. It courses in the posterior interventricular groove to reach the apex and supplies the adjacent parts of both ventricles and part of the interventricular septum. It gives a branch to the atrioventricular node.

The right coronary artery typically supplies the right ventricle, posterior part of the left ventricle, the right atrium and interatrial septum including the sinuatrial and atrioventricular nodes.

Left Coronary Artery

Left coronary artery arises from the left posterior aortic sinus behind the pulmonary trunk. It lies in the atrioventricular sulcus between the pulmonary trunk and the left auricle and anastomoses with the posterior interventricular branch at the back of the heart.

Its branches are:
Unnamed branches that supply the aorta, pulmonary trunk, and the left auricle, atrium and ventricle.

Anterior interventricular artery arises as the coronary artery enters the anterior interventricular groove. It descends towards the apex of the heart, turns around the apex and ascends in the posterior interventricular groove (reaching the posterior interventricular branch of the right coronary artery). It supplies both ventricles and most of the interventricular septum.

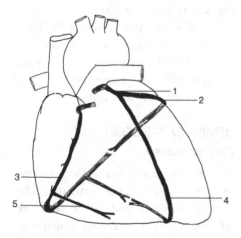

FIG. 31-14 Coronary arteries. (1) Left coronary a; (2) Circumflex a; (3) Right coronary a; (4) Anterior interventricular a; (5) Posterior interventricular a.

Circumflex artery is the direct continuation of the left coronary artery usually courses in the left part of the coronary groove finally anastomosing with the right coronary artery. It provides branches to the left atrium and ventricle. Its origin and area of distribution can be very variable. It may give branches to the atrioventricular and or sinuatrial nodes.

The left coronary artery typically supplies most of the left ventricle and left atrium. The anterior interventricular branch supplies most of the interventricular septum (and atrioventricular bundle and its branches). It may be the sole supplier of the sinuatrial and atrioventricular nodes.

Variations

The posterior interventricular artery may arise from the left coronary via the circumflex branch of the left coronary artery (10%). The circumflex artery may arise independently from the right aortic sinus rather than from the left. The anterior interventricular artery may arise independently from the left aortic sinus. The right coronary artery may arise from the pulmonary trunk.

VEINS OF THE HEART

Venae cordis minimae are small veins that begin in the heart.

Anterior cardiac veins lie in front of the right atrium and ventricle. They have independent openings into the right atrium.

Coronary sinus is a large venous channel lying in the atrioventricular

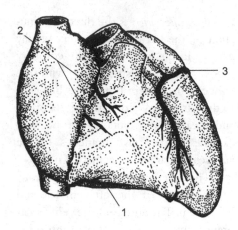

FIG. 31-15 Coronary vein. (1) Small cardiac v; (2) Anterior cardiac v; (3) Great cardiac v.

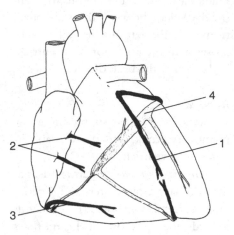

FIG. 31-16 Coronary veins. (1) Great cardiac v; (2) Anterior cardiac v; (3) Small cardiac v; (4) Coronary sinus.

sulcus between the left atrium and ventricle. It is a continuation of the great cardiac vein and opens into the right atrium, where its orifice has a rudimentary unicuspid valve. It receives blood from most of the named cardiac veins.

Tributaries are of the coronary sinus are the following:

Great cardiac vein begins at the apex of the heart, ascends in the anterior interventricular sulcus with the anterior interventricular branch of the left coronary artery then with the circumflex branch, and curves around the heart in the coronary sulcus, to continue as the coronary sinus.

Middle cardiac vein begins at the apex of the heart, passes on the back of the heart in the posterior interventricular groove, and joins the coronary sinus near its termination.

Small cardiac vein courses along the lower border of the anterior surface of

the heart, curves around its right margin to enter the right end of the coronary sinus near its termination.

Oblique vein passes across the back of the left atrium, it joins the coronary sinus. It is the remnant of the distal sinus horn — the more proximal part becomes the coronary sinus.

NERVES OF THE HEART
Cardiac Plexuses

This system controls the force or frequency of the heart contractions. The cardiac plexus consists of an admixture of sympathetic and parasympathetic (vagus) nerves. The plexus is divided into two parts:

Superficial plexus lies in the concavity of the aortic arch in front of the ligamentum arteriosum. This plexus consists of nerves from the inferior cervical cardiac branch of the left vagus nerve and the superior cervical cardiac branch of the left sympathetic trunk.

Deep plexus lies on the tracheal bifurcation. It consists of cardiac branches from all cervical sympathetic ganglia of both sides except the superior left, together with superior and inferior cervical and thoracic cardiac branches of the right vagus nerve and superior cervical and thoracic branches of the left vagus nerve.

Branches from the superficial plexus to the heart are distributed to

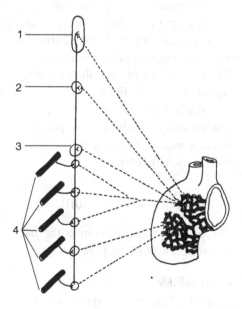

FIG. 31-17 Cardiac plexus. (1) Superior cervical cardiac n; (2) Middle cervical cardiac n; (3) Inferior cervical cardiac n; (4) Thoracic cardiac n.

the left half of deep, left anterior pulmonary and right coronary plexuses.

Branches from the deep plexus are distributed to the right atrium, right anterior pulmonary and right and left coronary plexuses, left atrium, left anterior pulmonary and left coronary plexus.

Branches of the coronary plexuses accompany the arteries and innervate the areas supplied by the arteries. Vagal fibers are cardioinhibitory while sympathetic fibers are cardioaccelerating, vasodilating and sensory.

CONDUCTING SYSTEM OF THE HEART

This system is responsible for the orderly sequence of the cardiac cycle. It consists of specialized cardiac muscle fibers that connect "pacemaker" regions of the heart with general cardiac muscle fibers. Intrinsic rhythmic contraction of cardiac muscle fibers is regulated by specialized nodal cells (pacemakers).

The conducting system consists of the sinuatrial node, atrioventricular node, atrioventricular bundle that has two limbs and a subendocardial plexus of Purkinje cells.

Impulses begin in the sinuatrial node, spread across atrial muscle to reach the atrioventricular node. From here, impulses travel in the atrioventricular bundle to cross the fibrous skeleton and then to Purkinje fibers isolated from surrounding myocardium by a connective tissue sheath to distribute in ventricular myocardium in particular to reach

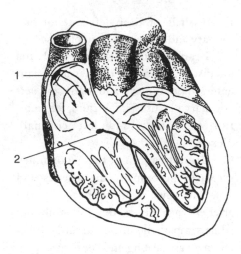

FIG. 31-18 Conducting system. (1) Sinoatrial node; (2) Atrioventricular node.

papillary muscles slightly ahead of the general myocardium and prevent valves from being everted.

Sinoatrial node is a small mass of specialized myocardial cells (P cells or nodal myocytes) located in the medial wall of the right atrium close to the opening for the superior vena cava. Myocytes of the node are very small and differ from atrial muscle in the nature of their action potential and dictates the rate of heart beat. It is said to have "pacemaker" characteristics. The sinuatrial node is usually supplied by a branch of the right coronary artery and it receives mostly parasympathetic nerves that slows its autorhythmicity.

Atrioventricular node is a similar smaller mass of nodal cells located in the interatrial septum just above the opening of the coronary sinus. Nodal cells have an intrinsic autorhythmicity like those of the sinuatrial node but slower and are normally stimulated by atrial depolarization. They contact atrial muscle and Bundle or P cells of the atrioventricular bundle. Slowed contraction through the node allows complete contraction of the atria before the beginning of ventricular contraction.

Atrioventricular bundle consists of modified cardiac muscle cells (Purkinje cells) located beneath the endocardium that have a rapid conduction rate. The bundle ascends to the membranous part of the interventricular septum. It then divides at the lower border of the membranous septum into right and left crura which continue onto the ventricular walls.

The right crus enters the septomarginal trabecula to reach the ventricular wall and anterior papillary muscle. Its fibers then form a subendocardial plexus of Purkinje fibers.

The left crus courses beneath the endocardium reaching the papillary muscles and also form a plexus in the wall of the left ventricle.

5. Posterior Mediastinum

The posterior mediastinum is the narrow region behind the pericardium.

BOUNDARIES ·

Anterior: Posterior surface of pericardium.
Posterior: Bodies of the fifth to twelfth vertebrae.

Superior: A line between the base of manubrium and the fourth thoracic vertebra.
Inferior: Diaphragm.

CONTENTS
Thoracic Aorta

Thoracic aorta is the continuation of the aortic arch. It extends from the left side of the lower border of the fourth thoracic vertebra to the lower border of the twelfth thoracic vertebra. It is continuous with the abdominal aorta at the aortic opening of the diaphragm.

Branches:

Posterior intercostal and subcostal arteries: Nine pairs supply the intercostal spaces.
Superior phrenic artery: Small branches to the posterior diaphragm.
Esophageal branches: At least two branches descend to help supply the esophagus.
Pericardial branches: A few small vessels to the back of the pericardium.
Mediastinal branches: Supply the lymph nodes of the posterior mediastinum.
Left bronchial arteries: These two arteries descend laterally to the back of the left bronchus and accompany it into the lung. The right bronchial artery usually arises from the first right aortic intercostal artery.

Esophagus

Esophagus enters the posterior mediastinum to the right of the aorta and descends from the fifth to the seventh thoracic vertebrae. It curves gradually forward and to the left until it lies in front of the aorta and traverses the diaphragm through the esophageal hiatus at the level of the ninth thoracic vertebra. It is narrowest at its beginning, compressed slightly as it passes behind the left bronchus and constricted again as it passes through the diaphragm.

Veins of the esophagus drain primarily into inferior thyroid veins in the neck, azygos and hemiazygos veins in the chest and left gastric vein in the abdomen. Thus connections can be established between the portal system below and the caval system and azygos system above.

Vagus Nerves

The right vagus passes between the first part of the subclavian artery and subclavian vein, down by the right brachiocephalic vein to the back of the root of the right lung where it forms the right posterior pulmonary plexus. It emerges from the plexus as two cords that pass to the back of the esophagus to form a plexus (the esophageal plexus) with the nerve of the opposite side. The right nerve passes through the esophageal hiatus of the diaphragm to reach the back of the stomach and joins splenic and celiac plexuses.

The left vagus descends between the left subclavian and carotid arteries and behind the left brachiocephalic vein (where it is crossed by the phrenic nerve) then to the left of the arch of the aorta to the posterior surface of the root of the left lung where it forms the left posterior pulmonary plexus.

It emerges from the plexus as one or two cords, and passes to the front of the esophagus. It passes through the diaphragm onto the anterior surface of the stomach joining the hepatic plexus.

At the lower end of the esophagus, the part of the plexus in front of the esophagus is known as the anterior vagal trunk and that behind the esophagus is the posterior vagal trunk but both have an admixture of left and right vagus nerves.

The esophageal plexus also receives branches from the greater splanchnic nerve and the thoracic sympathetic trunk. It supplies the smooth muscle and glands of the lower two-thirds of the esophagus.

The upper third of the esophagus contains skeletal muscle in its wall and is supplied by the recurrent laryngeal nerves. The wall of the middle third contains skeletal and smooth muscle fibers and the wall of the lower third contains only smooth muscle.

Thoracic Sympathetic Trunk

Thoracic sympathetic trunk is the continuation of the cervical sympathetic trunk that enters the thorax by passing by the side of the vertebrae over the neck of the first rib. It descends beneath the parietal pleura in front of the intercostal neurovascular bundle to traverse the diaphragm behind the medial arcuate ligament and to continue into the abdomen as the lumbar sympathetic trunk.

The sympathetic trunk consists of preganglionic fibers ascending or descending to reach their synaptic stations.

Postganglionic fibers pass medially to viscera which they reach as a plexus on the walls of a blood vessel or occasionally as independent fibers or laterally to join the ventral rami of every spinal nerve (31 pairs) as a grey ramus communicans. Some of these run with the spinal nerves to dorsal rami and reach blood vessels, arrector pili muscles and sweat glands of the posterior body wall.

There may be twelve separate ganglia on the sympathetic trunk but as the first may be fused with the inferior cervical ganglion (to form the cervicothroacic or stellate ganglion) and others also may be fused, there are most commonly eleven ganglia.

Chain ganglia contain synapses of preganglionic axons with cell bodies of postganglionic neurons, cell bodies of postganglionic (postsynaptic) neurons and fibers (sensory or postganglionic fibers) passing through the ganglia with or without synapsing in the ganglion.

Branches arising from the ganglia are:

Rami communicantes carry preganglionic sympathetic fibers to and postganglionic fibers from spinal nerves. Sensory nerves from viscera pass through the sympathetic trunk into rami communicantes thus reaching spinal nerves and dorsal roots.

Esophageal branches follow arteries to form the esophageal plexus with fibers from the left and right pulmonary plexuses, branches from upper thoracic ganglia and branches from both vagus nerves.

Aortic branches participate in forming the **aortic plexus**. The upper part of the aortic plexus is formed on the aorta by roots from the upper five sympathetic ganglia and afferent fibers from the vagi. The lower part receives additional branches from the splanchnic nerves.

Cardiac branches are direct branches from the upper four or five ganglia as well as direct cardiac nerves from cervical ganglia contribute to the cardiac plexus on the tracheal bifurcation.

Fibers from the left superior cardiac sympathetic nerve and inferior cervical cardiac of the left vagus and branches from the deep cardiac plexus form the superficial cardiac plexus on the concavity of the arch of the aorta in front of the ligamentum arteriosum.

Fibers from cardiac cervical ganglia (except the left superior nerve) and all the cardiac branches of the vagi abd recurrent laryngeal nerves (except the inferior cervical of the left vagus) form the deep cardiac plexus between the arch of the aorta and the trachea.

Pulmonary branches follow the posterior intercostal arteries to reach the posterior pulmonary plexus.

Greater splanchnic nerve are preganglionic sympathetic fibers that arise from the fifth to ninth ganglia as large roots that unite to form a nerve that descends medially over the bodies of the vertebrae on the lateral side of the azygos or hemiazygos veins respectively. It pierces the crus of the diaphragm and ends in the celiac ganglia and plexus.

Lesser splanchnic nerve arises from the tenth and eleventh ganglia, descends laterally to the greater splanchnic nerve and pierces the crus of the diaphragm to end in the celiac plexus.

Lowest splanchnic nerve arises from the twelfth ganglion, enters the abdomen by piercing the crus or alongside the sympathetic trunk and to end in the renal plexus.

Azygos Vein

Azygos vein developed from the supracardinal veins draining the dorsal body wall into the proximal part of the posterior cardinal veins. A cross communication develops between the two supracardinals, the left one looses its contact with the posterior cardinal forming the hemiazygos vein. The right supracardinal and proximal portion of the right posterior cardinal forms the azygos vein. The vena azygos is connected to the back of the inferior vena cava at the level of the renal veins but the connection may disappear and the vena azygos is then formed by the union of the right ascending lumbar and right subcostal veins.

The posterior body wall drains posteriorly into the azygos system that joins the superior and inferior vena cavae directly or indirectly.

The azygos system of veins have extensive connections with the valveless

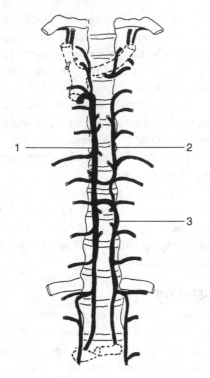

FIG. 31-19 Azygos system of veins. (1) Azygos v;
(2) Accessory hemiazygos v; (3) Hemiazygos v.

vertebral system of veins so they connect veins of the trunk and neck with
intracranial veins.

The azygos vein usually arises from
the right ascending lumbar vein joining
the right subcostal vein. At the level
of the renal vein, it ascends on the right
crus of the diaphragm and enters the
thorax piercing the right crus or by
passing through the aortic hiatus. It
continues up on the side of the vertebral bodies to the fourth thoracic vertebra where it arches anteriorly at the
sternomanubrial plane over the root of

the lung to enter the superior vena cava
near its termination.

The azygos vein is a right sided
channel largely responsible for draining
the posterior body walls left and right of
the thorax and abdomen. Some channels start on the left side and shunt their
blood across the midline and all end in
the right atrium.

The azygos vein receives all posterior intercostal veins except the below
tributaries:

The supreme intercostal vein on
each side drains forward over the apical
pleura to end in the brachiocephalic
vein.

The left second to the fourth intercostal veins (like the same on the right)
form a common trunk, the left superior
intercostal vein. Unlike the right trunk,
the common trunk on the left side terminates outside the azygos system (into
the left brachiocephalic vein.

Right subcostal and lower eight
posterior intercostal veins.

Right superior intercostal vein from
the second to fifth intercostal spaces.

Phrenic vein from the back of the
diaphragm.

Esophageal, pericardial and mediastinal branches.

Right bronchial vein from the lung.

Hemiazygos Vein

Hemiazygos vein, like the azygos vein,
arises from the junction of the left subcostal and ascending lumbar veins. It
ascends by piercing the left crus of the

diaphragm to enter the thorax. It ascends on the side of the vertebral bodies to the level of the ninth thoracic vertebra, where it curves sharply to the right, passing behind the aorta, esophagus, and thoracic duct to terminate in the azygos vein.

Principle tributaries are the left subcostal and lower four left posterior intercostal veins. Other tributaries are the left mediastinal veins and the lower esophageal veins.

Accessory Hemiazygos Vein

Accessory hemiazygos vein is a longitudinal channel in the upper part of the thorax on the left side. It arises from the medial end of the fourth or fifth intercostal space, and descends on the side of the vertebral bodies to the level of the eighth thoracic vertebra where it curves sharply to the right and joins the azygos vein.

Tributaries are the left bronchial veins and the fifth to eighth left posterior intercostal veins.

Thoracic Duct

Thoracic duct is a lymphatic channel that arises at the junction of the intestinal, lumbar and descending intercostal trunks from a sac, the cisterna chyli, that lies on the surface of the second lumbar vertebra to the right of the abdominal aorta. The duct enters the thorax through the aortic hiatus and ascends alongside the vertebral column between the aorta and the azygos vein. At the level of the fourth to sixth thoracic vertebra, it inclines to the left and continues along the left side of the esophagus deep to the arch of the aorta. It ascends into the root of the neck where it joins the left subclavian vein in the angle of junction with the internal jugular vein.

The tributaries of the thoracic duct are:

Channels from posterior mediastinal lymph nodes and small intercostal nodes.

Left bronchiomediastinal trunk, left subclavian trunk and left jugular trunk.

The thoracic duct receives all of the lymphatics below the diaphragm making it the main collecting trunk of the entire lymphatic system.

On the right side, the right bronchomediastinal, jugular, and subclavian trunks terminate in the brachiocephalic, internal jugular and subclavian veins respectively. The right jugular and subclavian trunks may unite to form the right lymphatic duct that descends across the front of the first part of the subclavian artery to end in right subclavian vein at its junction with internal jugular vein.

LYMPHATICS

Parietal lymphatics of the thorax drain to the below nodes.

Sternal nodes lie along the internal thoracic artery. They receive lymph from the breast, anterior thoracic wall, and upper abdominal wall and efferent vessels contribute to the bronchomedi-astinal trunk.

Intercostal nodes lie near the heads of the ribs, one or two for each inter-costal space. They receive lymph from the five lower spaces of the posterolat-eral thoracic wall and efferent vessels reach the thoracic duct.

Phrenic nodes are located in three groups on the thoracic aspect of the diaphragm:

Anterior nodes receive lymph from the upper surface of liver, diaphragm and anterior abdominal wall and effer-ent vessels and efferent vessels reach sternal nodes.

Middle nodes lie on the surface of the diaphragm close to the phrenic nerves. They receive lymph from the middle part of diaphragm and efferent vessels reach the anterior phrenic nodes.

Posterior nodes are located on the back of the crura next to the aorta. They receive lymph from the posterior diaphragm and efferents pass to poste-rior mediastinal nodes.

Visceral lymphatics of the thorax drain to the below nodes:

Anterior mediastinal nodes are located in the superior mediastinum. They receive lymph from the thymus, peri-cardium, heart and, pleura. Efferents join those from the trachea, bronchi and lungs to form the bronchomediastinal trunk.

Posterior mediastinal nodes sur-round the lower thoracic esophagus and receive lymph from the esophagus, peri-cardium and lower lobes of the lungs. Efferents drain into the thoracic duct and descending intercostal lymphatic trunks.

Tracheobronchial nodes form an inferior group in the angle of bifurca-tion of the trachea and a superior group in the angle between the trachea and bronchus on each side. They receive lymph from the visceral pleura, heart lower part of the trachea, bronchi and lung. Efferent channels ascend to the trachea.

Pleura, Lungs and Diaphragm

On each side of the mediastinum, two serous sacs enclose and invest the lungs. The visceral layer covers the lung and the parietal layer lines the inner surface of the chest wall, upper surface to the diaphragm, sides of the pericardium and superior mediastinum. The visceral and parietal layers become continuous in front of and behind the root of the lung. Below the root of the lung, a fold, the pulmonary ligament extends downward along the medial surface of the lung. It allows for the distension of the pulmonary veins. In development, the lungs expand into the medial aspect of the pleural cavities pushing mesothelium of the mediastinal pleura ahead hence the lung buds remain covered with mesothelium and each sac remains a closed sac.

Chapter 32

Pleura, Lungs and Diaphragm

Chapter Outline

The pleura is a thin, slippery serous membrane that lines the thoracic wall, mediastinum and diaphragm.

1. Layers

The pleura is lined by the parietal pleura and the visceral pleura.

Parietal pleura clothes the thoracic wall. It consists of mesothelium, subserosa of dense connective tissue containing elastic fibres and containing sensory nerve endings. It is much thicker than visceral pleura.

Visceral (pulmonary) pleura is thinner but also comprises the mesothelium and subserosa whose elastic fibers radiate into interlobular septa intimately associated with the lung and follows its fissures.

Parietal pleura comprises:

Costal pleura covers the ribs and intercostal muscles.

Mediastinal pleura covers the mediastinal surface.

Diaphragmatic pleura covers the thoracic surface of the diaphragm.

Cervical pleura covers the projection of the pleural cavity above thoracic inlet (extending to the neck of the first rib posteriorly) forming the scupula in the neck. Here, it is reinforced by a thickened sheet of connective tissue attached to the inner surface of the first rib — the suprapleural membrane.

The two layers of pleura are continuous on the mediastinal surface as a narrow tube, the upper part of which envelopes the structures of the root of the lung and the collapsed lower part of which forms the pulmonary ligament.

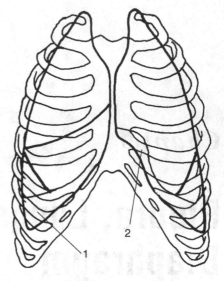

FIG. 32-1   Pleura from in front. (1) Costodiaphragmatic recess; (2) Costomediastinal recess.

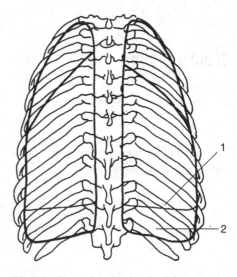

FIG. 32-2   Pleura from behind. (1) Inferior border of lung; (2) Inferior reflection of pleura.

Between the parietal and visceral layers of pleura is a potential space, the pleural cavity. It is filled with a capillary thin film of serous fluid that provides lubrication allowing the visceral surface to slide over the parietal surface. The fluid layer also provides surface tension between the layers keeping the lung inflated.

2. Lines of Pleural Reflection

These are lines along which the costal pleura which become continuous with the mediastinal pleura anteriorly and posteriorly, and the diaphragmatic pleura inferiorly. They are important clinically to determine whether penetrating wounds have or have not entered the lung or pleural cavity, and the location of accumulation of fluid.

Posterior reflection where the mediastinal pleura becomes continuous with the costal pleura posteriorly. It extends from the cupula vertically to just below the twelfth rib (where the pleural cavity is unprotected by the twelfth ribs).

Anterior reflection where the costal pleura becomes continuous with the mediastinal pleura anteriorly.

The anterior border of the pleural cavity differs slightly on each side. The right and left anterior lines pass behind the sterno-clavicular joints, and meet in the median plane at or above the sternal angle.

The *right reflection* continues downward in the midline to the back of the xiphoid process while the left deflects laterally to the margin of the sternum (fourth costal cartilage) and continues downward to the sixth costal cartilage.

The right reflection descends lower than the left anteriorly (and is unprotected by ribs or cartilage in the right xiphicostal angle).

The *inferior reflection* is where the costal pleura becomes continuous with the diaphragmatic pleura. On both sides this line passes obliquely across the eighth, tenth and twelfth ribs crossing the eighth in the midclavicular line, tenth in the midlateral line and twelfth at the neck of the twelfth rib.

At two points, potential spaces are formed by separation of the parietal pleural layers. The margins of the lungs alternatively advance or recede from these spaces in respiration.

PLEURAL SPACES

Costomediastinal recess is the part of the pleural cavity between the anterior reflection of costal to mediastinal pleura.

Costodiaphragmatic recess is outlined by the inferior reflection of costal to diaphragmatic pleura.

The **neurovascular system** of the pleura is supplied by:
Arteries: Bronchial and pulmonary arteries supply the visceral pleura and intercostal and pericardiophrenic arteries supply the parietal pleura.
Lymphatics: From the visceral pleura drain to lymph nodes in the hilum.

Those draining the parietal pleura drain to lymph nodes in the back of the thorax (which in turn drain to axillary nodes), root of the neck and along the internal thoracic vessels.

Veins: via pulmonary veins.

Nerves: Supplying the parietal pleura pass with the phrenic, intercostal, vagal, and sympathetic nerves. The parietal pleura is extremely sensitive. Sensory nerves in the phrenic nerve supply the mediastinal pleura and central diaphragmatic pleura. Sensory nerves supplying the visceral pleura pass with the vagus and sympathetic nerves via the pulmonary plexus. The visceral pleura is insensitive and some of these nerves may be involved in respiratory reflexes.

Lungs

Chapter Outline

1   Surfaces and Borders
2. Fissures and Lobes
3. Root of the Lung
4. Bronchial Tree — Bronchopulmonary Segments

The lungs are the essential organs of respiration. Each lung is covered by pleura and is attached to the heart and trachea by its root and the pulmonary ligament. It is otherwise free in a pleural cavity. The lungs are elastic, spongy and pliable, they expand and contract in inspiration and expiration. The color of the lungs varies from pink in the newborn to dark and mottled in adults from the accumulation of carbon particles with aging or exposure to a carbon containing atmosphere.

1. Surfaces and Borders

The lungs are conical in shape and have the following surfaces and borders.

Apex is rounded and rises above the anterior end of the first rib to the neck of the rib. It is covered by a cupola of the pleura and grooved by the subclavian artery.

Base is concave and rests on the diaphragm. The base of the right lung is separated by the diaphragm from the liver and the left lung from the liver, spleen and stomach.

Costal surface is a large and convex surface conformed to the curvature formed by the ribs, sternum and costal cartilages.

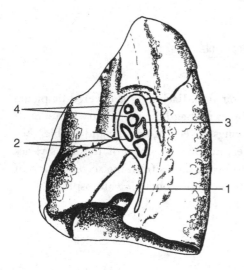

FIG. 32-3 Medial surface of right lung. (1) Pulmonary ligament; (2) Pulmonary v; (3) Bronchus; (4) Pulmonary a.

Mediastinal surface bears the impressions of the heart and other structures covered by mediastinal pleura.

RELATIONS OF THE RIGHT LUNG

Pericardium and heart, upper end of the inferior vena cava, a superior vena cava, thymus, end of the ascending aorta and beginning of the aortic arch, right phrenic nerve and pericardiacophrenic artery, arch of the vena azygos, trachea, right vagus, side of the esophagus, trachea and vena azygos (behind the esophagus).

RELATIONS OF THE LEFT LUNG

Pericardium and heart, pulmonary trunk and ascending aorta, descending aorta, lower part of the esophagus, arch of the aorta, left subclavian, esophagus, thoracic duct, left recurrent laryngeal nerve, left common carotid artery, left phrenic and pericardiacophrenic artery and left vagus nerve.

Anterior border separates the costal from the mediastinal surfaces. On the left side, it has a deep indentation, the cardiac notch which leaves part of the pericardium uncovered by lung. The lingula is a small tongue shaped part of the left lung between the cardiac notch and the oblique fissure. It corresponds to the middle lobe of the right lung.

Posterior border is a poorly-defined line between the costal surface and the posterior part of the medial surface.

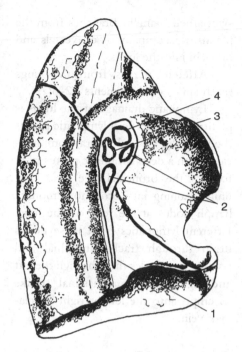

FIG. 32-4 Medial surface of left lung. (1) Pulmonary ligament; (2) Pulmonary v; (3) Bronchus; (4) Pulmonary a.

Inferior border is sharp margin that separates the diaphragmatic surface from the costal and mediastinal surfaces. It occupies the costodiaphragmatic recess in respiration.

2. Fissures and Lobes

Fissures of visceral pleura divide the lungs into lobes. Both lungs are divided by an oblique fissure that extends from the posterior border somewhat below the apex to the inferior border near the midclavicular line. The right lung has a horizontal fissure that extends from near the middle of the anterior border to the oblique fissure.

Right lung consists of superior, middle and inferior lobes.

Left lung consists of superior and inferior lobes.

The number of lobes, however may vary, being increased by the presence of an extra fissure or decreased by fusion.

The oblique fissure on the right side begins at the level of the head of the fifth rib and follows the line of the sixth rib ending near the sixth costochondral junction. This line approximates the line of the medial border of the scapula when the arm is abducted and the hand placed on the back of the head.

The horizontal fissure begins from the oblique fissure near the midaxillary line at the level of the sixth rib reaching the anterior border at the level of the fourth costal cartilage.

3. Root of the Lung

The root of the lung is a wedge shaped area above and behind the cardiac impression on the mediastinal surface of each lung. The root is enveloped in pleura and is bounded below by the pulmonary ligament. It contains structures that enter or leave the lung at the hilus.

These are:
Principal (primary) bronchus lies behind the blood vessels.

Pulmonary artery is a single vessel which divides into branches that accompany the bronchi.

The **pulmonary veins** are found on both side of the lungs paired veins take an independent course within the lung.

Bronchial arteries come off the aorta. Two branches supply the left lung. For the right lung, the artery arises from or within the superior left bronchial branch, or from the first right aortic intercostal artery. They supply the bronchi, pulmonary vessels, and the pleurae.

Bronchial veins originate near the hilus and drain into the azygos vein on the right side and the accessory hemiazygos vein on the left.

In the anteroposterior plane, the upper of the two veins lie in front, the pulmonary artery and bronchus are behind and the bronchial vessels most posterior. Vertically, the arrangement is (from above downwards) superior lobar bronchus, pulmonary artery, principal bronchus, and lower pulmonary vein on the right side. On the left side is the pulmonary artery, principal bronchus and lower pulmonary vein.

Pulmonary autonomic plexuses consist of branches of the vagus and sympathetic nerves via the interconnected anterior and posterior pulmonary plexuses. The anterior plexus lies on the front of the root and is formed mainly by anterior pulmonary branches of the vagus and fibers from the cardiac plexus. The posterior plexus is much larger and located on the back of the root of the lung. It is formed by a trunk of the vagus and branches from the second to the fourth thoracic sympathetic ganglia. Branches from the plexuses accompany blood vessels and bronchi into the lungs.

Afferent nerves from the lungs reach spinal cord segments T2–6(7).

Lymphatic drainage of the lungs is plentiful. A superficial lymphatic plexus (beneath the visceral pleura) communicates with a deep lymphatic plexus (that follows the bronchial tree) and they unite draining into bronchopulmonary lymph nodes at the root of the lung. Efferent lymphatics from these nodes drain through tracheobronchial and paratracheal nodes then through the left and right bronchomediastinal trunks to empty into the left or right subclavian vein.

4. Bronchial Tree

In fetal development, the early entodermal lung bud gives rise to the specialized lining epithelium of the conducting and respiratory passages of the lungs. It grows into the surrounding mesoderm that contributes the supporting tissues. In addition, the lungs expand into the coelom (pleural cavities).

The lung bud bifurcates (at 4 mm) into two primary bronchi. These branch in a dichotomous fashion forming some 17 generations of divisions by birth. An additional six generations branch during early postnatal growth. Within the lung tissue, the main bronchi divide into secondary or lobar bronchi on each side. These in turn divide into tertiary or segmental bronchi that run to discrete regions of lung known as bronchopulmonary segments (see below).

The larger subdivisions of the bronchial tree act as conducting tubes and airconditioners and their walls are reinforced to retain patency during respiration. These tubes are the trachea, primary and secondary bronchi, bronchioles and terminal bronchioles.

More distally are respiratory tubes where gaseous exchange occurs with the vascular system. These tubes (terminating in sacs) are respiratory bronchioles, alveolar ducts and finally alveolar sacs.

In the adult, the trachea bifurcates at the level of the intervertebral disc between T4 and T5 into right and left main bronchi. Each of these bronchi descends into the hilus of the lung. Three lobar bronchi arise on the right side and two on the left; each divides into segmental bronchi. The portion of the lobe of the lung supplied by a segmental bronchus is a self-contained, functionally independent unit of lung tissue known as a bronchopulmonary segment. There are usually ten bronchopulmonary segments on the right side and eight on the left.

With respect to the distribution of bronchopulmonary segments, the left and right lungs differ in that:

(1) The right upper and middle lobar bronchi are partially fused on the left side to form the left upper lobe bronchus.

(2) Apical and posterior segmental bronchi divide late (arise from a common stem) on the left side.

(3) The anterior basal and medial basal bronchi divide later on the left side.

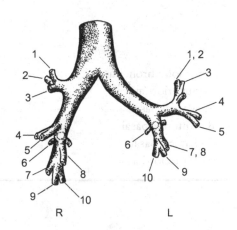

FIG. 32-5 Bronchopulmonary segments. Right lung. (1) Apical; (2) Posterior; (3) Anterior; (4) Lateral; (5) Medial; (6) Apical; (7) Anterior; (8) Medial basal; (9) Lateral basal; (10) Posterior basal. Left lung. (1, 2) Apicoposterior; (3) Anterior; (4) Superior lingular; (5) Inferior lingular; (6) Apical; (7) Medial basal; (8) Anterior basal; (9) Lateral basal; (10) Posterior basal.

Table 1. Summary of the Bronchial Tree

Trachea

Left principal bronchus	Right principal bronchus
Superior lobe bronchus	Superior lobe bronchus
Apicoposterior	
1. Apical	1. Apical
2. Posterior	2. Posterior
3. Anterior	3. Anterior
4. Superior lingular	
5. Inferior lingular	
	Middle lobe bronchus
	4. Lateral
	5. Medial
Inferior lobe bronchus	Inferior lobe bronchus
6. Superior (apical)	6. Superior (apical)
7. Medial basal	7. Medial basal
8. Anterior basal	8. Anterior basal
9. Lateral basal	9. Lateral basal
10. Posterior basal	10. Posterior basal

Diaphragm

Chapter Outline

1. Structure
2. Openings of the Diaphragm
3. Structures also Traversing the Diaphragm

The diaphragm is the convex musculofascial partition separating the thoracic and abdominal cavities.

1. Structure

- **Origin:** Extensive, muscular, from the circumference of the thoracic outlet; namely: the back of xiphoid process by a pair of short, narrow slips (the sternal part), inner surfaces of the lower six costal cartilages by wide slips (costal part), lateral arcuate ligament (a thickening of the fascia over quadratus lumborum), the medial arcuate ligament (the fascia over psoas major) and the bodies of the upper lumbar vertebrae (the lumbar or vertebral part).

The part of the diaphragm arising from the lumbar vertebrae form two muscular crura or bundles. The right crus extends from the upper three or four vertebrae and the left crus extends from the upper three lumbar vertebrae.

Median arcuate ligament is a fibrous arch uniting the left and right crura and forming the aortic opening.

- **Insertion:** Crescentic central tendon divided into right, left, and median lobes.
- **Action:** Chief muscle of respiration. It descends in inspiration drawing the central tendon down and increasing the volume of the thorax (and decreasing the abdominal volume).
- **Nerve supply:** Phrenic nerve (C3–C5) separately to each side however the halves contract synchromously. There is some innervation of the crura from lower thoracic and upper lumbar nerves and sensory innervation by intercostal nerves at the margin.

- **Arterial supply:** Pericardiacophrenic, musculophrenic, superior and inferior phrenic arteries.

Vertebrocostal trigone is the triangular gap in the origin of the diaphragm between the twelfth rib and the lateral arcuate ligament. It is closed by a thin sheet of fibrous tissue.

2. Openings in the Diaphragm

The aortic opening is under the crura and behind the median arcuate ligament opposite the twelfth thoracic vertebra. The aorta passes behind the diaphragm. The opening transmits the aorta, thoracic duct and azygos vein.

Esophageal opening is slung in the right crus opposite the 10th thoracic vertebra. It transmits the esophagus, esophageal branches of the left gastric artery and vein, and the anterior and posterior vagal trunks.

Foramen for the inferior vena cava lies between the right and median lobes of the central tendon. It transmits the inferior vena cava, right phrenic nerve, and lymphatics from the liver.

3. Structures also Traversing the Diaphragm

Superior epigastric vessels pass behind xiphoid process and 7th costal cartilage.

Musculophrenic vessels pass between 7th and 8th costal cartilages.

Hemiazygos vein passes through the left crus.

Subcostal vessels and nerve pass behind lateral arcuate ligament.

Lower five intercostal vessels and nerve pass behind slips from 7th to 12th cartilages. Sympathetic trunk pass behind medial arcuate ligament.

Three splanchnic nerves pass through crura.

Left phrenic nerve traverses the left dome of the central tendon.

Abdomen

The abdomen is that part of the trunk that lies between the thorax and the pelvis. The abdominal cavity is separated from the thorax by the diaphragm and from the pelvic cavity by an arbitrary line passing through the arcuate lines of the bony pelvis. The abdominal cavity is often considered to comprise the abdominal cavity and the cavity of the lesser pelvis since these cavities are continuous.

The abdomen contains most of the digestive organs, part of the urogenital system, the spleen, suprarenal glands and parts of the autonomic plexuses.

The lesser pelvis contains the urinary bladder, abdominal parts of the ureters, sigmoid colon, rectum and some coils of small intestine as well as the internal genitalia.

The abdominal walls are mostly arranged in layers that accommodate for expansion in pregnancy, ingestion of foods and storage of fat.

Chapter **33**

Anterior Abdominal Wall

Chapter Outline

1. Horizontal Planes of the Abdomen

The following planes vary with individuals but anatomical relationships to the planes are clinically useful.

Transpyloric plane lies midway between the jugular notch and pubic symphysis. It passes through the lower border of the first lumbar vertebra. In the plane, lie the pylorus, inferior margin of the liver, neck of the gallbladder, hili of the kidneys, anterior end of the spleen, origin of the superior mesenteric artery, origin of the portal vein, neck of the pancreas and the conus medullaris (end of the spinal cord).

Subcostal plane lies tangential to the lower border of the skeleton of the thorax. It passes through the upper border of the third lumbar vertebra.

Supracristal plane passes through the highest level of the iliac crests and the lower part of the body of the fourth lumbar vertebra. The abdominal aorta bifurcates in this plane (and the plane is a useful landmark for performing a lumbar puncture).

Intertubercular plane passes through the highest level of the iliac crests from in front (tubercles of the ilium). It passes through the body of the fifth lumbar vertebra and the inferior vena cava begins in this plane.

The umbilicus and transumbilical plane is variable but usually lies at the level of the disc between the bodies of the third and fourth lumbar vertebrae.

The abdomen is divided topographically into right and left, upper and lower quadrants by the vertical median plane and horizontal umbilical plane. It is also frequently divided into nine regions by two horizontal planes (subcostal and transtubercular) and two vertical planes (right and left lateral planes).

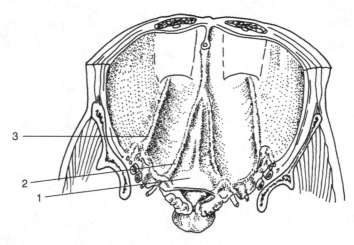

FIG. 33-1 Anterior abdominal wall from within. (1) Median umbilical fold; (2) Medial umbilical fold; (3) Lateral umbilical fold.

The lateral planes run sagittally through the midinguinal point (midway between the anterior superior iliac spine and pubic symphysis). The regions defined are laterally, left and right hypochondriac, lumbar and inguinal regions and medially, the epigastric, umbilical and hypogastric regions.

The anterior abdominal wall consists of three sheets of muscle, fleshy at the sides but aponeurotic in front and behind. In addition, there is a wide vertical muscle (rectus abdominis) on each side of the midline, in front, enclosed in an aponeurosis.

2. Fascia

SUPERFICIAL FASCIA

Superficial fascia consists of two distinct layers (particularly in the lower abdominal wall) and contains a variable amount of fat.

Layers of Superficial Fascia

1. An adipose layer.
2. A deeper, more collagenous membranous layer. The membranous layer is attached below to the iliac crest, fascia lata of the thigh a finger's breadth below the inguinal ligament and the pubic tubercle. It continues in front of the pubic symphysis below and between the two pubic tubercles and is attached to the inferior ischiopubic rami and posterior border of the perineal membrane.

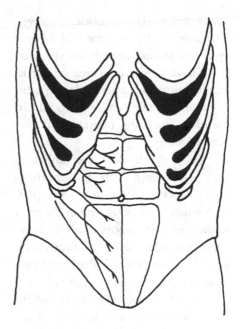

FIG. 33-2 Cutaneous distribution of thoracoabdominal nerves.

Contents

Cutaneous nerves are the anterior and lateral cutaneous branches of the seventh to eleventh intercostal nerves (T7 to 11) and the anterior cutaneous branches of the subcostal (T12) and iliohypogastric (L1) nerves.

T10 supplies the skin around the umbilicus. Three nerves, (T9, 8, and 7) supply the region above and three nerves (T11, 12 and L1) supply the region below. T7 to L1 run between internal oblique and transversus muscles supplying them. T7–T12 (not L1) pierce rectus abdominis supplying it.

Cutaneous arteries are branches of the lower intercostal arteries that accompany the lateral cutaneous nerves. Branches of the superior and inferior epigastric arteries accompany the anterior cutaneous nerves. Three superficial inguinal arteries (superficial epigastric, external pudendal and superficial circumflex iliac) are branches of the femoral artery.

Below the level of the umbilicus, cutaneous veins such as superficial inguinal veins accompany corresponding arteries and below end in the great saphenous vein. Above the umbilicus, superficial veins drain to thoracoepigastric veins that reach the lateral thoracic vein of the axillary system.

Superficial epigastric veins and lateral thoracic veins anastomose and can unite the veins of the upper and lower halves of the body.

Veins from the umbilical region communicate with paraumbilical veins and reach the portal vein via the ligamentum teres.

Cutaneous lymphatics from the upper part of the abdomen drain into axillary nodes and lymphatics from the lower part of the abdomen drain into inguinal nodes.

DEEP FASCIA

Deep fascia is the extremely thin investing layer of the external oblique muscle. It continues medially over the external oblique aponeurosis to reach the linea alba. Below, it is prolonged onto the spermantic cord as the external spermatic fascia.

3. Muscles

Abdominal muscles flatten the anterior abdominal wall, compress the viscera and help flex and rotate the vertebral column. They help to maintain the position of abdominal organs, raise intra-abdominal pressure (in expelling urine, feces, vomiting), assist in forced expiration (sneezing, coughing or phonation). They are innervated by the seventh to twelfth intercostal, iliohypogastric and iliolumbar nerves.

External Oblique

Fibers of external oblique are directed downwards and forwards.

- **Origin:** Lower eight ribs (along a line a handsbreadth from the costal margin) interdigitating with serratus anterior and latissimus dorsi muscles.
  1. The posterior border is free and fleshy between the last rib and the iliac crest.
  2. A small triangular gap between external oblique and latissimus dorsi is the lumbar triangle. It is the site of lumbar hernias and may be surgical approach to the kidney.
  3. Between the pubic tubercle and the anterior superior iliac spine, the free lower border of the external oblique aponeurosis is the inguinal ligament.
- **Insertion:** Fleshy fibers pass anteroinferiorly (external intercostals) and become aponeurotic

and completely cover rectus abdominis. It reaches the anterior half of the outer lip of the iliac crest, xiphoid process, linea alba, pubic crest, and pectineal line.

The pubic crest forms the base of a triangular deficiency in the aponeurosis, the superficial inguinal ring.

Internal Oblique

Internal oblique is upper muscle fibers that radiate upward and forward, lower fibers radiate downwards towards the pubic crest and between, fibers run almost horizontally.

- **Origin:** Behind by aponeurosis from the thoracolumbar fascia, below by fleshy fibers from the anterior two-thirds (intermediate lip) of the iliac crest, and the lateral half of inguinal ligament.
- **Insertion:** Posterior (muscular) fibers insert into the lower borders of the lower six costal cartilages.

The aponeurotic part divides to form anterior and posterior laminae of the rectus sheath blending in the midline with its fellow of the opposite side to form a raphe, the linea alba. Fibers arising from the inguinal ligament arch medially over the spermatic cord and unite with the aponeurosis of transversus abdominis forming the conjoint tendon. This passes in front of the rectus abdominis to insert into the crest and medial part of the pectineal line of the pubis. The rectus sheath is therefore

deficient behind and its lower edge forms the arcuate line.

Transversus Abdominis

Most of its muscle fibers run horizontally (transversely).

- **Origin:**
  (1) Posteriorly fleshy fibers arise from the aponeurosis formed by fusion of the three layers of the thoracolumbar fascia.
  (2) Inner surfaces of the sixth to twelfth costal cartilages interdigitating with the diaphragm.
  (3) Anterior two-thirds of inner lip of the iliac crest.
  (4) Lateral third of the inguinal ligament.
- **Insertion:** Xiphoid process, linea alba, pubic crest and pectineal line.

Linea alba is a tendinous band (raphe) extending from the xiphoid process to the pubic symphysis. It is formed by the interlocking insertions of the aponeuroses of the three flat abdominal muscles.

Rectus Abdominis

Rectus abdominis is a long, thin, wide muscle lying ventrally alongside the midline. It is enclosed within the rectus sheath. Three tendinous intersections cross the muscle, one at the level of the umbilicus, one at the xiphoid, and one between and attached to the anterior layer of the rectus sheath. They are an indication of the segmental origin of the muscle.

- **Origin:** Pubic crest and symphysis.
- **Insertion:** Front of the xiphoid process and fifth to seventh costal cartilages.

Transversalis fascia is the internal fascial lining of transversus. It is separated from the peritoneum by a thin layer of extra peritoneal areolar tissue. It is continuous posteriorly with the fascia lining the posterior abdominal wall. In the inguinal region, the femoral vessels are covered anteriorly by transversalis fascia that becomes the anterior layer of the femoral sheath (the posterior layer arises from the iliac fascia).

Pyramidalis

Pyramidalis is a small, triangular muscle in front of the lower part of rectus abdominis. It may be absent.

- **Origin:** Pubic crest.
- **Insertion:** Linea alba below the umbilicus.
- **Action:** Tightens the linea alba.
- **Nerve supply:** Subcostal nerve (T12).

RECTUS SHEATH

The aponeuroses mostly of transversus and internal oblique form a sheath around the rectus abdominis and stop the rectus from bowstringing.

The aponeurosis of internal oblique splits at the lateral border of rectus into anterior and posterior laminae that ensheath the rectus and fuse in the

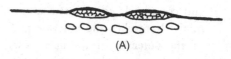

(A)

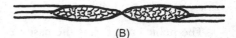

(B)

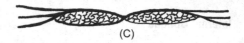

(C)

FIG. 33-3 Transverse section of rectus sheath. (A) Above costal margin; (B) At an intermediate level; (C) Below the arcuate line.

midline forming the linea alba. The posterior lamina is reinforced by the aponeurosis of transversus abdominis and the anterior lamina is reinforced by the aponeurosis of external oblique.

The composition of the sheath varies at three levels:

(1) **Above the costal margins**, the posterior layer lacks a contribution from internal oblique and transversus but external oblique forms a strong anterior lamina. Posteriorly, rectus lies on costal cartiages five to eight and the xiphoid process.

(2) **Below the arcuate line**, the sheath lacks a posterior lamina — the aponeurosis of internal oblique joins the aponeurosis of transversus and passes in front of rectus as the conjoined tendon (falx inguinalis). The aponeurosis of external oblique reinforces (fuses with) the anterior layer.

(3) **At an intermediate level,** internal oblique aponeurosis splits at the lateral edge of rectus into anterior and posterior laminae that enclose the rectus and join with their fellow at the linea alba. The aponeurosis of external oblique fuses with the antrior lamina and the aponeurosis of transversus fuses with the posterior layer.

The **contents of sheath** include the rectus abdominis and pyramidalis which are attached to the pubic crest, superior and inferior epigastric vessels (which anastomose in the sheath), terminal parts of the seventh to eleventh intercostal nerves and lymphatics.

Arcuate line (or linea semicircularis) is the curved lower free edge of rectus sheath where the aponeurosis of transversus ends and passes onto the anterior wall of the rectus sheath.

Linea semilunaris is the curved line at the lateral edge of the rectus where the aponeuroses of internal oblique and transversalis meet.

4. Inguinal Region

The inguinal region is the triangular region below the transtubercular plane bounded by the inguinal ligament, the lateral margin of rectus muscle.

Superficial fascia consists of a fatty layer containing cutaneous vessels and nerves and a membranous layer that passes down over the inguinal ligament to join the fascia lata of the thigh.

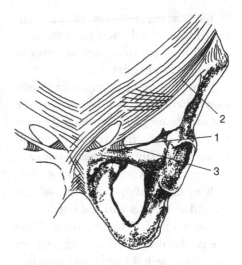

FIG. 33-4 Inguinal ligament. (1) Lacunar ligament; (2) Inguinal ligament; (3) Pectineal ligament.

Aponeurosis of external oblique gives rise to three ligaments:

Inguinal ligament is the thickened, inwardly rolled lower border of the aponeurosis of the external abdominal oblique extending from the anterior superior iliac spine to the pubic tubercle and crest. The deep fascia of the thigh is attached to the inguinal ligament.

Lacunar ligament (pectineal part of the inguinal ligament). The triangular backward and upward expansion of the medial part of the inguinal ligament attached to the pubic tubercle at its apex and extended onto the pectineal line. Its base forms a gutter-like floor to the medial margin of the femoral ring.

Pectineal ligament is a fibrous, lateral extension along the iliopectineal line from the base of the lacunar ligament.

It receives the insertion of the conjoined tendon and fibers from the linea alba. It is an anchor in surgical repair of some inguinal hernias.

Internal oblique originates in part from the lateral two thirds of the inguinal ligament. Its aponeurosis fuses with that of the transversus abdominis to form the falx inguinalis (conjoined tendon), which is attached in the midline with its fellow of the opposite side, to the end of the linea alba and inferiorly to the pecten pubis. The lowermost fibers overlie the deep inguinal ring and are continuous with the cremaster muscle.

The fibers of cremaster are disposed in downwardly directed loops whose attachments are continuous at both ends with those of internal oblique. Cremaster is found only in the male and causes elevation to the testis. It is supplied by the gentiofemoral nerve (L1).

In the pubic region, the transversus abdominis fuses with internal oblique contributing to the formation of falx inguinalis. Muscle fibers arise from the lateral third to one half of the inguinal ligament and arch over the spermatic cord from its deep surface above the deep inguinal ring. Fibers become aponeurotic and form the deeper layer of the conjoined tendon.

Some fibers, an inconstant thickening arising from the lower edge of transversus, arch down to the inguinal ligament just medial to the deep inguinal ring. This band of variable strength, is known as the interfoveolar ligament.

Transversalis fascia is the internal lining of the abdominal wall. Medially, it forms the posterior wall of the rectus sheath and the inguinal canal. In the inguinal region, it bounds the internal opening of the inguinal canal (deep inguinal ring) and is continuous with the internal spermatic fascia, the innermost layer of the covering of the spermatic cord. It is carried into the thigh below the inguinal ligament, as the anterior wall of the femoral sheath. Laterally, it is continuous with the iliac fascia and the lateral part of the inguinal ligament.

THE INGUINAL CANAL

The inguinal canal is an oblique passage about 4 cm long through the lower part of the anterior abdominal wall leading from the extraperitoneal layer internally to the subcutaneous fatty layer externally. It is situated above the medial half of the inguinal ligament and directed medially, downward and forward.

The inguinal canal has two openings, two walls, a roof and a floor:
Superficial inguinal ring is a triangular opening in the aponeurosis of the external oblique just above the pubic tubercle and the medial end of the inguinal ligament. The margins of the ring are the medial and lateral crura held together by intercrural fibers. The ring is closed by the thin external spermatic fascia.

Deep inguinal ring is a circular aperture in the transversalis fascia located about 1.5 cm above the mid-inguinal point (the point on the inguinal ligament equidistant from the pubic symphysis and the anterior superior iliac spine).

The boundaries of the inguinal canal are:
Anterior: Is formed by the aponeurosis of external abdominal oblique muscle reinforced in its lateral part by medial fibers of internal oblique.
Posterior: Is formed throughout by fascia transversalis reinforced medially by falx inguinalis.
Roof: Arched lower edges of transversus and internal oblique muscle.
Floor: Attachment of fascia transversalis to the inguinal and lacunar ligaments.

Contents of the Inguinal Canal

In the male

The canal consist spermatic cord, vestige of the processus vaginalis, terminal parts of ilioinguinal nerve (L1) and genital branch of genitofemoral nerve (L2), cremasteric artery and vein.

The spermatic cord is composed of the ductus deferens and artery to the ductus, testicular artery, pampiniform plexus of veins, testicular lymphatics, the cremasteric artery, genitofemoral nerve and deferential plexus (with the artery to the ductus) and testicular plexus (with the testicular artery).

In the male, the spermatic cord becomes enveloped by some layers of the anterior abdominal wall during the course of the descent of the testes from

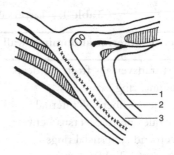

FIG. 33-5 Coverings of the spermatic cord and corresponding anterior abdominal layers. (1) External spermatic fascia; External oblique aponeurosis; (2) Cremaster m and fascia, Internal oblique; (3) Internal spermatic fascia, Transversalis fascia.

the abdominal cavity into the scrotum. These layers and the corresponding coverings of the spermatic cord are shown in the above table:

In the female

The inguinal canal transmits the round ligament of uterus and accompanying vessels and the ilioingual nerve (L1).

Nerve Supply

The spermatic cord is supplied by the genital branch of the genitofemoral nerve, and testicular sympathetic plexus joined by the pelvic plexus accompanying the deferential artery.

The anterior part of the scrotum is supplied by L1 through the anterior scrotal branches of the ilio-inguinal nerve (L1) and genital branch of the genitofemoral nerve (L1, 2).

The posterior part of the scrotum is supplied by S3 through posterior scrotal branches of the perineal nerve (branch of the posterior femoral cutaneous

Table 1.   Layer of Anterior Abdominal Wall

Layer of anterior abdominal wall	Coverings of spermatic cord
Fascia transversalis from the deep inguinal ring	Internal spermatic fascia
From fleshy fibers of internal oblique and transversus between deep and superficial rings	Cremaster muscle (a single areolar membrane in which cremaster fibers are spread out)
From external oblique aponeurosis at the external inguinal ring	External spermatic fascia

nerve) and perineal branch of the posterior femoral cutaneous nerve (S1, 2, 3).

5. Arteries

The superficial and deep arteries supplying the abdominal wall arise from the femoral and external iliac arteries respectively. The external iliac artery becomes the femoral artery at the level of the inguinal ligament.

SUPERFICIAL

Superficial epigastric artery arises from the femoral artery about a centimeter below the inguinal ligament, passes through the saphenous opening in the thigh and ascends in front of the inguinal ligament to reach the umbilicus. It anastomoses with the superficial branches of the inferior epigastric artery.

Superficial circumflex iliac artery arises from the femoral artery near the superficial epigastric artery. It

runs laterally toward the iliac crest and anastomoses with the deep circumflex iliac and superior gluteal arteries.

Superficial external pudendal artery arises from the medial side of the femoral artery, pierces the fascia over the saphenous opening in the thigh, crosses the spermatic cord and ascends toward the pubic tubercle. It supplies superficial structures in the lower anterior abdominal wall and external genitalia anastomosing with the internal pudendal artery.

DEEP

Musculophrenic artery is a lateral terminal branch of the internal thoracic artery. It enters the abdomen between slips of the diaphragm from the ninth and tenth costal cartilages and passes along the deep surface of the costal margin. It supplies the anterior abdominal wall, the abdominal surface of the diaphragm and anterior intercostal arteries to the seventh to ninth intercostal spaces. It anastomoses with the inferior epigastric artery (of the external iliac) and last two intercostal arteries.

Superior epigastric artery is the other (medial) terminal branch of the internal thoracic artery. It enters the abdomen between the sternal and costal slips of the diaphragm and descends between the rectus and posterior layer of the rectus sheath to ramify in the muscle and provide cutaneous branches and anastomoses with the inferior epigastric branch of the external iliac artery.

Inferior epigastric artery is one of the two branches of the external iliac artery arising near the midinguinal point. It ascends past the medial margin of the deep inguinal ring, pierces the transversalis fascia and courses vertically passing deep to the arcuate line, where it sinks into the rectus abdominis muscle. It anastomoses with the superior epigastric artery and providing collateral circulation between the subclavian and external iliac arteries.

The **pubic branch** descends medially behind the inguinal ligament then along the free border of the lacunar ligament to reach the back of the pubis and anastomose with the pubic branch of the obturator artery. It anastomoses with the pubic branch of the obturator artery and may be the source of the obturator artery.

Cremasteric artery passes into the deep inguinal ring and through the inguinal canal. It supplies cremaster and anastomoses with the testicular artery in the male. In the female, it is the small artery of the round ligament.

Deep circumflex iliac artery is the other (lateral) branch of the external iliac artery given off just above the inguinal ligament. It courses laterally ascends above the inguinal ligament then above the anterior superior iliac spine, where it gives rise to an ascending branch and pierces the transversalis fascia. Continuing along the iliac crest, it supplies the adjacent muscles and anastomoses with the lumbar arteries behind and iinferior epigastric artery in front.

Deep external pudendal artery arises from a common trunk with the superficial external pudendal artery. It pierces the fascia lata of the thigh, passes behind the sermatic cord and supplies the scrotum (labia) anastomosing with the scrotal (labial) branches of the internal pudendal artery.

6. Deep Nerves

Thoracoabdominal and intercostal nerves leave the 7th to 11th intercostal spaces deep to the cartilages and enter the abdominal wall by passing through the slips of origin of the diaphragm and transversus abdominis muscle. The nerves course downward and forward between the transversus and internal oblique to the rectus sheath, they pierce to emerge as the anterior cutaneous nerves. They supply transversus, internal oblique, external oblique, rectus and lateral cutaneous nerves that become cutaneous in the midaxillary line.

Subcostal nerve (T12) enters the abdomen behind the lateral arcuate ligament of the diaphragm. It courses below and parallel to the lowest intercostal nerve, provides a lateral cutaneous branch to the side of the gluteal region. Muscular branches to parts of transversus, internal oblique, rectus and piriformis terminate as an anterior cutaneous branch.

Iliohypogastric nerve (T12 or L1) courses behind the lower part of the kidney, pierces the aponeurosis of the transversus and internal oblique, to pierce the latter in front of the anterior superior iliac spine. It provides a lateral cutaneous branch to the side of the gluteal region. It pierces the external oblique a little above the superficial ring to supply the skin over the pubis and terminates in an anterior cutaneous branch.

Ilioinguinal nerve (T12 or L1) takes the same course as the iliohypogastric nerve but runs a little below it. After piercing internal oblique, it passes through the inguinal canal and emerges from the superficial inguinal ring to innervate the upper and medial thigh and the scrotum or mons pubis (through anterior scrotal (labial) branches). It has no lateral cutaneous branch.

Lymphatics drain the anterior abdominal wall in two directions from the umbilicus down toward superficial inguinal nodes and upwards toward axillary nodes.

Lymphatics from the deep surface of the abdominal wall follow the main abdominal vessels and can be divided into four groups:

1. Following inferior epigastric vessels — to external iliac nodes.
2. Following deep circumflex iliac vessels — to external iliac nodes.
3. Following lumbar vessels — to nodes of the lumbar chain.
4. Following superior epigastric vessels — to sternal nodes.

Chapter **34**

Peritoneum

Chapter Outline

1. Introduction
2. Peritoneal Cavity
3. Anterior Parietal Peritoneum

Peritoneum is a large continuous sheet comprising a smooth serous membrane that lines the abdominal walls and is reflected onto abdominal organs. It lines the peritoneal cavity and permits free, smooth movement between organs and the body wall.

1. Introduction

The peritoneum that lines the internal abdominal walls is known as the parietal peritoneum, and it is reflected on the viscera as the visceral peritoneum. Parietal peritoneum is supplied by sensory and vasomotor nerves from the adjacent body wall and responds to stimulation with intense pain. Visceral peritoneum is insensitive.

Some viscera are held against the posterior abdominal wall to a variable extent. Such organs are said to be retroperitoneal in location. Others are almost completely invested in peritoneum but remain connected to the dorsal body wall by duplications called mesenteries, omenta, and ligaments and are said to be intraperitoneal in location.

DEVELOPMENT

The organization and attachments of the viscera in the abdominal cavity is complicated in the adult but has developed from a relatively simple pattern in the embryo that should be followed from its initial pattern.

With closure of the abdominal wall during development, the two apposing layers of splanchnic mesoderm fuse and form a double layered membrane known as the primitive mesentery. This forms a pathway for vessels and nerves from the dorsal body wall to the organs. The alimentary canal is established as a tube suspended in the midline of the abdominal cavity by a ventral and dorsal mesentery.

The gut tube was attached throughout its length to the midline by a dorsal mesentery. The pharynx and upper esophagus lack a mesentery as they lie above the region where there is no coelom. The dorsal mesentery is given regional names — mesoesophagus, dorsal mesogastrium (greater omentum), mesoduodenum, mesentery proper (of the jejunum and ileum) mesocolon and mesorectum.

A ventral mesentery remains only in the midline and only above the umbilicus.

Initially, the gut tube traverses the abdominopelvic cavity in the midline straight from the diaphragm to the floor of the pelvis. It is divided for descriptive reasons into (cranial and caudal parts of) the foregut, midgut and a hindgut.

The cranial part of the foregut extends from the buccopharyngeal membrane at the back of the oral cavity to the tracheobronchial diverticulum and gives rise to the pharyngeal pouches, structures in the floor of the pharynx, pharyngeal clefts and arches and the respiratory system.

The caudal part of the foregut extends from the respiratory diverticulum to the origin of the liver bud and gives rise to the esophagus, stomach, duodenum, liver and gall bladder and pancreas.

The midgut extends from the origin of the liver bud to the junction of the right two thirds and left one third of the transverse colon. It gives rise to the distal part of the duodenum, jejunum, ileum, caecum and appendix, ascending colon and proximal part of the transverse colon.

The hindgut extends from the junction of the right two thirds and left one third of the transverse colon to the cloacal membrane.

The stomach has a ventral and dorsal mesentery. The lesser curvature of the stomach faces the ventral mesentery and the greater curvature faces the dorsal mesentery.

The liver develops as an outgrowth of the entoderm of the caudal part of the foregut and increasingly protrudes between the leaves of the ventral mesentery (ventral mesogastrium) but retains its contact with the diaphragm (septum transversum) at the bare area of the liver. The ventral mesogastrium is thereby divided by the liver into an anterior segment (the falciform ligament) and a posterior part (the lesser omentum).

The ventral mesentery does not persist caudal to the upper duodenum. The free border of the ventral mesentery contains the common bile duct, portal vein and hepatic artery and forms the epiploic foramen that leads into the lesser peritoneal sac.

The pancreas develops from two separate pancreatic buds, one developing in the ventral mesentery close to the liver and the other develops in the dorsal mesentery.

The spleen develops between the leaves of the dorsal mesogastrium and taps the blood supply to the stomach. The part of the mesogastrium between the dorsal midline and the spleen is called the lienorenal ligament and the part between the spleen and the stomach is known as the gastrosplenic ligament.

ROTATION OF THE STOMACH

The stomach is originally identified as a spindle shaped dilatation of the foregut. It appears to rotate firstly ninety degrees in a clockwise direction along its longitudinal axis. This brings the left vagus nerve to supply the front of the stomach and right vagus to supply the posterior wall of the stomach. The dorsal mesogastrium appears to be swept to the left helping to form the omental bursa (lesser sac).

The proximal and caudal ends of the stomach appear to rotate in an anteroposterior axis so that the caudal or pyloric end is displaced upward and to the right while the cranial end is displaced downward and to the left. The greater curvature faces downward and the lesser curvature faces upwards. As a result, the spleen is carried to the left and the pancreas is pushed to the right where it is pressed against the dorsal body wall and fuses with the dorsal body wall.

Rotation of the stomach appears to be the result of unequal growth of the gut wall in the region of the stomach and the change in position of surrounding organs, particularly the liver. Formation of the lesser peritoneal sac (omental bursa) is also the result of a burrowing process into the right side of the dorsal mesogastrium rather than its folding and creates a subdivision, the infracardiac bursa that extends cranially between the right lung and esophagus (mostly obliterated) and the caudal subdivision, the omental bursa.

The dorsal mesentery in the region of the small intestine enlarges and changes its orientation in parallel with the rapid elongation of the intestinal tube. The small intestine at first appears as a simple loop in the midline with cranial and caudal limbs attached at its apex through the body stalk to the remnant of the yolk sac by the vitelline duct.

Elongation of the intestinal loop occurs at a rate faster than elongation of the trunk so that the loop rotates (around the axis of the superior mesenteric artery) and herniates into the umbilical cord. The cranial loop is carried to the right and caudal limb the left (anti-clockwise rotation from in front).

The rotation appears to be caused by an enlarging left umbilical vein.

The protruded intestinal loop remains herniated for about three weeks but then begins to re-enter the abdomen in an orderly progression.

The re-entry of the intestine occurs when space becomes available at a time when the mesonephroi regress. Liver growth is relatively reduced and the size of the abdominal cavity is increased.

The proximal part of the jejunum is carried to the left side and the ileum later to the right hand side. The caecum is the last to leave the umbilical cord and slants to the upper right hand side.

Parts of the dorsal mesentery fuse with the posterior abdominal wall. When the colon reaches its definitive position, the mesenteries are pressed against the posterior abdominal wall and fuse with it so the ascending and descending colons are anchored to the posterior abdominal wall. The mesentery of the transverse colon covers the duodenum and fuses with its mesentery forming the greater omentum. The definitive attachment of the transverse colon extends from the hepatic flexure of the ascending colon to the splenic flexure of the descending colon.

The line of attachment of the mesentery proper to the posterior abdominal wall extends from the line of entry of the duodenum into the peritoneal cavity to the ileocecal junction.

The right side of the mesentery of the gut is involved in secondary fusions with the dorsal body wall.

(a)   of the dorsal mesogastrium,

(b)   of the dorsal mesoduodenum,

(c)   of the mesocolon in the region of the ascending colon and,

(d)   of the mesocolon in the region of the descending colon.

In the adult, the dorsal mesentery is attached to the posterior abdominal wall in a line that extends from the second part of the duodenum, along the anterior border of the pancreas to the anterior surface of the left kidney.

The **root of the mesentery** extends from the duodenojejunal junction, crosses the third part of the duodenum, aorta, inferior vena cava, right ureter, genitofemoral nerve, gonadal vessels and psoas to end at the ileocecal junction. Between its layers are jejunum and ileum, at its free border, superior mesenteric vessels, lymph glands and nerve plexuses from sympathetic and parasympathetic (vagus) nerves.

Peritoneum forms the mesentery of the sigmoid colon. It is attached along an inverted V shaped line (whose apex lies over the left sacroiliac joint).

TERMS USED TO DESCRIBE PROCESSES INVOLVING THE PERITONEUM

Omenta are broad sheets or duplications of peritoneum connecting the stomach with another organ.

The **greater omentum** arises from the dorsal mesogastrium formed by passing downward then folding upward of the two layers of the dorsal mesogastrium in front of the transverse colon and loops of small intestine. The leaves fuse to form a single sheet hanging from the greater curvature of the stomach. The greater omentum has three parts, a lower, apron like part, the gastrocolic ligament, a left part, the gastrolienal (gastro splenic) ligament and an upper part, the gastro-phrenic

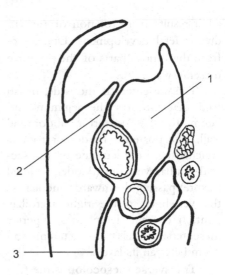

FIG. 34-1 Sagittal section of abdominal cavity showing disposition of peritoneum. (1) Omental bursa; (2) Lesser omentum; (3) Greater omentum.

ligament. These are attached to the transverse colon, spleen and diaphragm respectively.

The **lesser omentum** (ventral mesogastrium) extends from the porta hepatis and fissure for the ligamentum venosum on the liver to the lesser curvature of the stomach and first part of the duodenum. The right border of the lesser omentum is free and forms the anterior boundary of the opening into the lesser sac. The left border is attached to the diaphragm between the openings for the esophagus and inferior vena cava.

The **primitive mesentery** is a duplication of splanchnic mesoderm connecting the intestine to the posterior abdominal wall. The position of the organs and peritoneal folds in the adult results from rotation of the gut during fetal development. Derivatives from the various parts of the primitive mesentery are:

The mesentery is the wide, broad fold that attaches the jejunum and ileum to the posterior abdominal wall. The posterior border or root is attached near the left side of the second lumbar vertebra (duodeno-jejunal flexure) passing downwards and across the vertebrae to the right sacroiliac joint. It contains lymphatics, superior mesenteric vessels and jejunum and ileum between its layers.

Transverse mesocolon is the fold connecting the transverse colon to the posterior abdominal wall along the lower border of the pancreas. It forms part of the posterior wall of the lesser sac.

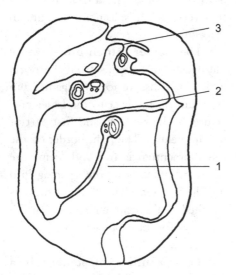

FIG. 34-2 Attachment of mesentry to posterior abdominal wall. (1) Root of mesentery; (2) Transverse mesocolon; (3) Lesser omentum.

Sigmoid mesocolon is the dorsal mesentery of the sigmoid colon. It is attached along an inverted V the apex of which lies in front of the left ureter. It contains lymphatics and the sigmoid colon. The lower left colic vessels enter the lateral limb and the superior rectal vessels enter the medial limb.

Mesoappendix is a triangular fold of peritoneum attaching the appendix to the terminal left layer of the mesentery of the ileum. The appendicular artery (branch of the ileocolic artery) often runs in its free edge.

LIGAMENTS

Ligaments are peritoneal reflections between organs or between the body wall and organs.

Phrenicolienal ligament is the portion of the dorsal mesogastrium extending from the spleen to the diaphragm.

Phrenicocolic ligament is a transversely placed fold of dorsal mesogastrium at the left colic flexure where the colon is attached to the peritoneum over the kidney. The ligament contacts the lower pole of the spleen.

Ligaments of the Liver

The **coronary ligament** is formed by the layers of peritoneum that reflect on to the diaphragm from the upper and lower margins of the bare area of the liver. It consists of a superior (anterior) layer that is continuous at the left with the right layer of the falciform ligament and an inferior (posterior) layer is reflected onto the right kidney and diaphragm and continuous with the right layer of the lesser omentum.

Triangular ligaments: The layers of the coronary ligaments meet at the right corner of the bare are and are reflected onto the diaphragm as the right triangular ligament. The left triangular ligament is formed by the lower layer of the falciform ligament and lesser omentum.

The **falciform ligament** is the long wide triangular anterior portion of the ventral mesogastrium connecting the diaphragm with the liver. It extends from the umbilicus to the inferior border of the liver (notch for the ligamentum teres) and encloses the ligamentum teres (umbilical vein) and some paraumbilical veins.

Gastrohepatic ligament is the part of the lesser omentum that extends between the liver and stomach.

Hepatoduodenal ligament is the part of the lesser omentum that extends between the liver and duodenum. It is continuous with the gastrohepatic ligament.

Gastrolienal (gastrosplenic) ligament is the part of the dorsal mesogastrium between the stomach and spleen. It contains the distal parts of the short gastric and gastroepiploic vessels.

FOLDS

Folds are peritoneal reflections with a free edge.

Rectouterine fold is formed by the posterior layer of the broad ligament of the uterus passing backward from the cervix. It forms the lateral boundary of the rectouterine pouch.

Lateral umbilical folds are folds in the parietal peritoneum that extend from the origin of the inferior epigastric vessels to the lower border of the posterior wall of the rectus sheath (arcuate line). They contain the inferior epigastric vessels.

Medial umbilical fold are folds in the parietal peritoneum that extend from the pelvis to the umbilicus. They contain the obliterated umbilical arteries.

Median umbilical fold is a fold of parietal peritoneum extending from the apex of the bladder to the umbilicus. It contains the urachus, in the adult a fibrous cord that joined the bladder to the allantois in the umbilical cord.

2. Peritoneal Cavity

The peritoneal cavity is the potential space between the visceral and the parietal peritoneum. It is completely closed in the male but communicates with the outside in the female via the uterine tubes. The cavity is divided into two parts:

Greater sac is the major part of the peritoneal cavity subdivided by the greater omentum and the transverse colon into supra- and infracolic compartments.

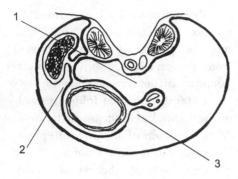

FIG. 34-3 Transverse section showing disposition of peritoneum at the level of the stomach. (1) Epiploic foremen; (2) Gastrosplenic ligament; (3) Lesser omentum.

Lesser sac (omental bursa) is a diverticulum from the greater sac that lies mostly behind the stomach and lesser omentum. Its anterior wall consists of the caudate lobe of the liver, the lesser omentum, the posterior surface of the stomach, and the anterior two layers of the greater omentum. The posterior wall consists of the transverse mesocolon and the posterior two layers of greater omentum. The lesser sac consists of a series of recesses.

RECESSES

- *Vestibule*: Narrow canal extending from the epiploic foramen.
- *Superior omental recess*: Between the caudate lobe and the diaphragm.
- *Lienal recess*: Between the spleen and the stomach.
- *Inferior omental recess*: Between the layers of the greater omentum.

Epiploic foramen is the opening between the greater and lesser sacs. Its boundaries are:

- *Anterior*: Free border of the lesser omentum containing the portal vein, hepatic artery and bile duct (portal triad).
- *Posterior*: Inferior vena cava and right crus of diaphragm.
- *Inferior*: First part of duodenum.

3. Anterior Parietal Peritoneum

On the peritoneal surface of the umbilical region, five ridges of peritoneum are produced by the underlying cords, which are remnants of fetal tubes.

Median umbilical fold is produced by the urachus. It extends from the bladder to the umbilicus.

Medial umbilical folds are produced by the obliterated umbilical arteries. They converge from the sides of the bladder to the umbilicus.

Lateral umbilical folds are produced by the inferior epigastric arteries. They extend from the side of the deep inguinal ring to the arcuate line.

Ligamentum teres or round ligament of the liver is produced by the underlying obliterated umbilical vein and lies in the margin of the falciform ligament, that extends from the umbilicus to the inferior border of the liver.

PERITONEAL SPACES, FOSSAE AND GUTTERS

Mesenteries divide the abdominal cavity into two major compartments and

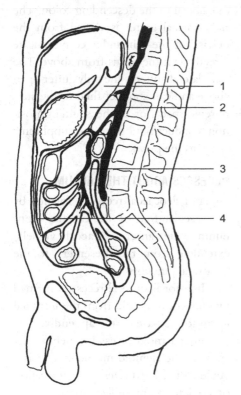

FIG. 34-4 Sagittal section of abdomen showing peritoneum. (1) Lesser sac (omental bursa); (2) Lesser omentum; (3) Mesentery; (4) Greater omentum.

several blind fossae. The dorsal mesogastrium and lesser omentum separate the greater and lesser sacs. The gastrocolic ligament, fusion of the greater omentum and transverse mesocolon divides the greater sac into a supracolic compartment and an infracolic compartment.

Recesses, spaces and fossae are potential sites for infections and collection of fluids.

ABOVE THE GREATER OMENTUM

Right subphrenic space is situated above and in front of the liver and beneath the diaphragm. Left subphrenic space is situated below the liver within the lesser sac. Subphrenic spaces are separated by the falciform ligament.

Hepatorenal recess or pouch is situated between the right lobe of the liver, right kidney and right colic flexure.

The hepatorenal recess is the lowest part of the abdominal cavity in the supine subject (a common position of the unconscious patient) and the site of accumulation of extravasated fluids.

BELOW THE GREATER OMENTUM
Duodenal Recesses

Superior duodenal recess is located to the left of the distal end of the duodenum behind the superior duodenal fold. The inferior mesenteric vein lies behind this fold.

Inferior duodenal recess is located to the left of the distal end of the duodenum behind the inferior duodenal fold. It may extend to lie in front of the left colic artery and inferior mesenteric vein.

Paraduodenal fossa is located to the left of the ascending part of the duodenum behind the paraduodenal fold whose free end contains the inferior mesenteric vein and ascending branch of the left colic artery.

Retroduodenal fossa is uncommon but may be the largest of the duodenal recesses. It is located behind the horizontal and ascending parts of the duodenum in front of the aorta and is bounded on both sides by the duodenoparietal folds.

Caecal Recesses

Superior ileocaecal recess lies between the vascular fold of the caecum and mesentery of the ileum.

Inferior ileocaecal fossa is located between the ileocaecal fold and front of the mesoappendix.

Retrocaecal fossa is located behind the caecum and on either side by caecal folds that pass from the sides of the caecum to the dorsal body wall. The vermiform appendix frequently occupies this recess.

Intersigmoid fossa lies behind the sigmoid mesocolon and its mouth opens at the apex of the inverted V shaped mesocolon where the left ureter crosses the left common iliac vessels.

Pelvic Fossae

Rectovesical fossa in the male lies between the rectum and urinary bladder.

Rectouterine fossa in the female lies between the uterus and rectum.

Paracolic Gutters

Alongside the attachments of the mesentery and ascending and descending colons to the posterior abdominal wall are four "gutters", one lateral to the right of the ascending colon, one between the root of the mesentery and the ascending colon, one medial to the attachment of the descending colon and one lateral to the descending colon. The right medial gutter is closed from the pelvis but the remainder communicate directly with the pelvis from above. The right lateral gutter is the only gutter open above and can conduct fluid to the subphrenic space, or in the standing position, conduct fluid from the subphrenic space to the pelvis.

RECESSES ABOUT THE CAECUM

Superior ileocaecal recess is formed by the vascular fold of the caecum (containing the anterior caecal vessels) extending from the mesentery to the ileocaecal junction.

Inferior ileocaecal recess is formed by the ileocaecal fold from the terminal ileum to the base of the appendix.

Intersigmoid recess lies behind the apex of the sigmoid mesocolon. It is a guide to the left ureter and bifurcation of the left common iliac artery.

The disposition and continuity of peritoneum can be traced in horizontal and vertical section.

The greater and lesser sacs traced horizontally at the level of the opening into the lesser sac (the epiploic foramen).

Peritoneum can be traced from the falciform ligament of the liver. It extends over the abdominal wall and diaphragm then over the lateral part of the left kidney. It is reflected laterally behind splenic vessels to the spleen, forming the posterior layer of lienorenal ligament.

The peritoneum covers the spleen as far as its hilum then passes to the stomach, forming anterior layer of gastro-splenic ligament. It passes over the anterior wall of the stomach into anterior layer of lesser omentum. It turns around the common bile duct forming anterior edge of opening into lesser sac (where the lesser sac commences).

Passing from right to left: within the lesser sac, peritoneum forms the posterior layer of the lesser omentum. It passes over the posterior surface of the stomach then forms the posterior layer of the gastro-splenic ligament, reaching hilum of spleen.

The peritoneal layer then forms the anterior layer of lieno-renal ligament and now passing from left to right continues over the left kidney, aorta, and inferior vena cava, forming the posterior boundary of opening into lesser sac.

It continues over the right kidney to the liver and can be traced over the inferior surface of the liver to the right border and around onto the anterior surface to the falciform ligament.

The two sacs traced vertically.

From the porta hepatis two layers of peritoneum can be seen to pass to the lesser curvature of stomach where they separate. One layer passes in front and the other layer passes behind stomach, thus enclosing it. The layers rejoin at the greater curvature of the stomach, forming the anterior layers of the greater omentum.

The layers pass down in front of and beyond the transverse colon then bend upwards and backwards, and separate to enclose transverse colon. They continue to the posterior abdominal wall as the transverse mesocolon. Opposite the lower border of the pancreas, they separate, one layer passing upwards and the other downwards.

The ascending layer passes over the upper surface of the pancreas and posterior part of the diaphragm then onto the posterior surface of the liver to the porta hepatis.

The descending layer passes along the superior mesenteric vessels, around the jejunum and ileum, and back to the vertebral column forming the mesentery. It passes downwards in front of the vertebral column and lower part of the aorta and encloses the pelvic colon, forming the pelvic mesocolon.

The descending layer passes forwards in the male to the bladder, forming the recto-vesical pouch and posterior false ligaments. In the female, it passes to the vagina and uterus, forming the posterior ligaments of uterus and recto-vaginal pouch. The layer extends over the uterus to the bladder, forming the uterovesical pouch and posterior vesical ligaments.

The descending layer passes over the bladder to the anterior abdominal wall, covering the urachus and obliterated umbilical arteries as far as umbilicus, covers the under surface of diaphragm and is reflected over the upper surface of the liver and passes around the anterior border of the liver to its under surface as far as the porta hepatis.

LIGAMENTS OF THE UTERUS

Broad ligaments each pass from the side of the uterus to the lateral pelvic wall. They contain the round ligament of uterus, uterine tube, the ovary and its ligament, and branches of ovarian and uterine vessels.

Anterior ligaments define the margins of the utero-vesical pouch.

Recto-uterine folds define the margins of the recto-vaginal pouch.

Infundibulo pelvic ligament is the upper part of the lateral margin of the broad ligament. They extend from the infundibulum of the uterine tube and upper or tubal end of ovary to lateral pelvic wall. In its free lateral margin are the ovarian vessels, nerves and lymphatics.

Chapter 35

Gastrointestinal Tract

Chapter Outline

1. Embryology
2. Abdominal Part of Esophagus
3. Stomach
4. Small Intestine
5. Large Intestine
6. Arteries
7. Portal Venous System
8. Nerves
9. Lymphatics of the Intestine

The gastrointestinal tract or alimentary canal conducts food (the esophagus), digests food and resorbs its breakdown products (stomach, intestines and associated glands), and eliminates its indigestible material. Products of digestion pass through the mucosal epithelium to the bloodstream and lymphatic capillaries while a muscular layer mechanically moves the intestinal contents.

1. Embryology

The digestive tract develops from an entodermal lined cavity located on the ventral surface of the embryo. With head and tail folding, this cavity is divided into an intraembryonic **primitive gut**, an extraembryonic **yolk sac** and **allantois**.

The cephalic part of the digestive tube is a blind ending tube known as the **foregut**. It is commonly divided into a cranial part extending from the buccopharyngeal membrane to the tracheobronchial diverticulum (which gives rise to the respiratory apparatus), and a caudal part extending from the tracheobronchial diverticulum to the liver diverticulum (anterior intestinal portal).

The derivatives of the foregut are supplied by branches of the celiac trunk.

The **midgut** extends from the liver bud to the junction of the right two thirds of the transverse colon and left third of the transverse colon (the posterior intestinal portal) and gives rise to the distal part of the duodenum, jejunum, ileum, caecum, appendix, ascending colon and proximal two thirds of the transverse colon. At its middle part, the midgut is temporarily connected to the yolk sac at the umbilicus. The derivatives of the midgut are supplied by branches of the superior mesenteric artery.

The **hindgut** extends from the posterior intestinal portal to the cloacal membrane and gives rise to the distal

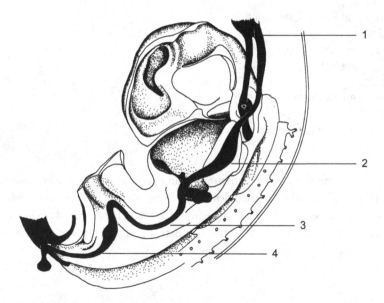

FIG. 35-1 Sagittal section of 9 mm embryo showing the gut tube (black) and its derivatives. (1) Tracheobronchial diverticulum; (2) Stomach; (3) Midgut; (4) Hindgut.

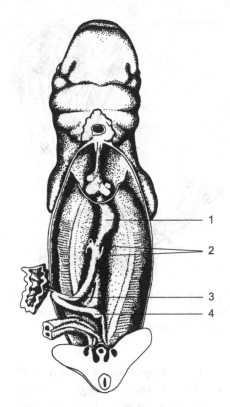

FIG. 35-2 Dissection of 9 mm embryo from in front showing the digestive tube. (1) Stomach; (2) Pancreatic diverticulum; (3) Midgut; (4) Hindgut.

third of the transverse colon, descending colon, sigmoid colon, rectum and upper anal canal. The derivatives of the hindgut are supplied by branches of the inferior mesenteric artery.

2. Abdominal Part of the Esophagus

In the adult, a short segment of the esophagus protrudes into the abdominal cavity extending from the esophageal opening of the diaphragm to the cardiac opening of the stomach at its lesser curvature. The segment of the esophagus partly within the thorax, at the diaphragm and of the abdominal part (the lower esophageal sphincter) slows the passage of swallowed food and acts as a barrier to reflux from the stomach. Nevertheless there is no anatomical sphincter at the cardioesophageal junction. The "sphincter" is the result of the circular smooth muscle layer being tonically contracted, a functional external sphincter from the right crus of the diaphragm (the phrenicoesophageal ligament) and valve like mucosal folds within the lumen. In addition, there is an acute angle formed between the esophagus and upper cardiac part of the stomach extended within the lumen of the stomach as a large fold closing the entrance when there is increased intragastric pressure.

The **neurovascular system** of the abdominal esophagus is supplied by:

Arteries: Phrenic and esopageal branches of the left gastric artery (of the celiac trunk).

Veins: The submucosa of the esophagus contains an extensive plexus of large blood vessels. Esophageal veins drain into the azygos and hemiazygos veins.

Veins of the lower esophagus anastomose with branches of the left gastric vein of the stomach.

Since the veins have no valves, an anastomosis can provide a shunt between the systemic caval system and the portal system in cases of portal hypertension.

Lymphatics flow: From the thoracic part of the esophagus drain to phrenic, mediastinal and tracheal lymph nodes while those from the abdominal esophagus drain to the upper gastric nodes.

Nerves: The gastroesphageal junction is supplied by motor splanchnic nerves and vagal trunks to smooth muscle. Sensory fibers in the abdominal esophagus mediate pain in reaction to reflux of acid gastric contents.

3. Stomach

The stomach is a cylindrical, dilated, distensible part of the alimentary canal located in the upper left part of the abdomen. It varies in position, shape and size and is related to the dome of the diaphragm and left lobe of the liver and posteriorly to a group of structures comprising the "stomach bed". It presents anterior and posterior surfaces, and a concave lesser curvature and a convex greater curvature that extend from the cardiac opening to the pyloric opening.

PARTS

The stomach presents three parts or regions based on the histological composition of its mucosa (cardiac, body and pyloric parts) however, external macroscopic features separating additional regions are poorly defined.

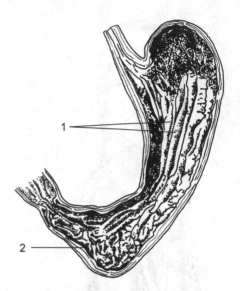

FIG. 35-3 Stomach showing the lining mucous membrane. (1) Longitudinal ridges (rugae); (2) Pyloric canal.

Cardiac opening (or ostium) is the opening of the esophagus into the stomach. On the right side it is continuous with the lesser curvature but on the left side, there is a notch between the opening and the greater curvature — the cardiac incisure.

Cardiac part adjoins the esophagus and is characterized by the presence of cardiac glands in its mucosa.

Fundus is the part of the stomach above the level of the esophageal opening. It usually contains swallowed air and is visible in radiographs. Its mucosa is similar to that of the body of the stomach.

Body is the part of the stomach between the fundus and the pyloric antrum. It is not necessarily separated from the pyloric antrum histologically at the angular notch.

Pyloric part or **pylorus** is characterized by the presence of pyloric glands in its mucosa and is subdivided into three regions:

Pyloric antrum is the slightly dilated proximal part of the stomach below the body. It leads into:

Pyloric canal of about 2–3 cm long. It is the cylindrical distal part of the stomach between the antrum and the thickened circular muscle layer comprising the pyloric sphincter.

The internal structure of the stomach comprises gastric mucosa that is thrown into folds or rugae arranged like coarse honeycombed corrugations. In the region of the lesser curvature, rugae run longitudinally forming gutters along which fluid may pass from the esophagus directly into the duodenum.

PERITONEAL ATTACHMENTS

Both surfaces of the stomach are covered with peritoneum. At the curvatures, the peritoneum forms duplications that connect to other organs. These are:

Lesser omentum extends from the first 3 cm of the duodenum, the lesser curvature of the stomach (gastrohepatic ligament), and from the diaphragm (gastrophrenic ligament) to the margins of the porta hepatis of the liver. The right border is free and forms the anterior boundary of the epiploic foramen. Here, it encloses the bile ducts, proper hepatic artery, right and left gastric vessels, autonomic plexuses, lymph nodes and most of the portal vein.

The **epiploic foramen** is a short peritoneal canal that unites the lesser and greater peritoneal sacs. It is located the portal vein, hepatic artery and bile duct in the free edge of the lesser omentum and in front of the inferior vena cava and right crus of the diaphragm. Above is the caudate lobe of the liver and below, the first part of the duodenum.

Greater omentum is a peritoneal fold consisting of four layers. The anterior two layers hang from the greater curvature of the stomach in front of the transverse colon almost to the pelvis then turn back as the posterior two layers. These layers reach the transverse colon and fuse with its mesentery (transverse mesocolon). This new layer extends to the lower border of the pancreas, where it divides to pass upwards and downward over the structures of the posterior abdominal wall. The four layers of the greater omentum extending below the transverse colon are fused.

Gastroepiploic vessels from the celiac trunk lie between the anterior two layers near the margin of the greater curvature of the stomach.

Gastrophrenic ligament is the part of the greater omentum above the spleen that joins the greater curvature to the diaphragm.

Gastrosplenic ligament is the part of the dorsal mesogastrium that extends between the upper third of the greater curvature to the hilus of the spleen. It is continuous with the gastrophrenic ligament above and with the anterior two layers of the greater omentum below.

Lienorenal ligament passes from the left kidney to the spleen. The right or anterior layer is formed by the lesser sac and the posterior layer by the greater sac.

The gastrosplenic and lienorenal ligaments contribute to forming the left boundary of the lesser sac.

RELATIONS OF THE STOMACH

Above: Lesser omentum and gastric vessels.
Below: Greater omentum and gastroepiploic vessels.
Anterior: The gastric area of the left lobe of the liver, anterior abdominal wall and diaphragm.
Posterior: The "stomach bed" separated by the omental bursa from:

Left crus of the diaphragm, pancreas, left kidney and suprarenal gland, transverse colon, splenic artery and spleen.

BLOOD SUPPLY TO THE STOMACH

They drain directly or indirectly from the celiac trunk, and is derived from the embryologic artery of the foregut.

Left gastric artery is the smallest of the direct branches of the celiac trunk that passes up behind the lesser sac to reach the cardiac end of the stomach. It turns to the right along the lesser curvature supplying it and anastomosing with the right gastric artery.

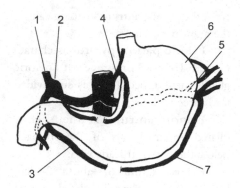

FIG. 35-4 Blood supply of the stomach. (1) Hepatic a; (2) Right gastric a; (3) Right gastroepiploic a; (4) Left gastric a; (5) Splenic a; (6) short gastric a; (7) Left gastroepiploic a.

Right gastric artery is a branch of the common hepatic artery (a branch of the celiac trunk). It passes along the lesser curvature of the stomach from right to left anastomosing with the left gastric.

Short gastric arteries are branches of the splenic artery that run between the layers of the gastrosplenic ligament to supply the fundus.

Left gastroepiploic artery, also a branch of the splenic, arises near the spleen runs in the gastrosplenic ligament to the left end of the stomach then along the greater curvature between layers of the greater omentum. It supplies both surfaces of the stomach and some of the greater omentum anastomosing with the right gastroepiploic and middle colic arteries.

Right gastroepiploic artery is a branch of the gastroduodenal branch of the common hepatic artery. It runs along the greater curvature of the stomach from right to left supplying part of the stomach (gastro) and greater omentum (epiploic).

Gastric veins accompany the gastric arteries either into the portal vein itself or its two main tributaries, the splenic and superior mesenteric veins. The right and left gastric veins open into the portal vein. The right gastric vein receives a small tributary, the prepyloric vein that is a surgical landmark to the gastroduodenal junction. The left gastroepiploic and short gastric veins open into the splenic vein and right gastroepiploic drains into the superior mesenteric or middle colic vein.

NERVES SUPPLY TO STOMACH

Gastric nerves from the vagi (parasympathetic) and branches of the coeliac plexus (sympathetic) supply the stomach.

Parasympathetic fibers contract the muscular wall, relax the pyloric sphincter and control the cephalic phase of gastric secretion.

The **anterior (left) vagal trunk** forms a large hepatic branch that runs in the lesser omentum and supplies the liver and pyloric part of the stomach.

The **posterior (right) vagal trunk** gives a large celiac branch that passes with the left gastric artery to reach the celiac plexus. It supplies the posterior surface of the body of the stomach and intestine.

Sympathetic fibers such as the preganglionic neurons are situated in the spinal cord at T6–T10 and their fibers pass in spinal nerves or white rami into or through (splanchnic nerves) sympathetic chain ganglia or prevertebral plexuses (in particular the celiac plexus). Postganglionic fibers run with blood vessels.

Sympathetic nerves constrict blood vessels, inhibit motor activity and secretion and cause contraction of the pyloric sphincter.

Afferent fibers from the stomach are concerned with reflex activity (nerves travel with vagal nerves) and pain (nerves travel with sympathetic nerves).

LYMPHATICS DRAINING THE STOMACH

Three lymphatic drainage areas correspond closely to the arterial territories.

These are:

- Upper two thirds — left gastric artery.
- Right two thirds of the lower third — gastroepiploic of the splenic.
- Left third of the lower third — right gastroepiploic of the hepatic.

Lymph nodes concerned with drainage of the stomach are:

- Hepatic group (hepatic, cystic, subpyloric) — on the hepatic artery and its branches.
- Gastric group (superior and inferior) — on the left gastric artery.
- Pancreaticolienal — on the splenic artery and its branches.

Efferents of these nodes drain to nodes around the celiac trunk in front of the aorta.

Lymphatics drain to four groups of lymph nodes but there is a free anastomosis between vessels in the stomach wall. The watershed is defined by a line parallel with the greater curvature two thirds of the way down the anterior surface of the stomach. Lymph from the region above the line drain toward the lesser curvature to left gastric nodes along the lesser curvature. Below the line, lymph from the area supplied by the short gastric vessels pass to pancreaticosplenic nodes while the remainder of the greater curvature drains to right gastroepiploic and pyloric nodes. From all of these nodes, lymph passes to the cisterna chyli.

4. Small Intestine

DUODENUM

Duodenum is the first and shortest segment of the small intestine (about 25 cm long) extending from the pyloric sphincter to the duodenojejunal junction. It is C-shaped with its convexity to the right around the head of the pancreas. It is retroperitoneal, has no mesentery (except for the hepatoduodenal part of the lesser omentum from the first 3 cm). It is attached to the body wall except in its first part and so it is relatively immobile.

First, or superior part, is about 5 cm in length extends from the pylorus to the right side of the body of the first lumbar vertebra in front of the pancreas. It is located intraperitoneally and supported by the hepatoduodenal ligament. Internally, it lacks circular folds and from radiographic appearance, has

been termed the duodenal cap. It passes in front of the common bile duct, common hepatic artery and portal vein.

Second, or descending part, about 8–10 cm in length passes downward at the level of the second lumbar vertebra. It is secondarily retroperitoneal situated in front of right renal vessels, behind the transverse colon and below the right lobe of the liver. The lumen contains permanent circular folds (plicae circulares) that increase in size and complexity towards the jejunum. This segment receives the hepatopancreatic ampulla and the secondary pancreatic duct.

Third, or horizontal part, is 10 cm in length at the level of the third lumbar vertebra is also secondarily retroperitoneal crossing the right iliopsoas, inferior vena cava and aorta.

Fourth, or ascending part, is about 2.5–5 cm in length passing upwards to the level of the second lumbar vertebra along the left side of the aorta to terminate at the duodenojejunal flexure.

The structure of the duodenum is similar to that of the remainder of the small intestine but its wall is thicker and it has the richest pattern of corrugations of the mucous membrane (plicae circulares).

Duodenojejunal flexure is the sharp curvature in the small intestine held in place by the suspensory muscle of the duodenum. This is a fibromuscular band ascends behind the pancreas from the fourth segment of the duodenum to the right crus of the diaphragm.

DUODENAL RECESSES

The peritoneum between the posterior abdominal wall and the lateral sides of the duodenum may be pinched up to form folds that cover the fossae.

The **inferior duodenal recess** is the largest fold running parallel with the lower part of the duodenum.

The **superior duodenal recess** is also arranged alongside the lower duodenum but opens downwards and is related medially to the inferior mesenteric vein.

The **paraduodenal recess** is located between the other two recesses behind the inferior mesenteric vein and left colic artery.

These recesses can be the sites of hernias during development.

Since the duodenum proximal to the opening of the bile and main pancreatic ducts is derived from the foregut and the remainder from the midgut, the blood supply comes from the vessels of these two regions — the celiac trunk and the superior mesenteric artery.

The **neurovascular system** of the duodenum is supplied by:

Arteries: **Common hepatic artery** (from the celiac trunk) passes above the first part of the duodenum, gives small branches. The gastroduodenal branch of the common hepatic passing behind the duodenum divides at its lower border into superior pancreaticoduodenal and

right gastroepiploic arteries. The right gastric branch of the gastroduodenal artery also supplies the first part. The supraduodenal artery is a branch of the common hepatic (gastroduodenal or right gastric) and passes to the anterior surface of the first part.

The second part of the duodenum is supplied by the **superior pancreaticoduodenal branch** of the gastroduodenal artery above the level of the duodenal papilla.

The remainder of the duodenum is supplied by the **inferior pancreaticoduodenal artery** (from the superior mesenteric artery).

There is a free anastomosis between branches of the inferior pancreaticoduodenal and jejunal arteries in the wall of the small intestine.

Veins: Terminate directly or indirectly into the portal vein or its tributaries, the splenic or superior mesenteric veins.

Nerves: Sympathetic nerves through splanchnic nerves, the celiac plexus and parasympathetic fibers through vagus nerves through the celiac and superior mesenteric plexuses. Nerve plexuses in the walls of the duodenum are continuous with those in the stomach and jejunum.

Lymphatics: Anterior and posterior groups of lymphatic channels anastomose with one another. The anterior collecting vessels drain to nodes ventral to the pancreas then to nodes associated with the hepatic artery. Posterior collecting vessels pass to nodes behind the head of the pancreas that drain to superior mesenteric nodes.

The root of the mesentery is the attachment of the mesentery to the posterior abdominal wall. It is about 15 cm long and begins at the duodenojejunal flexure. It passes obliquely downwards to about 1 cm short of the ileocecal junction. It crosses the horizontal (third part) of the duodenum, aorta, inferior vena cava, right psoas muscle and ureter and gonadal vessels in front of psoas. The superior mesenteric artery and vein enter and leave the mesentery as the root passes over the horizontal part of the duodenum.

JEJUNUM AND ILEUM

Jejunum is the part of the small intestine that extends for about 3 m from the duodenojejunal junction at the left side of the second lumbar vertebra to the ileum. There is no distinct line of demarcation at the junction of the jejunum with the ileum.

Ileum is the continuation of the small intestine that extends for about 4 m terminating in the right iliac fossa at the caecum.

The characteristics of the jejunum and ileum are:
(1) The jejunum lies in upper left part of abdominal cavity;
(2) Walls of the jejunum are thicker and more vascular than those of the ileum;
(3) The lumen of the jejunum is wider than that of the ileum;

(4) The circular folds (plicae circulares) on the internal surface are large and closer set in the jejunum and rudimentary in the ileum;
(5) Aggregate nodules of lymphoid tissue (Peyers patches) are few in the jejunum but numerous in the internal surface of the ileum.

The **neurovascular system** of the small intestine is supplied by;

Arteries: Jejunal and ileal arteries arise from the left side convexity of the superior mesenteric artery (the artery of the midgut). Branches form a series of anastomosing arcades from which straight (non-anastomosing) arteries run to the intestinal wall. There may be a weak anastomosis within the intestinal wall.

Veins: Accompany the arteries and terminate in the superior mesenteric vein then in the portal vein.

Lymphatics: Vessels begin as central lacteals in the cores of villi join a mucosal plexus then form collecting vessels. These accompany the superior mesenteric vessels and drain into lymph nodes.

Mesenteric nodes located along the marginal artery close to the intestinal wall, intermediate nodes along the stems of mesenteric arteries; and then into main nodes near the roots of the mesenteric arteries beside the aorta. The efferent trunks ultimately reach the thoracic duct.

Nerves: Supplying the small intestine are vagal (parasympathetic), sympathetic and sensory fibers from the celiac and superior mesenteric plexuses. Sensory (pain and reflex regulating) fibers are derived from thoracic nerves reaching the superior mesenteric plexus through splanchnic nerves.

Sympathetic (postganglionic) fibers arise from neurons in the celiac and superior mesenteric plexus. They slow peristalsis and constrict blood vessels.

Parasympathetic (preganglionic) fibers from the vagus pass through the celiac plexus and synapse on postganglionic neurons in myenteric and submucous plexuses in the wall of the intestine. They promote peristalsis and glandular secretion.

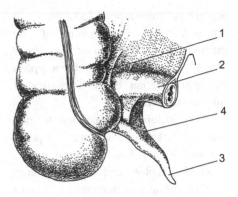

FIG. 35-5 Caecum and appendix. (1) Ileocaecal fold; (2) Ileum; (3) Appendix; (4) Mesoappendix.

5. Large Intestine

The large intestine includes the caecum, appendix and colon (ascending, transverse, descending and sigmoid), the rectum and anal canal.

Caecum is the beginning of the colon located in the right iliac fossa. It is a blind pouch that is enveloped by peritoneum but has no mesentery.

Ileocecal valve is a horizontal slit-like projection of the ileum into the posteromedial wall of the junction of the caecum. Folds of mucous membrane and muscle fibers, above and below the opening guard the valve.

Frenulae of the ileocaecal valve arise as continuities of the folds and extend from both ends of the valve

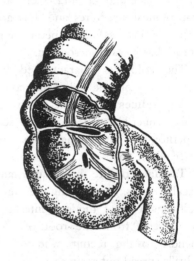

FIG. 35-6 Ileocaecal valve.

surround the intestine forming the caecocolic junction.

Vermiform appendix extends from the posteromedial wall of the caecum but may lie in a considerable range of locations (subcaecal, retrocecal or retrocolic). It is enveloped by peritoneum and has its own triangular mesentery (mesoappendix), which is attached to the left lamina of the mesentery. The appendicular artery (from the ileocolic artery) that supplies the appendix, lies in the free edge of the mesoappendix.

The **colon** is about one to two meters long and divided into four parts, the ascending colon, transverse colon, descending colon and sigmoid colon. The ascending and descending parts of the colon, being retroperitoneal, are fixed to the posterior abdominal wall. The transverse and sigmoid colons are intraperitoneal structures and thus are completely invested with peritoneum and each has its own mesocolon.

The colon is characterized by appendices epiploicae and taeniae coli.

Appendices epiploicae which are fat-filled finger-like projections enclosed in peritoneum projecting from the surface of the colon.

Taeniae coli are three equidistant muscular bands, representing the outer longitudinal coat of the large intestine. They extend from the root of the appendix to the rectum where they gradually spread out again as a continuous layer. Because the teniae are shorter than the colon, they produce pouches or sacculations, the haustra.

ASCENDING COLON

Ascending colon extends from the ileocolic opening to the hepatic flexure. Peritoneum covers its anterior and lateral surfaces. It lies on iliacus, quadratus lumborum and lateral part of the right kidney.

The ascending colon is supplied by ileocolic artery, the terminal branch of the superior mesenteric artery. At the junction of the ileum and caecum, the ileocolic divides into an ascending colic artery to the basal part of the ascending colon, anterior and posterior cecal arteries to the caecum, the appendicular artery to the appendix and ileal branches to the terminal ileum.

Anastomoses between the ileocolic, right, middle and left colic arteries result in a continuous channel at the margin of the intestine that extends from the caecum to the rectum known as the marginal artery.

TRANSVERSE COLON

Transverse colon extends from the hepatic flexure to the splenic flexure. It is attached by the transverse mesocolon to the pancreas.

The **neurovascular system** of the transverse colon is supplied by

Arteries: The middle colic artery is the first major branch of the superior mesenteric artery after the inferior pancreaticoduodenal artery lies in the transverse mesocolon and contributes to the marginal artery. Its right branch anastomoses with the ascending branch of the right colic artery and the left branch anastomoses with the ascending branch of the left colic artery.

Nerves: The transverse colon is the most caudal part of the intestines. It is innervated by the vagus nerves.

DESCENDING COLON

Descending colon extends from the splenic flexure to the sigmoid colon at the pelvic brim. Peritoneum invests its anterior and lateral surfaces and it is related posteriorly to the left kidney, quadratus lumborum and iliacus. It is about 25 cm long and smaller in diameter than the ascending colon.

The **neurovascular supply** of the descending colon is supplied by:
Arteries: The left colic branches of the inferior mesenteric artery (the artery of the hindgut). Ascending branches usually anastomose with the left branch of the middle colic and the descending branch anastomoses with the sigmoid branch of the inferior mesenteric artery.
Nerves: Parasympathetic innervation from pelvic splanchnic nerves S2 to S4 and sympathetic innervation from lumbar splanchnic nerves L1 and L2. Afferent nerves travel with lumbar splanchnic nerves.

Sigmoid colon extends from the medial border of psoas to the level of the third sacral vertebra where the rectum begins. It has an extensive mesentery (sigmoid mesocolon) whose line of attachment forms an inverted V whose apex lies over the left ureter and division of the left common iliac arteries and defines the intersigmoid recess.

The **neurovascular system** of the sigmoid colon is supplied by:
Arteries: Sigmoid and rectosigmoid branches of the inferior mesenteric artery. Sigmoid arteries anastomose with the descending branch of the left colic artery and the rectosigmoid supplies the terminal sigmoid colon and the superior rectum. It may or may not anastomose with the superior rectal artery.
Lymphatics: Scattered lymph nodes lie on the surface of the colon. Drainage channels follow arteries to reach regional nodes. Ileocolic, right colic and middle colic nodes drain into superior mesenteric nodes and then into the intestinal trunk that empties into the thoracic duct. The descending and sigmoid colons drain into inferior mesenteric nodes, para-aortic nodes and then into the thoracic duct. The descending colon may drain to celiac nodes via channels along the inferior mesenteric vein or to pelvic nodes via channels alongside the superior rectal vessels.

RECTUM AND ANAL CANAL

Rectum is continuous with the sigmoid colon near the third sacral vertebra. It has three curvatures (flexures), upper and lower to the right and middle to the left. The lower part of the rectum is slightly dilated forming the rectal ampulla. The rectum becomes continuous with the anal canal at the anorectal junction (pelvic diaphragm) formed by the levator ani muscles.

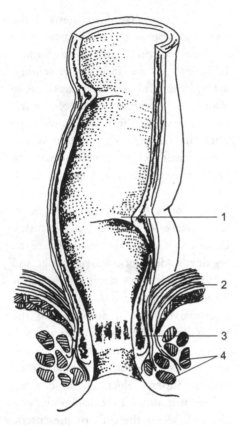

FIG. 35-7 Lower rectum and anal canal. (1) Transverse rectal fold; (2) Levator ani m; (3) Internal anal sphincter; (4) External anal sphincter.

Three sickle shaped, transverse folds of mucous membrane and submucosa project into the lumen of the rectum resembling plicae circularis. They may afford shelf like support of rectal contents.

Anal canal is the terminal 4 cm of the alimentary tract. It projects down and back to end at the anus. A right angled turn backwards at the anorectal junction is maintained by a sling of the puborectalis part of levator ani. Behind is the anococcygeal body, a mass of fibrous tissue separating the anal canal from the coccyx. In front is another fibrous body, the perineal body that separates the lower anal canal from the membranous urethra and bulb of the penis (or lower end of the vagina).

Internally, six to ten vertical folds (anal columns) marking the upper 15 mm are joined at their lower end by small crescentic folds (anal valves). Anal sinuses are pockets behind the valves and anal ducts or glands open into the sinuses.

The *pecten* is the smooth area below the anal valves and the anal verge then merges with the skin of the anus. The intersphincteric line is located between the internal and external anal sphincters.

RECESSES AND GUTTERS

Peritoneal recesses are formed around the intestines by folds of peritoneum and may be the site of herniation, collection of fluids or entrapment of a segment of intestine.

Retrocaecal recess extends upward behind the caecum.

Superior ileocecal recess lies behind the vascular fold of the caecum and contains the anterior caecal vessels. It extends from the mesentery to the ileocecal junction.

Inferior ileocecal recess between the ileocecal fold and the mesentery of the appendix.

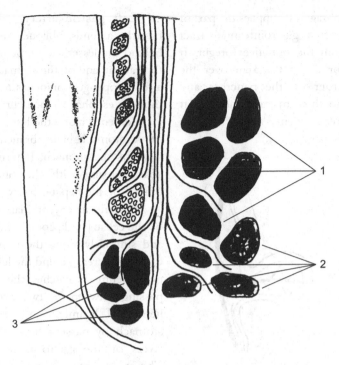

FIG. 35-8 Section of anal canal. (1) Deep part of external sphincter; (2) Superficial part of external sphincter; (3) Subcutaneous part of external sphincter.

Intersigmoid recess is situated behind the apex of the sigmoid mesocolon. It is less common in adults than children.

Paracolic gutters are established between the posterior abdominal wall and primary or secondary attachments of the mesenteries of the colon. As a result, they dictate the direction of flow of intra-abdominal collections of fluids and are clinically important.

Right lateral gutter is located lateral to the ascending colon and caecum. This is the only gutter open above and conducts fluid from the hepatorenal pouch to the pelvis

Left lateral gutter is located lateral to the descending colon and sigmoid colon.

Right medial gutter lies between the root of the mesentery and ascending colon. It is closed above and below.

Left medial gutter lies between the root of the mesentery and descending colon.

6. Arteries

CELIAC TRUNK

The celiac trunk arises from the front of the abdominal aorta between the crura of the diaphragm. It supplies the part of the abdominal gastrointestinal tract derived from the primitive foregut. It extends for about 0.5–3 cm over the upper border of the pancreas and divides into three major branches: left gastric artery, splenic artery and the common hepatic artery.

(A)

(B)

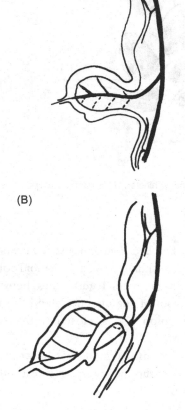

FIG. 35-9 Rotation of the gut during development. (A) Primitive unrotated position; (B) First stage. Original cranial limb is forced into 90 degrees counter clockwise position around the superior mesenteric a.

Left gastric artery, the smallest branch, ascends obliquely to the left behind the lesser sac of peritoneum to the cardiac end of the stomach. It gives a few esophageal branches to the lower thoracic esophagus, then curves sharply and descends along the lesser curvature of the stomach, giving branches to both sides of the stomach. It terminates by anastomosing with the right gastric branch of the hepatic artery.

Splenic artery, the largest branch of the celiac trunk, courses horizontally and sinuously along the upper border of the pancreas behind the lesser sac of peritoneum. It reaches the spleen by passing between the two layers of the lienorenal ligament and reaches the stomach by passing between the two layers of the gastrosplenic ligament. The splenic artery supplies the spleen and part of the pancreas and stomach.

Its branches are:

Pancreatic arteries pass to the body of the pancreas as the artery runs along the body.

Short gastric arteries pass in the gastrosplenic ligament to the fundus of the stomach and anastomose with the arteries of the esophagus.

Left gastroepiploic artery arises near the spleen. It runs in the gastrosplenic ligament, then along the greater curvature of the stomach between the two layers of the greater omentum. It supplies the left inferior region of the stomach and part of the greater omentum and ends by anastomosing with the right gastroepiploic artery (of the hepatic artery).

Splenic branches form the branching termination of the splenic artery. They pass through the lienorenal ligament dividing near the spleen to enter the hilum of the spleen.

Common hepatic artery courses to the right along the upper border of the pancreas to the first part of the duodenum and then up between layers of the lesser omentum anterior to the opening of the lesser sac. It ascends as the proper hepatic artery to the porta hepatis.

Its branches are:

Right gastric artery runs to the left along the lesser curvature of the stomach giving branches to both surfaces of the pyloric portion of the stomach and anastomoses with the left gastric artery.

Gastroduodenal artery descends behind the first part of the duodenum, ending at its lower border by dividing into:

Superior pancreaticoduodenal artery divides into anterior and posterior divisions that loop around the head of the pancreas, providing branches to both the duodenum and pancreas. Its divisions anastomose with similar divisions of the inferior pancreaticoduodenal artery (of the superior mesenteric artery).

Right gastroepiploic artery runs to the left along the greater curvature giving branches to the stomach and greater omentum. It anastomoses with the left gastroepiploic artery (branch of the splenic artery).

Supra and retroduodenal arteries may arise from the gastroduodenal artery to supply the superior and inferior aspects of the first part of the duodenum respectively.

Right and left hepatic arteries are terminal branches of the proper hepatic artery from the common hepatic artery. They enter the substance of the liver, the right branch supplying the right lobe, right half of the caudate lobe and the gall bladder. A small cystic branch descends along the cystic duct to supply the gallbladder.

The left hepatic artery supplies the left lobe, the quadrate lobe and the left half of the caudate left lobes of the liver.

SUPERIOR MESENTERIC ARTERY

The superior mesenteric artery, the artery of the midgut, supplies the intestine from immediately distal to the entrance of the bile duct into the duodenum to the junction of the proximal two thirds of the transverse colon and distal third. It arises from the front of the abdominal aorta, immediately below the origin of the coeliac trunk and behind the pancreas and passes between the pancreas and over the third part of the duodenum to enter the root of the mesentery. It passes to the right and

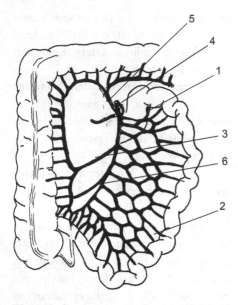

FIG. 35-10 Superior mesenteric artery. (1) Jejunal branches; (2) Ileal branches; (3) Right colic a; (4) Inferior pancreaticoduodenal a; (5) Middle colic a; (6) Ileocolic a.

terminates in the right iliac fossa by anastomosing with a branch of the ileocolic artery.

Its branches are:

- **Inferior pancreaticoduodenal artery** divides into anterior and posterior branches that anastomose with corresponding divisions of the superior pancreaticoduodenal artery (of the gastroduodenal branch of the common hepatic artery). It supplies the descending and inferior parts of the duodenum and head of the pancreas.
- **Jejunal and ileal arteries** is about 12–15 intestinal arteries arise from the left side of the superior

mesenteric artery and pass in the mesentery toward the jejunum and ileum. They divide after about 5 cm and anastomose with a neighboring branch forming a series of arcades. Straight, terminal vasa recti are given off the arcades to reach the intestine. There are more arcades in the region of the ileum than of the jejunum.

- **Middle colic artery** arises at the lower border of the pancreas. It runs forward in the transverse mesocolon and divides into right and left branches that supply the transverse colon. These branches anastomose with the right colic artery (from the superior mesenteric) and left colic (from the inferior mesenteric) arteries respectively.
- **Right colic artery** arises below the duodenum (but may be absent). It curves to the right dividing into ascending and descending branches that supply the ascending colon and anastomose with the middle colic (of the superior mesenteric artery) and ileocolic arteries (of the superior mesenteric artery) respectively.
- **Ileocolic artery** is the lowermost (terminal) of the major branches of the superior mesenteric artery descends to the caecum and appendix. It divides into an ascending branch to the ascending colon and descending branches to the lower part of the ileum that anastomose with the right colic artery and the end of the superior mesenteric

artery respectively. The ileocolic artery also provides ileal and anterior and posterior cecal branches the latter of which gives an appendicular branch.

INFERIOR MESENTERIC ARTERY

Inferior mesenteric artery is the artery of the hindgut supplying the region from the distal third of the transverse colon to the upper part of the anal canal. It arises from the left side of the aorta, a little above its bifurcation, opposite the third lumbar vertebra. It descends to the left, behind the parietal peritoneum, crosses left psoas, left common iliac vessels and ureter to reach the back of the rectum to terminate as the superior rectal artery. It supplies the left half of the transverse colon, descending colon, sigmoid colon, rectum and anal canal.

Its branches are:

Left colic artery passes laterally dividing into ascending and descending branches that anastomose with the middle colic artery (of the superior mesenteric) and an ascending branch of the sigmoid artery (of the inferior mesenteric) respectively. It supplies the descending colon.

Sigmoid arteries are about three arteries that descend in the sigmoid mesocolon, each dividing into ascending and descending branches that form arcades from which vasa recti arise. Sigmoid arteries anastomose primarily with the left colic arteries (of the inferior mesenteric artery) and supply the sigmoid colon.

Superior rectal artery is the terminal branch of the inferior mesenteric artery where it crosses the left common iliac artery. It enters the pelvic mesocolon and gives two branches that supply the wall of the rectum and anal canal. These anastomose with the middle and inferior rectal arteries (of the internal pudendal artery).

The anastomosis between ileocolic, right, middle and left colic arteries result in a vascular channel at the margin of the intestine from caecum to rectum. The channel is usually continuous but may be deficient in places.

7. Portal Venous System

Veins of the portal system have no valves and collect blood from the digestive tract. They form a trunk, the portal vein, that enters the liver and breaks up into small branches (perilobular vessels) in its substance then into sinusoids within lobules of the liver. Blood then collects in the center of the lobules in intralobular veins, hepatic veins (a central and two lateral) immediately joining the inferior vena cava.

The liver is perfused by blood from two sources, the common hepatic artery (20%) and portal vein (80%) — the former providing 80% of the liver's oxygen requirements.

Portal vein is formed by three unpaired veins, the splenic (cf celiac), superior and inferior mesenteric veins returning blood delivered to the intestines by the celiac, superior mesenteric and inferior mesenteric arteries and its unpaired glandular derivatives (spleen and pancreas but not the liver).

The portal vein is formed by the union of the splenic vein and the superior mesenteric vein behind the neck of the pancreas in front of the right crus of the diaphragm. It passes up behind the first part of the duodenum between the layers of the lesser omentum and continues up behind and between the bile duct and hepatic artery to the porta hepatis. Here it divides into right and left branches that enter the liver and ramify. Capillaries reunite into hepatic veins that terminate in the inferior vena cava.

Its main tributaries are the ligamentum teres, splenic vein, superior mesenteric vein and inferior mesenteric vein.

Ligamentum teres is the obliterated umbilical vein connected to the branch of the portal vein to the left lobe of the liver in front and the ligamentum venosum (ductus venosus) behind. These are remnants of a fetal connection between the placenta and inferior vena cava.

Splenic vein drains the region supplied by the celiac artery. It is formed by the union of a number of branches at the hilus of the spleen that unite and the main vessel passes behind the pancreas below the splenic artery to join the superior mesenteric vein. Its tributaries include pancreatic, short gastric, left gastroepiploic and inferior mesenteric veins.

Superior mesenteric vein drains the region supplied by the superior mesenteric artery which includes the entire small intestine, caecum, ascending and right half of the transverse colon. It ascends in the root of the mesentery to the neck of the pancreas, where it joins the splenic vein behind the upper border of the pancreas to form the portal vein.

Inferior mesenteric vein drains the region of the hindgut supplied by the inferior mesenterid artery — the upper two thirds of the anal canal, rectum, sigmoid colon, descending colon and left half of the transverse colon. It travels with the inferior mesenteric artery near the duodenojejunal flexure then behind the pancreas joins the splenic vein. Tributaries include the superior and inferior left colic and superior rectal veins.

If the portal vein is slowly obstructed, portal blood may reach the inferior vena cava via anastomotic veins (portacaval anastomosis).

These channels are located at:
The upper end of the gastrointestinal tract by esophageal branches of the left gastric vein anastomosing with the esophageal branches of the azygos and hemiazygos veins.
Lower end of the gastrointestinal tract by anastomosis of the superior rectal vein with the middle and inferior rectal

veins (branches of the internal iliac vein).

Paraumbilical veins run in the falciform ligament to anastomose with the superior and inferior epigastric veins.

Branches of the colic and splenic veins anastomose with branches of the renal vein and veins of the body wall.

These channels may enlarge and rupture.

8. Nerves

VAGUS NERVE

The anterior and posterior vagal trunks are formed from the esophageal plexus on the esophagus and pass through the esophageal opening of the diaphragm.

Anterior trunk (from the left vagus nerve) courses on the anterior surface of the stomach and provides gastric branches to the stomach as far as the pylorus and hepatic branches that pass through the lesser omentum on the hepatic artery to reach the liver, gallbladder and bile duct. A celiac branch courses through the celiac plexus and superior mesenteric plexus to supply the small intestine and large intestine as far as the splenic flexure.

Posterior trunk (from the right vagus nerve) courses over the posterior surface of the esophagus and stomach. Gastric branches supply the posterior surface of the stomach above the pylorus. Celiac branches pass through the celiac plexus and superior mesenteric plexus to reach the small intestine and large intestine approximately as far as the splenic flexure.

SYMPATHETIC FIBERS

A group of interconnected plexuses is located in front of the abdominal aorta and its branches (prevertebral plexuses). The coeliac plexus consists of the two celiac ganglia that lie on the crura of the diaphragm, and a dense network of preganglionic vagal and sympathetic fibers.

The intermesenteric plexus from which are derived the subordinate plexuses around the major and minor branches of the aorta. The principal branches are the superior and inferior mesenteric, aorticorenal, and phrenic plexuses.

Splanchnic (preganglionic) nerves terminate in the upper plexuses. Rami (postganglionic) pass from these plexuses with the arteries to the abdominal organs.

9. Lymphatics of the Intestine

Visceral lymph drainage begins in lacteals in the mucosa of the intestine. The regions drained correspond approximately to the areas supplied by major vessels of the gastrointestinal tract. Lymph nodes are interposed on a common plan.

Distal nodes are situated near the intestinal wall between small vessels entering the gut. They are located in the intestinal wall interconnect by lymphatic channels and lymph from the intestine may bypass nodes of this group.

Intermediate nodes situated along larger branches of larger branches of major visceral vessels.

Of intermediate nodes draining to superior mesenteric nodes, mesenteric nodes drain the small intestine, ileocolic nodes drain the caecum and appendix, right colic nodes drain the ascending colon and middle colic nodes drain the transverse colon. Superior mesenteric nodes receive lymph from the duodenum and pancreas. These nodes and the celiac nodes give rise to the intestinal trunk that drains into the cisterna chyli.

Of intermediate nodes draining to inferior mesenteric nodes, the descending and sigmoid colons drain to left colic nodes and the rectum drains to superior rectal nodes. There are anastomotic connections with middle rectal nodes that drain along internal vessels to inguinal and hypogastric nodes. Inferior rectal nodes drain along the external iliac vessels to the inguinal and hypogastric nodes.

Proximal nodes are situated along the stem of the three major visceral branches of the abdominal aorta. They drain into the intestinal trunk that in turn drains into the cisterna chyli.

Lymphatic drainage of the rectum and anal canal represent a watershed. The upper rectum drains to lymph nodes associated with the superior rectal artery. The lower rectum and anal canal above the level of the mucocutaneous junction drain in lymphatics accompanying the middle rectal vessels to internal iliac nodes. Below the mucocutaneous junction, lymphatics drain via cutaneous and subcutaneous channels to inguinal nodes.

Chapter 36

Liver, Biliary Apparatus, Pancreas, and Spleen

Chapter Outline

1. Liver
2. Gallbladder
3. Biliary Ducts
4. Pancreas
5. Spleen

The liver and pancreas are digestive glands associated with the functions of the gastrointestinal tract. The spleen is an important lymphoid organ interposed in the vascular system.

1. Liver

EMBRYOLOGY

The liver begins to develop as a ventral outgrowth of the entoderm of the distal foregut (the hepatic diverticulum) in the region of the anterior intestinal portal. Cells from the outgrowth penetrate the septum transversum. The cranial part of the diverticulum gives rise to cords of liver cells and bile ducts while the caudal part gives rise to the gall bladder and cystic duct.

The hepatic diverticulum forces its way into the septum transversum that forms most of the diaphragm. The part of the septum between the liver and anterior abdominal wall becomes drawn out as the falciform ligament and the part between the liver and foregut is also drawn out becoming the lesser omentum.

Surface mesoderm over the liver becomes the visceral peritoneum except where the liver contacts the diaphragm that is devoid of peritoneum — the bare area of the liver.

The fate of two major embryonic veins is closely associated with the development of the liver.

Vitelline veins carry blood from the yolk sac to the sinus venosus. To reach the sinus, they pass through the septum transversum where cords of liver cells invade the veins and an extensive network of liver sinusoids is formed. Right and left vitelline veins grow asymmetrically, the right dominating as the hepatocardiac part of the inferior vena cava. An anastomosis of both vitelline veins develops into a single vessel, the portal vein.

Umbilical veins carry oxygenated blood from chorionic villi in the placenta to the sinus venosus. They originally pass on each side around the developing liver to reach the sinus venosus but connect with the hepatic sinusoids as the liver enlarges. Segments of the umbilical veins adjacent to the liver disappear but the distal left umbilical vein persists.

With increasing placental circulation, a direct channel (the ductus venosus) forms between the left umbilical vein and right hepatocardiac channel largely bypassing liver sinusoids.

After birth, the left umbilical vein is obliterated forming the ligamentum teres and the ductus venosus is obliterated forming the ligamentum venosum.

SURFACE

The liver is a four sided pyramidal shaped organ located in the uppermost part of the abdomen (right and left hypochondriac and epigastric regions). It is the largest gland in the body producing an exocrine secretion (bile) and with an endocrine function (by releasing products directly into the blood stream and carrying out reactions of carbohydrate, protein and lipid metabolism). It has a transient hemopoetic function in the fetus.

The diaphragmatic (superior) surface is smooth and shaped to fit the diaphragm. It is divided descriptively

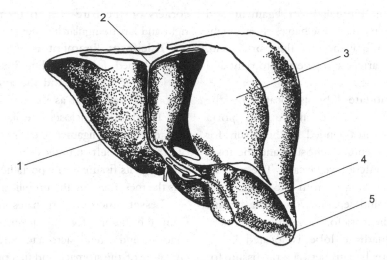

FIG. 36-1   Liver from behind. (1) Gastric impression; (2) Caudate lobe; (3) Bare area; (4) Renal impression; (5) Colic impression.

into unequal lobes by a fold of peritoneum — the **falciform ligament**.

The visceral (inferior) surface is concave and in contact with the stomach and pylorus, duodenum and right colic flexure, right kidney and right suprarenal gland. It is covered with peritoneum except where the gall bladder is attached and at the porta hepatis and fissure for the ligamentum venosum that give attachment to the lesser omentum. It is separated from the diaphragmatic surface by the sharp inferior border.

The visceral surface has an H-shaped division that consists of:

Left sagittal limb — contains fissures for the ligamentum teres (obliterated umbilical vein) and ligamentum venosum (fibrous remnant of the ductus venosus). It joins the left extremity of the porta hepatis.

Right sagittal limb — contains fossae for the gallbladder and the inferior vena cava.

Cross bar — the porta hepatis, is the hilus of the liver bounded by part of the lesser omentum. It transmits the portal vein, hepatic artery, hepatic duct, nerves and lymphatics.

LOBES

Right and left lobes are separated on the diaphragmatic surface by the falciform ligament. On the visceral surface, the lobes are separated by the fissure for the ligamentum venosum behind and the ligamentum teres in front.

Right and left functional lobes are defined by a boundary forward from the gallbladder and inferior vena cava to the right of the falciform ligament and median plane. These halves receive right and left branches of the portal vein, hepatic artery and give rise to right and left hepatic ducts.

Caudate lobe is located on the posterior surface above the porta hepatis and bounded by the fissure for the ligamentum venosum and the fossa of the inferior vena cava. The caudate process extends from the right lower border and separates the porta hepatis from the fossa for the inferior vena cava.

Quadrate lobe is located below the porta hepatis between the fissure for the ligamentum teres and the fossa for the gallbladder.

The quadrate and most of the caudate lobes belong functionally to the left half of the liver.

The inferior surface presents the impressions of the abdominal organs with which the liver comes in contact. These impressions are named after the structure involved.

The liver is clothed completely with visceral peritoneum except for the large convex area, the bare area, where it rests against the diaphragm. The peritoneum is reflected at the liver as a number of ligaments.

LIGAMENTS

Ligaments are five in number of which four are composed of peritoneum.

Coronary ligament is the single layer of peritoneum reflected from the margins of the bare area onto the diaphragm. The margins meet in the

corners of the bare area to form the right and left triangular ligaments.

Falciform ligament is the long, sickle shaped fold extending between the lobes of the liver and the anterior abdominal wall as far as the umbilicus. The free border encloses the ligamentum teres (round ligament), the remnant of the fetal left umbilical vein that extends in its fissure to the porta hepatis. It is the mesentery of the umbilical vein.

Lesser omentum originates in part from the fissure for the ligamentum venosum and extends across to the lesser curvature of the stomach and first part of the duodenum (ventral mesogastrium).

NEUROVASCULAR SUPPLY

The neurovascular system of the liver is supplied by:

Artery: Common hepatic artery.

Veins: Portal and hepatic veins.

The liver has a double blood supply. The portal vein carries venous blood from the gastrointestinal tract to liver sinusoids. The common hepatic artery supplies ducts, blood vessels and areolar tissue of the parenchyma and ends in the sinusoids.

Nerves: Vagal and sympathetic nerves are vasomotor from the hepatic plexus and celiac plexus to the liver and biliary passages.

2. Gallbladder

The gallbladder is a pear-shaped organ is about 6 cm–10 cm long situated in the

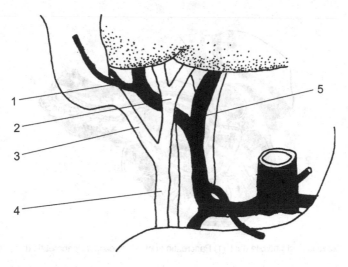

FIG. 36-2   Relationship of structures at the lesser omentum (pedicle). (1) Right hepatic a; (2) Common hepatic duct; (3) Cystic duct; (4) Common bile duct; (5) Left hepatic a.

fossa for the gallbladder on the isceral surface of the liver between the ight and quadrate lobes. It may be embedded in the substance of the liver or suspended by a short mesentery. It has a body, the main part of the gallbladder, a fundus, the blind lower end of the body that projects beyond he lower margin of the liver and a narrow upper end, the neck that joins the cystic duct.

The **neurovascular system** of the gallbladder is supplied by:

Artery: Cystic artery usually as a single stem arising from the right hepatic artery but cystic arteries are quite variable and may arise from the left or proper hepatic artery.

Vein: Cystic vein drains into the superior mesenteric vein. Small veins join those from the pancreas and duodenum and empty into the liver.

Nerves: Vagus and sympathetic nerves from the hepatic plexus. The polypeptide cholecystokinin- pancreozymin (CCK-PZ) produced by I cells in the wall of the duodenum, jejunum and ileum promotes contraction of the gallbladder (and secretion of pancreatic acinar cells).

Afferent (pain) fibers especially innervate the biliary passages. They travel with splanchnic nerves and pain is referred to the upper right quadrant or epigastrium or may be referred to the right scapula.

Lymphatics: Superficial collecting lymphatics drain from the middle posterior surface with the inferior vena cava to nodes around its termination. Those from the remainder of the posterior surface drain to hepatic nodes near the

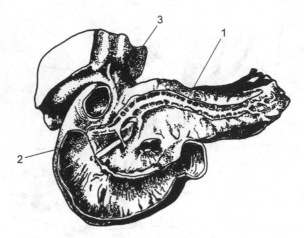

FIG. 36-3   Pancreatic duct and bile duct. (1) Pancreatic duct; (2) Accessory pancreatic duct; (3) Bile duct.

porta hepatis. From the posterior of the left lobe, vessels pass upward through the esophageal opening to paracardial nodes while the remaining convex part of the right lobe travel with the inferior phrenic artery to celiac nodes. Deep collecting lymphatics ascend with hepatic veins to nodes around the inferior vena cava or descend to end in hepatic nodes in the porta hepatis.

3. Biliary Ducts

Biliary ducts conduct bile from the gallbladder to the duodenum. The ducts consist of cystic ducts, common hepatic duct and the bile duct.

Cystic duct is a very short, twisted canal extends from the neck of the gallbladder downwards, backwards and to the left to the common hepatic duct at or below the porta hepatis. Spiral smooth muscle bundles at the neck continue into the cystic duct where they throw the mucous membrane of the duct into a series of 5–12 crescentic folds forming the spiral valve.

Common hepatic duct is formed in the porta hepatis by the union of the right and left hepatic ducts from the right and left lobes of the liver.

Bile duct (choledodochus) is formed by the union of the cystic and common hepatic ducts. It descends in the free border of the lesser omentum in front of the portal vein and lateral to the hepatic artery. It passes behind the first part of the duodenum to a groove on the back of the head of the pancreas then obliquely through the posteromedial wall of the second part of the duodenum. It expands near its termination to form an ampulla, that receives the pancreatic duct and passes obliquely through the posteromedial wall to form the greater duodenal papilla on the inner surface of the duodenum.

The intramural part of the common bile duct is the narrowest part of the duct and calculi lodge at this point.

Hepatopancreatic ampulla (duodenal papilla) is formed by the confluence of the common bile duct and main pancreatic duct.

Hepatopancreatic sphincter comprises of various parts.

* Sphincter choledochus — the portion of the sphincter surrounding the terminal bile duct within the wall of the descending part of the duodenum.
* Papillary sphincter — surrounds the ampulla.
* Pancreatic sphincter — surrounds the pancreatic sphincter.

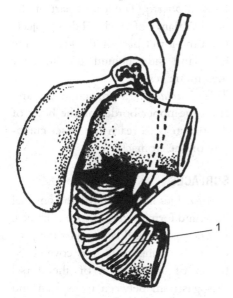

FIG. 36-4 Interior of second part of duodenum and biliary tract. (1) Circular folds.

The **neurovascular system** of the biliary ducts is supplied by:

Arteries: Cystic artery (from the right hepatic artery), a branch of the hepatic artery and the retroduodenal branch of the gastroepiploic artery from the common hepatic artery).

Veins: Enter adjacent portal vessels.

Nerves: Vagal parasympathetic nerves from the hepatic branch of the anterior vagal trunk stimulate contraction of the gallbladder and relaxation of the choledochal sphincter.

Sympathetic fibers inhibit gallbladder contraction but hormonal control (CCK-PZ) is more important.

Afferent fibers travel with sympathetic nerves. Pain is referred to the right epigastrium or hypochondrium often extending to the right scapula.

Lymphatics: From the gallbladder, hepatic ducts and upper bile duct drain to nodes in the porta hepatis, to the cystic node (at the junction of common hepatic and cystic ducts) and nodes at the anterior border of the omental foramen. The lower end of the bile duct drains to pancreaticosplenic nodes.

4. Pancreas

The pancreas is an accessory digestive (exocrine) gland and combined endocrine gland that lies transversely across the posterior abdominal wall behind the stomach and between the duodenum and spleen.

The pancreas begins to develop from two out-pocketings from the endodermal lining of the gastrointestinal tract on opposite sides of the duodenum.

The **dorsal pancreas** arises in the dorsal wall of the duodenum opposite the hepatic diverticulum. It grows in the dorsal mesogastrium more rapidly than the ventral pancreas.

The **ventral pancreas** appears ventrally in the caudal angle between the hepatic diverticulum and the duodenum. It remains smaller than the dorsal pancreas and is carried from the duodenal wall by the lengthening common bile duct. The bile duct moves caudally. The ventral pancreas is carried into the dorsal mesentery as a result of unequal growth of the duodenal wall and interlocks with the dorsal pancreas but accounts only for the head and uncinate process while the dorsal pancreas accounts for most of the head of the pancreas.

The pancreas has a connective tissue framework (like salivary glands) and an areolar tissue capsule. It is subdivided by fibrous septa into lobules. Reticular tissue embeds individual alveoli. The exocrine part is glandular epithelium organized into a compound, tubuloalveolar gland. The duct system drains into the duodenum by a main excretory duct. The endocrine part consists of scattered epithelial masses (pancreatic islets) whose cells produce the hormone insulin secreted into capillary channels.

PARTS OF THE PANCREAS

Head: Expanded part lies in the curve of the duodenum. It lies in front of the inferior vena cava and the two large renal veins.

Neck: The constricted part of the pancreas that lies in front of the superior mesenteric vessels. Behind the neck, the superior mesenteric vein and splenic vein combine to become the portal vein.

Body: The main part of the pancreas continues to the left and is triangular in cross section with anterior, posterior and inferior surfaces.

Tail: The blunt end of the pancreas extends into the lienorenal ligament close to the hilus of the spleen. It contains the majority of the *pancreatic islets*.

PROCESSES

Uncinate process: (The lowest part of the head) A flange of the head that projects from the lower part of the head to the left hand side behind the superior mesenteric vessels.

Tuber omentale: Is a projection upward, from the upper border of the body of the pancreas to reach the lesser curvature of the stomach.

SURFACES

Anterior: Lies in the floor of the omental bursa and forms part of the stomach bed.

Posterior: is flat and lies against the posterior abdominal wall. It is covered by fusion of the portion of the dorsal mesogastrium between the spleen and dorsal midline to the dorsal body wall.

It crosses the aorta, crura of the diaphragm, superior mesenteric vessels, splenic vein, left kidney and suprarenal gland and left renal vessels.

Inferior: Lies in relation to the duodeno-jejunal flexure, splenic flexure of the colon and small intestine.

BORDERS

Anterior: The ridge along which the transverse mesocolon is attached.

Superior: Extends from the tuber omentale and carries the convoluted splenic artery.

Inferior: Separates the inferior from the posterior borders and is closely related to the inferior pancreatic artery (usually from the dorsal pancreatic artery).

PANCREATIC DUCT

Pancreatic ducts develop from axial ducts of ventral and dorsal pancreatic buds. The dorsal duct arises directly from the duodenal wall but the base of the ventral pancreatic duct is carried up the common bile duct. When the dorsal and ventral primordia fuse, the ventral duct taps the dorsal duct so that the distal segment of the dorsal duct and entire ventral duct combine to form the pancreatic duct. The proximal segment of the dorsal duct remains as the accessory pancreatic duct remaining as a tributary of the main duct.

The adult pancreas contains two ducts.

Pancreatic duct begins in the tail receiving small subsidiary ducts in its course through the body of the pancreas. In the head, it curves down and behind and then to the right to join the common bile duct and terminates obliquely in the duodenum at the greater duodenal papilla.

Accessory pancreatic duct is variable in development representing the proximal part of the dorsal pancreatic duct. It frequently extends from the first curve of the main duct through the head of the pancreas to open in the lesser duodenal papilla located superior to (above) the greater duodenal papilla.

NEUROVASCULAR SUPPLY

The neurovascular system of the pancreas is supplied by:

Arteries: From numerous branches of the splenic artery (from the celiac trunk) but the head is supplied by the superior pancreaticoduodenal artery (from the gastroduodenal branch of the common hepatic artery) and inferior pancreaticoduodenal artery (from the superior mesenteric artery).

There is a free anastomosis between all of the vessels supplying the pancreas.

The superior pancreaticoduodenal artery from the gastroduodenal artery bifurcates into anterior and posterior arcades that supply the head of the pancreas and the duodenum.

The inferior pancreaticoduodenal artery from the superior mesenteric artery also bifurcates into anterior and posterior arcades that supply the head of the pancreas and duodenum.

The uncinate process is part of the head and receives blood from the inferior pancreaticoduodenal artery.

The pancreaticoduodenal vessels form an anastomosis with each other as they lie between the head and duodenum but small branches of these and of the splenic artery anastomose in the substance of the pancreas.

The great pancreatic artery, caudal pancreatic artery and dorsal pancreatic artery all arise from the splenic artery.

Veins: Tributaries from the head of the pancreas empty into the main pancreaticoduodenal vein that passes into the portal vein. Veins from the body empty into pancreatic veins that pass into the splenic vein and those from the tail empty into the caudal pancreatic vein that pass into the splenic vein.

Nerves: Vagal and sympathetic fibers via the celiac branch of the posterior vagal trunk and celiac plexus. Parasympathetic nerves stimulate exocrine secretion.

Lymphatics: Drain the acini and stroma of the gland and reach pancreaticosplenic and aortic nodes. Islet tissue does not appear to have a lymphatic drainage.

5. Spleen

EMBRYOLOGY

The spleen develops as an accumulation of mesenchymal cells on the surface of the dorsal mesogastrium that enlarges and bulges into the abdominal cavity. With formation of the omental bursa, a portion of the dorsal mesogastrium between the spleen and dorsal midline fuses with the dorsal body wall. The remaining part connecting the left kidney with the spleen is the lienorenal ligament while the part connecting the spleen to the stomach is the gastrosplenic ligament. The spleen in effect taps the blood supply of the stomach from the aorta.

The spleen is the largest lymphoid organ within the body and is located on the left side of the body between the upper part of the stomach and the diaphragm interposed in the blood stream. It has its own true fibroelastic capsule and is enveloped by peritoneum. Many trabeculae pass from the capsule to the interior. There is along deep hilus where blood vessels enter and leave the trabeculae. The parenchyma (splenic pulp) is of two distinct types.

SPLENIC PULP

White pulp surrounds and follows the arteries like a sheath. At intervals, it

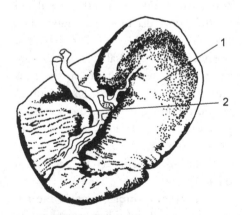

FIG. 36-5 Spleen; (1) Gastric surface; (2) Renal surface.

thickens into ovoid masses known as splenic nodules.

Red pulp is more abundant occurring in irregular masses known as pulp cords.

Splenic sinuses are sausage-shaped channels that intervene between arteries and veins. Sinuses and terminal blood vessels are embedded in a common mass of red pulp.

SURFACES

Diaphragmatic: Is convex and smooth separating the spleen from the 9th, 10th and 11th ribs, lung and pleura;

Visceral: Is triangular in outline. It has gastric, renal, and colic impressions related to the stomach, left kidney and splenic flexure of the colon respectively. A central fissure, the hilus in the lower medial part of the spleen is pierced by the splenic vessels and nerves. The tail of the pancreas may reach the hilus.

BORDERS

Superior: Notched showing the remains of fetal lobulation.

Inferior: Base.

NEUROVASCULAR SUPPLY

The neurovascular system of the spleen is supplied by:

Arteries: Terminal branches of the *splenic artery* (from the celiac trunk) that runs a tortuous course along the superior border of the pancreas, pass into the lienorenal ligament and divide into the hilus.

Veins: Although not a digestive organ, the spleen drains into the portal vein. Veins leave the spleen from the hilus and join the splenic vein that joins the superior mesenteric vein behind the neck of the pancreas to form the portal vein. Short gastric veins and the left gastroepiploic veins directed to the right between the layers of the greater omentum drain the greater curvature of the stomach.

Nerves: Autonomic nerves reach the spleen from the celiac plexus. Postganglionic sympathetic nerves supply smooth muscle in the capsule, trabeculae and splenic vessels in the pulp.

Lymphatics: Drain the capsule and trabeculae and drain to adjacent nodes.

Chapter 37

Kidneys, Ureters, and Suprarenal Glands

Chapter Outline

1. Kidneys
2. Ureter
3. Suprarenal Glands

The kidneys are the essential organs of excretion. The ureters convey urine from the kidneys to the bladder. Suprarenal glands are endocrine organs.

1. Kidneys

Kidneys are paired, bean-shaped, retroperitoneal organs which lie one on each side of the vertebral column on the ventral surface of quadratus lumborum and diaphragm lateral to psoas and the vertebral column. The right kidney is slightly (2 cm–8 cm) lower than the left and related posteriorly to the 12th rib and anteriorly to the liver, duodenum and hepatic flexure of the colon anteriorly. The left kidney is related to the 11th and 12th ribs posteriorly and the pancreas, spleen and splenic flexure of the colon interiorly.

Each kidney has a "true" fibrous capsule and is enclosed, with the suprarenal gland by an adipose capsule

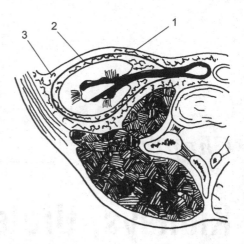

FIG. 37-2   Renal fascia in cross section through posterior abdominal wall. (1) Renal fascia; (2) Pararenal fat; (3) Perirenal fat.

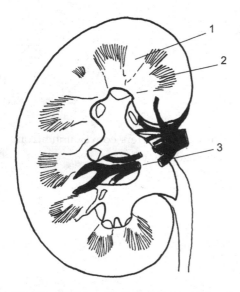

FIG. 37-1   Frontal section through kidney. (1) Renal column; (2) Pyramid; (3) Interlobular a.

in a pocket of extraperitoneal tissue. This tissue is condensed by movements of the kidney during respiration to form the renal fascia. The pocket is open below but closed medially and above:

Perirenal fat forms an adipose capsule around the kidney within the pocket formed by renal fascia.

Renal fascia is derived from condensation of extraperitoneal fat. It has an anterior layer passing over the kidney that fuses medially with the connective tissue anterior to the great vessels and a posterior layer that fuses with psoas fascia posterior to the great vessels. The fascial layers enclose the inferior vena cava and aorta between them in the median plane.

Pararenal fat is adipose tissue outside to the renal fascia.

MEDULLA AND CORTEX

Medulla that consists about 15 conical renal pyramids, whose apices converge toward the renal sinus. Two or three pyramids commonly fuse at their apices to end on a common renal papilla (8–18) which project into the lumina of minor calyces.

Cortex arches over the base of pyramids as cortical arches or lobules and cortical tissue extends between pyramids as renal columns. The cortex is divided into outer and inner (juxtamedullary) zones in which the structure of nephrons differ. With a hand lens, the cortex can be seen to be divided into alternate radially oriented areas, medullary rays (pars radiata) that contain concentrations of proximal tubules, arched tubules and collecting ducts and convoluted parts (pars convoluta or labyrinth) which contain a high proportion of renal corpuscles, proximal and distal convoluted tubules.

The medial border of the kidney is indented at the hilus that expands into a large cavity, the renal sinus. The major renal vessels and nerves leave and enter from the hilus and the ureter leaves from the hilus.

Within the renal sinus, the renal pelvis (upper expanded end of the ureter) divides into two or three major calices. Each of these ramifies into seven to fourteen minor calices, cup-shaped indentations that receive the collecting tubules of renal papillae.

FIG. 37-3 Cast of renal pelvis and calyces.

DEVELOPMENTAL NOTE

The metanephros forms the definitive kidney in the human. Like the mesonephros, it has a double origin, Secretory units develop in a caudal mass of intermediate mesoderm, the metanephric blastema collecting ducts develop from a branching bud of the mesonephric duct.

The **collecting system** arises in the fourth week as a dorsomedial bud, the ureteric bud, from the mesonephric duct close to its entrance to the cloaca. The ureteric bud penetrates the metanephric blastema where it splits into cranial and caudal parts, the major calyces. These continue to branch for 13 or more generations. Major calyces absorb the proximal ducts up to the third or fourth generations establishing minor calyces. Fifth and later generations form definitive collecting tubules.

The **secretory system** arises from tissue of the metanephric blastema. The distal part of each newly formed tubule is covered with a tissue cap. Parts of the cap separate from the main cap forming cell clusters on each side of the tubule. These clusters develop a vesicle that gives rise to the excretory tubule (or nephron). The proximal end of the nephron becomes invaginated forming the glomerular (Bowman's) capsule. The distal end of the nephron connects with and then opens into one of the collecting tubules.

BLOOD SUPPLY TO KIDNEY

Renal arteries arise from each side of the abdominal aorta at the level between L1 and L2. Both arteries pass transversely to the hilus over the crura of the diaphragm and psoas major. The right artery lies behind the inferior vena cava and the head of the pancreas. Renal arteries not only supply the parenchyma of the kidney but filter blood receiving one fifth of cardiac output for renal filtration. The left artery lies behind the left renal vein and the body of the pancreas. Each renal artery provides an inferior suprarenal artery that ascends to the suprarenal gland, ureteric branches to the upper part of the ureter (and

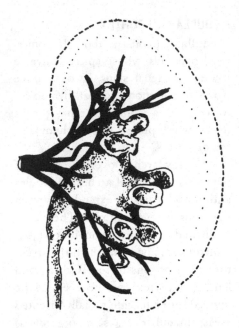

FIG. 37-4 Frontal section through kidney showing the relationship of branches of the renal artery and renal pelvis and calyces.

occasionally gonadal and inferior phrenic arteries).

Segmental blood supply: Renal arteries divide into segmental branches within the renal sinus. An anterior branch (to the renal pelvis) divides into four branches that supply the apical (superior), upper anterior superior), middle (anterior inferior) and lower (inferior) segments of the kidney. The posterior branch supplies the posterior segment of the kidney.

Segmental arteries are functional end arteries.

Segmental arteries give rise to interlobar arteries. Each interlobar artery enters a renal column and give arcuate arteries at the base of pyramids at the junction of cortex and medulla. Arcuate arteries give interlobular branches which course in medullary rays. Interlobular arteries supply afferent glomerular arterioles.

Renal veins pass in front of the arteries transversely to empty into the inferior vena cava. The right vein usually has no tributaries. The left vein is longer than the right, crosses the aorta and receives the left gonadal vein, left suprarenal vein and left inferior phrenic vein and communicates with the azygos vein.

NERVE SUPPLY TO KIDNEY

Sympathetic nerves from T12 and L1 and L2 run in the least splanchnic and lumbar splanchnic nerves. They synapse in the aorticorenal ganglia and through the renal plexus.

Afferents are referred to T12 to L2 dermatomes (lumbar and inguinal regions and anterosuperior thigh).

2. Ureter

A muscular, retroperitoneal duct conveys urine by peristaltic contractions from the kidney to the bladder. The renal pelvis, formed by union of the major calyces. The renal pelvis issues from the hilus where it lies behind the renal vessels and descends along the medial border of the kidney narrowing toward the lower end of the kidney where it becomes the ureter proper. Half of the 25 cm tube is within the abdomen and the other half in the pelvis. It descends medial to psoas major muscle to enter the pelvis crossing the origin of the external iliac artery. It enters the posterolateral angle of the bladder obliquely.

The ureter is constricted (a) at the ureteropelvic junction, (b) where it crosses the iliac vessels, and (c) where it joins the bladder. Kidney stones can be arrested at either of these positions.

In the male, the ureter crosses the ischial spine, obuturator nerve and anterior branches of the internal iliac artery then turns medially below the ductus deferens to reach the bladder. In the female, it crosses the cervix below the uterine artery lateral to the upper vagina to enter the posterior bladder. It is vulnerable to injury or improper ligation during hysterectomy.

The **neurovascular system** of the ureter is supplied by:

Arteries: The renal pelvis and upper ureter are supplied by the renal artery. The middle ureter is supplied by the testicular (ovarian) artery. The pelvic ureter is supplied by the vesicle arteries. Vessels ascend, descend and anastomose on the ureter but the blood supply is delicate and can be easily interrupted during surgery.

Veins: Correspond to the arteries.

Nerves: Arise from the lesser splanchnic nerves (T12) that synapse in the aorticorenal ganglion (region of the renal pelvis) lumbar splanchnic nerves (L1 and L2).

Pelvic splanchnic nerves (S2–S4) supply parasympathetic innervation to the entire ureter.

Visceral afferents from the upper ureter carry pain fibers referred to the iliac crest. Those from the middle third are referred to the inner thigh and leg while those from the lower part it is referred to the suprapubic area.

3. Suprarenal Glands

Suprarenal glands are small endocrine glands located on either side of the celiac artery on the superomedial aspect of the kidneys on the crura of the diaphragm. Each is composed of a yellow cortex that produces mineralcorticoids (essential for salt metabolism), glucocorticoids (affecting metabolism particularly of connective tissue) and male and female sex hormones. Each also has a deep red medulla that secretes epinephrine into the blood stream in response to cholinergic stimulation. The right gland is triangular; the left is semilunar in shape.

The **neurovascular system** of the suprarenal glands is supplied by:

Arteries: Arise from three sources (a) *superior suprarenal arteries* from the inferior phrenic artery that descend to the gland; (b) *middle suprarenal artery* from the abdominal aorta that arise at the level of the renal arteries; and (c) *inferior suprarenal artery* from the renal artery that run laterally to the gland.

Veins: One suprarenal vein terminates on the right side in the inferior vena cava and on the left side in the renal vein.

Nerves: Arise from the celiac plexus and lower ends of splanchnic nerves as well as the upper lumbar sympathetic trunk. These fibers are preganglionic destined for the suprarenal medulla.

Chapter 8

Posterior Abdominal Wall

Chapter Outline

The posterior abdominal wall extends from the posterior attachment of the diaphragm to the iliac crests. It comprises the five lumbar vertebrae and their discs, psoas major and minor, quadratus lumborum, ilium and iliacus, the diaphragm, erector spinae, latissimus dorsi and the origins of internal oblique and transversus abdominis from the throracolumbar fascia. The aorta, inferior vena cava, kidneys, suprarenal glands, caecum, ascending colon and descending colon lie against the posterior abdominal wall.

1. Fascia

Superficial fascia contains a variable amount of fat beneath the skin, cutaneous tributaries of the lumbar vessels, and cutaneous branches of the dorsal rami of the lower thoracic nerves.

Deep fascia of the posterior abdominal wall is dense fascia covering muscles of the wall and anchors peritoneum and retroperitoneal viscera.

Iliac fascia covers the Iliacus muscle. It is attached to the bone at the margins of Iliacus and inguinal ligament. It is prolonged into the femoral sheath.

Psoas fascia covers the surface of psoas muscle. It is attached to vertebral bodies, fibrous arches on the sides of the vertebral bodies and transverse processes of the lumbar vertebrae.

Psoas fascia can retain pus originating in vertebral bodies that then follows psoas major to its insertion and points in the groin (psoas abscess).

Thoracolumbar fascia is a retinaculum retaining underlying muscles that act on the vertebral column. The posterior layer (lamella) encloses the extensor muscles of the vertebral column and extends from the base of the skull to the sacrum. In the lumbar region, it comprises thick ensheathing layers. The anterior and middle layers enclose quadratus lumborum and extending laterally from the transverse processes of the lumbar vertebrae give origin to the internal oblique and transversus.

Anterior lamella extends from the front of the iliolumbar ligament to the inferior border of the twelfth rib. Medially, it is attached to the front of the lumbar transverse processes and laterally it blends with the middle lamella along the lateral border of quadratus lumborum (where transversus and internal oblique take origin).

Middle lamella extends from the back of the iliolumbar ligament and iliac crest up to the twelfth rib. Medially, it is attached to the tips of the lumbar transverse processes and laterally blends with both posterior and anterior lamellae.

Posterior lamella lies over the erector spinae muscles. Medially, it is attached to the tips of the lumbar spines and supraspinous ligaments of all sacral, lumbar and thoracic vertebrae.

Quadratus lumborum muscle is wider below than above and so extends laterally beyond the erector spinae below but not above. This relationship is of significance in posterior surgical approaches to the abdominal cavity.

2. Muscles of the Posterior Abdominal Wall

Quadratus Lumborum

Quadratus lumborum is rectangular and ensheathed by the anterior and middle lamellae of the thoracoabdominal fascia.

- **Origin:** Upper border of the Iliolumbar ligament, posterior

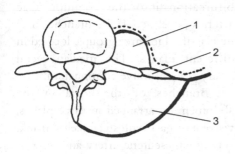

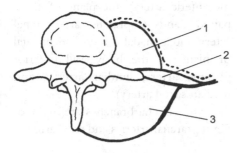

FIG. 38-1 Transverse section through thoracolumbar fascia. (1) Psoas major; (2) Quadratus lumborum m; (3) Erector spinae m.

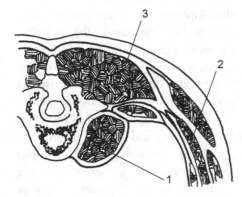

FIG. 38-2 Cross section of lumbar fascia and muscle.

quarter of the inner lip of the iliac crest and tips of the lower two or three lumbar transverse processes.

- **Insertion:** Medial half of the inferior border of the twelfth rib and the tips of upper lumbar transverse processes.
- **Action:** Fixes the last rib during inspiration.
- **Nerve supply:** L2 and L3 (anterior primary rami).

Psoas Major

Psoas major is a muscle large and fusiform. It lies on the side of the vertebral column of the lumbar region extending under the inguinal ligament into the thigh.

- **Origin:** Lumbar vertebrae and fibrous arches (over lumbar vessels and rami communicantes) on the sides of the bodies and front of the lumbar transverse processes.
- **Insertion:** Lesser trochanter of the femur.
- **Action:** Flexor and medial rotator of the extended thigh. It is a postural muscle (with antagonists) maintaining the trunk balanced on the lower limb.
- **Nerve supply:** L2 and L3.

Psoas Minor

This muscle is a long, slender muscle frequently absent.

- **Origin:** Disc between the twelfth thoracic and first lumbar vertebra and adjacent vertebral bodies.
- **Insertion:** Pecten pubis near the iliopubic eminence.

- **Action:** Helps flex the vertebral column.
- **Nerve supply:** L1.

The important relations of psoas major muscle.

Psoas fascia enclosing the muscle.

Lumbar nerves forming the lumbar plexus in the substance of the muscle.

Ureter, gonadal vessels and inferior vena cava lying on it.

Front of the hip joint in the thigh separated by a bursa.

Lies behind the femoral artery in the thigh.

Ilacus

Iliacus joins psoas major muscle to form the iliopsoas muscle.

- **Origin:** Upper part of the floor of the iliac fossa and adjacent ala of the sacrum.
- **Insertion:** Tendon of psoas major muscle and lesser trochanter of the femur.
- **Action:** Iliopsoas muscle flexes and rotates the thigh medially. After flexion it rotates the thigh laterally. It helps flex the vertebral column.
- **Nerve supply:** Femoral nerve (L2 and L3).

3. Arteries

Abdominal aorta extends in the median plane from its point of entry to the abdomen through the diaphragm (level of the disc between T12 and L1) inclining to the left to its point of bifurcation into the common iliac arteries (level of L4). Its branches may be classified in three groups, located in different planes. The median sacral artery is its primitve continuation.

Branches of the aorta in the abdomen are arranged in three planes, visceral or gastrointestinal (celiac trunk, superior mesenteric artery and inferior mesenteric artery), the plane of three paired glands (suprarenal artery, renal artery and gonadal artery) and parietal or plane of the body wall (inferior phrenic artery, lumbar arteries and median sacral artery).

The visceral branches are the middle suprarenal arteries and renal arteries.

PARIETAL GROUP

Inferior phrenic arteries arise from the abdominal aorta beside the celiac trunk. They diverge across the crura of the diaphragm to divide into a medial branch that branches on the underside of the diaphragm, and a lateral branch, which ramifies on the posterolateral body wall. The inferior phrenic arteries provide superior suprarenal arteries to the suprarenal glands and branches to the esophagus and twigs to the liver.

There are four pairs of **lumbar arteries** that lie in series with the posterior intercostal arteries. They arise from the back of the aorta at the levels of the bodies of the upper four lumbar vertebrae (the fifth pair known as iliolumbar arteries arise from the internal iliac arteries). They run laterally behind

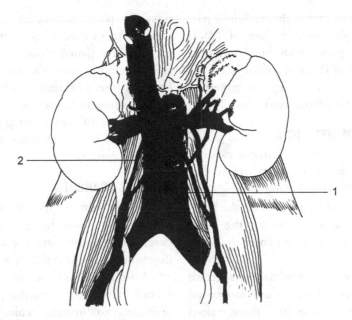

FIG. 38-3 Blood vessels and other structures in contact with the posterior abdominal wall. (1) Abdominal aorta; (2) Inferior vena cava.

the psoas major muscle and sympathetic trunk and in front of quadratus lumborum to terminate by branching between the transversus abdominis and internal oblique.

From each lumbar artery, a posterior branch arises that accompanies the dorsal ramus of its corresponding nerve and a ventral branch.

The *posterior branch* divides into muscular branches to the back muscles and a spinal branch to the contents of the vertebral column.

The *ventral branch* supplies adjacent muscle and nerves and anastomoses with segmental arteries above and below.

Median sacral artery is a small branch arising from the bifurcation of the aorta. It passes in the median plane over the fifth lumbar vertebra and middle of the sacrum to reach the coccyx.

4. Veins

Inferior vena cava is formed by the junction of two common iliac veins to the right and in front of L5 behind the right common iliac artery. It passes upward to the right side of the aorta to become embedded in a groove in the posterior border of the liver. The inferior vena cava receives hepatic veins and passes through the opening for the inferior vena cava in the tendinous part of the diaphragm (at the level of T8). It enters the pericardium then lower posterior part of the right atrium where the orifice is protected by an imperfect valve of the inferior vena cava.

DEVELOPMENTAL NOTE

A longitudinal vein (posterior cardinal vein) appears on each side of the vertebral column travelling through the abdomen and thorax to join the anterior cardinal vein forming the common cardinal vein and enter the sinus venosus.

The posterior cardinal veins receive parietal veins (lumbar and intercostal veins), veins from the three paired glands (suprarenal, renal and gonadal) but not veins from the three unpaired glands (liver, pancreas and spleen).

Cross communications are established between left and right posterior cardinal veins shunting venous blood predominantly to the right side of the body. These form the left common iliac vein and left renal vein (the segments of the left posterior cardinal vein between the left common iliac and left renal vein and between the veins draining the paired glands on the left side and azygos vein disappear).

A communication is also established above the renal vein between the right posterior cardinal vein and the right vitelline vein.

As a result, the inferior vena cava has a (post renal) segment lying on a posterior plane developing from the right posterior cardinal vein and draining the three paired glands; a segment that developed from the anastomosis between the posterior cardinal and vitelline veins; and a prehepatic segment on the plane of and draining the gastrointestinal tract and three unpaired glandular derivatives.

PARIETAL BRANCHES

Iliolumbar vein emerges at the medial side of the psoas major muscle. It descends to empty into the common iliac vein on the same side.

Lumbar veins (four or five segmental pairs) pass from the posterior abdominal wall medially behind quadratus lumborum and on the transverse processes of the vertebrae.

The first and second lumbar veins end in the ascending lumbar vein that passes through the aortic opening or through the left crus of the diaphragm to join the subcostal veins and form the azygos or hemiazygos veins.

The third and fourth lumbar veins pass forwards between psoas muscle and the vertebral bodies to empty into the inferior vena cava.

The fifth lumbar vein empties into the iliolumbar vein or common iliac vein.

Ascending lumbar vein is a vertical anastomotic channel (the abdominal equivalent of the azygos or hemiazygos veins) located over the lumbar transverse processes. Below it connects with the common or internal iliac veins and above pass through the crura or aortic opening of the diaphragm and joins the subcostal vein to form the azygos vein.

5. Nerves

LUMBAR NERVES

The ventral rami of lumbar nerves pass into the substance of psoas major in front of the transverse processes. There, L1 receives a branch from T12. After providing muscular branches mostly to psoas, quadratus lumborum and intertransversari, L1 to L4 divide to form the roots of the lumbar plexus — upper part of the lumbosacral plexus — that pass through psoas. The five lumbar roots cross the root of the transverse process of the vertebra next below but the lower division of L4 joins L5 to form the lumbosacral trunk and crosses the ala of the sacrum.

LUMBAR PLEXUS

The nerves of the lumbar plexus are derived from anterior primary rami of T12 and L1 to L4. The posterior division of the plexus is equivalent to the lateral branch of a spinal nerve. The anterior division of the plexus is equivalent to the continuation of a spinal nerve distal to the lateral branch.

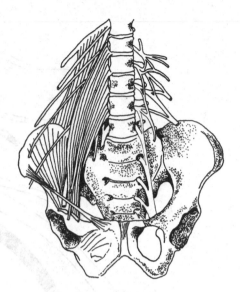

FIG. 38-4   Lumbar plexus *in situ.*

The branches of the plexus are:

Iliohypogastric nerve (L1) appears from at the lateral border of psoas major, crosses quadratus lumborum to the iliac crest. It the pierces transversus abdominis and divides into a lateral (iliac) branch to the skin of the upper gluteal region and an anterior (hypogastric) branch supplying internal oblique and transversus abdominis and is sensory to the pubic (hypogastric) region.

Ilioinguinal nerve (L1) emerges from the lateral border of psoas over quadratus lumborum and iliacus to the iliac crest. It pierces transversus abdominis and internal oblique (supplying them) then accompanies the spermatic cord through the inguinal canal, emerges from the superficial inguinal ring and

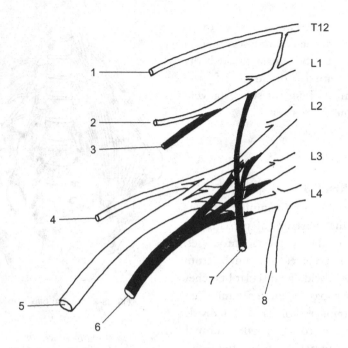

FIG. 38-5 Lumbar plexus. (1) Subcostal n; (2) Iliohypogastric n; (3) Ilioinguinal n; (4) Lateral femoral cutaneous n; (5) Femoral n; (6) Obturator n; (7) Genitofemoral n; (8) Lumbosacral trunk.

supplies the skin of the upper medial thigh, root of the penis (or mons pubis) and anterior scrotum (or superior labia majora).

Genitofemoral nerve (L1, L2) emerges on the front of psoas major, and descends behind the psoas fascia to a variable level and divides into a genital branch and a femoral branch.

The genital branch descends to cross the external iliac artery then passes through the deep inguinal ring into the inguinal canal with the spermatic cord. It supplies the cremaster muscle, and the skin of the scrotum in the male or the round ligament and skin of the labium majus in the female.

The **cremasteric reflex** is the contraction of the cremaster muscle with raising of the testis within the scrotum following stroking of the upper, medial thigh.

The femoral branch is a lateral branch descends beneath the inguinal ligament in the anterior wall of the femoral sheath lateral to psoas and emerges through the saphenous opening to supply a small area of skin on the upper front of the thigh (L1). It provides the afferent limb of the cremasteric reflex.

LATERAL FEMORAL CUTANEOUS NERVE

The nerve is derived from the posterior division — from the loop between L 2 and L3. Emerges through the lateral margin of psoas at the iliac crest. It crosses iliacus covered by iliac fascia, passes beneath the lateral end of the inguinal ligament close to the anterior superior iliac spine and divides into anterior and posterior cutaneous branches — sensory to the upper lateral aspect of the thigh.

FEMORAL NERVE

Femoral nerve arises from the posterior division of L2 to L4 — given off posterior to the roots of the obturator nerve. The largest branch of the lumbar plexus emerges at the lower lateral border of psoas major. It descends behind the iliac fascia, in the groove between the psoas and iliacus to pass behind the inguinal ligament into the thigh. It gives muscular branches to iliacus in the iliac fossa and is motor to the anterior aspect of the thigh and sensory to the anterior and medial aspects of the thigh. Through its saphenous branch, it is sensory to the anterior and medial aspects of the leg.

The femoral nerve supplies the afferent and efferent limbs of the knee-jerk reflex (a transient stretch of the patellar tendon produces a brief contraction of quadratus femoris).

OBTURATOR NERVE

This nerve arises from the anterior division of L2 to L4 — anterior to roots of the femoral nerve. It emerges at the medial border of psoas lying medial to the femoral nerve and passes in front of the sacroiliac joint into the pelvis. It enters the thigh through the obturator foramen and supplies the adductor group of thigh muscles and may be sensory to the medial aspect of the thigh.

ACCESSORY OBTURATOR NERVE (L3, L4)

This nerve is present in only about 10% of cases. It descends along the medial side of psoas major and over the superior pubic ramus into the thigh under pectineus. It supplies pectineus and provides a branch to the hip joint and communicates with the anterior branch of the obturator nerve.

LUMBAR SYMPATHETIC TRUNK

The lumbar sympathetic trunk enters the abdomen behind the medial arcuate ligament and descends on the side

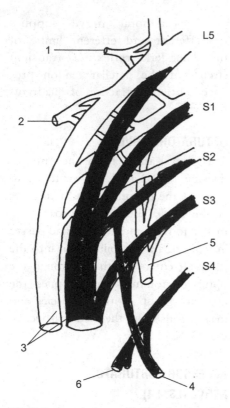

FIG. 38-6 Sacral plexus. (1) Superior gluteal n; (2) Inferior gluteal n; (3) Sciatic n; (4) Pudendal n; (5) Posterior femoral cutaneous n; (6) Perineal branch of S4.

of the bodies of the lumbar vertebrae. It usually contains about four ganglia. The upper two ganglia receive white rami, all give grey rami to lumbar nerves. Rami pass under fibrous arches of psoas.

In addition to the usual rami communicantes associated with sympathetic ganglia, lumbar splanchnic nerves arise from both the ganglia and the trunk. These are two to four short rami which join the intermesenteric and hypogastric plexuses.

The sympathetic supply to the abdominal viscera is distributed by means of the plexuses and nerves along the abdominal aorta and its branches. Preganglionic supply comes from T5–1T2. They enter the sympathetic trunk but do not relay in chain ganglia. They leave the chain in three separate nerves, the greater splanchnic nerve (T5–T9), lesser splanchnic nerve (T10–T11) and lowest splanchnic nerve (T12) which descend piercing the diaphragm.

Pelvis

The pelvis is the part of the trunk below and behind the abdomen. It is directly continuous with the abdominal cavity but for topographic purposes, it is often divided into greater and lesser pelves by an oblique line passing through the sacral promontory, arcuate line of the ilium, pectineal line of the pubis and upper margin of the pubic symphysis. The line is called the terminal line or pelvic brim.

The greater pelvis is related to the abdomen and largely comprises the iliac fossae.

The lesser pelvis is the true pelvis and contains the pelvic viscera. It has a pelvic inlet or superior pelvic aperture circumscribed by the terminal line (linea terminalis).

The pelvic outlet or inferior pelvic aperture is diamond shaped and bounded by the back of the pubis, ischiopubic rami, ischial tuberosities, sacrotuberous ligaments and tip of the coccyx.

The bony pelvis is a bony ring to which the lower limb is attached and is largely covered internally and externally by muscles.

Chapter **39**
_____

Pelvic Viscera

Chapter Outline

1. Pelvic Peritoneum
2. Urinary Bladder
3. Urethra
4. Rectum
5. Arteries
6. Veins
7. Nerves

The pelvic viscera are the two organs used for the collection of the body wastes prior to elimination (urinary bladder and rectum) and the pelvic genital organs.

1. Pelvic Peritoneum

The pelvic viscera are embedded in extraperitoneal connective tissue (endopelvic fascia) and only partially covered by peritoneum as the viscera lie largely below it. From the anterior abdominal wall behind the pubis, peritoneum reflects onto the superior surface of the bladder.

In the male, peritoneum passes over the upper ends of the seminal vesicles and then over the rectovesical pouch (the hollow in the peritoneal cavity between bladder and rectum) onto the rectum. The pararectal fossae are on either side of the rectum and the paravesical fossae on either side if the bladder is distended.

In the female, the hollow of the pelvic cavity is divided by the elevation from the pelvic floor (the broad

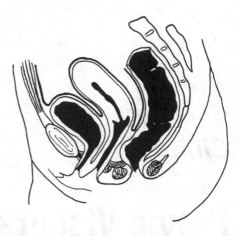

FIG. 39-2 Sagittal section of pelvis in female showing relationship of pelvic viscera, peritoneum and peritoneal pouches.

ligament) that covers the uterus, uterine tubes and ovaries. Peritoneum leaves the anterior abdominal wall, crosses the bladder and onto the vesicular surface of the uterus (defining the vesicouterne pouch) then onto the intestinal surface of the uterus down over the vaginal fornix and passes over the rectouterine pouch onto the rectum.

Peritoneum covers only the upper two thirds of the rectum and the rectum is distinguished from the sigmoid colon by its lack of a complete peritoneal investment (the lower third is retroperitoneal).

2. Urinary Bladder

The urinary bladder is a hollow, muscular organ lying in the anterior half of the pelvis behind the pubic symphysis between the diverging walls of the pelvis and behind the rectovesical

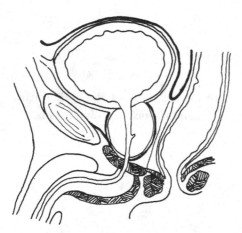

FIG. 39-1 Sagittal section of pelvis in male showing relationship of pelvic viscera, peritoneum and peritoneal pouches.

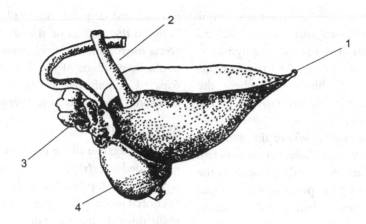

FIG. 39-3 Lateral view of bladder, prostate and seminal vesicles. (1) Median umbilical ligament; (2) Ureter; (3) Seminal vesicle; (4) Prostate.

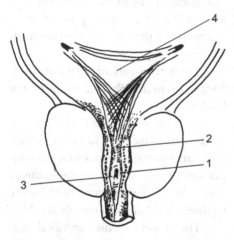

FIG. 39-4 Dissection of bladder and urethra from in front in male. (1) Prostatic utricle; (2) Urethral crest; (3) Seminal colliculus; (4) Trigone.

septum. It is covered with endopelvic fascia.

Posteriorly, the bladder is related in the male to the ductus deferens and seminal vesicles and through the rectovesical septum to the rectum.

In the female, the bladder is placed in front of the uterus and vagina.

The bladder serves as a reservoir for the urinary wastes. When empty, its superior surface is flat but when filled, it rises into the abdominal cavity and may reach the umbilicus.

PARTS

When empty, the bladder presents four surfaces and four angles:

Surfaces

Superior: Covered by peritoneum.
Inferolateral (2): Form the sides of the bladder.
Posterior: Form the base or fundus of the bladder.

Angles

Each of the four angles is the point of attachment of a duct.

Anterior (*apex*) where the superior and inferolateral surfaces meet behind the margin of the pubic symphysis.

Posterolateral (2): At the superolateral part of the fundus where the ureters attach and are crossed by the ductus deferens.

Inferior (*neck*): Where the inferolateral surfaces meet below. It rests on and is firmly attached to the prostate in the male, and on the pelvic diaphragm and into the genital hiatus and urogenital diaphragm in the female.

The body of the bladder is located between the apex in front and base behind.

The muscular coat of the bladder is composed of bundles of smooth muscle collectively known as the **detrusor muscle**. There is a triangular sheet of muscle in the trigone between the mucosa and detrusor. Some fibers run between the orifices of the ureters and some form a median thickening — the uvula. Some fibers of the detrusor pass forward as pubovesicalis on each side and some backwards as rectovesicalis.

ATTACHMENTS

The bladder, like all the pelvic viscera, is embedded in extraperitoneal connective tissue, the endopelvic fascia. The neck is anchored to the pelvic diaphragm in the male through the prostate but in the female, it rests on the pubococcygeal part of levatores ani. At the neck of the bladder, localized thickening of superior fascia of the pelvic diaphragm create medial and lateral puboprostatic ligaments of the male (or the pubovesical ligaments of the female).

Three remnants of fetal structures contribute minimally to fixation of the bladder.

The median umbilical ligament extends from the apex to the umbilicus. It is a remnant of the urachus, the connection of the bladder to the allantois. The fold of peritoneum raised by the ligament is the median umbilical fold.

The second medial umbilical ligament extends from the bladder to the umbilicus and contain the obliterated distal parts of the umbilical arteries. The fold of peritoneum raised by the ligament is the medial umbilical fold.

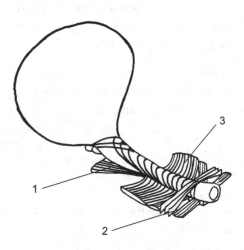

FIG. 39-5 Muscles of external urinary sphincter. (1) Rectourethralis; (2) Deep transverse perinei; (3) Levator ani.

INTERIOR OF THE BLADDER

The mucous membrane is loosely attached to its musculature and wrinkled when the bladder is empty but smooth when dilated.

Trigone in the region of the fundus is bounded by the orifices of the ureters at the upper angles, and by the internal urethral orifice, at the neck. The mucous membrane is bound to the underlying muscular coat and is always smooth in the trigone.

Interureteric fold is a fold of mucous membrane that connects the upper angles of the trigone. It is formed by an underlying bundle of muscle fibers prolonged medially from the termination of the ureters.

Uvula is an elevation just above the internal urethral orifice. It is formed by the underlying middle lobe of the prostate.

NEUROVASCULAR SUPPLY

The neurovascular system of the bladder is supplied by:

Arteries: Superior vesical arteries are branches of the patent part of the umbilical arteries that supply the upper part of the bladder.

The artery of the ductus deferens (branch of the inferior vesical artery) supplies the base of the bladder.

Inferior vesical arteries (from anterior trunk of internal iliac arteries) supplies the ductus deferens and lower part and neck as well as side of prostate and seminal vesicles.

In the female, the base of the bladder is also supplied by the vaginal artery

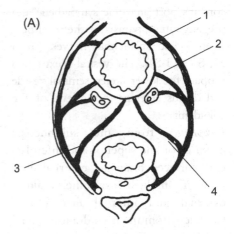

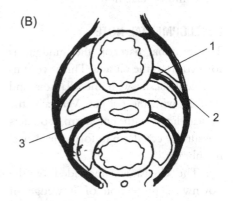

FIG. 39-6 Neurovascular sheaths (A) in male. (1) Superior vesical a; (2) Artery of ductus deferens; (3) Sacrogenital fold; (4) Inferior vesical a; (B) In female; (1) Inferior vesical a; (2) Lateral ligament of bladder; (3) Uterosacral and cardinal ligaments.

(also from anterior trunk of the internal iliac artery).

Veins: Drain downwards to the vesical or prostatic plexus drains into the internal iliac vein.

Nerves: The vesical plexus is a continuation of the hypogastric plexus and contains postganglionic sympathetic and preganglionic parasympathetic fibers from pelvic splanchnic nerves (from S2, S3 and S4). The vesical plexus also supplies the lower ureter, seminal vesicle and ductus deferens to the level of the epididymis.

Lymphatics from the submucosal plexuses of the superior and inferolateral surfaces drain to external iliac nodes. From the base, channels drain to external and internal iliac nodes. Channels from the neck drain to sacral and common iliac nodes.

DEVELOPMENTAL NOTE

The terminal part of the hindgut is formed from the **cloaca**. This is continued cephalically with the allantois and caudally with the tail gut. Ventrally, the cloaca, lined with entoderm, contacts the surface ectoderm to form the cloacal membrane.

The cloaca is subdivided by the downward projection of a wedge of mesenchyme, the urorectal septum, that divides the cloaca into dorsal rectum and ventral bladder and urogenital sinus.

The bladder is continuous with the allantois and its caudal end receives two common stems of mesonephric ducts and ureters.

Growth is associated with absorption of the posterior wall of the bladder. The stems of the two duct systems are incorporated so that four ducts acquire separate openings into the bladder.

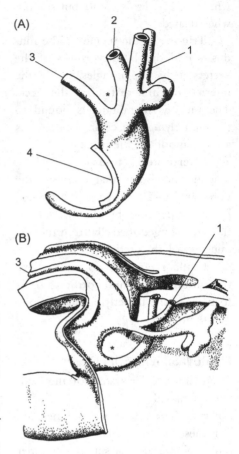

FIG. 39-7 Model of cloacal region (A) at 5 mm. (1) Mesonephric duct; (2) Hindgut; (3) Allantois; (4) Cloacal membrane. (B) At 6 weeks.

These displace, the two ureters coming to lie apart rostrally and the two mesonephric ducts lie close to one another caudally. This absorption outlines the trigone in adult anatomy. Mesoderm exposed in the absorption is covered by entodermal epithelium.

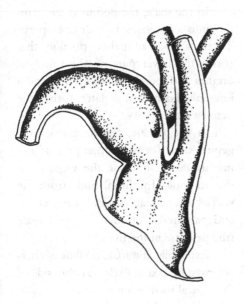

FIG. 39-8 Partial division of cloaca at 8 mm.

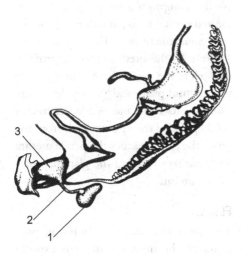

FIG. 39-9 Mesonephros at 8 mm. (1) Definitive kidney (metanephros); (2) Ureteric bud; (3) Urogenital sinus.

3. Urethra

MALE

In the male, the urethra carries urine from the bladder or components of seminal fluid as it receives the ejaculatory ducts, secretion of the prostate, bulbourethral glands and urethral glands. It extends from the internal urethral orifice at the neck of the bladder to the external urethral orifice, at the tip of the penis (20 cm). It is divided into three parts traversing the prostate, urogenital diaphragm and the penis.

In the female, the urethra extends from the internal urethral orifice to the external urethral orifice at the front of the vaginal opening (4 cm), developmentally, the equivalent of the prostatic and membranous parts of the male.

Prostatic urethra is the part of the urethra that extends from the neck of the bladder at the apex of the trigone passing through the prostate.

Urethral crest is a narrow longitudinal elevation on its posterior wall of the prostatic urethra.

Prostatic sinuses are the grooves on either side of the urethral crest into which the ducts of the prostate open.

Seminal colliculus is a median eminence of the urethral crest.

Prostatic utricle is a blind pouch directed upward and backward from the seminal colliculus. It is a remnant of the fused end of the paramesonephric ducts and homologue in the male of the uterus and upper vagina in the female.

Ejaculatory ducts open on each side of the prostatic utricle. The ducts are formed just above the base of the prostate by the union of ducts of the seminal vesicles with the narrowed ends of the ductus deferentes ane are about 2 cm long.

Membranous urethra is the part of the urehra that extends from the lower end of the prostate and pierces the urogenital diaphragm where it is surrounded by fibers of sphincter urethrae.

The bulbourethral glands lie behind and on either side of the membranous urethra and their ducts perforate the inferior fascia of the urogenital diaphragm to enter the spongy urethra.

Spongy urethra is about 15 cm long extending from the bulb of the penis, through the corpus spongiosum to the external urethral orifice. The mucosa presents a number of openings of urethral glands. The lumen is larger in the bulb and widens again in the glans as the navicular fossa. The lumen is transverse except at the external meatus where it is vertical and this, together with the navicular fossa, helps create a compact urinary stream.

DEVELOPMENTAL NOTE

At about the 5 mm stage, two parts of the primitive urogenital sinus can be destinguished — the vesicourethal canal, above the point of entry of the mesoephric duct, and the urogenital sinus below. The vesicourethral canal gives rise to the bladder and upper part of the urethra.

In the male, the definitive urogenital sinus comprises two distinct parts. There is a small pelvic portion that forms the lower part of the prostatic urethra and membranous urethra and a long phallic part that later forms the penile urethra.

In the female, the definitive urogenital sinus forms a small part of the urethra, lower fifth of the vagina and the vestibule. Epithelial buds from the wall of the cranial urethra form urethral and paraurethral glands in the female (the prostate in the male).

Externally, the male phallus is characterized by cloacal folds on either side of the cloacal membrane and a rostral cloacal eminence. With division of the cloacal membrane into rostral urogenital membrane and caudal anal membrane, the phallus comprises a rostral genital tubercle and genital folds on either side of the urogenital membrane. Entoderm at the bottom of the urethral groove proliferates while the urethral folds bridge over the urethral plate, and the urethral plate canalizes to form the penile urethra. A solid epithelial cord grows back from the tip of the glans, reaches the penile urethra and canalizes forming the glandular part of the urethra.

FEMALE

The prostatic and membranous portions of the male urethra correspond to the female urethra. The female urethra extends from the neck of the bladder through the pelvic diaphragm to terminate in front of the vaginal opening (about 4 cm). In its course, it is intimately associated with the anterior wall of the vagina. The homologue of the prostate in the female is a cluster of paraurethral glands at each side of the lower part of the urethra. Their ducts open into the vestibule on the sides of the external urethral orifice, and are known as the paraurethral ducts. Urethral glands open into the urethra.

The **neurovascular system** of the urethra is supplied by:

Arteries: That supply the structures traversed by the urethra also supply the urethra.

In the prostate, the urethra is supplied by branches of the inferior vesical and middle rectal arteries.

The membranous part of the urethra is supplied by the artery of the bulb of the penis (from the internal pudendal artery).

The spongy part of the urethra is supplied by the urethral artery and branches of the deep and dorsal branches of arteries of the penis.

Veins: Drain into the prostatic plexus and then into the internal pudendal vein.

Nerves: The prostatic part is supplied by the prostatic plexus which continues as the cavernous nerves to supply the membranous part. Branches of the internal pudendal nerve supply the spongy part.

Lymphatics: Drain along the internal pudendal vessels to reach internal iliac nodes. Some drain to external iliac nodes.

MICTURITION

The normal process of micturition is controlled by at least three centers in the central nervous system, a micturition center in the frontal cortex that controls voluntary aspects, a pontine micturition center in the pons that coordinates the activities of the sphincters and bladder so that they work in synchrony and a sacral reflex center that is a primitive voiding center responsible for bladder contraction.

Peripherally, the bladder receives autonomic nerves. **Sympathetic** adrenergic (A1) receptors are located in the trigone, bladder neck and urethra and contribute to continence by contraction of the smooth muscle of the bladder neck. Adrenergic B2 receptors are located in the bladder neck and body of the bladder. They are inhibitory and relax the bladder neck (and body to permit storage).

Parasympathetic cholinergic receptors are located in the body, trigone, neck and urethra.

The bladder and internal urethral sphincter are primarily under sympathetic control. Sympathetic activity allows the bladder to increase capacity without increasing resting pressure being exerted by the detrusor muscle. Sympathetic activiy stimulates the internal urethral sphincter to remain closed, it inhibits parasympathetic stimulation and inhibits the micturation reflex.

Immediately preceeding parasympathetic stimulation, sympathetic influence on the internal sphincter is suppressed and the internal sphincter is relaxed.

The **pudendal nerve** is inhibited and the external sphincter opens and the pelvic diaphragm (puborectalis) is inhibited facilitating voluntary micturition. This is aided by voluntary raising of intra-abdominal pressure.

At the termination of micturition, strong contraction of the external urethral sphincter, pelvic floor (and bulbocavernosus) expel residual urine from the urethra and continence is restored. **Detrusor muscle** relaxes in the absence of stretch.

4. Rectum

The terminal part of the large intestine lies in the pelvis. It extends from the level of S3 where the sigmoid mesocolon disappears to the tip of the coccyx or where the puborectalis forms a sling surrounds the rectum and the rectum enters the anal canal. As it descends, the rectum curves slightly to the right and then to the left. These external curves correspond to three internal transverse rectal folds (valves) in the inner layers of the wall that project inward.

Peritoneum covers only the anterior and lateral surfaces of the upper third and the anterior surface of the middle third of the rectum.

Ampulla is the widest part of the rectum and is capable of considerable distension.

Feces are stored primarily in the sigmoid colon. Movement of feaces into the ampulla is associated with the urge to defecate.

The rectum pierces the pelvic diaphragm to continue as the anal canal.

Puborectalis arises from the pubis and skirts the edge of the genital hiatus to form a sling with its fellow around and behind the rectum. This rectal sling holds the anorectal junction forward in a 90° turn and is the most important factor in fecal continence.

The **neurovascular system** to the rectum is supplied by:
Arteries: Superior rectal artery (branch of the inferior mesenteric), middle rectal artery (branch of the internal iliac artery), inferior rectal (branch of the

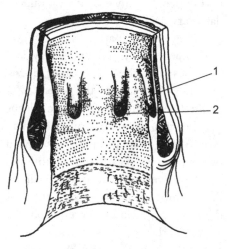

FIG. 39-10 Longitudinal section of anal canal. (1) Anal column; (2) Anal sinus (crypt).

internal pudendal artery), and the median sacral arteries.

These arteries are involved in an extensive anastomosis in the submucosa.

Veins: The superior rectal veins drain into the portal system, the inferior rectal and medial sacral veins drain to the systemic system.

Nerves: Sympathetic (L1 and L2) inferior mesenterid and hypogastric plexuses and pelvic splanchnic nerves (S2–S4). Sympathetic activity inhibits muscle activity but facilitates sphincter constriction. Parasympathetic activity stimulates activity but relaxes the internal sphincter.

Lymphatics: Of the rectum drain across the ischioanal fossa to internal iliac nodes; and across the surface of levator ani to internal iliac nodes. Posteriorly, behind the rectum to sacral nodes and upwards along superior rectal vessels to nodes at the bifurcation of the left common iliac artery then along vessels accompanying the great vessels to nodes at the origin of the superior mesenteric artery.

RELATIONS OF THE RECTUM

Valuable clinical information can be gained by inserting the index finger into the anal canal and then into the rectum and palpating structures related to their walls. Resistance is encountered first by the sphincters and then puborectalis muscle.

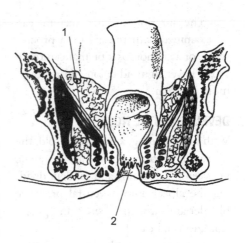

FIG. **39-11** Lower rectum and anal canal. (1) Levator ani m; (2) Pectinate line.

The following structures can be felt per rectum:

Anterior in the male: membranous urethra (when catheterised), prostate, seminal vesicles (when distended), bladder (when full), bulbourethral glands (when enlarged) and ductus deferens (when displaced and enlarged).

Anterior in the female: cervix and external os of the uterus, vagina, body of the uterus (when retroverted), rectouterine fossa and under certain pathological conditions, the ovary, uterine tube and broad ligament.

Lateral in both sexes: ischial tuberosity, ischial spine, sacrotuberous ligament, iliac lymph nodes (if enlarged) and abnormal structures in the ischioanal fossa.

Posterior in both sexes: pelvic surface of the sacrum and coccyx.

The interior of the rectum can be examined with the aid of a proctoscope or a sigmoidoscope that can reach to the lower sigmoid colon and rectosigmoid junction.

DEVELOPMENTAL NOTE

With body folding, a transverse fold, the urorectal septum, forms at the angle between the allantois and hindgut dividing the cloaca into a ventral primitive bladder and urogenital sinus and posterior anorectal canal.

A single nerve (the pudendal) supplies the muscles into which the cloacal sphincter is divided. The internal pudendal artery supplies the entire area so the bladder and rectum come to have a common nerve supply (pelvic splanchnic nerves and hypogastric plexus).

DEFECATION

Defecation is prevented primarily by **puborectalis**, the most medial part of levator ani ,that loops behind the anorectal junction drawing it forward and forming a sling that holds the junction at 90° and making it an effective valve. The mass of fat in the ischioanal fossa is also compressed in standing by tonic activity of muscles of the buttocks.

Normally, the sigmoid colon contains feces and the rectum is empty.

When the volume of feces stored in the sigmoid colon is sufficiently large, mass movement empties their contents into the rectum causing dilation of the rectal ampulla and the "call to stool" is perceived.

Intermittent rectal distension initiates the rectoanal inhibitory response resulting in internal anal sphincter relaxation and concurrent external anal contraction.

Increased intra-abdominal pressure and intra-rectal pressures are made to exceed those of the anal canal resulting in a co-ordinated reflex relaxation of the internal and external sphincters as well as of puborectalis. This relaxation (and adoption of the sitting position) increases the anorectal junction.

Relaxation of the pelvic floor muscles results in descent of the perineal floor and also increases the anorectal angle. Intra-abdominal pressure is directed onto the bolus expelling it.

Upon completion, a closing reflex is activated in which the internal sphincter and puborectalis transiently contract to restore the anorectal angle. This allows the internal anal sphincter to recover its tone thus closing the anal canal.

5. Arteries

Median sacral artery is a small, unpaired branch of the aorta that arises from the back of the aorta just above its bifurcation. It descends over L4 and L5 and the pelvic surface of the sacrum and provides the small pair of fifth lumbar arteries and branches to the rectum.

Superior rectal artery is the continuation of the inferior mesenteric artery that crosses the pelvic inlet and becomes the superior rectal artery. It courses in the sigmoid mesocolon giving branches to the lower sigmoid colon. It divides into two branches (left and right) that descend on either side of the rectum. These subdivide about 15 cm from the end of the anus, pierce the muscularis muscle, and descend as vasa recta in the submucosa to form a series of loops at the lower end of the rectum.

The superior rectal artery anastomoses with the middle and inferior rectal arteries but there may be a gap or poor anastomosis between the lowest sigmoid branch and superior rectal branch of the inferior mesenteric artery important in surgical resection.

Internal iliac artery arises from the common iliac artery in front of the sacro-iliac joint. It terminates near the upper margin of the greater sciatic foramen by dividing into anterior and posterior trunks.

Branches of the posterior trunk are all distributed to the pelvic wall.

Iliolumbar artery ascends behind the common iliac vessels and beneath psoas major. It divides into:

Iliac branch supplies iliacus muscle and provides a large nutrient artery to the ilium.

Lumbar branch supplies psoas major and quadratus lumborum muscles. It provides a spinal branch that passes through the lumbosacral intervertebral foramen to the cauda equina.

Lateral sacral arteries descend lateral to the anterior sacral foramina and branches supply the contents of the sacral canal. Their terminal branches pass through the posterior foramina to supply the muscles and skin on the back of the sacrum.

Superior gluteal artery is the largest branch and a continuation of the posterior trunk. It courses between the lumbosacral trunk and the first sacral nerve and passes out of the pelvis, above piriformis, through the greater sciatic foramen into the gluteal region.

Branches of the anterior division are either **parietal or visceral**.

VISCERAL BRANCHES

The **umbilical artery** is almost completely obliterated. It forms a prominent fold, the medial umbilical fold, on the anterior abdominal wall. The internal iliac on each side originally passed through the umbilical cord to supply the placenta. At birth, severing of the placenta causes the umbilical artery to obliterate as far back as the branches to the ventral part of the cloaca (now the bladder) — the superior vesical arteries.

From its patent proximal end the following also arise a few arteries:

Superior vesical artery supplies the upper part and sides of the bladder and the terminal part of the ureter.

Artery of the ductus deferens accompanies the ductus deferens as far as the testis, where it anastomoses with the testicular artery. It supplies the back of the bladder and the seminal vesicles.

Inferior vesical artery may arise in common with the middle rectal artery or separately. It supplies the fundus of the bladder, prostate, seminal vesicles, and ureter and anastomoses with the superior and middle vesical arteries.

Middle rectal artery may arise separately from the anterior trunk or in common with the inferior vesical or internal pudendal arteries. It supplies the middle portion of the rectum, the prostate, seminal vesicles, and ductus deferens in the male or to the vagina in the female.

Uterine artery is a large vessel that may be a separate branch of the anterior trunk or arise in common with the vaginal or middle rectal artery. It descends medially to the cervix on the upper surface of the pelvic diaphragm then it ascends tortuously between the layers of the broad ligament on the side of the uterus.

The uterine artery crosses above the ureter.

It divides into a large ascending branch that supplies the uterus, medial part of the uterine tube and ovary that anastomoses with the ovarian artery (from the abdominal aorta) and a descending branch that supplies the upper part of the vagina (anastomosing with the vaginal artery).

The uterine artery is homologous to the artery to the ductus deferens in the male.

Vaginal artery (homologous to the inferior vesical artery in the male) supplies branches to the front and back of the vagina, base of the bladder and lower rectum.

PARIETAL AND LIMB ARTERIAL BRANCHES

Inferior gluteal artery is the larger of the two terminal branches of the anterior division of the internal iliac. It passes back between S1 and S2 nerves (sometimes S2 and S3) and leaves the pelvis through the lower part of the greater sciatic foramen below piriformis to enter the gluteal region. It supplies branches to piriformis, coccygeus and levator ani.

Outside of the pelvis, the inferior gluteal artery supplies muscular branches to gluteus maximus and lateral thigh rotators, a companion artery to the sciatic nerve, the capsule of the hip joint, and anastomotic branches to the trochanteric fossa (trochanteric anastomosis) and cruciate anastomosis.

Internal pudendal artery is the other terminal branch of the anterior division. It passes down and back and leaves the pelvis through the greater sciatic foramen, passing below piriformis. Crossing the ischial spine, it re-enters the pelvis through the lesser sciatic foramen and enters the pudendal canal on the medial side of the ischial tuberosity. It perforates the perineal membrane and supplies the external genitalia.

Obturator artery descends forward on the side wall of the pelvis, leaving the pelvis through the obturator canal on the upper part of the obturator foramen. Its iliac branches supply iliacus and the bladder. A pubic branch ascends on the pelvic surface of the pubis and anastomoses with the inferior epigastric artery.

6. Veins

Median sacral vein is formed by a pair of venae comitantes of the median sacral artery and terminates in the left common iliac vein.

Superior rectal vein originates from capillaries in a submucous plexus in the rectal wall. They ascend from the pelvis to become the inferior mesenteric vein of the portal system.

Internal iliac vein begins by the union of venae comitantes corresponding to branches of the internal iliac artery (except the umbilical and iliolumbar arteries). The visceral veins are derived from plexuses that drain the viscera. These are the vesical, prostatic, pudendal, and rectal plexuses. The internal iliac vein ascends joining the external iliac vein to form the common iliac vein.

7. Nerves

LUMBOSACRAL TRUNK

Lumbosacral trunk is a thick cord formed on the ala of the sacrum by the lower division of the ventral ramus of L4 that joins the ventral ramus of L5. It enters the pelvis to join the sacral plexus.

Sacral plexus is formed from the lumbosacral trunk (L4 and L5) and the ventral rami of S1 to S4. The nerves forming this plexus emerge from the anterior sacral foramina and lie on the anterior surface of piriformis behind the internal iliac artery

and rectum. The plexus divides into ventral and dorsal divisions.

Ventral divisions supply developmentally ventral muscles that become adductors and flexors of the thigh, flexors of the leg and intrinsic muscles of the foot. Dorsal divisions supply developmentally dorsal muscles that become extensors and abductors of the thigh, extensors and evertors of the leg and foot.

Movements of the hip are controlled from spinal segments L2–L5, the knee L3–S1 and ankle L4–S2.

VENTRAL BRANCHES

Ventral (preaxial) branches supply ventral structures — ventrum of the thigh, ventrum of the leg, ventrum of the foot and the perineum. There are seven ventral branches:

Nerves to Levator Ani and Coccygeus (S3, S4) arise from a common loop and descend on the pelvic surface of these muscles to supply them.

Nerve to Quadratus Femoris (L4, L5, S1) and Inferior Gemellus leaves the pelvis through the greater sciatic foramen beneath the sciatic nerve to supply the quadratus femoris and gemellus inferior muscles in the gluteal region.

Nerve to Obturator Internus and Superior Gemellus (L5, S1, S2) leaves the pelvis medial to the sciatic nerve but lateral to the internal pudendal vessels and pudendal nerve below the piriformis muscle. It supplies obturator internus and gemellus superior in the gluteal region.

Muscular branches to the pelvic diaphragm and external anal sphincter (S3, S4) the perineal branch of S4 pierces the coccygeus muscle and enters the ischio-anal fossa under the edge of coccyx. It supplies the superficial part of the anal sphincter that arises from the coccyx.

Pelvic splanchnic nerves (S2, S3, S4) are parasympathetic nerves that arise from S2 to S4, the same spinal segments as the pudendal nerve and supply the corresponding involuntary sphincters of the bladder and anal canal. They join the visceral (inferior hypogastric) sympathetic plexus for distribution to the pelvic organs.

DORSAL BRANCHES

Dorsal (postaxial) branches supply dorsal structures, muscle dorsal to the plexus (piriformis), skin and muscles of the buttock, the dorsum of the leg and foot.

Nerve to Piriformis (S1, S2) immediately enters the pelvic surface of piriformis.

Superior Gluteal nerve (L4, L5, S1) leaves the pelvis above the piriformis muscle to enter the gluteal region. It supplies gluteus medius and minimus and ends by supplying tensor fascia lata.

Inferior Gluteal nerve (L5, S1, S2) is formed one segment lower than the superior gluteal nerve. It leaves the pelvis below piriformis superficial to the sciatic nerve to enter gluteus maximus.

Posterior Femoral Cutaneous nerve (S1, S2) leaves the pelvis below piriformis and passes through the gluteal region piercing fascia lata to the back of the knee. It provides cutaneous nerve supply to the lower buttock, perineum and back of the thigh.

Perforating cutaneous nerve (from the back of **S2, S3** is so named because it perforates the sacrotuberous ligament) and curves around the lower border of gluteus maximus to supply the skin of the lower medial side of the buttock.

TERMINAL BRANCHES

Sciatic nerve (L4, L5, S1, S2, S3) is formed at the lower border of piriformis by the union of tibial and common fibular components and leaves the pelvis at the lower margin of the piriformis muscle entering the buttock lying on the ischium.

The tibial component, the larger, medial and ventral component supplies the ventrum of the thigh, leg and foot.

The common fibular component is smaller, lateral and dorsal and may arise separately from the plexus in front of or piercing piriformis. It supplies the dorsum of the leg and foot (and exceptionally supplies a muscle of the ventrum of the thigh — short head of biceps femoris).

Pudendal nerve (S2, S3, S4) leaves the pelvis medial to all structures emerging below piriformis. It passes over the ischial spine and enters the pudendal canal for distribution to the perineum (including the external genitalia, sphincter urethrae and sphincter ani externus).

Coccygeal plexus is formed by the union of ventral rami of parts of S4, S5 and the coccygeal nerve that unite on the pelvic surface of the coccygeus muscle.

Anococcygeal nerves arise from the plexus, pierce coccygeus and the sacrotuberous ligament to supply the skin of the post-anal region.

Sacral sympathetic trunk or the lumbar sympathetic trunk crosses the margin of the ala of the sacrum to enter the pelvis medial to the anterior sacral foraminae. They descend on the pelvic surface of the sacrum just medial to the pelvic sacral foramina and terminate on the coccyx by the union of both trunks in a small swelling, the ganglion impar.

Somatic branches are given to all sacral nerves and visceral branches leave the upper ganglia to join pelvic plexuses.

SYMPATHETIC PLEXUSES

The **aortic plexus** extends into the pelvis receiving fibers from the lower lumbar splanchnic nerve becoming the superior hypogastric plexus. This divides into two (left and right) hypogastric nerves, in front of the sacrum. These descend on each side of the rectum, where they are joined by pelvic splanchnic nerves to become the superior hypogastric plexus.

Subdivisions of the sympathetic nerve plexus accompany the visceral branches of the internal iliac arteries to supply the pelvic organs and are named accordingly.

Middle rectal plexus accompanies the middle rectal artery. **Prostatic plexus** goes through the hypogastric plexus via a pelvic plexus. **Vesical plexus** supplies the bladder. **Uterovaginal plexus** follows the uterine artery to supply the uterus, ovary, vagina, urethra and cavernous bodies of the clitoris.

Chapter 40

Pelvic Genital Organs

Chapter Outline

In the male, spermatozoa are produced in the testis. They then pass into the epididymis where they mature and are stored. On emission, spermatozoa pass sequentially through the ductus deferens, ejaculatory duct, urethra to the external urethral orifice. Other genital organs produce nutrient secretions that comprise seminal fluid.

Semen is a mixture of spermatozoa and secretions of seminal vesicles, the prostate and bulbourethral glands. The secretion of seminal vesicles contains fructose and prostaglandins. The secretion of the prostate is alkaline and neutralizes the acidity of vaginal secretions.

In the female, ova are produced in the ovaries and pass down the uterine tube to the cavity of the uterus. The uterine tubes also convey spermatozoa to the ovum and fertilization occurs in the uterine tube. The fertilized ovum reaches the cavity of the uterus where it implants.

1. Genital Organs in the Male

The male genitalia comprise (primary sex organs) the testes, a series of ducts leading from the testes (epididymis); ductus deferens, ejaculatory ducts; and glands that contribute their secretion to seminal fluid (accessory sex organs) — a single prostate gland, paired seminal vesicles and bulbourethral glands. The testes, being intrapelvic in the fetus are considered with the pelvic genital organs.

TESTES

Testes are paired, oval, white organs, located within the scrotum. Each is covered by the following.

Tunica vaginalis is a remnant of the processus vaginalis, an evagination of the peritoneal cavity into the scrotum. The visceral layer of tunica vaginalis is equivalent to the visceral layer of peritoneum. It is not present where the epididymis and beginning of the

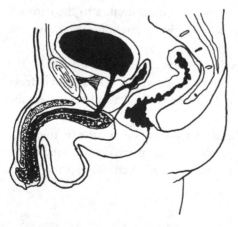

FIG. 40-1 Male genital system.

spermatic cord attach to the upper posterior border of the testis. The parietal layer of the tunica vaginalis is equivalent to the parietal layer of peritoneum.

Tunica albuginea is the tough fibrous coat of the testis that projects into the testis as an internal median ridge, the mediastinum testis.

Septulae testis are fine septa that extend from the mediastinum and subdivide the testis into over 200 conical compartments. Each compartment contains two or more convoluted seminiferous tubules.

Straight seminiferous tubules of each tubule enter the mediastinum from the seminiferous tubules and anastomose in it forming a network, the rete testis.

Efferent ducts (about 6–12) emerge from the mediastinum into the head of the epididymis.

The **neurovascular system** of the testis is supplied by:

Arteries: **Testicular artery** arises from the aorta below the renal arteries join the spermatic cord and enter the testis through its mesentery (mesorchium). It also gives a branch to the epididymis.

Veins: Venous drainage of the testis collects into the **pampiniform plexus** that surrounds the testicular artery. This plexus functions as a counter current heat exchanger insuring that the temperature in the testis is some 4° cooler than in the abdomen. The plexus gives rise to the testicular vein that terminates in the inferior vena cava on the right side and left renal vein on the left side.

Nerves: Vasomotor from the autonomic aortic and renal plexuses.

EPIDIDYMIS

Epididymis is a C-shaped body situated along the posterior border of the testis. It contains a highly coiled duct of the epididymis that begins from the efferent ductules and is continuous at the tail with the ductus deferens.

The *head* is the enlarged upper end of the epididymis and with the body is a major segment for storage and where maturation of spermatozoa takes place.

The *body* is attached to the posterior margin of the testis and separated from the testis by the sinus of the epididymis.

The *tail* is the lower end of the epididymis that becomes thicker and leads into the ductus deferens.

The **neurovascular system** of the epididymis is supplied by.

Arteries: Testicular artery (a branch of the aorta) that anastomoses with the artery of the ductus deferens and cremasteric artery (from the inferior epigastric artery).

Veins: Join the pampiniform plexus.

Lymphatics: From the testis and epididymis run with testicular vessels to lumbar aortic nodes.

Nerves: **Inferior hypogastric plexus** branches along the ductus deferens.

DEVELOPMENT

Testes develop as longitudinal ridges medial to the mesonephric ridges that are proliferations of celomic epithelium and underlying condensed mesenchyme.

Primordial germ cells first appear among entodermal cells of the yolk sac and migrate through the mesentery of the hindgut to reach the genital ridges.

Celomic epithelium of the genital ridges proliferate into underlying mesenchyme becoming primitive sex cords. They penetrate deeply into the medulla as medullary cords.

Cords give rise to seminiferous tubules, straight tubules and the rete testis.

Interstitial mesenchyme produces hormone producing interstitial cells.

Appendix testis is a small vestigial body on the upper end of the testis. It is the remnant of the upper end of the paramesonephric duct.

Appendix epididymis is a small vestigial body on the head of the epididymis. It is the remnant of the mesonephric duct.

DUCTUS DEFERENS

Ductus deferens is the continuation of the duct of the epididymis. It begins at the tail of the epididymis and ascends from the scrotum in the spermatic cord. At the deep inguinal ring, it courses beneath the peritoneum along the side wall of the pelvis and curves medially passing over the ureters. It expands into the ampulla behind the bladder. Behind the neck of the bladder, the ductus deferens narrows and joins the duct of the seminal vesicle to form the ejaculatory duct.

The **neurovascular system** of the ductus deferens is supplied by:

Arteries: The artery of the ductus deferens from the superior vesical branch of the anterior trunk of the internal iliac artery also supplies the seminal vesicle and ejaculatory duct anastomoses with the testicular artery.

Veins: Veins join the prostatic an vesical venous plexuses.

Nerves: Autonomic nerves from the superior and inferior hypogastric plexuses.

SPERMATIC CORD

Extends from the deep inguinal ring through the inguinal canal into the scrotum ending at the posterior border of the testis.

The cord consists of the following:
- Ductus deferens (lies posteriorly in the cord).
- Artery of the ductus deferens, a branch of the umbilical artery.
- Cremasteric artery, a branch of the inferior epigastric artery.
- Testicular artery, a branch of the aorta.
- Testicular veins from the pampiniform plexus of veins.
- Testicular lymphatics that drain to lumbar nodes.
- Genital branch of the genitofemoral nerve (L1–L2).

- Remnants of the processus vaginalis of peritoneum.
- Autonomic nerves from the testicular and deferential plexuses.
- The spermatic cord is also enveloped by layers derived from the abdominal wall.

Layer of scrotum
- Skin.
- Dartos muscle (smooth muscle attached to the skin).

Coverings of cord
- External spermatic fascia cf aponeurosis of external oblique muscle. A thin outer covering attached to the crura at the superficial inguinal ring.
- Cremaster muscle and fascia derived from internal oblique muscle within the inguinal ring.
- Internal spermatic fascia derived at the deep inguinal ring from transversalis fascia — the innermost layer.

Constituents of cord
- Areolar tissue — extraperitoneal fatty tissue.
- Tunica vaginalis — a double layered serous membrane.

SEMINAL VESICLES

Seminal vesicles are pear-shaped, branched, sacculated pouches that lie on the back of the bladder. Their narrow duct joins the termination of the ductus deferens to form the ejaculatory duct.

The **neurovascular supply** of the seminal vesicles is by:

Arteries: **Artery of the ductus deferens** from the umbilical artery and anastomoses with the testicular artery.

Veins: Join the prostatic and vesical plexuses.

Nerves: L2 to L4 via the hypogastric plexus.

Ejaculatory ducts are formed just above the base of the prostate by the union of the ductus deferens and the duct of the seminal vesicle. They penetrate the base of the prostate obliquely and open into the prostatic sinus in the prostatic urethra on both sides of the utricle.

PROSTATE

Prostate is an almond shaped gland that lies behind the pubic symphysis immediately below the bladder. It has an apex, its lowermost part that lies on the urogenital diaphragm and a base or superior

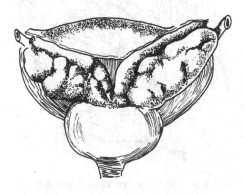

FIG. 40-2 Posterior surface of bladder showing prostate and seminal vesicles.

surface below the bladder. Its inferolateral surface is convex and separated from the pelvic fascia by a plexus of veins.

The anterior surface is separated from the pubic symphysis by a retropubic pad of fat.

The posterior surface is flat and triangular with a median sulcus between the right and left lobes.

The glandular tissue is embedded in a fibromuscular stroma.

Secretion of the prostate is eliminated via 20–30 prostatic ducts that open into the prostatic sinuses.

The gland is perforated in the center by the (prostatic) urethra and from behind by the ejaculatory ducts. In cross section, it has two lateral lobes and a single median lobe. The median lobe separates the urethra in front from the ejaculatory ducts behind.

"False" *capsule* consists of condensed endopelvic fascia encloses the gland. It is continuous with the superior fascia of the urogenital diaphragm and contains the prostatic venous plexus that receives the deep dorsal vein of the penis and drains into the vesical venous plexus.

Medial puboprostatic ligaments fix the prostate and neck of the bladder. They pass from the back of the pubes to the fascia at the base of the prostate and bladder.

The **neurovascular system** of the prostate is supplied by:

Arteries: **Inferior vesical artery** from the internal iliac artery and middle rectal arteries from the superior rectal artery.

Veins: Venous drainage is to a plexus between the true and false capsules that empty into the vesicoprostatic plexuss then into the **internal iliac vein**.

Enucleation of diseased prostate is directed toward the removal from the compressed periphery rather than from the line of the false capsule.

Nerves: **Prostatic plexus** to blood vessels and smooth muscle.

Lymphatics: Drain to internal iliac nodes, some to external iliac nodes or some to sacral nodes.

Bulbourethral glands are paired, small glands embedded in the substance of the sphincter urethrae behind the membranous urethra. Their ducts pass through the bulb of the penis and open into the spongy urethra.

2. Pelvic Organs of the Female

The female internal genitalia comprise the uterus, uterine tubes, ovaries, and upper vagina.

UTERUS

Uterus is an inverted pear-shaped, muscular organ flattened from before backwards located between the rectum and the bladder. It is formed by the fusion of the paired paramesonephric (Mullerian) ducts and is dependent upon estrogen as well as somatotrophin so is larger in maturity than immaturity or senility.

Cervix is the lowest narrow part that projects postero-inferiorly through the upper anterior wall of the vagina. It has supra- and intravaginal parts.

The vaginal part is tapered and rounded and at its lower end, it presents the ostium uteri. It has anterior and posterior lips and posteriorly faces the posterior vaginal wall.

The anterior surface of the supravaginal part lies below the floor of the vesicouterine pouch and closely related to the bladder. The posterior surface is covered with peritoneum and leaves the cervix to cap the posterior vaginal fornix.

The external os is the opening of the cervical canal into the vagina.

The cervical canal is the segment between the internal os and external os.

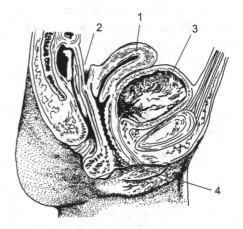

FIG. 40-3 Sagittal section through female pelvis. (1) Uterus; (2) Rectouterine pouch; (3) Bladder; (4) Vagina.

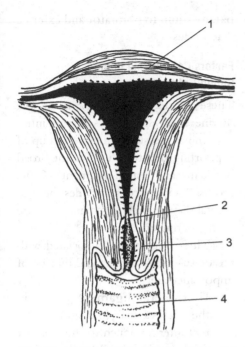

FIG. 40-4 Frontal section through uterus and upper part of vagina. (1) Fundus; (2) Isthmus; (3) Cervix; (4) Vagina.

It contains glands that produce a protective mucus plug during pregnancy.

Body is the major part of the uterus and is connected with the cervix at the constricted isthmus (the internal os). The shape of the cavity of the body is flattened and triangular.

Uterine horns are the point of entry of the uterine tubes on each side.

Fundus is the upper rounded and blunt region above the uterine horns.

The vesical surface of the uterus is covered with peritoneum as far as the isthmus. Peritoneum passes down the intestinal surface as far as the vagina.

The lateral borders of the uterus are connected to the broad ligament.

Positions of the Uterus

Normal position of the uterus is anteflexed with the body bent forward on itself at the level of the internal os.

Anteversion refers to the angle between the cervix and vagina that is normally about 90°.

Other positions are: retroflexed when the axis of the body passes upward and backwards; and retroverted where the axis of the cervix with the vagina is upward and backward.

The **neurovascular system** to the uterus is supplied by:

Arteries: **Uterine arteries,** branches of the internal iliac artery, reach the lateral side of the cervix and turn upwards between the layers of the broad ligament.

Uterine arteries run in the transverse cervical (cardinal) ligament and cross the ureters superiorly where they may be involved in surgery.

The uterine arteries give ascending (tubal) branches to the uterine tubes and that anastomose with the ovarian arteries and descending (vaginal) branches which in turn anastomose with those of the opposite side and perineal branches of the internal pudendal arteries.

Veins: The uterine plexus in the broad ligament drains to uterine and ovarian veins then drains into pelvic venous plexuses.

Nerves: **Autonomic innervation** is by the uterovaginal plexus, an extension of the inferior hypogastric plexus (T12–L2, L3). Parasympathetic nerves leave spinal nerves in pelvic splanchnic nerves (S2–S4) or nervi eregentees.

Visceral afferents from the body of the uterus and uterine tubes pass with T12–L2 afferents and are referred to the middle of the back, inguinal and pubic regions and anterior thigh. Afferents from the cervix pass along the nervi eregentes (S2–S4) and may be referred to the perineum, gluteal region, posterior thigh and leg.

Lymphatics: The fundus and upper body drain along with the blood supply to lumbar (aortic) nodes. The lower part of the body drains to external iliac nodes. The cervix drains along the uterine vessels to internal iliac nodes, along internal pudendal vessels to internal iliac nodes and in the parametrium to obturator and external iliac nodes.

Factors Contributing to the Support of the Uterus

These include fascial sheaths of the uterine and vaginal vessels and intra-abdominal pressure. An upper group of supporting structures include the broad ligament, and round ligament of the uterus. A lower group includes the urogenital diaphragm and tendinous centre of the perineum.

Three "ligaments" associated with the uterus in its support and may be of importance are:

- **Uterosacral ligament** passes in the rectovaginal fold around the rectouterine pouch to the front of the second or third sacral vertebra. It is continuous with the posterior layer of the broad ligament at the cervix.

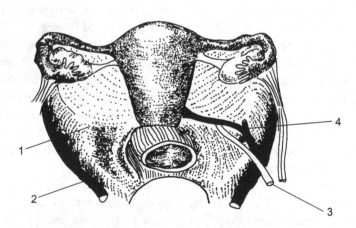

FIG. 40-5 Uterosacral and lateral cervical ligaments from above and behind. (1) Upper edge of lateral cervical ligament; (2) Uterosacral ligament; (3) Ureter; (4) Uterine a.

- **Transverse cervical (cardinal) ligament** passes laterally from the cervix in the lowest part of the broad ligament to superior fascia of the pelvic diaphragm. It contains the uterine artery.
- **Pubocervical ligament** is the anterior continuation of the pubocervical ligament that reaches the pubis in the floor of the retropubic space.

BROAD LIGAMENT

Broad ligament is the mesentery of the embryonic paramesonephric duct. It extends from the side of the uterus to the side wall of the pelvis. The uterine tube is enclosed in the medial four-fifths of the upper free end of the ligament. The suspensory ligament of the ovary is enclosed in the lateral fifth. The mesovarium, attached to the posterior surface divides the broad ligament into two parts. The upper part is known as the mesosalpinx.

Peritoneal folds are continuations of the broad ligament.

Mesosalpinx supports the uterine tube. It begins at the junction of the mesovarium and superior limit of the broad ligament and contains the tubal branch of uterine vessels.

Mesovarium supports the ovary. It is the posterior extension of the broad ligament.

Suspensory ligament of the ovary is an elevation of peritoneum between the mesovarium and the pelvic brim. It contains ovarian vessels and nerves.

Contents of the Broad Ligament

- *Uterine tube* in the upper free border on both sides.
- *Ovary* attached by mesovarium to the posterior surface.
- *Proper ligament of the ovary* attached from the medial pole of the ovary to the horn of the uterus just below the point of entry of the uterine tubes.
- *Round ligament of the uterus* extending from the lateral border of the uterus near the point of attachment of the ligament of the ovary to and then through the inguinal canal to the labia majora.
- *Uterine vessels* in the base of the broad ligament in the transverse cervical ligament.
- *Ovarian vessels* in the suspensory ligament of the ovary and mesovarium.
- *Lymphatics and nerves.*
- *Vestigal structures* such as round ligaments of the ovary homologous to the gubernaculum of the testis in the male; remnants of the mesonephric ducts such as epoopheron vestigal mesonephric tubules (homologous to the epididymis), paroopheron vestigal mesonephric tubules (homologous to the para-didymis), Gärtners ducts in the mesometrium adjacent to the uterus are vestigal mesonephric ducts (homologous to the ductus deferens) and retii ovarii adjacent to the hilum of the ovary (homologous to the retii testis).
- Round ligament (of the uterus) is a fibrous cord also known as the

ligamentum teres (uteri), is the remnant of the lower part of the gubernaculum of the ovary. It extends between layers of the broad ligament from the superolateral angle of the uterus to the deep inguinal ring. It passes through the inguinal canal, emerges through the superficial ring, and terminates in the labium majus.

* *Rectouterine fold* extends from the sacrum to the cervix of the uterus.

Uterine Tubes

Uterine tubes are fibromuscular tubes covered by peritoneum that convey ova from the ovaries to the cavity of the uterus. Each tube is divided into numerous parts.

Uterine part the segment located within the wall of the uterus and extends in the cavity of the uterus as the uterine opening.

Isthmus is the segment with thicker walls than the ampulla.

Ampulla is the longest and widest part of the tube. It is partly tortuous and its walls are thin.

Infundibulum is the distal funnel shaped segment of the uterine tube that begins at the pelvic opening of the uterine tube.

Fimbriae are a number of finger-like processes that project from the margins of the infundibulum.

OVARIES

Ovaries are the female gonads. They are white glands that lie against the lateral pelvic walls in the ovarian fossa between the ureter behind and the broad ligament in front. The suspensory ligament of the ovary extends from the tubal end of the ovary to the pelvic wall and the ligament of the ovary from its uterine end. The ovary is enveloped by peritoneum that attaches to the broad ligament by the mesovarium.

The **neurovascular system** of the ovary is supplied by:

Arteries: **Ovarian artery** from the aorta and ovarian and tubal branches of the uterine artery (branch of the internal iliac artery).

Veins: A pampiniform plexus from which the ovarian veins arises to terminate assymetrically in the inferior vena cava on the right side and the renal vein on the left.

Nerves: Branches from the aortic and renal plexuses but their function is unclear. Visceral afferents are referred to spinal segments T12–L2.

Lymphatics: Drain to aortic nodes situated near the renal artery.

The two ligaments associated with the ovary are:

Suspensory ligament of the ovary (infundibulo-pelvic ligament) extends from the tubal end of the ovary to the lateral pelvic wall, it lies at the upper margin of the uterine tube and contains ovarian vessels, lymphatics and nerves.

Ligament of the ovary is a prominent cord that extends from the uterine end of the ovary to the lateral side of the uterus, just below the entrance of the uterine tube. It consists of a fibromuscular cord enclosed by the broad ligament.

VAGINA

Vagina is an organ of copulation and birth canal that lies partly in the lower

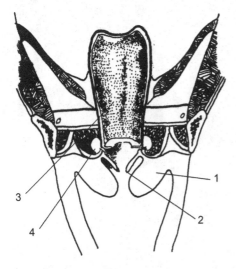

FIG. 40-6 Coronal section through vagina showing relationship of external genital organs from in front. (1) Labium majus; (2) Labium minus; (3) Vagina; (4) Greater vestibular gland.

part of the pelvis and partly in the perineum. Its walls are in contact with each other except where the cervix projects into the vaginal cavity at its upper end. This forms the constricted anterior, lateral and posterior fornices.

The **neurovascular system** of the vagina is supplied by:

Arteries: Vaginal branch of the uterine artery, branches of the middle rectal artery, and internal pudendal arteries.

Veins: Through the vagiinal plexus to internal iliac veins.

Lymphatics: From the upper third to external and internal iliac nodes, from the middle third to internal iliac nodes, and from the lower third to internal iliac nodes and superficial inguinal nodes.

Nerves: From sympazthetic and pelvic splanchnic nerves through the uterovaginal plexus. Afferents travel with nervi erigentes to S2–S4 spinal segments. The lower part of the vagina is innervated by the pudendal nerve.

Chapter 41

Pelvic Walls

Chapter Outline

1. Pelvic Muscles
2. Pelvic Fascia

The diamond shaped pelvic outlet is closed by the obturator internus and piriformis muscles laterally and by the levator ani and coccygeus muscles behind and below. The urogenital diaphragm contributes above and in front.

1. Pelvic Muscles

Two muscles, levator ani and coccygeus are limited to the pelvis. Two others, piriformis and obturator internus, line the pelvic wall and also act on the femur (short lateral rotators). The lesser pelvic walls comprise from before backwards, posterior surface of the body of the pubis, obturator internus, piriformis, and the pelvic surface of the sacrum and coccyx.

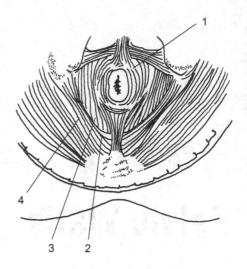

FIG. 41-2 Ischioanal fossa (muscles). (1) External anal sphincter; (2) Puborectalis m; (3) Pubococcygeus m; (4) Iliococcygeus m.

Obturator Internus

Obturator internus contributes to the lining of the pelvic wall. It leaves the pelvis through the lesser sciatic foramen.

- **Origin:** Within the pelvis, from the entire margin of obturator

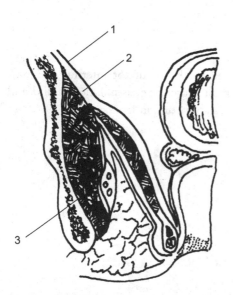

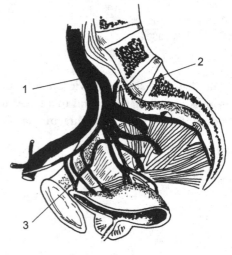

FIG. 41-3 Internal iliac artery. (1) External iliac a; (2) Internal iliac a; (3) Obturator a.

FIG. 41-1 Coronal section of pelvis showing pelvic fascia. (1) Obturator fascia; (2) Tendinous arch of levator ani; (3) Pudendal canal.

foramen, inner surface of the obturator membrane and pelvic surface of the coccyx. The muscle bends at right angles around the lesser sciatic notch, a bursa intervening.

- **Insertion:** Medial surface of the greater trochanter of the femur above the trochanteric fossa.

In the gluteal region, superior and inferior gemelli blend with the tendon of obturator internus.

- **Action:** Lateral rotator of the thigh and assists in abduction of the thigh.
- **Nerve supply:** Nerve to obturator internus and superior gemellus (preaxial) from the sacral plexus (L5, S1 and S2).

Piriformis

Piriformis: A flat and triangular muscle. Leaves the pelvis through the greater sciatic foramen.

- **Origin:** Pelvic surface of the second to fourth sacral vertebrae between and lateral to the pelvic sacral foraminae.

- **Insertion:** Upper border and medial surface of the greater trochanter of the femur.
- **Action:** Lateral rotation and abduction of the thigh.
- **Nerve supply:** Directly from ventral rami of S1 and S2.

Piriformis is an important landmark. The superior gluteal nerve and vessels leave the pelvis above the upper border of piriformis. The sciatic nerve and nerve to quadratus femoris, inferior gluteal nerve and vessels, posterior femoral cutaneous nerve, nerve to obturator internus, internal pudendal vessels and pudendal nerve leave the pelvis below the lower border of piriformis.

Ischiococcygeus

Ischiococcygeus is triangular in shape lying behind and on the same plane as levator ani. It forms the posterior third of the muscular part of the pelvic diaphragm.

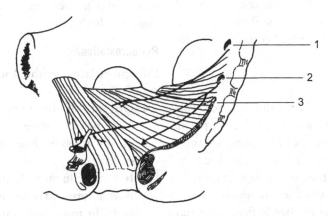

FIG. 41-4   Nerves of the pelvic floor. (1) S3; (2) S4; (3) S5.

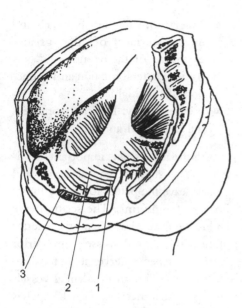

FIG. 41-5 Pelvic muscles. (1) Puborectalis m; (2) Iliococcygeus m; (3) Pubococcygeus m.

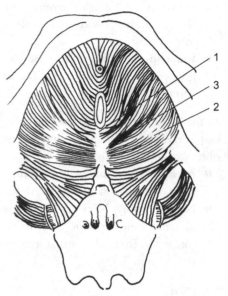

FIG. 41-6 Pelvic diaphragm from below. (1) Puborectalis m; (2) Iliococcygeus m; (3) Pubococcygeus m.

- **Origin:** Pelvic surface of the spine of the ischium and sacrospinous ligament (an aponeurotic part of the muscle).
- **Insertion:** Lateral margin of the lower sacrum and coccyx.
- **Action:** Draws the coccyx forward elevating the pelvic floor.
- **Nerve supply:** Ventral rami of S4 and S5.

Levator Ani

Levator ani forms the anterior two thirds of the muscular part of the pelvic floor. It comprises two parts, an anterior pubic part and a posterior iliac part:

Pubic part is the most important functional part of levator ani. It is made of three sets of muscle fibers that run backwards and medially. Puboprostaticus is the most medial set of fibers (pubovaginalis in the female). Puborectalis is an intermediate set of fibers in both sexes and pubococcygeus is the most lateral set in both sexes.

Puboprostaticus

Puboprostaticus is the most medial set of fibers.

- **Origin:** (Of the anterior or pubic part) pelvic surface of the body of the pubis with a slight lateral extension.
- **Insertion:** In the male, puboprostaticus is inserted into the perineal body. In the female, pubovaginalis

crosses the side of the vagina to reach the perineal body. Some fibers blend with the muscular walls of the vagina and with muscle fibers from the contralateral side form a partial sphincter for the vagina.

Puborectalis

Puborectalis is the intermediate set of fibers of levator ani.

- **Origin:** Dorsum of the pubis. Its fibers pass across the side of the rectum to loop around the anorectal junction.
- **Insertion:** Joins its fellow from the opposite side behind the rectum. Some fibers become continuous with the deeper part of the external anal sphincter and outer longitudinal coat of the rectum.

Pubococcygeus

Pubococcygeus is the thickest and most lateral portion of the levator ani.

- **Origin:** Superior pubic ramus lateral to the obturator canal.
- **Insertion:** Lowest part of the sacrum, anococcygeal raphe and the coccyx.

Iliac part of levator ani comprises iliococcygeus and pelvis diaphragm.

Iliococcygeus

Iliococcygeus is thin and its muscular fibers form the most posterior (iliac) part of levator ani.

- **Origin:** A tendinous band (the tendinous arch of the levator ani muscle)

which runs from the pubic part of the muscle to the ischial spine.

- **Insertin:** Anococcygeal raphe and coccyx in the midline.
- **Action:** Levator ani supports and raises the pelvic floor and resists increased intraabdominal pressure as in forced expriation and defecation.
- **Nerve supply:** Branches of S3 and S4 and inferior rectal nerve (branch of the internal pudendal nerve).

Pelvic Diaphragm

Levator ani and coccygeus and their fascia form the pelvic floor, that separates the pelvic cavity from the perineum. From above downwards, the diaphragm includes the superior fascia of the pelvic diaphragm, levator ani and ischiococcygeus and the inferior fascia of the pelvic diaphragm.

The pelvic diaphragm is perforated by the urethra and the rectum in both sexes. In addition, the vagina perforates the pelvic diaphragm of the female.

2. Pelvic Fascia

The pelvic viscera, vessels, and muscles are covered with areolar tissue. This fascia is distinguished as three layers.

VISCERAL LAYER

Visceral layer is an endopelvic fascia or layer of extraperitoneal connective tissue invests the bladder, the rectum, and internal genital organs. The endopelvic

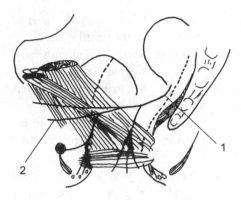

FIG. 41-7 Nerve supply to external sphincter and puboanal sling; (1) Pudendal n; (2) Perineal n.

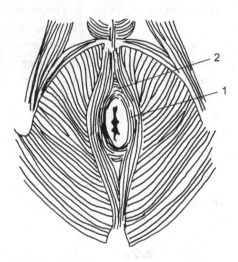

FIG. 41-8 External anal sphincter from below. (1) Subcutaneous part; (2) Superficial part.

fascia has an attachment to the superior fascia of the pelvic diaphragm at the sides and in front of the bladder and prostate. This attachment is along a line of thickening of the superior fascia known as the arcus tendineus fascia pelvis. It begins posteriorly from the arcus tendineus levator ani and extends forward on pubococcygeus to the lower border of the pubis.

FASCIAL LIGAMENTS

Fascial ligaments arise from fused thickenings of visceral pelvic fascia that extend from the broad tendinous arch of pelvic fascia and superior fascia of the pelvic diaphragm. These converge medially onto pelvic organs giving them some passive stability. Thickenings are reinforced by smooth muscle fibers to form the "ligaments".

The **puboprostatic ligament** (pubovesical ligament in the female) is the sheet of endopelvic fascia sweeping medially from the arcus tendineus fascia pelvis to invest the prostate and bladder.

In the female, these fascial ligaments are known anteriorly as the pubovescical ligament, laterally as the cardinal (transverse cervical) ligament and posteriorly, as the uterosacral ligaments. These can be palpated per rectum.

PARIETAL LAYER

Parietal layer is the downward continuation of transversalis fascia into the pelvis following fusion with the periosteum covering linea terminalis and back of the pubis. This layer is divided into two parts.

Obturator fascia is the part of parietal fascia covering the intrapelvic part of the obturator internus except at the lesser sciatic notch, where the muscle leaves the pelvis. Tendinous arch of levator ani is the thickened boundary between the intrapelvic and extrapelvic parts of the obturator fascia. It extends from the pubis to the ischial spine. At the arch, the fascia splits into superior and inferior fasciae of the pelvic diaphragm that clothes levator ani and ischeococcygeus.

The superior fascia of the pelvic diaphragm is a continuity of the transversalis fascia. It covers the superior surface of levator ani and ischiococcygeus and passes backward and over the sacral plexus, piriformis and lateral front of the sacrum.

The inferior fascia of the pelvic diaphragm lines the inferior surface of levator ani and posterior surface of ischiococcygeus. It fuses in front and medially with the fascia of the perineal region at the urogenital hiatus.

Piriform fascia is the part of parietal fascia that covers the intrapelvic portion of piriformis and is attached to the front of the sacrum and the side of the greater sciatic foramen.

Bones and Joints of the Pelvis

Chapter Outline

The hip bones are large irregular bones articulating with each other in front and with the sacrum and coccyx behind to form the bony pelvis. Internally, the part of the bony pelvis above the arcuate line (the superior pelvic aperture or pelvic brim) is known as the false (or greater) pelvis, and the part below the arcuate line is the true (or lesser) pelvis.

The greater pelvis is part of the abdominal cavity and is lined by the greater peritoneal sac extended into the lesser pelvis to cover terminal parts of the gastrointestinal tract, urinary system and female internal reproductive organs.

Each of the hip bones consists developmentally of three bones, the ilium, ischium, and pubis that unite at the cavity for the head of the femur (the acetabulum). From a single cartilaginous model, three separate major (and several secondary minor) ossification centers appear prenatally in the third, fourth and fifth months. At birth, the major centers are still separated by a Y-shaped cartilage at the center of the acetabulum. Fusion into a single bone occurs between 15 and 16 years.

1. Ilium

The Iliums consists of a fan shaped wing or ala and an irregular handle-like body. The iliopubic eminence marks the region of fusion of the ilium with the ischium.

Crest is a prominent border of the ala or wing of the ilium. The margins of the long, arched, upper border of the iliac crest are known as the external, intermediate and internal lips. The outer lip gives attachment to the fascia lata of the thigh, external abdominal oblique

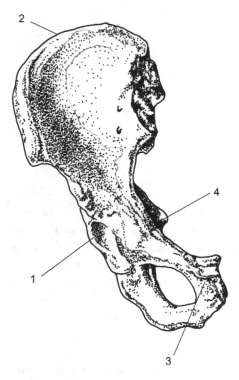

FIG. 42-1 Right hip bone — anterior aspect. (1) Acetabulum; (2) Iliac crest; (3) Body of pubis; (4) Ischial spine.

anteriorly and latissimus dorsi posteriorly. The internal lip is the upper limit of the iliac fossa and anteriorly gives attachment to the transversus abdominis and posteriorly, quadratus lumborum. The intermediate line gives attachment to the internal abdominal oblique muscle.

Anterior superior iliac spine marks the anterior end of the iliac crest. It gives attachment to the inguinal ligament and sartorius muscle.

Posterior superior iliac spine is less well marked and located at the posterior end of the iliac crest. It gives attachment to the sacrotuberous ligament, posterior sacroiliac ligament and multifidus muscle.

Tubercle of the crest is a thickening of the outer lip about 6 cm behind the anterior superior iliac spine. It is a palpable landmark of the horizontal transtubercular plane at the level of L5, the level of the beginning of the inferior vena cava.

Anterior border extends from the anterior superior iliac spine to the iliopubic eminence. The anterior inferior iliac spine subdivides the border and gives rise to the straight head of rectus femoris muscle.

Posterior border extends from the posterior superior spine to the margin of the greater sciatic notch. The posterior inferior iliac spine at the inferior extremity of the auricular surface divides this border and marks the upper end of the greater sciatic notch.

Gluteal (external) surface is the curved wide surface between the anterior and posterior borders. Three curved lines subdivide this area: Inferior gluteal line extends from above the anterior inferior spine above the acetabulum towards the deepest part of the sciatic notch; Anterior (or middle) gluteal line extends from behind the anterior superior spine towards the sciatic notch; and Posterior gluteal line extends from in front of the posterior superior spine downward to the sciatic notch.

Gluteus minimus arises between the anterior and inferior lines, gluteus medius between the anterior and posterior lines and the small iliac part of gluteus maximus from the region behind the posterior gluteal line. The reflected head of rectus femoris arises from the area immediately above the acetabular rim.

INNER SURFACE

Iliac fossa is the smooth internal concavity of the ala bounded by the crest above and the arcuate line below. The fossa gives rise to iliacus muscle.

Auricular surface is the ear shaped rough region behind the iliac fossa for articulation with the first two segments of the sacrum. The rough iliac tuberosity behind and above the auricular surface gives attachment to the interosseous ligament.

2. Ischium

Ischium is the posterior and inferior V-shaped part of the hip bone. It has

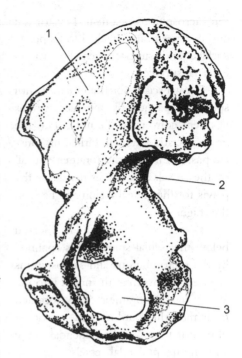

FIG. 42-2 Right hip bone — medial aspect. (1) Iliac fossa; (2) Greater sciatic notch; (3) Obturator foramen.

three parts, a body that forms part of the acetabulum and greater sciatic notch; a tuberosity which is a roughened area on the posterior and inferior aspect of the body; and a ramus, the thin part on the inferior aspect continuous with the pubic ramus.

BODY

Body has an upper thickened part fused with the pubis and ilium at the acetabulum.

Ischial spine projects from the posterior border dividing it into greater and lesser sciatic notches. The spine receives the sacrospinous ligament. Levator ani and coccygeus are attached to its pelvic surface and superior gemellus to its external surface.

Greater sciatic notch, above the ischial spine, is spanned by the sacrospinous ligament converting the notch into the greater sciatic foramen. Piriformis muscle passes through the foramen and all of the vessels and nerves that leave the pelvis for the gluteal region and back of the thigh.

Lesser sciatic notch is located below the ischial spine and is spanned by the sacrotuberous and sacrospinous ligaments converting the notch into the lesser sciatic foramen. The foramen transmits the tendon and nerve of obturator internus, the pudendal nerve and internal pudendal vessels.

TUBEROSITY

Tuberosity is located at the end of the posterior border. It has an upper quadrilateral part that gives origin to the semimembranosus, long head of biceps and semitendinosus muscles and a lower triangular part that gives origin to the adductor magnus muscle.

RAMUS

Ramus extends forward and medially from the tuberosity to join the inferior ramus of the pubis forming the pubic arch or ischiopubic ramus. The outer surface gives origin to the obturator externus and part of adductor magnus muscle.

3. Pubis

Pubis has a body, a superior and a inferior rami.

BODY

Body has three surfaces, a smooth pelvic surface that supports the bladder; a rough femoral surface for the attachment of muscles; and an oval symphyseal surface bearing ridges to which fibrocartilage of the symphysis is attached.

The upper border of the body, the pecten pubis or pubic crest, has the pubic tubercle at its lateral end. The tubercle gives attachment to the medial end of the inguinal ligament. Rectus

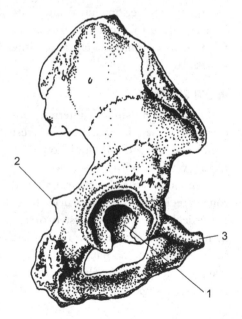

FIG. 42-3 Right hip bone — lateral aspect. (1) Acetabulum; (2) Ischial spine; (3) Pubic tubercle.

abdominis and pyrimidalis muscles arise from the pubic crest.

The pubic symphysis, pubic crest and pectineal line constitute the continuous pubic part of the arcuate line.

Superior ramus

Superior ramus extends from the body backward, upward and laterally to the anterior part of the acetabulum. Its medial end is fused with the body of the pubis. Its lateral end is fused with the ilium and the ischium to form the acetabulum.

Obturator crest descends from the pubic tubercle to the acetabular notch and gives attachment to the pubofemoral ligament.

Obturator groove on the pelvic surface is an oblique groove (converted *in vivo* by the obturator membrane into a canal) that lodges the obturator nerve and vessels.

Inferior Ramus

Inferior ramus descends, backward and laterally from the body of the pubis to fuse with the ramus of the ischium.

The subpubic angle is the angle between inferior rami in the articulated pelvis and is used to assess the adequacy of the pelvic outlet to accommodate the fetal head.

Acetabulum is the large cup shaped fossa on the outer surface of the hip bone that articulates with the head of the femur at the hip joint. It lies at the site of fusion of the three parts of the hip bone. The fossa is surrounded by the superior pubic ramus of the pubis in front, body of the ischium below and behind, and ilium above and behind.

Acetabular notch is the inferior deficiency in the acetabulum. The notch is converted into a foramen by the strong transverse acetabular ligament. Through this foramen, the acetabular artery (from the obturator artery) enters to supply the fat in the floor of the acetabulum and give rise to the vessels of the ligament of the head of the femur.

Obturator foramen is bounded by the pubis and ischium and their rami. The obturator membrane is attached to the margin of the foramen but not to the obturator sulcus. Obturator externus muscle is attached to the external surface of the obturator foramen and membrane.

4. Joints of the Pelvis

The joints of the pelvis include the sacroiliac joint, pubic symphysis, lumbosacral, and sacrococcygeal articulations. The latter two are considered with the vertebral column.

Sacroiliac joint

Sacroiliac joint between the hip bone and sacrum is particularly strong and reinforced behind and in front by heavy ligaments and further strengthened by accessory ligaments (sacrotuberous, sacrospinous and iliolumbar).

- **Type:** Plane synovial joint with limited movement.
- **Articulating elements:** Auricular surfaces of the sacrum and the ilium. Surfaces are covered by cartilage and their irregular surfaces result in partial interlocking.
- **Ligaments:**

Capsule is attached medially to the sacrum and laterally to the ilium.

Ventral sacroiliac ligament comprises thin short bands connecting the ala and pelvic surfaces of the ilium and pelvic surface of the sacrum and adjacent ilium.

Dorsal sacroiliac ligament fills the depression between the posterior inferior spine of the ilium and tuberosity of the ilium and spreads out to reach the intermediate sacral crest.

The *interosseous sacroiliac ligament* is short, strongest fibers deep to the dorsal ligaments connect the tuberosities of the ilium and sacrum.

PUBIC SYMPHYSIS

- **Type:** A secondary cartilaginous joint.
- **Articulating elements:** Medial surface of the bodies of the right and left pubic bones.
- **Ligaments:**

Superior pubic ligament extends across the superior surface of the symphysis. It extends to the pubic tubercles.

Arcuate pubic ligament extends below the symphysis to the inferior pubic rami.

Between the arching fibers and the transverse ligament of the perineum is a small gap that transmits the deep dorsal vein of the penis or clitoris.

Interpubic disc is a thick mass of fibrocartilage connecting the articulating elements. It frequently has a sagittal cleft in the adult.

- **Movements:** Minimal movement between elements.

LIGAMENTS BETWEEN SACRUM AND ISCHIUM

Sacrospinous ligament is a triangular ligament that extends from the lowest lateral part of the sacrum and upper coccyx across the inner aspect of the sacrotuberous ligament to the apex of the ischial spine. It is coextensive with the coccygeus muscle and converts the lesser sciatic notch into the lesser sciatic foramen.

Sacrotuberous ligament extends from the greater part of the dorsal surfaces of the sacrum and coccyx and posterior border of the ilium above the greater sciatic notch. Fibers pass downward and laterally converging to attach to the tuberosity of the ischium converting the greater sciatic notch into the greater sciatic foramen.

43

Perineum

Chapter Outline

1. Urogenital Triangle of the Male
2. Urogenital Triangle of the Female
3. Anal Triangle

The perineum is the diamond shaped space below the pelvic diaphragm — the entire pelvic outlet. It has the following boundaries: Anterior: arcuate pubic ligament below the pubic symphysis; Anterolateral: ischiopubic ramus to the tuberosity of ischium; Posterolateral: sacrotuberous ligament (from the tuberosity to the coccyx); Posterior: coccyx.

It is divided topographically by an imaginary transverse line between the anterior parts of the ischial tuberosities. The urogenital triangle containing the external genitalia and lower part of the urinary tract is anterior to the transverse line and the anal triangle containing the anal canal and anus is posterior to the line.

1. Urogenital Triangle in the Male

The male urogenital triangle is arranged in five layers.

Skin has a median raphe which extends from the anus through the scrotum.

Superficial perineal fascia is a subcutaneous layer divided into a superficial fatty layer and a deep membranous layer.

Superficial layer is continuous with the fatty layer of the anterior abdominal wall, fatty superficial layer in the thighs, with a similar fatty layer in the anal triangle and with the dartos layer of the scrotum.

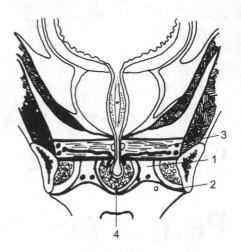

FIG. 43-2 Coronal section of male perineum. (1) Superficial perineal space; (2) Deep perineal fascia; (3) Deep perineal space; (4) Bulb of penis.

Deep (or membranous) layer is attached behind to the posterior border of the perineal membrane and at the sides, to the ischiopubic rami. It is continuous anteriorly with the dartos layer of the scrotum, membranous fascia of the anterior abdominal wall and the superficial fascia of the penis.

Deep perineal fascia covers approximately the same triangular area as the superficial layer and has similar attachments. It is continuous, anteriorly, with the thin fascia covering the external oblique muscle. Laterally, it is continuous with the fascia lata of the thigh. It invests the superficial perineal muscles and dips between them to become associated with the inferior fascia of the urogenital diaphragm. It extends over the penis as the deep penile fascia.

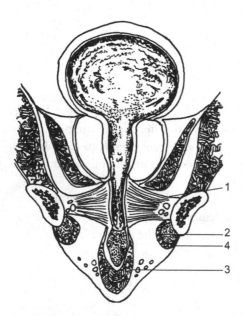

FIG. 43-1 Coronal section of pelvis and perineum. (1) Urogential diaphragm; (2) Crus of penis; (3) Bulbocavernosus m; (4) Ischiocavernosus m.

Subcutaneous perineal pouch is the space enclosed by the membranous layer of superficial fascia, its attachments and the perineal membrane.

It has the following boundaries:
Roof: Membranous layer of superficial fascia.
Floor: Perineal fascia.
Above: The pouch is open above.
Contents: Testes in the male.

Superficial perineal pouch is the space between the deep perineal fascia and the inferior fascia of the urogenital diaphragm.

It has the following boundaries:
Roof: Perineal fascia.
Floor: Perineal membrane (attached to the ischiopubic rami).

The pouch is closed posteriorly by the attachments of the fascia and membrane to each other. It extends as a subcutaneous triangular pouch lying external to the urogenital diaphragm with an outlet in front and above on each side of the pubic symphysis leading to the superficial inguinal space deep to the anterior abdominal wall and upper part of the thigh. Lateral attachments of the membranous layer of superficial fascia are distal to the inguinal sulcus along the hip flexion crease to fascia lata.
Contents: The superficial perineal pouch contains the root of the penis/clitoris and associated superficial perineal muscles, branches of the internal pudendal vessels, and the pudendal nerve.

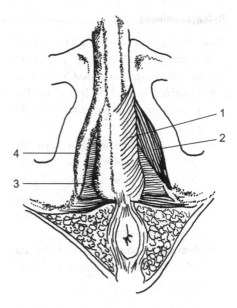

FIG. 43-3 Muscles of the superficial perineal space. (1) Bulbospongiosus m; (2) Ischiocavernosus m; (3) Bulb of penis; (4) Crus of penis.

There are four superficial muscles in the superficial perineal space. All receive their blood supply from the internal pudendal artery and their innervation from the perineal branch of the pudendal nerve.

Ischiocavernosus

Ischiocavernosus covers each crus of the clitoris/penis.

- **Origin:** Ramus of the ischium near the tuberosity.
- **Insertion:** Superficial aspect of the crus penis and sides of the dorsum of the adjacent part of corpus cavernosum.
- **Action:** Compresses the crus. Helps to maintain erection of the penis by compressing the crus.

Bulbospongiosus

Bulbospongiosus is a midline muscle that covers the bulb of the penis — two symmetrical parts are joined by a raphe.

- **Origin:** Perineal body and median fibrous raphe on the bulb of the penis.
- **Insertion:** Inferior fascia of the urogenital diaphragm, dorsum of the corpus spongiosum, and deep fascia of the dorsum of the penis. Fibers of the latter insertion cross over corpus cavernosum.
- **Action:** Compress the bulb and helps empty the urethra after micturition.

Superficial Transverse Perinei

Superficial transverse perinei are narrow, thin muscles and may be absent.

- **Origin:** Medial side of the ramus of the ischium near the tuberosity. It crosses the posterior margin of the perineal membrane to meet its fellow at the perineal body.
- **Insertion:** Perineal body.
- **Action:** Tethers the perineal body in the midline i.e. it tethers the visceral canals.

The tendinous center of the perineum (perineal body) is a fibromuscular node located in the midline between the anal canal and the urogenital apparatus at the rectovesical (rectovaginal) septum. In the female, it is situated between the lower part of the vagina in front and anal canal behind. In the male, it is situated between the bulb of the

penis in front and anal canal behind. It is formed where fibers of some perineal muscles, perineal fascia and anal muscles converge and interlace.

In the female, the perineal body supports the posterior wall of the vagina and is an indirect support for the pelvic organs. Here, seven muscles meet and fuse: These are, bulbospongiosus, external anal sphincter, two superficial transverse perinei, sphincter urethrovaginalis and two levator ani (pubovaginalis part) muscles.

In the male, eight muscles meet and fuse at the perineal body. These are, bulbospongiosus, external anal sphincter, two superficial perinei, two deep transverse perinei and two levator ani (puboprostatic part) muscles.

Deep perineal pouch is the space between the inferior and superior fascias of the urogenital diaphragm.

It has the following boundaries:
Roof: Perineal membrane.
Floor: Fascia coextensive laterally with the obturator fascia. Medially in the region of the urogenital hiatus, this fascia is fused with the inferior fascia of the pelvic diaphragm in both males and females.

The pouch is open above.
Contents: In the female: External urethral sphincter (surrounding the urethra — inner muscle layer smooth, outer muscle layer striated), and sphincter urethrovaginalis. The pouch transmits the vagina.
In the male: External urethral sphincter surrounding the intermediate part of

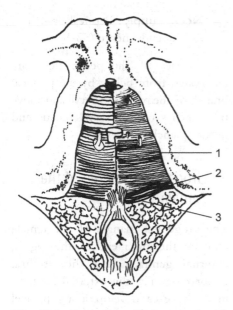

FIG. 43-4 Urogenital diaphragm (male). (1) Inferior fascia of urogenital diaphragm; (2) Perineal body; (3) Deep transverse perinei.

the pelvic surface of the sphincter urethrae and the deep transverse perinei muscles.

The **deep perineal space** contains the deep perineal muscles, membranous part of the urethra, bulbourethral gland, internal pudendal vessels, artery of the bulb of the penis, and dorsal nerve of the penis.

The deep perineal muscles, together with covering fascial layers, comprise the urogenital diaphragm.

Inferior fascia of the urogenital diaphragm (or perineal membrane) is a strong sheet covering the lower surface of the sphincter urethrae and the deep transverse perinei muscles. Its lower surface is largely covered by the root of the penis. Its posterior border is fused

the urethra (inner muscle layer smooth, outer muscle layer striated) and two transverse perineal muscles.

At a deeper level, deep to the fascia forming the floor of the deep perineal pouch lie the anterior parts of the ischioanal fossae on both sides of the urogenital hiatus.

The deep perineal pouch and its coverings form a protective wall for the urogenital hiatus.

Superior fascia of the urogenital diaphragm is a fascial layer is derived from the parietal pelvic fascia, continuous laterally with the fused obturator and lunate fasciae and attached to the ischiopubic rami on each side. It covers

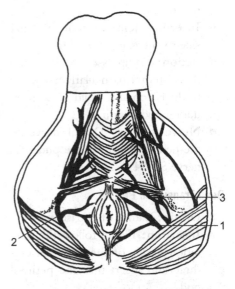

FIG. 43-5 Nerves and vessels of the urognital and anal regions in the male. (1) Inferior rectal n; (2) Inferior rectal a; (3) Perineal n.

with the posterior border of the superior fascia and the perineal body. The fascia is thickened anteriorly to form the transverse perineal ligament. An oval gap separates the membrane from the arcuate pubic ligament and transmits the dorsal vein of the penis.

Sphincter Urethrae

Sphincter urethrae surrounds the lower urethra, neck of the bladder and enters the prostate.

- **Origin:** Inferior fibers of sphincter urethrae encircle the membranous urethra, attach to the inferior ramus of the pubis and continue into the prostate. Superior fibers form a U-shape merging into the smooth muscle layer of the bladder and also continuing into the prostate.
- **Insertion:** Joins fibers of the other side to encircle the urethra.
- **Action:** Compresses the membranous urethra particularly when the bladder contains fluid. It relaxes during micturition.
- **Nerve supply:** Dorsal nerve of the penis (branch of the pudendal nerve S2–S4).

Deep Transverse Perinei

- **Origin:** Ramus of the ischium.
- **Insertion:** Fibers follow the posterior margin of the sphincter urethrae and insert into the perineal body.
- **Action:** Helps fix the perineal body.

- **Nerve supply:** Perineal branch of the pudendal nerve (S2–S4).

Muscles of the pelvic floor are innervated from below by the perineal branch of the pudendal nerve and above by branches of the sacral plexus and pelvic splanchnic nerves.

2. Urogenital Triangle of the Female

The urogenital triangle of the female contains the lower end of the vagina, external genitalia and the urethra. Embryologically, the external genitalia in both sexes develop along parallel lines and there are homologies between the parts. Labia majora are homologs of the scrotum, labia minora represent

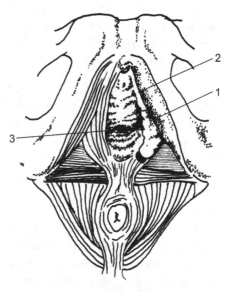

FIG. 43-6   Female perineum. (1) Crus of clitoris; (2) External urethral orifice; (3) Vagina.

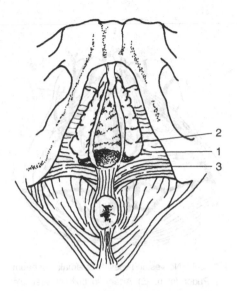

FIG. 43-7 Urogenital diaphragm — female. (1) Inferior fascia of urogenital diaphragm; (2) Bulb of vestibule; (3) Deep transverse perinei m.

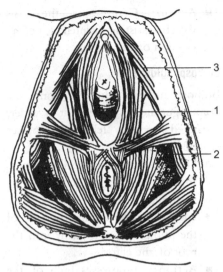

FIG. 43-8 Female perineum. (1) Bulbospongiosus m; (2) Superficial transverse perinei m; (3) Ischiocavernosus m.

skin and subcutaneous tissue at the side of the penile urethra and the penis in the male is homologous to the clitoris in the female.

The five layers of the urogenital triangle in the female are:

Skin is folded to shape the external genital organs.

Superficial perineal fascia consists of two layers: Superficial and deep.

The superficial fatty layer extends into the labia majora.

The deep membranous layer has the same attachments posteriorly and laterally as the male. The layer unites anteriorly to the vagina to become continuous with the membranous layer of the subcutaneous fascia of the abdomen.

Deep perineal fascia has the same attachments as in the male.

Superficial perineal space has the same boundaries as in the male. It encloses the superficial perineal muscles, the greater vestibular glands, and branches of the internal pudendal vessels and the pudendal nerve.

The superficial perineal muscles are:

Ischiocavernosus

Ischiocavernosus is smaller than in the male. Each overlie the crura of the clitoris.

- **Origin:** Medial side of the ischial tuberosities and ischial rami.
- **Insertion:** Margin of the pubic arch and medial and lower aspects of each crus of the clitoris.

- **Action:** Compresses the crus thereby assisting maintenance of an erection of the clitoris.

Bulbospongiosus

Bulbospongiosus differs from the homologue in the male as it is widely separated from its fellow and passes around the vagina.

- **Origin:** Perineal body. Fibers pass forward covering the bulbs of the vestibule.
- **Insertion:** Partly into the side of the pubic arch and partly into the root of the clitoris.
- **Action:** Compresses bulb of the vestibule and acts as a sphincter of the vaginal opening.

Superficial Transverse Perinei

Superficial transverse perinei is similar to the same muscle in the male.

- **Origin:** Anterior part of ischial tuberosities.
- **Insertion:** Perineal body.
- **Action:** Helps to fix the perineal body.

Deep perineal pouch has the same boundaries as the corresponding space in the male and is enclosed by the inferior and superior fascia of the urogenital diaphragm. It contains the deep perineal muscles, membranous part of the urethra, middle segment of the vagina, internal pudendal vessels, artery of the bulb of the vestibule, and the dorsal nerve of the clitoris.

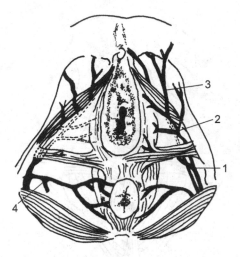

FIG. 43-9 Nerves and vessels of female perineum. (1) Pudendal n; (2) Artery to bulb of vestibule; (3) Posterior labial a and n; (4) Inferior rectal n; and a.

The deep perineal muscles are sphincter urethrae, deep transverse perinei and compressor urethrae.

Sphincter urethrae (external urethral sphincter)

This muscle surrounds the middle third of the urethra blending above with the smooth muscle of the neck of the bladder above and smooth muscle of the lower urethra and vagina below.

- **Origin:** Inferior ramus of the pubis. Fibers arch anterior to the urethra but do not pass posterior to the urethra.
- **Insertion:** Lateral wall of the vagina.

DEEP TRANSVERSE PERINEI

- **Origin:** Ischiopubic ramus.

- **Insertion:** Perineal body and lateral wall of the vagina.
- **Action:** With superficial transverse perinei, aids in tethering the perineal body and therefore the vagina, urethra and anus in the midline.

Compressor Urethrae

- **Origin:** Ischiopubic rami. Fibers pass forward.
- **Insertion:** Fibers meet their fellows from the opposite side ventral to the urethra below sphincter urethrae.
- **Action:** Produces compression and elongation of the urethra thus aids urinary continence.

Sphincter Urethrovaginalis

- **Origin:** Perineal body. Fibers pass forward on either side of the urethra and vagina.
- **Insertion:** Fibers of the opposite side ventral to the urethra below compressor urethrae.
- **Action:** Produces compression and elongation of the urethra thus aiding urinary continence.

3. Anal Triangle

This region is identical in both sexes. It is divided into two lateral ischioanal fossae and a median area containing the lower part of the anal canal surrounded by the external sphincter ani. Superficial fascia is thick, areolar and contains many fat cells. The fat pad extends into the ischioanal fossa between levator ani

and obturator internus. Deep fascia lines the ischioanal fossa

Ischioanal fossae are spaces on either side of the anal canal. Each fossa is wedge-shaped and is filled with fat thus accommodating the enlargement of the anal canal during the passage of feces.

Boundaries are:
Anterior: Posterior margin of the muscles of the urogenital diaphragm and prolonged above the muscles as a narrow recess.
Posterior: Sacrotuberous ligament.
Lateral: Obturator internus and its fascia, and the tuberosity of the ischium.
Medial: Inferior fascia of the pelvic diaphragm (levator ani) and external anal sphincter muscles more superficially.

The *anterior recess* of the ischioanal fossa is a narrow prolongation that may reach the pelvic surface of the pubis.

The *posterior recess* of the ischioanal fossa extends laterally and posteriorly beneath gluteus maximus.

Pudendal canal is an elongated cleft or fascial canal formed by a split in the obturator internus fascia on the side wall of the perineum. It starts at the lesser sciatic foramen above the urogenital diaphragm and ends near the anterior end of the lateral edge of the urogenital diaphragm. In part of its extent, the floor of the canal is the falciform process of the sacrotuberous ligament. The pudendal canal contains the pudendal nerve and internal pudendal vessels.

Anal canal is the terminal part of the alimentary canal. It begins below the pelvic floor and terminates by opening to the exterior at the anus. The upper five sixths is surrounded by circular muscular fibers thickened to form the internal anal sphincter.

The sphincter is supported externally by the longitudinal muscle coat that blends with fibers of puborectalis muscle. External to this fusion is the external anal sphincter.

External anal sphincter surrounds the lower two thirds of the anal canal and consists of three parts.

Subcutaneous part fibers are arranged as a flat circular band surrounding the anus. This lowermost part of the sphincter overlaps the internal sphincter.

Superficial part is intermediate in position and its muscular fibers are elliptical in arrangement. It extends from the top of the coccyx and anococcygeal ligament posteriorly, skirts the anus and decussates anteriorly in the tendinous centre of the perineum.

The superficial part is the only part of the external anal sphincter with a bony attachment (to the coccyx posteriorly).

Deep part is thick and circularly arranged surrounding the upper part of the anal canal. Fibers blend with those of puborectalis and decussate with superficial transverse perinei.

Anorectal ring is the surgically important thickening around the anal canal formed by fusion of fibers of puborectalis, longitudinal fibers of the anal canal reinforced by the internal anal sphincter and deep external sphincter. Division of the ring causes incontinence but division of sphincters below this ring does not.

INTERNAL STRUCTURE OF THE ANAL CANAL

On the basis of pattern of structure of mucosa, blood supply and autonomic nerve supply, the anal canal can be divided into an upper two thirds or visceral part and a lower third or somatic part. Skin of the margin of the anus is pigmented and contains hair follicles and glands (the cutaneous zone of the anus). Internal to this beginning at a clinically recognizable "white line" is a zone characterized by smooth hairless skin (the intermediate zone). True mucous membrane begins above the level of anal valves, remnants developmentally of a proctodeal sphincter.

Anal columns are from five to ten permanent vertical folds of mucous membrane that overlie veins and small arteries.

Anal valves are the mucosal folds that join the distal parts of anal columns and enclose anal sinuses. Beneath the valves pass communicating veins.

Pectinate line is the lower limit of the anal valves. It is the approximate line of division between the visceral and somatic parts of the anal canal.

Pecten is the centimeter wide zone below the anal valves lined with stratified squamous epithelium until the lining becomes continuous with the smooth skin of the surface epithelium.

The **neurovascular system** of the anal canal is supplied by:
Arteries: *Middle rectal branch of the internal iliac artery* from the side of the rectum. They anastomose with the superior and inferior rectal arteries and prostatic, vesical and vaginal arteries.

Inferior rectal arteries are branches of the internal pudendal arteries supplying the musculature of the anal canal and anus and much of its mucous membrane.
Veins: Drain into an internal venous plexus in the submucosa and an external plexus outside the muscular coat. The internal plexus drains mainly to superior rectal veins. Both plexuses drain to inferior rectal veins (branches of internal pudendal veins). Middle rectal veins drain the muscular coat of the upper third of the anal canal.

The communication between superior and middle rectal veins is a site of an anastomosis between the portal and systemic venous systems. It may dilate (hemorrhoids) because of little support from surrounding tissue and the absence of valves in the superior mesenteric veins. Hemorrhoids may form in obstruction by the liver in portal hypertension.

Hemorrhoids below the pectinate line are known as external hemorrhoids (piles) and are particularly painful because their covering epithelium is innervated

by branches of the pudendal nerve. Hemorrhoids above the pectinate line are known as internal hemorrhoids and are painless because their covering epithelium is innervated by visceral afferents.

Lymphatics: Ascending channels from the entire rectum and anal canal accompany superior rectal vessels to nodes (in the angle of division of the superior rectal vessels) then ascending to inferior mesenteric nodes (at the root of the inferior mesenteric artery). Drainage channels originating below the pelvic diaphragm follow inferior rectal vessels across the ischioanal fossa perforate the lateral edges of the diaphragm to reach internal iliac nodes. Channels from a cutaneous plexus around the anus follow the genitofemoral sulcus to reach superficial inguinal nodes.

Nerves: Preganglionic parasympathetic fibers synapse with postganglionic cells in the wall of the anal canal supplying smooth muscle in sphincter ani internus. Postganglionic sympathetic fibers supply blood vessels.

Annular muscle fibers (of the subcutaneous and deep parts of sphincter ani externus) are supplied by the inferior rectal nerves while the superficial part of sphincter ani externus is supplied by the perneal branch of S4.

Afferent nerves are visceral above the pectinate line are concerned with reflex control of sphincters and somatic sensation (mostly pain fibers) below the line.

Chapter 44

External Genitalia

Chapter Outline

1. External Genital Organs in the Male

Scrotum is a pouch of thin skin located below and behind the penis and below the pubic symphysis. It is divided into two compartments, each of which contains a testis, epididymis, lower part of the spermatic cord and coverings of the cord.

Layers of the scrotum are:

Skin scrotal skin is dark and has a median raphe, the superficial indication of division of the scrotum into two compartments.

Dartos muscle replaces the superficial and deep layers of the superficial fascial layer of general body skin in the region of the scrotum. It comprises

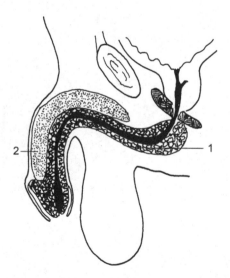

FIG. 44-1 Sagittal section of male external genitalia; (1) Corpus spongiosum; (2) Corpus cavernosum.

strands of smooth muscle and gives rise to the incomplete median scrotal septum.

The **neurovascular system** of the scrotum is supplied by:

Arteries: Poserior scrotal branches of the perineal artery, cremasteric branch of the inferior epigastric and superficial and deep external pudendal arteries, branches of the femoral artery.

Veins: Correspond to the arteries but the superficial external pudendal vein(s) end in the saphenous vein (not the femoral vein).

Lymphatics: Drain into the superficial inguinal lymph nodes.

Nerves: Ilioinguinal (ventral ramus of L1) nerve supplies the ventral scrotum and root of the penis. Medial and lateral scrotal nerves and perineal branch of the posterior femoral cutaneous nerves (S2–S4) supply the dorsal part of the scrotum.

Spinal cord segments L2–S2 are not represented in the scrotum. These segments indicate the origin of cutaneous nerves for the lower limb.

PENIS

Penis is the male organ of copulation. It is composed of three cylindrical bodies of vascular erectile tissue. Two of the bodies (corpora cavernosa penis) lie side by side on the dorsum. The central, ventral body (corpus spongiosum penis) has a conical extremity (the glans penis) formed by an expansion of the corpus spongiosum penis that fits over the blunt ends of the corpora cavernosa.

The corpus spongiosum and glans are perforated by the penile urethra.

The penis is suspended from the anterior abdominal wall in four legged mammals hence the anterior surface in humans is termed the dorsal surface and the posterior surface is termed ventral or urethral surface.

The penis is enclosed by:

Skin is thin, dark and devoid of hair except near the pubis.

Prepuce (foreskin) is a redundant fold of skin that overlaps the glans from its neck. Preputial glands located on the corona and neck of the glans produce a sebaceous secretion known as smegma.

Frenulum of the prepuce is a small median fold running from the deep surface of the prepuce to just below the external urethral orifice.

Superficial fascia (subcutaneous tissue) is loose thin, devoid of fat, and continuous with the dartos of the scrotum. It passes over the superficial perineal muscles to join the superficial perineal fascia.

Fundiform ligament is an elastic ligament attached to the penis near the junction of the body and root and lower part of linea alba and overlying membranous layer of covering subcutaneous tissue. It splits into two (left and right) parts that unite behind and insert into the septum of the scrotum.

Deep fascia is a fibrous tubular sheath enclosing the cavernous bodies. It extends from the groove between the shaft and corona of the glans of the penis to the ends of the erectile bodies and is continuous with the deep perineal fascia over muscles of the superficial perineal space.

Suspensory ligament of the penis below the fundiform ligament is a triangular fibrous band that extends between the pubic symphysis to the deep fascia on each side at the root of the penis.

The substance of the penis consists of masses of erectile tissue. Each tissue mass is enveloped by a fibroelastic coat, the tunica albuginea. It forms a septum where the corpora cavernosa are fused.

The penis is divided into three parts:

Root is the attached part of the penis located in the superficial perineal space. It consists of three masses of erectile tissue; two lateral crura and the median bulb of the penis.

Each crus is attached to its adjacent ischiopubic arch and deep perineal fascia and is covered by an ischiocavernosus muscle.

The bulb of the penis is the expanded, proximal, fused part of the corpus spongiosum situated between the crura in the superficial perineal space and adherent to the inferior fascia of the urogenital diaphragm. It narrows anteriorly to become the corpus spongiosum and is covered by a bulbospongiosus muscle.

Body is the free, pendulous part of the penis consisting of the paired corpora cavernosa and a single corpus spongiosum.

Each corpus cavernosum is the continuation of a crus, and both are fused and end bluntly covered by the glans.

The *corpus spongiosum* is the continuation of the bulb that lies in the groove along the under surface of the corpora. It is smaller than either corpus cavernosum but expands distally to form the glans penis.

Glans is the expanded end of the corpus spongiosum. It has a protuberant margin, or *corona* and a concavity that covers the blunted free ends of the corpora cavernosa.

Corona glandis is the prominent margin of the glans adjacent to the neck.

Neck of the glans is the constricted portion where the glans joins the body of the penis. The tip of the glans bears the external urethral orifice.

The **neurovascular system** of the penis is supplied by:
- *Arteries*: Three branches of the internal pudendal artery are on each side.
- **Artery of the bulb**, arises from the internal pudendal artery near the base of the perineal membrane, pierces the membrane reaching the bulb near its base. It supplies the bulb and the corpus spongiosum as well as the bulbourethral gland.

Deep artery of the penis, also a branch of the internal pudendal artery, pierces the perineal membrane to reach the crus and supplies the corpus spongiosum.

The deep artery possesses intimal cushions that regulate local blood flow.

In the flaccid state, almost all of the blood flows directly into dilated arteriovenous anastomoses to reach efferent veins while only a minimal amount of blood supplies the corpora cavernosa. Within the corpora, the deep artery divides into helicine arteries that empty directly into blood spaces of the erectile tissue and nutrient arteries that empty into trabeculae then into the blood spaces. Cavernous spaces are drained by veins that also have intimal cushions in their walls.

During erection, contraction of the arteriovenous anastomoses and dilation of the deep artery diverts blood into the cavernous bodies while the nutrient vessels and the draining veins are compressed. Overall blood flow is not reduced.

Dorsal artery of the penis, a branch of the internal pudendal artery, lies between the crus and pubic ramus, passes through the suspensory ligament of the penis and ends in the glans and prepuce. It supplies the fascia and skin of the penis.

Veins: The veins of the penis are located outside erectile tissue. The unpaired superficial dorsal vein of the penis drains the skin and fascia into the superficial external pudendal vein. The paired dorsal vein of the penis begin in the glans and drain the region deep to the deep fascia, pass below the pubic symphysis divide and empty into the prostatic plexus of veins.

Nerves: The **dorsal nerve of the penis** from the pudendal nerve pierces the perineal membrane and suspensory ligament of the penis to reach the dorsum. It gives off many branches and terminates in the glans.

The **deep branch of the perineal nerve** is mainly a motor nerve supplying the transverse perinei, ischiocavernosus, bulbospongiosus, sphincter urethrae, external anal sphincter and levator ani muscles. The nerve to the bulb is given off from the muscular nerve to bulbocavernosus. It enters the bulb supplying it, the corpus spongiosum and mucous membrane of the urethra as far as the glans.

Ilioinguinal nerve (L1) accompanies the spermatic cord through the inguinal canal and superficial inguinal ring and innervates the skin of the groin and root of the penis.

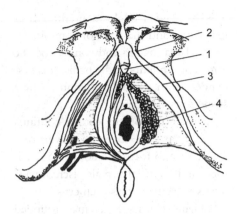

FIG. 44-2 External genitalia in female; (1) Glans of clitoris; (2) Body of clitoris; (3) Crus of clitoris; (4) Vagina.

2. External Genital Organs in the Female

External genitalia of the female consist of the mons pubis, labia majora, labia minora, clitoris and the vestibule of the vagina.

Mons pubis is a median elevation in front of the symphysis pubis produced by underlying fibrofatty tissue and is covered with hair after puberty.

Labia majora are a pair of elongated folds that extend back from the mons pubis and are separated by the median pudendal cleft. The folds are united in front by an anterior commissure at the mons and behind by the posterior commissure in front of the anus. They contain the termination of the round ligament of the uterus. The inner surfaces are smooth and hairless. Labia majora are homologous with the scrotum in the male.

Labia minora are a pair of thin folds is located on either side of the vagina. Posteriorly, in the virgin, the folds are joined by a slight transverse ridge of skin, the frenulum of the labia (fourchette). Anteriorly, each of the folds divides. The lateral pair forms a cover over the glans clitoris, the prepuce of the clitoris. Inferior to the clitoris, the medial portions fuse inferior to the clitoris to form the frenulum of the clitoris. Labia minora are homologs of the genital swellings that fuse in the male.

Clitoris is the female homologue of the penis consisting of erectile tissue but differing from the male penis in that it is not traversed by a urethra. It is located behind the anterior commissure of the labia majora and consists of the same components as the penis. It is suspended from the pubic symphysis by a suspensory ligament.

Two crura of the clitoris continue forward as the corpora cavernosa. These fuse to form the body of the clitoris and as in the male, ischiocavernosus muscle overlies each crus.

Glans clitoris is a small rounded elevation at the free end of the body formed by merging of the vestibular bulbs. It is homologous to the glans penis in the male.

Bulb of the vestibule corresponds to the bulb and corpus spongiosum of the penis in the male. It consists of paired elongated masses of erectile tissue located beneath a bulbospongiosus muscle on the sides of the vaginal opening. They unite in front to form a thin strand that extends along the lower surface of the body of the clitoris to the glans.

Prepuce or foreskin of the clitoris is formed by fusion of the lateral parts of the labia minora.

Frenulum of the clitoris is formed by the fusion of the medial parts of the labia minora.

Corpus spongiosum is only a thread of erectile tissue extending from the bulb of the vestibule to the glans and is not traversed by the urethra.

Vestibule is the developmental urogenital sinus bordered by the labia minora, frenulum of the clitoris and fourchette behind. It contains the openings of the vagina, urethra and ducts of the greater vestibular glands.

External urethral orifice is located between the clitoris and the vagina.

Only the lower third or part of the vagina below the pelvic diaphragm is in the perineum. The size and appearance of the vaginal orifice depends on the size and shape of the hymen. The hymen is a mucosal fold that partly closes the vestibule in the virgin. After rupture, its remnants persist as small nodules (carunculae hymenales).

The **vagina** is a fibromuscular sheath extending from the cervix of the uterus to the vestibule of the vagina. It is lined with a mucosa that forms transverse rugae on the anterior and posterior walls and anterior and posterior columns of rugae in the corresponding midlines.

Urethral carina is a keel-like ridge in the lower part of the anterior column of the vagina formed by the underlying urethra.

The long axis of the vagina is parallel to the plane of the pelvic inlet but perpendicular to the long axis of the cervix.

The vault of the vagina surrounds the vaginal part of the cervix forming the anterior, posterior and lateral fornices.

Relations of the vagina are important in clinical examination.
Anterior: Fundus of the bladder in its upper half and urethra in its lower half.

Posterior: Posterior fornix is capped by peritoneum of the rectouterine pouch and middle third of the rectum. The upper half of the remainder rests on the lower third of the rectum. The lower half is separated from the anal canal by the tendinous center of the perineum.

Superior: Lateral fornix is related to the parametrium of the broad ligament.

Sphincters of the vagina includes a weak circular muscular coat comprising pubovaginalis, sphincter urethrae and bulbosponiosae (from above down).

Greater vestibular glands are small, bean shaped glands homologous with the bulbourethral glands in the male. They lie on both sides of the lowest part of the vagina, enclosed by the bulbospongiosus muscle. Their ducts open on each side into the groove between the hymen and the labium minus. Their secretion is a mucoid lubricating fluid.

Lesser vestibular glands are smaller mucous secreting glands that also open into the vestibule. They are homologous to the urethral glands in the male.

The **neurovascular system** of the female external genitalia is supplied by:

Arteries: The internal pudendal arteries supply most of the external genitalia.

Anterior labial branches of the external pudendal arteries and posterior labial branches of the internal pudendal arteries supply the labia majora and minora.

Deep artery of the clitoris supplies the crura and corpora cavernosa of the clitoris.

Dorsal arteries of the clitoris supply the glans clitoris.

Artery of the bulb of the vestibule supplies the bulb of the labia.

Vaginal artery from the uterine artery or anterior part of the internal iliac artery also sends branches to the bladder, rectum and bulb of the vestibule.

Nerves: The **perineal nerve,** a branch of the internal pudendal nerve (S2, S3, S4) passes through the pudendal canal to the base of the perineal membrane where it gives two cutaneous branches (posterior labial/scrotal) and a motor branch, the deep perineal nerve that supplies the perineal muscles and three superficial muscles and twigs to levator ani and the external anal sphincter.

The **dorsal nerve of the penis** is also a branch of the pudendal nerve but a sensory nerve. It supplies the glans, prepuce, skin over the penis/clitoris and spongy urethra.

3. Superficial Perineal Arteries and Veins

These vessels supply the external genitalia. They reach the perineal region from the pelvis via the pudendal canal.

Pudendal canal is a tunnel in the substance of the obturator fascia. It courses from the lesser sciatic foramen, along the side of the ischioanal fossa and terminates in the deep perineal space at the side of the pubic arch.

Internal pudendal artery is one of the two terminal branches of the anterior division of the internal iliac artery. It leaves the pelvis through the greater sciatic foramen, enters the gluteal region, crosses the ischial spine, passes through the lesser sciatic foramen and enters the pudendal canal. It passes forward in the side wall of the ischioanal fossa to enter the deep perineal pouch. It pierces the deep perineal fascia and terminates by dividing into the deep and dorsal arteries of the penis (clitoris).

Its branches are:
Muscular branches supply pelvic and gluteal muscles.

Inferior rectal artery arises in the pudendal canal, pierces the fascia of the canal and supplies perianal muscles including levator ani and external anal sphincter.

Perineal artery passes below superficial transversus perinei muscles and supplies superficial perineal muscles.

Posterior scrotal (labial) branch pierce superficial and deep perineal fasciae and pass forward in the superficial perineal space. They supply the skin and dartos of the scrotum.

Artery of the bulb of the penis arises in the deep perineal space then passes medially through the urogenital diaphragm pierces the inferior fascia and supplies the bulb and bulbourethral gland.

Artery of the bulb of the vestibule also arises in the deep perineal space and passes medially through the urogenital diaphragm, pierces the inferior fascia and supplies the bulb of vestibule and greater vestibular gland.

Urethral artery arises ahead of the artery of the bulb, pierces the inferior fascia of the urogenital diaphragm and enters the corpus spongiosum of the glans penis (or clitoris). It supplies the corpus spongiosum.

Deep artery of the penis (clitoris) is a terminal branch of the internal pudendal artery, that pierces the inferior fascia of the urogenital diaphragm and enters the substance of the corpus cavernosum which it supplies.

Dorsal artery of the penis (clitoris) the other terminal branch of the internal pudendal artery, it pierces the inferior fascia of the urogenital diaphragm and passes between the crus and pubic ramus. It pierces the suspensory ligament and descends on the dorsum to the glans supplying it. The labial artery in the female supplies the labia.

Veins are venae commitantes of the arteries. In the male, the deep dorsal vein of the penis pierces the urogenital diaphragm between the arcuate and transverse ligaments and drains into the prostatic venous plexus.

In the female, the deep dorsal vein of the clitoris follows a similar path but drains into the vaginal venous plexus.

4. Superficial Perineal Nerves

The superficial perineal nerves are branches of the pudendal nerve (S2–S4) and supply most of the perineum. They are named according to the arteries that they accompany.

Pudendal nerve (S2, S3 and S4) is one of the two terminal branches of the sacral plexus. It passes through the greater sciatic notch between piriformis and coccygeus, winds around the sacrospinous ligament and re-enters the pelvis through the lesser sciatic notch. It enters the pudendal canal in the obturator fascia and divides into three branches.

Inferior rectal nerve passes across the ischioanal fossa, it divides to supply the levator ani and the external anal sphincter muscles and the skin around the anus.

Perineal nerve passes into the superficial perineal space (superficial to the perineal membrane). It divides into muscular branches that supply the transverse perinei, ischiocavernosus, bulbocavernosus, sphincter urethrae, external anal sphincter and levator ani muscles and posterior scrotal (labial) nerves that supply the dartos (labia majora) and skin.

The anterior scrotal skin is supplied by branches of the ilioinguinal (L1) and genitofemoral (L1 and L2) nerves.

Dorsal nerve of the penis (clitoris) passes into the deep perineal space (deep to the perineal membrane) then pierces the perineal membrane and suspensory ligament to reach the dorsum of the penis (clitoris), lateral to its artery. It gives branches to the corpus cavernosum and the glans.

Lower Limb

In its early embryonic position, the lower limb is abducted with the soles of the feet facing forward. The anterior surface is ventral and the posterior surface is dorsal. A line drawn through the long axis of the limb would define a preaxial (cephalic) border and a postaxial (caudal) border. The flexor aspect of the limb lies ventrally and the extensor aspect lies dorsally.

With further development, the lower limb rotates medially 90° so that the embryonic ventral, flexor compartment comes to lie posteriorly and the extensor compartment comes to lie anteriorly. The rotation appears as a torsion in the shaft of the femur and flexors and extensors of the hip are largely unaffected (flexors are anterior and extensors are posterior).

The arrangement of sensory spinal nerves in the adult reflects the change in orientation of the limb buds in development. Each sensory part of a spinal nerve innervates a strip of skin (a dermatome) in a sequence progressing down the trunk. As the limb elongates, the distribution of intermediate nerves migrates along the limb so that they no longer reach the surface of the proximal part of the limb. Dorsal and ventral axial lines separate adjacent areas of skin that are supplied by non-adjacent segments of the cord.

With axial rotation, the flexor aspect of the leg faces posteriorly and preaxial nerves are distributed along the anteromedial aspect of the leg. Spinal nerve segments on the anterior surface of the lower limb extend medially and distally and the big toe (hallux) receives a higher dermatome (L4) than the little toe (S1). The lower limb is an extension of the trunk and the lowermost dermatomes are located in the perineum not in the foot.

The upper and lower limbs have many structural similarities but these are analogous, not homologous and differences reflect functional specialization. Fusions within the pelvic girdle and greater stability of joints are adaptations to weight bearing. The leg is fixed in pronation and cannot be supinated while the foot is modified for support and propulsion at the expense of dexterity and intrinsic mobility.

The thigh extends from the iliac crest to the knee. It has one bone, the femur, whose proximal end articulates with the hip bone at the hip joint. The distal end of the femur articulates with the patella and the proximal end of the tibia at the knee joint.

The depression on the back of the knee is the popliteal fossa.

Front of the Thigh

Chapter Outline

1. Fascia
2. Muscles
3. Femoral Triangle
4. Adductor Canal

The front of the thigh extends from the inguinal ligament to the knee. It consists of a group of extensor muscles (of the leg) — iliopsoas, quadriceps femoris, sartorius, tensor fascia lata, pectineus and adductor longus. It also contains their related blood vessels and nerves.

1. Fascia

Superficial fascia consists of adipose and membranous layers that are continuous with the corresponding layers on the anterior abdominal wall.

The membranous layer is loosely attached to the deep fascia by areolar tissue. Immediately below the inguinal ligament, it is separated into two layers by the inguinal lymph nodes and the proximal end of the great saphenous vein.

The deeper part is continuous with the membranous layer of the abdominal wall and fuses with the deep fascia (fascia lata) a little below the inguinal ligament. It forms the lamina cribrosa covering the saphenous hiatus.

The outer part of the superficial fascia crosses the inguinal ligament and is continuous with the remainder of the superficial fascia of the thigh.

Cutaneous nerves are divided into two groups, according to their site of origin.

BRANCHES FROM THE LUMBAR PLEXUS

Ilioinguinal nerve (L1) emerges from the abdominal wall through the superficial inguinal ring accompanying the spermatic cord. It supplies the skin of the scrotum or labia and adjacent parts of the groin.

Femoral branch of the genitofemoral nerve (L1) enters the thigh by passing beneath the inguinal ligament. It pierces the deep fascia lateral to the saphenous opening and a little below the inguinal ligament to innervate the skin of the upper and front part of the thigh.

Lateral femoral cutaneous nerve (L2 and L3) enters the thigh through a fibrous sling beneath the lateral part of the inguinal ligament. It emerges from behind the lateral end of the inguinal ligament and divides into an anterior and posterior branches. The anterior branch pierces the deep fascia and supplies the skin of the anterolateral surface of the thigh. The posterior branch pierces the deep fascia immediately and descends backward to supply the skin on the lateral surface of the thigh to its midpoint.

BRANCHES FROM THE FEMORAL NERVE (L2, L3 AND L4)

Intermediate femoral cutaneous nerve pierces the deep fascia below the inguinal ligament and divides into medial and lateral branches that supply the skin down to the knee. It communicates with the genitofemoral and medial cutaneous nerve and gives a branch to the sartorius muscle.

Medial femoral cutaneous nerve is a branch of the anterior division of the femoral nerve. It gives branches that supply the middle third of the medial side of the thigh. At the apex of the femoral triangle it divides into:

Anterior branch pierces the fascia lata further down the thigh, lateral to the great saphenous vein to supply the lower medial side of the thigh.

Posterior branch descends on the posterior border of sartorius to pierce the deep fascia behind the great saphenous vein near the knee. It supplies the skin on the medial side of the leg.

Infrapatella branch of the saphenous nerve pierces the deep fascia on the medial side of the knee. It then curves downwards and forwards to innervate the skin below the patella.

Subsartorial plexus is formed by communication of the medial femoral cutaneous, obturator and saphenous nerves in the adductor canal.

Patellar plexus is a network of nerves consisting of terminal branches of the medial, intermediate and lateral femoral cutaneous nerves and the infrapatellar branch of the saphenous nerve. It is located in the fascia in front of the patella and the patellar ligament.

SUPERFICIAL INGUINAL ARTERIES

Four arteries arise from the femoral artery below the inguinal ligament.

Superficial circumflex iliac arteries pierce the deep fascia lateral to the saphenous opening. They extend below the inguinal ligament to the iliac crest and anastomose with the deep circumflex iliac, superior gluteal and lateral circumflex iliac arteries.

Superficial epigastric artery pierces the cribriform fascia, crosses the midinguinal point and ascends almost to the umbilicus supplying the anterior abdominal wall. It anastomoses with the vessel of the opposite side and the superficial inferior epigastric artery.

Superficial external pudendal artery pierces the cribriform fascia,

passes medially across the spermatic cord, or the round ligament, to supply the scrotum or labia majora. It anastomoses with the internal pudendal artery.

Deep external pudendal artery emerges more medially and passes behind the spermatic cord or round ligament to reach the skin of the scrotum or labia majora. It anastomoses with the scrotal or labial branches of the internal pudendal artery.

The femoral artery, the continuation of the external iliac artery, is conveniently divided into three segments: an upper third which is superficial in the femoral triangle; a middle third which is deep in the adductor canal; and a lower third which becomes the popliteal artery after passing through the adductor canal.

SUPERFICIAL INGUINAL VEINS

These veins correspond to the arteries and terminate in the great saphenous vein just before it pierces the cribriform fascia.

The **great saphenous vein** begins at the medial side of the dorsal venous arch of the foot and ascends in front of the medial malleolus with the saphenous nerve. It ascends along the medial side of the leg and behind the medial condyles of the tibia and femur. It ascends on the medial side of the thigh to the saphenous opening where it pierces the cribriform fascia and terminates in the femoral vein.

The great saphenous vein has numerous (at least 12) valves, communicates through the deep fascia with the deep veins, and has many unnamed tributaries, including the superficial inguinal veins.

SUPERFICIAL INGUINAL LYMPH NODES

The horizontal (or proximal) group of nodes lies in the superficial fascia along the line of fusion of the membranous layer of superficial fascia with the deep fascia of the thigh. They drain the areas supplied by superficial branches of the femoral artery, the lateral gluteal region, hypogastrium and perineum (external genitalia and anus).

The vertical (distal) nodes lie along the termination of the great saphenous vein and drain the skin of the lower limb and some of the perineum and gluteal regions.

Efferent vessels from the superficial group of nodes penetrate the cribriform fascia to terminate in the deep inguinal nodes and in external iliac nodes. These nodes also receive lymph from deep parts of the lower limb.

DEEP FASCIA OF THE THIGH

This fascia is known as the fascia lata. It surrounds the muscles of the thigh like a stocking attached (directly or indirectly) to all of the bony and ligamentous structures of the pelvis and of the knee.

Above, it is attached to the ischial tuberosity, side of the pubic arch, body of the pubis, inguinal ligament, and anterior iliac crest. Behind, the fascia lata is continuous with the gluteal fascia and through it to the remainder of the iliac crest, sacrum and coccyx and sacrotuberous ligament. Below the fascia is attached to the patella, patellar ligament and tibial condyles but posteriorly, it is continuous with the fascia of the back of the leg.

The fascia is thin medially but denser over the popliteal fossa where it is called the popliteal fascia.

Iliotibial tract is a wide, strong band-like thickening of the fascia lata on the lateral side of the thigh. It extends from the tubercle of the iliac crest to the lateral condyle of the tibia. Gluteus maximus is attached to it posteriorly at the level of the greater trochanter and tensor fascia lata muscle anteriorly below the trochanter.

Saphenous hiatus is an opening in the fascia lata a little below the medial end of the inguinal ligament. It is closed by the cribriform fascia, so named because it is pierced by several vessels, the great saphenous vein, superficial external epigastric and pudendal arteries and many lymphatics.

Falciform margin of the saphenous hiatus is the crescent shaped upper, lateral and lower edges of the saphenous hiatus that begins at the pubic tubercle and arches laterally to end behind the great saphenous vein. The medial margin is formed by the deep fascia over the pectineus muscle. It is not sharp and ascends behind the femoral vessels.

The **intermuscular septa**, lateral, medial and posterior septa are fibrous sheets that extend from the deep surface of the fascia lata to the linea aspera and supracondylar lines on the back of the femur.

The **lateral septum** is stronger and gives attachment to the vastus lateralis in front and short head of biceps femoris behind. The medial septum is weakest and continuous above with the pectineal fascia. These septa define three compartments containing muscles with a common function and nerve supply.

Anterior compartment contains flexors of the hip and extensors of the thigh supplied by the femoral nerve. Posterior compartment contains extensors of the hip and flexors of the knee supplied by the sciatic nerve. Medial compartment contains adductors of the hip supplied by the obturator nerve.

A fourth compartment contains abductors of the hip represented mainly by gluteus minimus, anterior part of gluteus medius and tensor fascia lata, all situated above the level of the greater trochanter. Below the trochanter, the compartment is represented by the iliotibial tract (insertion of tensor fascia lata and gluteus maximus).

Leg extensors are supplied by the femoral nerve, thigh adductors by the obturator nerve (except the leg flexor part of adductor magnus), leg flexors by the tibial component of the sciatic nerve (except short head of biceps femoris), and thigh abductors by the superior gluteal nerve.

Gluteus maximus is not an abductor and is supplied by the inferior gluteal nerve.

2. Muscles of the Front of the Thigh (Anterior Compartment)

EXTENSOR GROUP

Extensors of the knee are sartorius, quadriceps femoris and articularis genu. They are postaxial muscles and innervated by the femoral nerve (L2 and L3).

Sartorius

Sartorius is a ribbon-like muscle. It is the longest muscle in the body. It descends obliquely across the front and medial sides of the thigh.

- **Origin:** Anterior superior iliac spine and notch below it.
- **Insertion:** By a common aponeurosis (pes anserinus) formed by association with the tendons of gracilis, and semitendinosus into the upper, medial surface of the tibia. The common tendon is separated from the tibia by a bursa.
- **Action:** Flexes, abducts and laterally rotates the hip and flexes the knee.

Sartorius muscle forms the lateral border of the femoral triangle in the upper thigh and roof of the adductor canal in its middle third.

Quadratus femoris consists of four distinct parts having a common insertion, innervation and action. Rectus femoris arises from the ilium and has an action on the hip in addition to its action on the knee. All four parts end on the tibia through a common tendon attached to the base of the patella (a sesamoid bone in the tendon) and extended to the tibial tuberosity by means of the patellar ligament.

Quadratus femoris extends the knee, and the patella improves the angle of pull of the muscle on the tibial tuberosity.

Rectus Femoris

Rectus femoris is a bipennate fusiform muscle that lies in the middle of the front of the thigh superficial to the vasti.

- **Origin:** By two tendons that unite in front of the hip joint.
  Straight head from the anterior inferior iliac spine.
  Reflected head from above the upper brim of the acetabulum.
- **Action:** In addition to extension of the knee, flexion of the thigh at the hip joint.

Vastus Lateralis

This is the largest part of quadriceps forms the lateral side of the thigh.

- **Origin:** From the femur by a strong aponeurosis that extends from the root of the greater trochanter down the lateral side of the linea aspera, to the beginning of the lateral supracondylar ridge.

Vastus Medialis

Vastus medialis forms the medial side of the thigh.

- **Origin:** From the medial lip of linea aspera, distal half of the intertrochanteric line and medial intermuscular septum.

Vastus medialis is attached to the upper two thirds of the medial border of the patella and only slightly to the base. It draws the patella medially and acts to prevent lateral displacement of the patella in contraction of quadratus femoris.

Vastus Intermedius

Vastus intermedius lies under cover of the other vasti and rectus femoris.

- **Origin:** Anterior and lateral surfaces of the upper two thirds of the femoral shaft.

Articularis Genu

Articularis genu is a small bundle of muscle fibers considered to belong to the vastus intermedius muscle.

- **Origin:** Anterior surface of the lower part of the femur.
- **Insertion:** Synovial membrane of the knee joint.
- **Action:** Draws the synovial membrane upwards during extension of the knee.
- **Nerve supply:** Branch of the nerve to vastus intermedius muscle.

Pectineus

Pectineus is a flat quadrangular muscle located above and in the same plane as adductor longus. It belongs to the adductor group and is sometimes innervated by a branch of the obturator nerve but usually by anterior division of the femoral (L2, L3) or accessory obturator (L3) when present.

- **Origin:** Fleshy from the pectineal line and surface of the pubis.
- **Insertion:** By tendon in a line extending from the lesser trochanter to linea aspera.
- **Action:** Flexes and adducts the thigh and rotates it laterally.

Iliopsoas

Iliopsoas comprises iliacus and psoas major.

- **Origin:** Psoas major arises in the abdomen from the lumbar vertebrae, fibrous arches over lumbar vessels and sympathetic rami and front of the lumbar transverse processes.
- **Iliacus:** Arises in the pelvis from the greater part of the iliac fossa extending onto the sacrum.

When the muscles reach the thigh, psoas is located in the middle of the floor of the femoral triangle and below the hip joint at its insertion. Iliacus lies in the most lateral part of the floor of the femoral triangle in front of, then below the hip joint.

- **Insertion:** Psoas inserts by stout tendon into the lesser trochanter of the femur. Iliacus inserts by fleshy fibers into the tendon of psoas and lesser trochanter and below the joint.
- **Nerve supply:** L2, L3 and L4 directly or through the femoral nerve.
- **Action:** Flexes and medially rotates the extended thigh. Acts

with gluteus maximus to maintain balance of the trunk on the lower limb.

3. Femoral Triangle

Femoral triangle is a subfascial space on the front of the upper third of the thigh. It contains the main artery supplying the limb and branches of the nerve supplying muscles of the extensor (anterior) compartment.

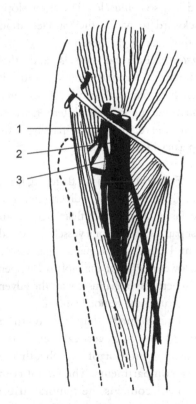

FIG. 45-1 Contents of femoral triangle and adductor canal; (1) Femoral n; (2) Femoral a; (3) Femoral v.

Its boundaries are:

Base: Inguinal ligament.

Lateral: Medial border of sartorius.

Medial: Medial border of adductor longus.

Apex: Where sartorius muscle crosses adductor longus muscle (leading into the adductor canal).

Roof: Fascia lata including the cribriform fascia covering the saphenous hiatus.

Floor: (From lateral to medial) anterior surfaces of adductor longus, pectineus, and iliopsoas muscles. The floor slopes backward to where the muscles attach to the back of the femur.

Contents: Femoral vessels and their branches; femoral nerve and its branches; femoral branch of the genitofemoral nerve; lateral femoral cutaneous nerve; lymphatics passing from superficial to deep nodes.

Femoral sheath is a flattened, fascial funnel formed by a diverticulum of the fascia transversalis that encloses the first part of the femoral vessels. It extends from behind the inguinal ligament to the lower margin of the saphenous opening where the fascia fuses with the adventitial coat of the femoral vessels.

The funnel shaped sheath is divided by two vertical septa that divide the interior of the sheath into three compartments. The lateral compartment contains the femoral artery, the intermediate compartment contains the femoral vein and the medial compartment (femoral canal) contains one or more deep lymph nodes and fat.

Femoral canal or medial compartment lies on the pectineus muscle beneath the cribriform fascia and upper margin of the saphenous hiatus and is conical in shape. Its oval upper end or mouth, the femoral ring, faces the peritoneal cavity and is closed by a plug of extraperitoneal connective tissue, the femoral septum. The canal allows for expansion of the femoral vein in standing but is largely obliterated thereby retaining abdominal contents.

The femoral canal is bounded medially by the concave margin of the lacunar ligament. The canal is the site of femoral hernia where extraperitoneal tissue with or without an intestinal loop may enter the canal and be trapped, (particularly on assuming upright posture), between the femoral vein laterally and lacunar ligament medially.

Femoral artery is the continuation of the external iliac artery appearing in the femoral triangle behind the inguinal ligament (at the mid inguinal point) in the lateral compartment of the femoral sheath. It descends almost vertically and slightly backward through the femoral triangle and the adductor canal then passes through the hiatus in the adductor magnus at the junction of the middle and lower thirds of the thigh and becomes the popliteal artery posteriorly in the popliteal fossa.

Branches of the femoral artery that arise in the femoral triangle are:

Superficial epigastric artery ascends through the saphenous opening passing upwards and superficially as far as the umbilicus to supply the epigastric region. It anastomoses with the inferior epigastric artery (branch of the external iliac artery).

Superficial external pudendal artery pierces the cribriform fascia and crosses the spermatic cord (or round ligament) supplying the lower abdominal wall and scrotum (labia). It anastomoses with the internal pudendal artery (branch of the internal iliac artery).

Superficial cicumflex iliac artery passes laterally towards the iliac crest supplying superficial tissues and anastomose with the deep circumflex iliac artery (branch of the external iliac artery).

Deep external pudendal artery arises from the medial side the femoral artery in the femoral sheath. It pierces the sheath and courses medially across pectineus and adductor longus muscle beneath the fascia lata. It passes behind the spermatic cord (round ligament) and is distributed to the skin of the external genitalia. It anastomoses with the scrotal (labial) branch of the internal pudendal artery.

Profunda femoris artery the largest branch of the femoral artery arising from the lateral and posterior side of the femoral artery shortly after it emerges from the femoral sheath. It descends curving medially to pass behind the femoral vessels. It leaves the femoral triangle between the pectineus and adductor longus muscles to enter the medial side of the thigh and ends by perforating adductor magnus in four branches that supply the hamstring muscles.

Its branches are lateral femoral circumflex artery arises from the lateral side behind rectus femoris dividing into three branches: ascending branch ascends beneath tensor fascia lata to supply the hip joint. It anastomoses with the inferior gluteal artery; transverse branch passes laterally around the femur to supply the gluteal muscles. It anastomoses with the medial circumflex artery in the cruciate anastomosis; descending branch descends along vastus lateralis to the knee. It anastomoses with the descending genicular artery (of the femoral), and superior lateral genicular (of the popliteal artery).

The *cruciate anastomosis* comprises vessels involved in the union of the medial and lateral circumflex arteries with the inferior gluteal artery (of the internal iliac) above and ascending branch of the first perforating artery below. The horizontal limb is formed by transverse branches of the medial and lateral circumflex femoral arteries and the vertical limb by descending branch of the inferior gluteal and ascending branch of the first perforating artery.

Profunda femoris artery ends in the lower third of the thigh by perforating adductor magnus and ending in vastus lateralis.

There are four perforating arteries.

First perforating artery arises at the lower border of pectineus. It perforates adductor magnus and supplies biceps and gluteus maximus. It anastomoses with the gluteal and circumflex arteries.

Second perforating artery arises at the middle of adductor brevis, perforates it and supplies the hamstring muscles and shaft of the femur. It anastomoses with the other perforating arteries.

Third perforating artery arises at the lower border of adductor brevis, perforates adductor magnus and supplies biceps.

Fourth perforating artery is the terminal part of the profunda femoris. It pierces adductor magnus and supplies short head of biceps and anastomoses with the popliteal and third perforating artery.

Medial femoral circumflex artery arises from the medial and posterior surface of the profunda femoris. It passes between pectineus and iliopsoas under the neck of the femur and supplies the upper part of adductor muscles and posterior hip and thigh.

Acetabular branch arises deep to adductor brevis and enters the hip joint beneath the transverse acetabular ligament to supply the hip joint. It anastomoses with the acetabular branch of the obturator artery.

Ascending branch arises at the lesser trochanter and passes along the upper border of quadratus femoris to the trochanteric fossa. It anastomoses with the superior and inferior gluteal arteries.

Transverse branch arises at the lesser trochanter and passes between the borders of quadratus femoris and adductor magnus supplying the hamstrings. It anastomoses with the first perforating artery, inferior gluteal and lateral circumflex in the cruciate anastomosis.

Femoral vein is the continuation of the popliteal vein above the adductor hiatus where it accompanies the femoral artery as it ascends through the adductor canal and upper two-thirds of the thigh. It passes through the femoral triangle posterior then medial to the femoral artery then lies lateral to the femoral canal ending behind the inguinal canal as the *external iliac vein*. Its tributaries are the long saphenous vein and veins corresponding to the branches of the femoral artery (except the superficial inguinal veins).

Femoral nerve (L2, L3 and L4) arises in the abdomen from the lumbar plexus. It enters the thigh behind the inguinal ligament, lateral to the femoral artery and under the iliac fascia breaking up almost immediately into its terminal branches.

Muscular branches:

Pectineus, quadriceps femoris, sartorius, vastus medialis and intermedius and rectus femoris.

Cutaneous branches:

The skin of the anterior abdominal wall appears developmentally to have slipped down over the ventral aspect of the thigh. Thus lumbar segments 1 to 4 are represented on the front of the thigh and spiral medially onto the medial surface of the thigh and leg. L4 and L5 segments supply the leg medially and laterally.

Intermediate femoral cutaneous nerve a branch of the femoral nerve pierces the deep fascia below the inguinal ligament and supplies the skin of the front of the thigh as far as the knee.

Medial femoral cutaneous nerve passes across the femoral artery and divides into an anterior branch that supplies the lower medial side of the thigh and a posterior branch that supplies the skin of the leg. It gives a branch to the subsartorial plexus.

Saphenous nerve is the longest branch of the femoral nerve. It accompanies the femoral artery through the femoral triangle and the adductor canal. Near the end of the adductor canal it gives an infrapatellar branch and then emerges at the surface by piercing the deep fascia on the medial side of the knee between the sartorius and gracilis muscles. It descends alongside the long saphenous vein to innervate the skin on the medial side of the leg and foot.

The subsartorial plexus is situated deep to sartorius. It is formed by communication between branches of the medial femoral cutaneous nerve, saphenous nerve and obturator nerve.

The patellar plexus in front of the knee is formed by communication between branches of the intermediate, medial, and lateral femoral cutaneous and saphenous nerves.

Deep lymph nodes drain the thigh one of three nodes are located deep to fascia lata along the medial side of the upper portion of the femoral vein often in the femoral canal.

Afferents drain the deep parts of the limb, the penis (or clitoris) and the superficial inguinal lymph nodes.

Efferents enter the abdomen alongside the external iliac artery to external iliac nodes, the lumbar trunk and then the cisterna chyli.

4. Adductor Canal

The adductor canal is an intermuscular, fascial compartment triangular in cross section, located in the middle third of the medial side of the thigh in the groove between vastus medialis and

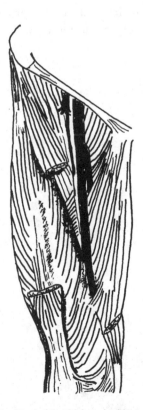

FIG. 45-2   Front of the thigh and adductor canal.

the front of the adductor muscles. It extends from the apex of the femoral triangle to the hiatus in the adductor magnus muscle.

Boundaries:

Anterolateral: Vastus medialis.

Anteromedial: Fascial bridge extending between vastus medialis and adductor longus and magnus and covered by sartorius.

Posteromedial: In its upper two-thirds; adductor longus and lower third; adductor magnus.

Contents: Femoral vessels, saphenous nerve, and nerve to vastus medialis.

The femoral artery provides a descending genicular artery just before it passes through the hiatus in the adductor magnus. This branch in turn divides immediately into: articular branch: descends medially deeply through the substance of vastus medialis, and then anastomoses with the medial superior genicular artery around the knee; saphenous branch: accompanies the saphenous nerve between gracilis and sartorius and joins the medial inferior genicular artery on the medial side of the knee. It supplies the medial aspect of the leg.

Nerve to vastus medialis is a branch of the femoral nerve that enters the femoral canal accompanying the descending genicular artery and then passes into the medial aspect of this muscle and gives a branch to the knee joint.

Chapter 46

Medial Side of Thigh

Chapter Outline

1. Muscles
2. Arteries
3. Veins
4. Nerves

The medial side of the thigh is that part from the medial part of the inguinal ligament to the side of the knee.

1. Muscles

ADDUCTOR GROUP

Adductor muscles of the thigh are separated from the extensor compartment by the medial intermuscular septum. They share an embryological origin with flexors (preaxial) being innervated by the obturator nerve (with the exception of pectineus and a part of adductor magnus) and have a common function of adduction, flexion, and medial rotation of the thigh.

The muscles are enclosed in a deep fascial sheath bounded by the medial intermuscular septum, fascia lata, a poorly defined posterior intermuscular septum between adductors and extensors whose deep apex is the linea aspera.

In the compartment, the adductors are arranged in three layers. The first of these contains pectineus, adductor longus, and gracilis. The second or intermediate layer contains adductor brevis and the third or deepest layer contains adductor magnus. The artery supplying the region is the profunda femoris.

Pectineus

Pectineus is a flat, quadrangular muscle that forms the medial part of the floor of the femoral triangle. It is continued below with the medial intermuscular septum that separates vastus medialis from the adductors.

- **Origin:** From the iliopectineal line and an area of the pubis in front of this.

- **Insertion:** Femur at the base of the lesser trochanter.
- **Nerve supply:** A branch of the femoral nerve (L2–L3, posterior division) that enters the lateral part of the muscle and also the obturator nerve (L2–L3, anterior division).

Pectineus is located at the boundary between the anterior and medial muscle groups supplied by the femoral and obturator nerves and is usually supplied by the femoral nerve but occasionally receives a twig from the obturator nerve. It is a key structure of the femoral triangle.

- **Action:** Adducts and medially rotates the femur and assists in flexion of the hip joint.

Adductor Longus

Adductor longus is triangular in shape. It lies in the same plane as pectineus and forms the medial boundary of the femoral triangle.

- **Origin:** By a narrow tendon from the medial superior ramus of the pubis between its crest and symphysis. It expands into a triangular muscular belly.
- **Insertion:** By a thin wide tendon into the lower two-thirds of the linea aspera.
- **Action:** Adducts the thigh and contributes to flexion and medial rotation of the thigh.

Gracilis

Gracilis is a thin superficial muscle on the medial side of the thigh and knee.

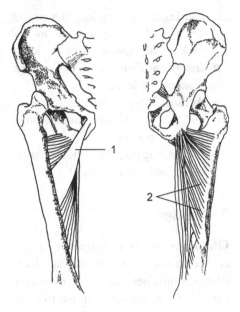

FIG. 46-1 Adductor muscles of the hip. (1) Adductor longus; (2) Adductor magnus.

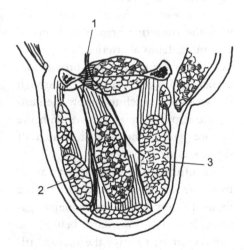

FIG. 46-2 Sagittal section through obturator foramen. (1) Obturator n; (2) Adductor longus m; (3) Adductor magnus m.

- **Origin:** Whole length of the inferior ramus of the pubic arch.
- **Insertion:** Medial upper surface of the tibia below the condyle and behind the sartorius muscle.
- **Nerve supply:** Obturator (the nerve of the pubis) L3–L4 anterior division.
- **Action:** Adducts the thigh and flexes the hip with the knee extended and assists in flexing the leg at the knee.

Adductor Brevis

Adductor brevis occupies the intermediate layer.

- **Origin:** Body and inferior ramus of the pubis between the gracilis and obturator externus.
- **Insertion:** Upper part of the linea aspera below the lesser trochanter.
- **Nerve supply:** Anterior division of the obturator (L2–L4) that passes down in front of the muscle (The posterior division passes down behind adductor brevis).
- **Action:** Adducts the thigh and flexes the hip.

Adductor Magnus

Adductor magnus shares the deepest layer with obturator externus in front of the pelvis and the neck of the femur. It is a composite of an adductor muscle (supplied by the obturator nerve) and a hamstring muscle (supplied by the tibial nerve).

Adductor Part

- **Origin:** Arises from the lateral half of the pubic ramus below the obturator foramen. Fibers of this part run almost horizontally.
- **Insertion:** Distal part of linea aspera and medial supracondylar line.
- **Action:** Adducts and laterally rotates the thigh and flexes the hip.
- **Nerve supply:** Obturator nerve (L2–L4, anterior division).

Extensor part (more posterior, hamstring)

- **Origin:** From the lower lateral part of the ischial tuberosity. Fibers from this part run almost vertically.
- **Insertion:** By a strong tendon into the adductor tubercle. The aponeurosis of insertion has, adjacent to the femur, four openings for perforating vessels (of the profunda femoris artery) and a large gap, the adductor hiatus through which the femoral artery passes and becomes the popliteal artery.
- **Action:** Adducts and medially rotates the thigh and aids the hamstrings in extending the hip.
- **Nerve supply:** Tibial part of the sciatic nerve (L4, L5 anterior division).

Obturator Externus

Obturator externus is flat and triangular.

- **Origin:** Medial half of the obturator membrane and adjacent edge of the obturator foramen. It

converges on a tendon that passes below and closely applied to the back of the capsule of the hip joint.

- **Insertion:** Floor of the trochanteric fossa of the femur.
- **Action:** Rotates the thigh laterally and aids in adduction and stabilizing the hip joint.
- **Nerve supply:** Posterior division of the obturator nerve (L3–L4).

2. Arteries

Obturator artery is a branch of the anterior trunk of the internal iliac artery that accompanies the obturator nerve through the obturator foramen to enter the thigh. Outside of the pelvis, it divides into anterior and posterior terminal branches at the upper border of obturator externus. Both branches supply neighbouring muscles.

The *anterior branch* anastomoses with the posterior branch and medial circumflex femoral arteries.

The *posterior branch* also provides an acetabular branch that passes through the acetabular notch to supply the head of the femur. It anastomoses with the inferior gluteal and medial branch of the obturator artery.

Profunda femoris artery arises from the lateral and back part of the femoral artery below the inguinal ligament. It curves downward, behind the femoral artery, to leave the femoral triangle by passing through its floor between the pectineus and adductor longus. It descends close to the femur, between adductor longus and magnus and terminates in the lower third of the thigh by piercing the adductor magnus. The terminal portion is also called the fourth perforating artery.

In the thigh, profunda femoris supplies the hamstrings (it is the artery of the adductor compartment).

Cruciate anastomosis is a cross-shaped arterial junction that is located over the greater trochanter where above, the inferior gluteal branches of the internal iliac artery anastomose with the circumflex femoral arteries and in turn below, anastomose with the first perforating artery of the profunda femoris. Perforating arteries of the profunda femoris artery form a series of anastomotic loops and the lowest communicates with the articular branches of the popliteal.

Components of the cruciate anastomosis are:
Horizontal limb: Transverse branches of the medial and lateral circumflex femoral arteries.

Vertical limb: Descending branches of inferior gluteal artery and an ascending branch of the first perforating artery.

3. Veins

Obturator veins are deep veins that accompany branches of the obturator artery. The united trunk pierces the obturator membrane, enters the pelvis, and terminates in the internal iliac vein.

Profunda femoris vein is the fourth perforating vein runs with its corresponding artery to join the femoral vein at a point lower than the origin of the artery. Its tributaries correspond to the branches of the profunda femoris artery. Circumflex branches enter the femoral vein directly.

Femoral vein arises from the hiatus in adductor magnus, ascends to the inguinal ligament where it becomes the external iliac vein. It receives muscular branches, the profunda femoris vein and the long saphenous vein. It has 4–5 valves.

4. Nerves

Obturator nerve (L2–L4, anterior division) arises from anterior branches of lumbar nerves L2–L4 in the abdomen, descends along the medial border of psoas major, behind the common iliac vessels then crosses the obturator internus and enters the thigh through the obturator foramen.

In the thigh, it divides into two branches that are separated by the adductor brevis muscle:

Anterior branch descends on adductor brevis behind pectineus and adductor longus to terminate in the subsartorial plexus. It provides articular and muscular branches.

Articular branches are given off near the obturator foramen. They pass through the acetabular notch to supply the hip joint.

Muscular branches innervate the gracilis and adductor longus and brevis.

Posterior branch supplies and pierces obturator externus then passes behind adductor brevis and reaches the anterior surface of the adductor magnus. It terminates as muscular and articular branches. The muscular branch innervating adductor magnus.

The articular branch passing with the popliteal artery to reach the knee joint.

The subsartorial plexus is located on the roof of the adductor canal beneath sarorius muscle. It contains branches of the saphenous nerve, medial femoral cutaneous and cutaneous branches of the obturator nerves. It supplies the skin of the region.

Chapter **47**

Gluteal Region

Chapter Outline

1. Fascia
2. Muscles
3. Deep Arteries
4. Deep Nerves

The gluteal region, or buttock, lies behind the pelvis from which blood vessels and nerves emerge to supply part of the lower limb. Above is the iliac crest and below, the lower border of gluteus maximus.

1. Fascia

Superficial fascia of the gluteal region is heavily laden with fat that gives the region its characteristic convexity. The crease for the hip joint is not related to the lower border of gluteus maximus. Superficial blood supply is derived from perforating branches of the gluteal arteries and lymphatic drainage is to the lateral group of superficial inguinal nodes.

Cutaneous nerves of the gluteal region are derived from posterior and anterior primary rami.

Posterior primary rami of L1–L3 slope downwards over the iliac crest to supply the skin of the upper buttock. Posterior primary rami of all five sacral nerves are cutaneous. S1–S3 supply the skin of the pudendal region while S4–S5 supply the skin over the coccyx.

Anterior primary rami supply the upper lateral skin (lateral cutaneous branches of the subcostal and iliohypogastric nerves, T12 and L1) and the lower part by the lateral femoral cutaneous nerve (L2).

Lateral femoral cutaneous nerve (from the nerve loop between L2 and L3) enters the thigh beneath the lateral part of the inguinal ligament where it divides. An anterior branch supplies the lateral anterior surface of the thigh and a posterior branch pierces the deep fascia to supply the skin over the greater trochanter.

Subcostal nerve (T12) lateral cutaneous branch pierces the deep fascia

immediately above the iliac crest. It descends across the gluteal region supplying it and the lateral side of the thigh to the greater trochanter.

Iliohypogastric nerve (T12) is the lateral cutaneous branch pierces the external and internal oblique muscles to reach the skin over the side of the buttock.

Lateral cutaneous branches of thoracoabdominal nerves become superficial passing through or between the digitations of serratus anterior. Each divides into a posterior branch that passes backward over latissimus dorsi and an anterior branch that passes downward and forward and medially.

Superior cluneal nerves are lateral cutaneous branches from the very small dorsal rami of L1 to L3. They pierce the thoracolumbar fascia at the lateral border of erector spinae muscles a little above the highest point of the iliac crest, cross the iliac crest and descend to supply gluteal skin over the greater trochanter almost to the gluteal fold.

Middle cluneal nerves are lateral cutaneous branches from the small dorsal rami of S1 to S3. They approach the surface in a line between the posterior superior iliac spine and the coccyx to supply the skin over the medial surface of the gluteal region.

Inferior cluneal nerves are gluteal branches of the posterior femoral cutaneous nerve (S1, S2, S3). They curve around the lower border of gluteus maximus to supply the skin over the lower buttock.

Perforating cutaneous nerves originate in the pelvis from the very fine rami of S2 and S3 of the sacral plexus. They perforate the sacrotuberous ligament and supply the skin over the lower buttock and medial fold of the buttock.

Deep fascia is thickest anterioriorly where it lies over the anterior part of gluteus medius. It splits into thin layers that enclose gluteus maximus and sends septa into the muscle that divide the muscle into coarse bundles.

2. Muscles

Gluteus Maximus

Gluteus maximus is the quadrilateral, coarsely fasciculated, most superficial muscle in the region and the heaviest muscle in the body.

- **Origin:** Upper part of the ilium behind the posterior gluteal line, dorsal surface of the sacrum, coccyx, and sacrotuberous ligament. It also arises from the deep fascia overlying gluteus medius (gluteal aponeurosis).
- **Insertion:** Lower fourth by fleshy fibres into the gluteal tuberosity, upper three fourths by an aponeurosis into the iliotibial tract.
- **Action:** Powerfully extends and abducts the thigh, and rotates it laterally. Insertion into the iliotibial tract produces extension of the thigh and holds the knee in extension in standing.

- **Nerve supply:** Inferior gluteal nerve (from L5, S1 and S2 of the sacral plexus).

Gluteus Medius

Gluteus medius lies anterior to and partly beneath gluteus maximus.

- **Origin:** Outer surface of the ilium between the middle and posterior gluteal lines and the overlying gluteal aponeurosis.
- **Insertion:** By a stong tendon into the lateral surface of the greater trochanter. A bursa separates the upper lateral surface of the trochanter from the tendon.
- **Action:** Abducts the thigh. It prevents adduction of the hip when the weight of the body is supported on the ipsilateral limb.
- **Nerve supply:** Superior gluteal nerve (L4, L5 and S1).

Gluteus Minimus

Gluteus minimus lies deep to gluteus medius.

- **Origin:** Lateral surface of the ilium between the anterior and middle gluteal lines.
- **Insertion:** Anterior surface of the greater trochanter.
- **Action:** Abducts the thigh. Prevents adduction of the hip when the weight of the body is supported on the ipsilateral limb.
- **Nerve supply:** Superior gluteal nerve (L4, L5, S1).

The following deep gluteal muscles are short lateral rotators of the thigh. They contribute to rotation of the hip when the foot is on the ground and increase the length of stride. They also contribute to stability of the hip joint.

Obturator Internus

Obturator internus tendon emerges from the pelvis through the lesser sciatic foramen.

- **Origin:** Internal (pelvic) surface of the hip bone and obturator membrane. Muscle fibres converge on a tendon that leaves the pelvis through the lesser sciatic foramen where it turns sharply forward into the gluteal region.

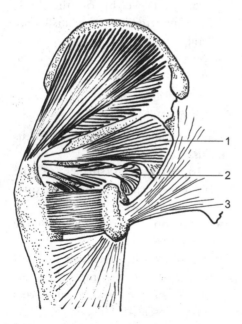

FIG. 47-1 Gluteal region. (1) Piriformis m; (2) Obturator internus m; (3) Quadratus femoris m.

- **Insertion:** Medial surface of the greater trochanter in front of the insertion of piriformis.
- **Action:** Abductor of flexed thigh; lateral rotator of extended thigh.
- **Nerve supply:** Nerve to obturator internus from the sacral plexus (L5, S1 and S2).

Gemellus Superior

Gemellus superior is a slip of muscle that arises from the upper margin of the lesser sciatic notch and that lies above and is attached to obturator internus.

- **Origin:** Ischial spine.
- **Insertion:** Tendon of obturator internus.
- **Action:** Lateral rotator of the thigh.
- **Nerve supply:** Nerve to obturator internus (L5, S1and L2).

Gemellus Inferior

Gemellus inferior is a similar slip of muscle lies below the obturator internus muscle and is attached to the lower margin of the lesser sciatic notch.

- **Origin:** Ischial tuberosity.
- **Insertion:** Tendon of the obturator internus muscle.
- **Action:** Lateral rotator of thigh.
- **Nerve supply:** Nerve to quadratus femoris (L4, L5 and S1).

Quadratus Femoris

Quadratus femoris is a thick quadrilateral muscle located below gemellus inferior and covering obturator externus.

- **Origin:** Upper lateral margin of ischial tuberosity.
- **Insertion:** Quadrate line (a line extending downward from the intertrochanteric crest).
- **Action:** Adductor and strong lateral rotator of the thigh.
- **Nerve supply:** Nerve to quadratus femoris (L4, L5 and S1) from the sacral plexus.

Tensor Fascia Lata

Tensor fascia lata is a fusiform muscle enclosed between two layers of fascia lata.

- **Origin:** Anterior part of outer lip of iliac crest.
- **Insertion:** Iliotibial tract.
- **Action:** Assists in flexion, abduction and medial rotation of thigh and assists gluteus maximus in tightening the iliotibial tract.
- **Nerve supply:** Superior gluteal nerve (L4, L5 and S1).

3. Deep Arteries

The deep arteries, branches of the internal iliac artery emerge from the pelvis through the greater sciatic foramen and enter the gluteal region above or below piriformis.

Superior gluteal artery is a continuation of the posterior division of the internal iliac artery. It appears in the gluteal region accompanied by its corresponding vein and divides immediately into two branches:

Superficial branch runs between gluteus maximus and medius and divides

immediately to supply gluteus maximus. It anastomoses with the inferior gluteal and medial circumflex arteries.

Deep branch runs forward between gluteus medius and minimus and divides into superior and inferior branches which accompany their corresponding nerves. The upper branch reaches the anterior superior iliac spine and anastomoses with the deep circumflex iliac artery. The inferior branch passes toward the greater trochanter and anastomoses with the lateral circumflex femoral artery.

Inferior gluteal artery is a branch of the anterior division of the internal iliac artery. It enters the gluteal region

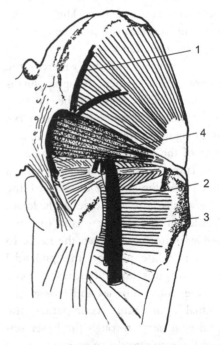

FIG. 47-2 Relations hips of piriformis. (1) Superior gluteal a and n; (2) Sciatic n; (3) Posterior femoral cutaneous n; (4) Pudendal n.

by passing through the lower part of the greater sciatic foramen between piriformis and superior gemellus with the sciatic nerve and internal pudendal artery. It lies between the ischial tuberosity and greater trochanter beneath gluteus maximus.

Muscular branches supply gluteus maximus and neighbouring lateral rotator muscles. They anastomose with the internal pudendal, medial circumflex femoral and obturator arteries.

Cutaneous branches accompany the posterior femoral cutaneous nerve to the skin of the buttock and back of the thigh.

Anastomotic branches descend across the lateral rotators. They join the cruciate anastomosis and supply the capsule of the hip joint.

Coccygeal branches pierce the sacrotuberous ligament to supply gluteus maximus and the skin over the coccyx.

Companion artery of sciatic nerve is a long nerve courses on or in the surface of the sciatic nerve and supplies it.

Internal pudendal artery is a branch of the anterior division of the internal iliac artery, emerges at the lower border of the greater sciatic foramen below piriformis between the nerve to obturator internus and the pudendal nerve. It gives small branches to some of the gluteal muscles and crosses the ischial spine and accompanies the pudendal nerve through the lesser sciatic foramen into the perineum.

4. Deep Nerves

Superior gluteal nerve (L4, L5 and S1) arises from the sacral plexus in the pelvis, it enters the gluteal region through the greater sciatic foramen passing above piriformis with the superior gluteal vessels. It passes forward between the gluteus medius and minimus muscles and divides into two branches.

The upper branch terminates in gluteus medius, while the lower branch innervates gluteus medius and minimus. It terminates in and supplies tensor fascia lata.

Inferior gluteal nerve (L5, S1 and S2) is a branch of the sacral plexus, emerging in the gluteal region below piriformis and ramifies immediately to innervate gluteus maximus.

Sciatic nerve (L4, L5, S1, S2 and S3), a terminal branch of the sacral plexus, enters the gluteal region passing below piriformis and crossing the gemelli and obturator internus to enter the thigh at the lower border of quadratus femoris. It has no branches in the gluteal region.

Posterior femoral cutaneous nerve (posterior branches from S1 and S2 and anterior branches from S2 and S3). This branch of the sacral plexus enters the gluteal region on or at the medial side of the sciatic nerve below piriformis. In the gluteal region, it gives the gluteal branch whereby several inferior clunial nerves wind around the lower border of gluteus maximus to innervate the skin of the lower lateral buttock; and the perineal branch which begins at the lower border of gluteus maximus, run medially over the hamstring muscles and innervate the skin of the scrotum or labium majus.

Nerve to quadratus femoris and **inferior gemellus** arises from the anterior branches of L4, L5 and S1. This branch of the sacral plexus also enters the gluteal region through the lower part of the greater sciatic foramen below piriformis, deep to the tendon of the obturator internus and gemelli. It gives branches to inferior gemellus and the hip joint, and terminates in the deep surface of quadratus femoris.

Nerve to obturator internus and **superior gemellus** arises from the anterior branches of L5, S1 and S2. This branch of the sacral plexus emerges in the gluteal region below piriformis medial to the sciatic nerve. It crosses the root of the ischial spine, and gives a branch to the superior gemellus. It leaves the gluteal region to reenter the pelvis through the lesser sciatic foramen to reach the perineal surface of obturator internus.

Pudendal nerve (anterior branches of S2, S3 and S4) is a terminal branch of the sacral plexus is the most medial structure to enter the gluteal region through the greater sciatic notch below piriformis. It passes over the ischial spine and sacrospinous ligament, enters the pelvis through the lesser sciatic foramen, and then enters the pudendal canal that leads it to the perineum.

Chapter 48

Back of Thigh and Popliteal Fossa

Chapter Outline

The back of the thigh extends from the gluteal fold to the upper border of the popliteal fossa. Posterior femoral muscles are the "hamstring" or flexor group of muscles. In common, they arise from the tuberosity of the ischium, cross and extend the hip joint (except short head of biceps) as well as the knee joint and are innervated by the tibial component of the sciatic nerve.

Short head of biceps has a supplemental origin from the femur supplied by the fibular division of the sciatic nerve. This suggests an evolutionary and embryologic origin from the extensor column of muscles.

1. Fascia

The **superficial fascia** of the back of the thigh contains a variable amount of fat.

Cutaneous nerves are branches of the medial (L2, L3), lateral (L2, L3), and posterior femoral cutaneous nerves (S1, S2, S3).

Cutaneous blood vessels reach the surface from deep vessels, after piercing the deep fascia

Deep fascia encloses the hamstrings in a tube bounded by the lateral intermuscular septum, fascia lata and a weak septum separating the adductor and flexor compartments. The apex of the compartment is the linea aspera.

2. Muscles

FLEXOR GROUP (HAMSTRINGS)

Flexor muscles of the thigh are innervated by the sciatic nerve.

Biceps Femoris

Biceps femoris is the most laterally situated muscle of this group. It is a secondary combination of a pre-axial muscle (long head) and a post-axial muscle (short head).

- **Origin:** *Long head*: By a common tendon with the semitendinosus muscle, from the upper medial part of the ischial tuberosity and from the lower part of the sacrotuberous ligament.

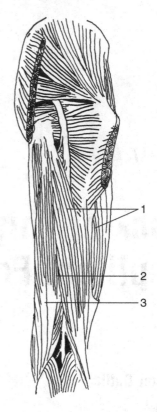

FIG. 48-1 Muscles of back of thigh. (1) Biceps femoris m; (2) Semitendinosus m; (3) Semi-membranosus m.

Short head: From the lateral lip of linea aspera, lateral, supracondylar line and lateral intermuscular septum.

- **Insertion:** At the knee, the combined tendon is split by the fibular collateral ligament. It ends on the head of the fibula, lateral condyle of the tibia and deep fascia of the leg.
- **Nerve supply:** Tibial part of the sciatic nerve (S1–S3)

- **Nerve supply:** Common fibular part of the sciatic
- **Action:** Flexes and laterally rotates of the leg (or the thigh medially if the limb is weight bearing). When the limb is fixed in weight bearing, biceps is active during forward bending and extends the pelvis on the thigh in raising from a forward bending position.

Semitendinosus

Semitendinosus is a fusiform shaped muscle occupying a groove on the surface of semimembranosus. It lies in the buttock, back of the thigh superficial to semimembranosus, and medial boundary of the popliteal fossa. Half way down the thigh, its muscular belly is replaced by a cord-like tendon.

- **Origin:** In common with long head of the biceps from the ischial tuberosity.
- **Insertion:** Medial surface of tibia, below the insertion of sartorius and gracilis.
- **Nerve supply:** Two branches of the tibial division of the sciatic (L4, L5, S1, S2).
- **Action:** Flexes the leg, extends the thigh and medially rotates the flexed leg.

Semimembranosus

Semimembranosus lies deep to semitendinosus. At its origin, it has a long flat tendon or membrane, hence its name.

- **Origin:** Upper outer part of ischial tuberosity.
- **Insertion:** Through three expansions (i) to a groove on the back of the medial tibial condyle; (ii) through the oblique popliteal ligament into the capsule of the knee joint; (iii) the fascia covering the popliteus muscle to the soleal line.
- **Nerve supply:** Tibial division of the sciatic (L4, L5, S1, S2)
- **Action:** Flexes of the leg and extends the thigh. Extends the trunk when the leg and thigh are fixed. Medially rotates the leg. Action on one joint depends on the relative position of the other joint.

Short head of biceps is supplied by the common fibular part of the sciatic nerve (L4, L5, S1) but the same segments as the tibial part that supplies the muscles of the flexor compartment.

3. Deep Arteries

The hamstring compartment receives its blood supply mainly from the profunda femoris artery by way of its perforating branches. These arteries pierce the adductor magnus giving branches to the hamstrings then pierce the lateral intermuscular septum to end in vastus lateralis.

Inferior gluteal artery gives branches to the upper part of the hamstrings.

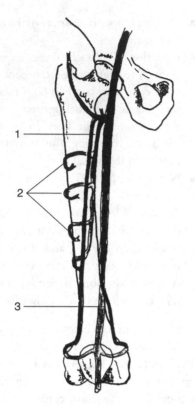

FIG. 48-2 Blood supply of thigh. (1) Descending branch of lateral femoral circumflex a; (2) Perforating a; (3) Descending genicular a.

Perforating and muscular branches of **profunda femoris artery** supply biceps femoris.

These arteries form a series of anastomoses along the back of the femur through their terminal arterioles.

4. Deep Nerves

Sciatic nerve enters the gluteal region through the greater sciatic foramen at the lower border of piriformis (although in 10% it may enter above or through piriformis). It runs down the midline of the thigh on the posterior surface of adductor magnus and crossed by long head of biceps. It terminates by dividing into the tibial and common fibular nerves near the upper apex of the popliteal fossa.

Its branches are:
Articular which arise proximally and enter the back of the capsule of the hip joint; *nerves to the hamstrings* (except the short head of biceps) are branches of the tibial division and leave the medial side of the sciatic nerve.

Posterior femoral cutaneous nerve (S2 and S3) is a branch of the sacral plexus emerges from under piriformis with the inferior gluteal artery. It descends in the midline of the thigh beneath the fascia lata on the long head of the biceps. In the popliteal fossa, it pierces the deep fascia and continues down to the middle of the leg in the superficial fascia, accompanied by the short saphenous vein.

In the gluteal region, it gives perineal branches to the skin of the upper medial side of the thigh and gluteal (inferior clunial) branches that wind around gluteus maximus to supply the skin over the lower buttock.

In the thigh, branches are cutaneous and they supply the skin on the back and medial side of the thigh and over the popliteal fossa.

In the leg, cutaneous branches supply the skin of the upper half of the back of the calf.

Popliteal Fossa

Chapter Outline

The popliteal fossa is a diamond-shaped space behind the lower part of the femur, the knee joint, and the upper part of the tibia.

1. Fascia

Superficial fascia contains a small amount of fat and the terminal branches of the posterior femoral cutaneous nerve, posterior branch of the medial femoral cutaneous nerve, sural communicating nerve, and upper part of the short saphenous vein.

Deep fascia of the popliteal fossa (popliteal fascia) is thin and strong.

2. Boundaries

The popliteal fossa has four sides, a roof and a floor.

Superior lateral: Biceps femoris.
Superior medial: Semitendinosus and semimemembranosus supported by sartorius and gracilis when the knee is bent.
Inferior lateral: Lateral head of gastrocnemius and plantaris.
Inferior medial: Medial head of gastrocnemius.
Floor: Popliteal surface of femur, oblique popliteal ligament and fascia covering popliteus.
Roof: Deep fascia (fascia lata) pierced by the posterior femoral cutaneous nerve and the short saphenous vein.

3. Contents

Popliteal artery is the continuation of the femoral artery from the hiatus in the adductor magnus muscle. It descends

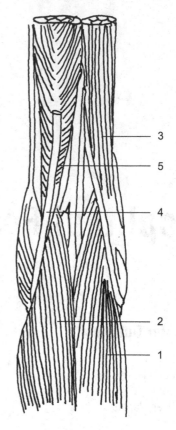

FIG. 48-3 Popliteal fossa — boundaries. (1) Lateral head of gastrocnemius m; (2) Medial head of gastrocnemius m; (3) Biceps femoris m; (4) Semitendinosus m; (5) Semimembranosus m.

vertically through the popliteal fossa to terminate by dividing at the distal borders of the popliteus muscle.

In addition to the terminal branches, the popliteal artery has muscular branches, genicular branches, and terminal branches.

Muscular branches supply the muscles of the back of the thigh and leg, and

Superior medial genicular artery crosses the medial head of gastrocnemius.

Inferior lateral genicular artery crosses the proximal tibiofibular joint then passes deep to the fibular collateral ligament.

Inferior medial genicular artery runs deep to the tibial collateral ligament.

The anastomosis around the knee joint is formed by the two superior and two inferior genicular arteries, descending branch of the lateral circumflex artery and descending genicular artery with the recurrent and circumflex fibular from the anterior tibial artery. There are cross connections superficial and deep to quadriceps.

The middle genicular artery (unpaired) passes directly forward piercing the oblique popliteal ligament to enter the knee joint supplying the cruciate ligaments.

Terminal branches are the anterior and posterior tibial arteries.

Popliteal vein is formed by the union of the venae comitantes of the anterior and posterior tibial and fibular arteries at the distal border of popliteus.

It has tributaries such as the short saphenous vein and veins corresponding to the arteries.

4. Deep Nerves

Tibial nerve (L4, L5, S1, S2 and S3) is the larger terminal branch of the sciatic

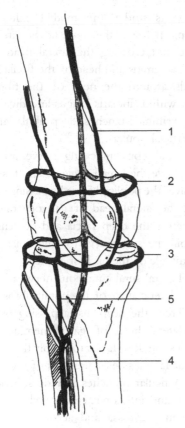

FIG. 48-4 Arterial anastomoses around the knee. (1) Descending genicular a; (2) Superior genicular a; (3) Inferior genicular a; (4) Anterior tibial a; (5) Recurrent genicular a.

provide cutaneous branches. They anastomose with muscular and perforating branches of the profunda femoris artery.

Genicular branches are the superior and inferior paired arteries that embrace the femoral condyles above and the tibial condyles below.

Superior lateral genicular artery crosses plantaris and lateral head of gastrocnemius.

nerve entering the popliteal fossa at its upper angle lateral to the popliteal vessels and descending as a separate nerve. The nerve crosses behind the vessels to leave the fossa at its lower angle at the lower border of popliteus passing deep to the fibrous arch of soleus to reach the back of the leg. It gives articular, muscular and a cutaneous branch.

The cutaneous branch is sural nerve which descends with the femoral artery through the adductor canal to reach the popliteal fossa. It runs vertically between the two heads of gastrocnemius and pierces the deep fascia below the knee (replacing the posterior femoral cutaneous nerve).

The muscular branches supply the muscles on the back of the leg.

There are three articular branches that arise proximally and accompany the medial and middle genicular arteries. They innervate the synovial membrane and cruciate ligaments of the joint.

Common fibular nerve (L4, L5, S1 and S2) is the smaller terminal branch of the sciatic nerve. It passes along the popliteal fossa from its upper angle and descends under the medial side of biceps. It leaves the fossa at the lateral angle and crossing the lateral head of gastrocnemius and head of the fibula. It winds around the neck of the fibula deep within fibularis longus and divides into terminal branches (deep fibular and superficial fibular).

Sural communicating nerve arises just above the lateral angle of the fossa. It leaves the popliteal fossa by passing out over the lateral head of gastrocnemius. It pierces the deep fascia and descends to join the sural nerve (of the tibial) at the middle of the back of the leg.

Lateral sural cutaneous nerve arises at the lateral angle of the popliteal fossa. It pierces the deep fascia passing over the lateral head of gastrocnemius. It pierces the deep fascia to supply the skin on the lateral side of the back of the leg.

Articular branches include superior lateral and inferior lateral branches that arise in the fossa and accompany the corresponding vessels to supply the capsule of the knee joint. The recurrent genicular branch originates outside the fossa.

Chapter 49

Front of Leg and Dorsum of Foot

Chapter Outline

1. Fascia

SUPERFICIAL

The superficial fascia is continuous with the superficial fascia of the thigh and the dorsum of the foot. It contains some fat, vessels and nerves over the muscles but little over the bones and joints.

Superficial Veins

Dorsal digital veins are two vessels which drain the medial side of the big toe (hallux) and the side of the dorsum of each toe. Each unite to form a dorsal metatarsal vein that terminates in the dorsal venous arch lying across the shafts of the metatarsals.

Small saphenous vein begins where the lateral dorsal digital vein of the little toe unites with the lateral end of the dorsal arch. It travels up the back of the leg with the sural nerve then between the heads of gastrocnemius and pierces the popliteal fascia ending in the popliteal vein. It communicates with deep veins and can anastomose with the great saphenous vein.

Great saphenous vein begins at the junction of the dorsal digital vein of the lateral side of the little toe with the dorsal venous arch. It passes in front of the medial malleolus with the saphenous nerve and bends behind the medial condyle of the femur. It ascends along the medial side of the thigh and receives the superficial circumflex iliac, superficial external pudendal and superficial epigastric veins before passing through the saphenous opening and joining the femoral vein. It has many valves and many connections with deep veins.

Cutaneous Nerves

Infrapatellar branch of the saphenous nerve is a branch of the saphenous nerve (from the femoral nerve) as it leaves the adductor canal. It pierces sartorius and fascia lata to suppy the skin in front of the patella and communicates with the lateral femoral cutaneous nerve to form the patellar plexus.

Lateral sural cutaneous nerve is a branch of the common fibular nerve (L4, L5, S1, S2) that supplies the skin on the middle third of the lateral back of the leg.

Cutaneous branches of superficial fibular nerve begin at the bifurcation of the common fibular nerve. The nerves pass deep to fibularis longus then between the fibular muscles and extensor digitorum longus, pierce the deep fascia. They divide into medial and intermediate dorsal cutaneous branches. These innervate the lower third of the leg, dorsum and medial side of the foot, and the first digit and adjacent sides of the second and third toes (medial branch) and adjacent sides of the third to fifth toes and skin over the lateral side of the ankle.

Terminal branches of the deep fibular nerve appear on the interosseous membrane at the ankle joint where the deep fibular nerve divides into medial and lateral (cutaneous) branches.

Medial branch travels with dorsalis pedis artery to reach and supply adjacent sides of the first and second toes.

Lateral branch passes laterally beneath extensor digitorum brevis (supplying it) and the tarsal metatarsal joints.

Sural nerve (S2) is the terminal and only cutaneous branch of the tibial nerve. It descends superficially between the heads of gastrocnemius and pierces the deep fascia being joined by the sural communicating nerve. It courses along the lateral side of the tendocalcaneus and supplies the lower lateral side of the leg and fifth toe.

Patellar plexus is formed in front of the knee by the communication between the anterior branch of the medial femoral cutaneous nerve, intermediate and lateral femoral cutaneous nerves and infrapatellar branch of the saphenous nerve.

Subsartorial plexus is formed deep to sartorius by the medial femoral cutaneos, saphenous and obturator nerves.

DEEP FASCIA AND COMPARTMENTS OF THE LEG

The deep fascia of the leg (crural fascia) is continuous with the fascia lata that envelopes the thigh. It is fused with the bony structures around the knee and with the periosteum of the subcutaneous surface of the tibia and fibula and malleoli. Intermuscular septa from the deep fascia subdivide the leg into three compartments.

Anterior intermuscular septum is attached to the anterior border of the fibula and separates the extensor muscles in the anterior compartment from the fibular muscles in the lateral compartment.

Posterior intermuscular septum is attached to the posterior border of the fibula and separates the fibular muscles in the lateral compartment from the flexor muscles in the posterior compartment.

Transverse intermuscular septum extends medially from the posterior intermuscular septum separating the superficial from deep groups of posterior or calf muscles. It is thick over popliteus where it is reinforced by fibers from tendon of semimembranosus muscle.

Muscles in the Anterior Compartment

Tibialis anterior, extensor hallucis longus, extensor digitorum longus and fibularis tertius are innervated by the deep fibular nerve.

Muscles in the Lateral Compartment

Fibularis longus and brevis are innervated by the superficial fibular nerve.

Muscles in the Posterior Compartment

Muscles in the posterior compartment are divided into a superficial group (triceps surae consisting of gastrocnemius, plantaris and soleus) and a deep group (popliteus, flexor hallucis longus, flexor digitorum longus and tibialis posterior). They are all innervated by the tibial nerve.

Deep fascia above the ankle and on the dorsum of the foot is thickened by transverse fibers to form bands, or retinaculae. These act as pulleys for the deep tendons and maintain their position close to the bones during muscle contraction — they prevent "bowstringing". Synovial sheaths surrounding tendons descend behind the retinacula.

RETINACULA

Two retinacula of the anterior compartment are superior extensor retinaculum and anterior extensor retinaculum

Superior Extensor Retinaculum

Fibers are attached to the anterior borders of the tibia and fibula just above the ankle joint. The band bridges all of the extensor (anterior compartment) muscles of the leg. A strong septum from its deep surface separates the

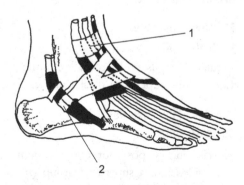

FIG. 49-1 Fibular retinacula and synovial sheaths of fibular muscles. (1) Superior extensor retinaculum; (2) Inferior fibular retinaculum.

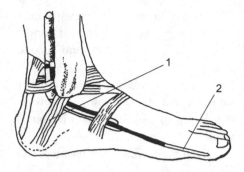

FIG. 49-2 Extensor retinacula. (1) Flexor retinaculum; (2) Flexor hallucis longus m.

underlying space so that tibialis anterior tendon is in its own synovial sheath is separate from a lateral space for the other long extensor tendons.

Inferior Extensor Retinaculum

sInferior extensor retinaculum is a Y-shaped band whose stem forms a loop for extensor tendons then branches into two bands. The upper band passes upwards and medially and is attached to the medial malleolus. The lower band passes downward and medially to the medial side of the foot to blend with the deep (plantar) fascia of the sole.

Tendons enclosed in their synovial sheaths, pass beneath this retinaculum in three tunnels contain, from lateral to medial, (1) extensor digitorum longus and fibularis tertius, (2) extensor hallucis longus, and (3) tibialis anterior.

2. Muscles

The muscles of the anterior compartment of the leg are extensors (dorsiflexors) of the foot with some inverting and others everting the foot. They are developmentally postaxial and innervated by the deep fibular nerve.

Tibialis Anterior

Tibialis anterior lies immediately against the lateral surface of the tibia.

- **Origin:** Upper half of the lateral surface of the tibia and interosseous membrane. Its tendon passes beneath the superior and inferior extensor retinacula surrounded by a synovial sheath.
- **Insertion:** Onto the medial side of the foot to the medial side of the first cuneiform and base of the first metatarsal.
- **Action:** Dorsiflexes and inverts the foot. With the foot on the ground, tibiallis anterior tilts the tibia forward and with its insertion into the same bones as fibularis longus, forms a sling for the longitudinal and transverse arches of the foot.
- **Nerve supply:** Deep fibular nerve (L4 and L5).

Extensor Digitorum Longus

Extensor digitorum longus is a pennate muscle on the lateral side of the anterior compartment.

- **Origin:** Lateral condyle of the tibia, upper anterior surface of the fibula and the interosseous membrane. Its tendon, enclosed in a synovial sheath, passes beneath the superior and inferior extensor retinacula and

it divides into four tendons in front of the ankle.

- **Insertion:** Into the base of the middle phalanges and the base of the distal phalanges of the lateral four toes.
- **Action:** Dorsiflexes (extends) and everts the toes and foot. When the foot is fixed, it draws the leg forward.
- **Nerve supply:** Deep fibular nerve (L5, S1).

Fibularis Tertius

Fibularis tertius is a lateral slip of extensor digitorum longus.

- **Origin:** Distal third of anterior surface of the fibula. The tendon passes under the extensor retinacula with extensor digitorum longus.
- **Insertion:** Dorsal base of fifth metatarsal.
- **Action:** Everts and dorsiflexes the foot.

Extensor Hallucis Longus

Extensor hallucis longus is located between tibialis anterior and extensor digitorum longus.

- **Origin:** Middle of the anterior surface of the fibula and adjacent interosseous membrane. The tendon arises on its superficial surface and passes beneath the superior and inferior extensor retinacula.
- **Insertion:** Base of the distal phalanx of the big toe or hallux.
- **Action:** Extends the hallux and dorsiflexes the foot.

- **Nerve:** Branch of the deep fibular nerve (L5, S1)

Extensor Digitorum Brevis

Extensor digitorum brevis lies on the lateral part of the dorsum of the foot, and is the only muscle on the dorsum of the foot forming a fleshy mass. Its most medial part is attached independently to the base of the proximal phalanx and may be considered a separate muscle, extensor hallucis brevis.

- **Origin:** Upper surface of the anterior process of calcaneus.
- **Insertion:** Tendon of extensor hallucis brevis inserts into the base of the proximal phalanx of the hallux. The other three tendons join the lateral side of the corresponding tendons of extensor digitorum longus.

The fifth digit lacks a short extensor tendon.

- **Action:** Extends metatarsophalangeal and interphalangeal joints of the four medial toes.
- **Nerve supply:** Branches of the deep fibular nerve (S1 and S2).

Muscles of the anterior compartment dorsiflex the foot of the free limb in walking allowing the foot to clear the ground in the swing phase of walking.

EXTENSOR EXPANSIONS

Tendons of the long and short extensors spread to form an extensor expansion on the dorsurn of the proximal phalanx. The expansion splits into three, a middle part that inserts into the base of the proximal phalanx and two collateral parts that unite and insert into the base of the distal phalanx. On the second, third and fourth toes, the expansions receive the insertions of tendons of the one lumbrical, two interossei and a tendon of extensor brevis. On the little toe, the expansion is joined by only one lumbrical and one interosseous muscle.

SYNOVIAL SHEATHS

Synovial sheaths allow the tendons to glide freely between underlying bones and ligaments and overlying retinacula. There are three sheaths in front of the ankle deep to the extensor retinacula:

Sheath of tibialis anterior extends from the upper edge of the superior retinaculum to the upper border of the inferior limb of the retinaculum.

Sheath of extensor hallucis longus begins above the inferior retinaculum and extends to beyond the cuneometatarsal joint of the hallux.

Sheath of extensor digitorum longus and fibularis tertius begins between the superior and inferior extensor retinacula and extends onto the dorsum of the foot to the base of the first metatarsal.

3. Deep Arteries

In the neurovascular plane of the anterior compartment, the anterior tibial vessels and the deep fibular nerve lie on the interosseous membrane beneath tibialis anterior.

Anterior tibial artery is a terminal branch of the popliteal artery arising from the popliteal artery in the back of the leg at the lower border of popliteus muscle. It passes between the two heads of origin of tibialis posterior over proximal border of the interosseous membrane medial to the neck of the fibula to enter the anterior compartment of the leg. It descends on the interosseous membrane between tibialis anterior and extensor digitorum longus then between tibialis anterior and extensor hallucis longus. Passing deep to the superior extensor retinaculum, it reaches the ankle joint midway between the malleoli and continues onto the foot as the dorsalis pedis artery

BRANCHES

Muscular branches arise at irregular intervals and supply the muscles of the anterior compartment.

Anterior tibial recurrent artery pierces tibialis anterior and supplies the anterior side of the knee and adjacent muscles. It anastomoses with the genicular branches of popliteal artery to participating in the genicular anastomosis.

Posterior tibial recurrent artery arises from the front of the popliteal artery and supplies the posterior part of the knee and popliteus muscle. It anastomoses with other genicular arteries to participate in the genicular anastomsis.

Medial anterior malleolar artery arises above the ankle joint and supplies the medial side of the ankle. It anastomoses with branches of the posterior tibial artery.

Lateral anterior malleolar artery proceeds posterior to extensor digitorum longus and fibularis longus to reach the lateral side of the ankle which it supplies. It anastomoses with posterior lateral malleolar artery and perforating branches of the fibular artery.

Dorsalis pedis artery is the continuation of the anterior tibial artery beyond the front of the ankle joint to the first interosseous space where it anastomoses with the lateral plantar artery to form the plantar arch (cf the radial artery).

Tarsal arteries, lateral and medial arise as the artery crosses the navicular. The lateral artery crosses under extensor digitorum to supply the tarsal joints. The medial arteries also supply tarsal joints.

Arcuate artery passes laterally in an arch over the base of the metatarsals deep to the extensor muscles. It anastomoses with the lateral tarsal and lateral plantar arteries. It gives rise to the second to fourth dorsal metatarsal arteries and branches into dorsal digital branches to three lateral metatarsal spaces. They supply the interossei and digital branches.

First dorsal metatarsal artery arises as the dorsalis pedis is about to pass into the sole. It lies over the dorsal first interosseous space, gives a branch to the medial side of the hallux then divides in the cleft between first and second toes to supply adjacent sides of the toes.

4. Deep Nerves

Common fibular nerve (from posterior branches of L4, L5 S1 and S2) branches from the tibial part of the sciatic nerve at the apex of the popliteal fossa. It follows the lateral margin of the tendon of biceps femoris (supplying the short head) to cross the lateral head of gastrocnemius and become superficial behind the head of the fibula. In the popliteal fossa, it gives lateral sural cutaneous and fibular communicating branches. It winds around the lateral side of the neck of the fibula deep to fibularis longus.

Terminal branches are deep fibular and superficial fibular nerves.

In the popliteal fossa, the common fibular nerve gives:
Articular branches to the knee and tibiofibular joint; cutaneous branches; lateral sural cutaneous nerve to the skin of the lateral back of the leg; fibular communicating nerve (with the medial sural cutaneous nerve) joining the sural nerve; and recurrent genicular nerve arises within fibularis longus and ascends with the anterior tibial recurrent artery to supply the part of tibialis anterior, tibiofibular and knee joints.

Deep fibular nerve arises from the division of common fibular nerve, it passes in front of the interosseous membrane piercing extensor digitorum longus. It descends lateral to the anterior tibial artery to the ankle where it divides into medial and lateral branches between the malleoli.

Muscular branches supply the muscles of the anterior compartment (tibialis anterior, extensor digitorum longus, fibularis tertius and extensor hallucis longus).

Articular branches accompany the anterior tibial recurrent artery to the knee joint.

Medial digital branch continues to the first intermetatarsal space, where it divides to supply the adjacent sides of the first and second toes and tarsometatarsal and metatarsophalangeal joints of the first digit.

Lateral digital branch passes laterally on the dorsum of the foot deep to extensor digitorum brevis innervating it and the adjacent tarsal and metatarsal joints.

Chapter 50

Lateral Side of Leg

Chapter Outline

1. Fibular Retinaculae
2. Fibular Muscles
3. Deep Nerve

The lateral compartment of the leg extends from the lateral side of the knee to the lateral side of the ankle. It contains two muscles, fibularis longus and fibularis brevis that arise from the fibula and evert and plantar flex the foot. They are innervated by the superficial branch of the common fibular nerve (L5, S1).

1. Fibular Retinaculae

Superior fibular retinaculum is a thickening of the fascia attached to the back of the lateral malleolus and lateral surface of the calcaneus. It retains the tendons of fibularis longus and brevis as they curve around the lateral side of the ankle. An osseofibrous tunnel retains a synovial sheath common to both tendons.

Inferior fibular retinaculum extends from the anterior part of the upper surface of the calcaneus to the lateral surface of the calcaneus. It binds the fibular tendons to the lateral side of calcaneus and provides a septum between the tendons. Fibularis brevis occupies the upper compartment and fibularis longus occupies the lower compartment.

2. Fibular Muscles

The two muscles in this group are innervated by the superficial fibular nerve and serve as evertors and plantar flexors of the foot in walking.

Fibularis Longus

Fibularis longus is a bipennate muscle arising more superficially and higher on the fibula than fibularis brevis

- **Origin:** Upper half of the lateral surface of the fibula, anterior and posterior intermuscular septa, and deep fascia over the muscle. Its tendon passes behind the lateral malleolus behind the tendon of fibularis brevis, below the fibular tubercle along the lateral surface of the

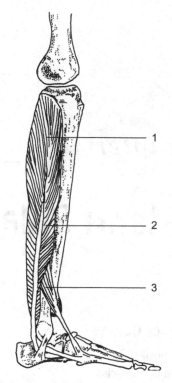

FIG. 50-1 Fibular muscles. (1) Fibularis longus m; (2) Fibularis brevis m; (3) Fibularis tertius m.

calcaneus to the base of the fifth metatarsal. It enters the groove on the plantar surface of the cuboid bone and crosses the sole obliquely.

The attachments of fibularis longus are the same as those of tibialis anterior.

- **Insertion:** Lateral side of the base of the first metatarsal and the base of the medial cuneiform.

A sesamoid of fibrocartilage develops in the tendon and protects the tendon from the tuberosity of the cuboid.

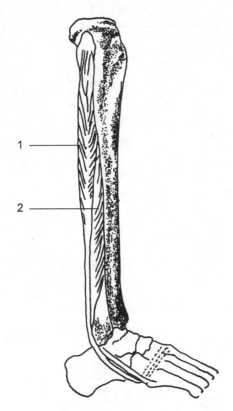

FIG. 50-2 Fibular muscles. (1) Fibularis longus m; (2) Fibularis brevis m.

The attachments of fibularis longus are to the same bones as the attachments of tibialis anterior.

Fibularis Brevis

Fibularis brevis lies deep to fibularis longus and is the smaller, shorter fibular muscle.

- **Origin:** Lower two-thirds of the lateral surface of the fibula and anterior and posterior intermuscular septa. Its tendon also passes under the superior fibular retinaculum anterior to fibularis longus tendon, behind the lateral malleolus in a common synovial sheath with fibularis longus and forward deep to the inferior fibular retinaculum over the lateral surface of the calcaneus.
- **Insertion:** Base (tuberosity) of the fifth metatarsal.

Synovial sheath extends above the lateral malleolus. It is a single synovial tube that encloses both the tendons fibularis longus and brevis. It divides between the superior and inferior fibular retinacula, and each tendon then has a separate sheath to its insertion.

3. Deep Nerve

Superficial fibular nerve is a terminal branch of the common fibular nerve. It begins on the lateral side of the neck of the fibula and descends in the substance of fibularis longus to the upper end of fibularis brevis, innervating both muscles. It passes obliquely over the anterior border of fibularis brevis and descends between it and the anterior intermuscular septum directly under the deep fascia. It pierces the fascia at the distal third of the leg and divides immediately into its lateral and medial terminal branches. The medial branch supplies adjacent sides of the first and second toes and the lateral branch supplies adjacent sides of the third, fourth and fifth toes.

Chapter 51

Back of Leg

Chapter Outline

1. Fascia
2. Muscles of the Back of the Leg
3. Deep Arteries
4. Deep Nerves

The back of the leg or posterior crural compartment extends from the lower border of the popliteal fossa to the back of the ankle. Component muscles are divided into superficial and deep groups by a transverse intermuscular septum. They are flexors of the foot and toes and supplied by the preaxial, tibial nerve (L4–S3 anterior division of the sciatic nerve).

1. Fascia

The superficial fascia of the back of the leg contains the following:

CUTANEOUS VEINS

Great saphenous vein arises from the dorsal venous arch on the dorsum of the foot to begin at the junction of the dorsal digital vein on the medial side of the big toe. It ascends in front of the medial malleolus accompanied by the saphenous nerve. It then bends behind the medial condyle of the femur and continues on the medial side of the thigh piercing the cribriform fascia of the saphenous opening then piercing the femoral sheath to end in the femoral vein.

The great saphenous vein communicates with deep veins particularly those within soleus and it has many valves.

Small saphenous vein arises from a plexus on the dorsum at the lateral side of the foot. It ascends behind the lateral malleolus along the midline of the back of the leg between the heads of gastrocnemius (with the sural nerve) to pierce the fascia over the popliteal fossa and terminate in the popliteal vein.

The small saphenous vein has several valves and communicates with deep veins of the foot and the great saphenous vein.

CUTANEOUS NERVES

Saphenous nerve is a terminal branch of the femoral nerve (L4). It becomes subcutaneous in the leg descending with the great saphenous vein alongside the medial side of the leg. It supplies the medial side of the calf and foot as far as the ball of the big toe (hallux).

Lateral sural cutaneous nerve arises from the common fibular nerve close to the head of the fibula and joins the sural branch of the tibial. It supplies the skin over the upper lateral side and back of the leg.

Sural nerve (from the union of the medial sural cutaneous nerve, a branch of the tibial nerve and the communicating fibular branch of the common fibular nerve) arises in the middle of the popliteal fossa. It descends under the deep fascia in the midline between the heads of the gastrocnemius muscle and pierces the fascia in the middle of the leg and descends alongside the small saphenous vein. It passes behind the lateral malleolus, and along the lateral side of the foot and fifth digit. It innervates the posterior aspect of the leg, the lateral side of the foot, and the fifth toe.

Posterior femoral cutaneous nerve is a mixed branch of the sacral plexus (posterior branches of S1, S2 and anterior branches of S2 and S3). It pierces the popliteal fascia and descends with the small saphenous vein to the middle of the calf. It innervates the skin over the popliteal fossa, back of the thigh and of the upper half of the back of the leg (calf).

The *posterior branch of medial femoral cutaneous nerve* (a branch of the anterior division of the femoral nerve) descends on the posterior border of sartorius piercing the fascia lata to join the subsartorial plexus with the saphenous and obturator nerves. It innervates the upper third of the back of the leg.

DEEP FASCIA

The deep fascia or crural fascia is continuous with the fascia lata of the thigh but attached to underlying bone and ligaments around the knee. The upper part is thickened by expansions of biceps, gracilis, semitendinosus and sartorius. It blends with the periosteum of the subcutaneous surface of the tibia. From its deep lateral surface, it gives an anterior and posterior intermuscular septum that pass to the fibula. The posterior compartment is subdivided by a transverse crural septum.

FLEXOR RETINACULUM

It is a thickened band of the deep transverse septum on the medial side of the ankle. It extends from the medial malleolus to the medial tubercle of the calcaneus, and holds down the tendons, nerves, and blood vessels that pass from the back of the leg to the sole of the foot. Septa extend from the deep aspect of the retinaculum to the tibia and ankle joint to form four tunnels, that contain the tendons of (from medial to lateral), tibialis posterior, flexor digitorum longus, posterior tibial vessels and the tibial nerve, and the tendon of flexor hallucis longus.

FASCIAL SEPTA

The posterior compartment of the leg is bounded anteriorly by the tibia, interosseous membrane, and fibula, and posteriorly by the investing deep fascia. The compartment is further subdivided by two septa.

The *outer septum* separates the deep flexors from the superficial flexors and extends from the medial border of the tibia to the posterior border of the fibula.

The *inner septum* (deep transverse fascia of the leg) covers tibialis posterior and separates it from the deep flexors. It extends between the medial border of the tibia and the posterior border of the fibula. Above, it is attached to the soleal line and below, it contributes to the flexor retinaculum.

2. Muscles of the Back of the Leg (of the Posterior Compartment)

SUPERFICIAL GROUP

The superficial group comprises the triceps surae (gastrocnemius, pantaris and soleus). Their tendons combine to form the tendocalcaneus. The muscles are innervated by the tibial nerve (L4–L5) and act as plantar flexors of the foot.

Gastrocnemius

Gastrocnemius is the most superficial layer forming the bulk of the calf. It is a two-headed muscle.

- **Origin:** Lateral head: Upper and posterior part of lateral condyle of the femur; Medial head: Popliteal

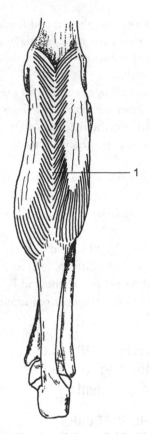

FIG. 51-1 Muscles of the calf (1). (1) Gastrocnemius m.

walking, it raises the heel against the body weight.

A bursa is located between the lateral head and the joint capsule.

Plantaris

The plantaris has a short belly and a long tendon. It lies above the lateral head of gastrocnemius but may be absent.

- **Origin:** Above the lateral condyle on the popliteal surface of the

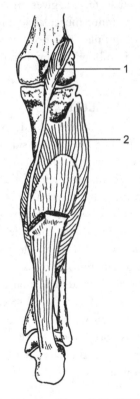

surface of the femur above the medial femoral condyle.

A bursa lies deep to each tendon of origin.

- **Insertion:** The heads converge on an aponeurosis that fuses with the underlying tendon of soleus to form the tendo calcaneus.
- **Action:** Plantar flexes and inverts the foot and flexes the knee. In

FIG. 51-2 Muscles of the calf (2). (1) Plantaris m; (2) Soleus m.

femur and oblique popliteal ligament. Its tendon descends between
gastrocnemius and soleus.

- **Insertion:** Medial side of tendocalcaneus or posterior calcaneus.

Soleus

Soleus is a thick, flat muscle that
extends beneath gastrocnemius.

- **Origin:** *Fibular origin*: From the
back of the head and upper third of
the shaft of the fibula; Tibial origin:
From the soleal line and its continuation along the medial border of
the tibia; *Tendinous origin*: From a
tendinous arch that extends from
the tubercle on the neck of the
fibula to the soleal line on the tibia.

The posterior tibial artery and tibial
nerve pass beneath the fibrous arch
from which soleus arises.

- **Insertion:** Tendo calcaneus and
posterior calcaneus.

DEEP GROUP

Muscles of the deep posterior compartment include a weak flexor of the
knee and three plantar flexors of the
foot and toes.

Popliteus

Popliteus is thin, flat, and triangular and
lies on the floor of the popliteal fossa.

- **Origin:** Medial two thirds of the
posterior shaft of the tibia above
the soleal line. Fibers run supero-
laterally to form a tendon that

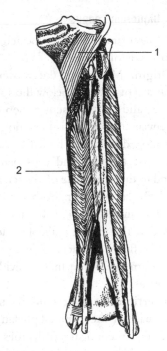

FIG. 51-3 Muscles of the calf. (1) Popliteus m;
(2) Flexor digitorum longus m.

grooves back of the lateral
condyle of the femur, and pierces
the capsule.

- **Insertion:** Sulcus below the lateral
femoral epicondyle and back of the
lateral meniscus.

- **Action:** Flexes the leg and rotates it
medially. With the foot on the
ground, the muscle laterally rotates
the femur unlocking the extended
knee to begin flexion of the leg.
A bursa beneath popliteus communicates with the knee joint.

- **Nerve supply:** Tibial nerve (L4,
L5, S1).

Flexor Digitorum Longus

Flexor digitorum longus is located on the tibial side of the leg.

- **Origin:** Middle of the posterior surface of the tibia below the soleal line and intermuscular septum between it and tibialis posterior. Its tendon descends behind the distal part of the tibia and runs behind and to the lateral side of the tendon of tibialis posterior. Both tendons curve into the sole by passing behind the medial malleolus, deep to the flexor retinaculum in a common groove but in their individual sheaths.
- **Insertion:** As four tendons into the base of the terminal phalanges of the second to fifth toes cf flexor digitorum profundus of the upper limb also giving origin to lumbricals.
- **Action:** Flexes the distal phalanges of the lateral four toes and assists in plantar flexion and inverstion of the foot.
- **Nerve supply:** Tibial nerve (S1, S2).

Quadratus plantae is inserted into the tendon of flexor digitorum longus in the sole of the foot near the point of division of the muscle into individual tendons to the digits.

Flexor Hallucis Longus

Flexor hallucis longus lies on the fibular side of the leg.

- **Origin:** Lower two-thirds of the posterior surface of the fibula and posterior intermuscular septum between it and posterior tibial and fibular muscles.

Its tendon passes behind the ankle joint, along a groove on the posterior surface of the tibia then on the back of the talus then beneath the sustentaculum tali. Beneath the flexor retinaculum, it is enclosed in a separate synovial sheath. It enters and crosses the sole of the foot diagonally above flexor digitorum longus.

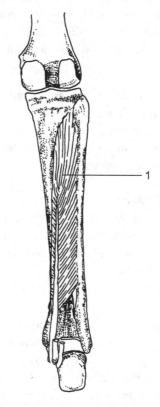

FIG. 51-4 Muscles of the calf (3). (1) Tibialis posterior m.

- **Insertion:** Base of the distal phalanx of the hallux.
- **Action:** Flexes the hallux and plantar flexes and inverts the foot. With the foot on the ground, it is a powerful "push off" muscle putting spring into the action of the foot.
- **Nerve supply:** Tibial nerve (S1, S2).

Tibialis Posterior

Tibialis posterior is the deepest muscle of this group lying between flexor digitorum longus and flexor hallucis longus.

- **Origin:** Lateral half of the back of the tibia below the soleal line; all of the posterior surface of the interosseous membrane except the lowest part; and upper two thirds of the medial surface of the fibula. Muscle fibers have a bipennate pattern. Its tendon descends medially to a groove on the back of the medial malleolus and enters the sole of the foot under the flexor retinaculum above the sustentaculum tali.
- **Insertion:** Mostly to the tuberosity of the navicular.
- **Action:** Inversion and plantar flexion of the foot. With the foot on the ground, the muscle maintains the longitudinal and transverse arches of the foot through continuity with ligaments to the cunieforms cuboid, sheath of fibularis longus and bases of the second to fourth metatarsals.
- **Nerve supply:** Tibial nerve (L5, S1).

SYNOVIAL SHEATHS

There are three sheaths behind the medial malleolus.

Sheath of tibialis posterior located immediately against the posterior and inferior surface of the medial malleolus extends from above the flexor retinaculum to just below the flexor retinaculum.

Sheath of flexor hallucis longus extends from a little above the flexor retinaculum to the middle of the sole.

Sheath of flexor digitorum longus begins just above the flexor retinaculum and extends onto the sole to where the tendons of flexor digitorum longus and flexor hallucis longus cross.

While separate in their fibrous compartments, the sheaths may communicate with each other below the flexor retinaculum.

3. Deep Arteries

Anterior tibial artery is a terminal branch of the popliteal artery at the lower border of popliteus. It gives rise to the circumflex fibular artery and the posterior tibial recurrent artery then passes to the front of the leg through a gap at the upper border of the interosseous membrane (between the origins of tibialis posterior) to reach the anterior compartment. It becomes the dorsalis pedis in front of the ankle.

Posterior tibial artery is the continuation of the popliteal artery beginning with the division of the popliteal

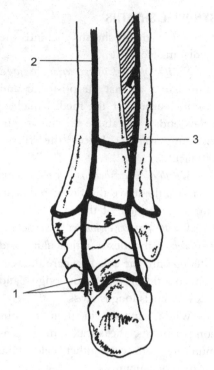

FIG. 51-5 Arterial anastomoses around the ankle. (1) Lateral and medial plantar a; (2) Posterior tibial a; (3) Fibular a.

branches supply to the skin on the medial side of the leg. Nutrient branch enters the nutrient foramen of the tibia. Communicating branch anastomoses with the communicating branch of the fibular artery. Medial calcaneal branch supply to the region behind the tendo calcaneus and around the heel. Terminal branches supplies lateral and medial plantar arteries.

Fibular artery is the largest branch of the posterior tibial artery arising a little below the origin of the posterior tibial artery. It descends obliquely and laterally behind soleus towards the fibula passing behind the ankle joint to terminate as a number of lateral calcaneal branches.

Branches of the fibular artery are:
Muscular branches to the muscles attached to the fibula (soleus, tibialis posterior, flexor hallucis longus, fibularis longus and brevis).

Nutrient branch enters the nutrient foramen of the fibula (passing downwards).

Perforating branch runs forward just above the inferior tibiofibular joint to the front of the leg.

Communicating branch begins at the same level as the perforating branch and anastomoses with the communicating branch of the posterior tibial artery.

Lateral calcaneal branches ramify behind the lateral malleolus anastomosing with the medial calcaneal branches of the posterior tibial artery.

artery at the lower border of popliteus. It descends through the fibrous arch in the origin of soleus and accompanied by the tibial nerve, descends in the deep posterior compartment. It passes behind the medial malleolus and under the flexor retinaculum into the sole to divide into its terminal branches (medial and lateral plantar arteries).

Muscular branches supply soleus and the deep flexors (flexor hallucis longus, popliteus, flexor digitorum longus and tibialis posterior). Cutaneous

The fibular artery is developmentally the major artery of the leg and is a large collateral vessel communicating distally with the anterior tibial artery.

4. Deep Nerves

Tibial nerve arises from anterior branches of L4, L5 and S1, S2 and S3 as a component of the sciatic nerve. In the thigh, it supplies the hamstring muscles. In the popliteal fossa, it gives rise to articular and muscular branches and contributes to the medial sural cutaneous nerve. In the leg, it continues through the popliteal fossa beneath the tendinous arch of soleus then beneath the transverse intermuscular septum overlying tibialis posterior. It terminates beneath the flexor retinaculum by dividing into medial and lateral plantar nerves.

Muscular branches innervate soleus and the deep flexors, tibialis posterior, flexor digitorum longus and flexor hallucis longus.

Medial calcaneal nerves arise in the lower part of the leg and innervate the skin of the heel and posterior sole of the foot.

Articular branches penetrate the deltoid ligament and innervate the ankle joint.

Terminal branches gives rise to medial and lateral plantar nerves.

Chapter 52

Sole of the Foot

Chapter Outline

1. Fascia
2. Deep Layers of Sole
3. Arteries
4. Nerves

The sole or plantar surface of the foot extends from the heel to the tips of the digits. Skin is thin on the toes and over the instep but thick over the heel and heads of the metatarsals.

1. Fascia

SUPERFICIAL

There is a considerable amount of fat in the subcutaneous tissue enclosed between tough, fibrous connective tissue strands that extend from the skin to the deep fascia. It is especially thick on the heel, the lateral margin of the foot, the ball of the foot, and the ends of the toes. The fascia also contains the medial calcanean nerves and arteries, lateral calcaneal branch of the sural nerve and cutaneous branches of the medial and lateral plantar nerves and arteries.

DEEP (PLANTAR FASCIA)

This fascia is thin on the medial and lateral sides of the sole of the foot where it covers the abductors of the large and small toes. It is thickened in the center (to form the plantar aponeurosis) that helps to maintain the longitudinal arch of the foot, and over the toes where it forms the fibrous flexor sheaths which retain the long digital flexors.

Plantar aponeurosis is a very thick layer of the deep fascia of the sole covering and giving attachment to flexor digitorum brevis. It is attached proximally to the medial process of the tuberosity of the calcaneus. In front, it widens and splits into five digital slips, one for each toe, near the heads of the metatarsals. Transverse fibers (the deep transverse metatarsal ligaments) interconnect these slips. Each slip divides into a thin superficial layer that terminates in the skin crease between the sole

and the toes and a deep layer inserting into the base of the proximal phalanges of the second to fifth toes.

Septa arise from the plantar aponeurosis dividing the sole into muscular compartments, the compartment of the big toe, compartment of the small toe and the central compartment.

Fibrous flexor sheaths are canals formed from the deep fascia of the toes containing the tendinous terminations of short and long flexor muscles. Sheaths are attached to the margins of the plantar ligaments of the joints, margin of the proximal and middle phalanges, to the plantar surface of the distal phalanges beyond the insertion of the flexor tendons, forming osseoaponeurotic canals. Opposite the bodies of the phalanges, the sheaths are strong and have transverse fibers (the annular part). Opposite the interphalangeal joints, they are thin and decussate (the cruciform part). The outer layer of the synovial sheaths that envelope the tendons serve as linings of the fibrous sheaths.

2. Deep Layers of Sole

Intramuscular septa pass upward into the sole from both the medial and lateral margins of the plantar aponeurosis separating flexor digitorum brevis and divide the sole into three compartments. Flexor digitorum brevis is located centrally below the plantar aponeurosis.

The sole is described in terms of four layers of muscles and tendons.

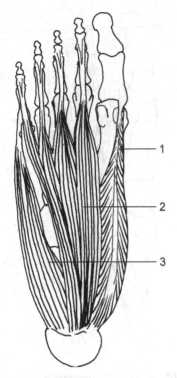

FIG. 52-1 Sole of foot — first layer of muscles. (1) Abductor hallucis m; (2) Flexor dogitorum brevis m; (3) Abductor digiti minimi brevis m.

SUPERFICIAL LAYER

The superficial (first) layer comprises the short common digital flexor (flexor digitorum brevis) and special abductors on each side (abductor hallucis and abductor digiti minimi).

Abductor Hallucis

This muscle covers the long tendons that enter the sole.

- **Origin:** Front of the medial tubercle of the calcaneus, plantar aponeurosis, and flexor retinaculum.

- **Insertion:** Medial sesamoid and base of the proximal phalanx of the big toe.
- **Action:** Aids in abduction and flexion of the hallux. It helps to maintain the longitudinal arch of the foot.
- **Nerve supply:** Medial plantar nerve (S1 and S2).

Abductor hallucis is known as the "gateway to the sole" through which the medial and lateral plantar nerves and vessels and tendons of the long digital flexors and of tibialis posterior pass to enter the sole.

Flexor Digitorum Brevis

Flexor digitorum brevis lies deep to the plantar aponeurosis.

- **Origin:** Summit of the medial process of calcaneus. At about level with the middle of the metatarsus, the muscle divides into four fleshy bellies to the four lateral toes. Each tendon enters a fibrous flexor sheath and splits to embrace a long flexor tendon (cf tendons of flexor digitorum superficialis in the fingers).
- **Insertion:** The split tendons reunite on its deep surface to be inserted into the margins of the middle phalanx of the lateral four toes.
- **Action:** Flexes the middle and proximal phalanges of the lateral four toes and helps to maintain the longitudinal arch of the foot.
- **Nerve supply:** Medial plantar nerve (S1 and S2).

Abductor Digiti Minimi

The muscle is lateral to the short common flexor.

- **Origin:** Both medial and lateral processes of the calcaneus and the plantar aponeurosis.
- **Insertion:** Lateral side of the base of the proximal phalanx of the little toe.
- **Action:** Abducts and flexes the little toe and helps to maintain the longitudinal arch of the foot.
- **Nerve supply:** Lateral plantar nerve (S1 and S2).

Three muscles of the first layer arise from the medial front of the tubercle of the calcaneus — the short common digital flexor and special abductors on each side.

MIDDLE LAYER

The middle (second) layer of muscles includes long digital flexors and associated muscles (and structures on the superficial neurovascular plane).

Flexor digitorum longus (tendon) passes behind the medial malleolus to enter the sole lateral to the tendon of tibialis posterior. It crosses the sole obliquely superficial to the tendon of flexor hallucis longus muscle, from which it receives a fibrous slip.

Flexor digitorum longus receives the insertion of the quadratus plantae muscle and divides into its four digital tendons, which enter fibrous flexor sheaths. Each tendon is enclosed in a common synovial sheath together with

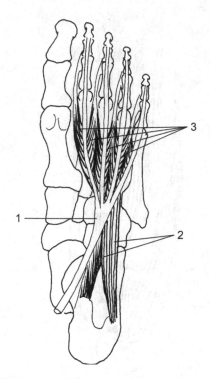

FIG. 52-2 Sole of foot — second layer of muscles. (1) Flexor digitorum longus m; (2) Quadratus plantae m; (3) Lumbrical m.

the tendon of the flexor digitorum brevis that it pierces to reach its insertion into the base of the distal phalanx (cf flexor digitorum profundus).

Quadratus Plantae

Quadratus plantae is a quadrangular, shaped accessory muscle to flexor digitorum longus serving to correct its oblique pull on the lateral four toes.

- **Origin:** *Medial head*: Muscular from the medial surface of the tuberosity of the calcaneus and medial border of the long plantar ligament;

Lateral head: By a tendon from the lateral edge of the tuberosity of the calcaneus.

- **Insertion:** Lateral margin of the tendon of the flexor digitorum longus muscle.
- **Action:** Assists flexor digitorum to flex the toes at all of their joints and helps to correct its oblique pull on the lateral four toes.
- **Nerve supply:** Lateral plantar nerve (S1 and S2).

Lumbricals

The four lumbricals are fusiform muscles are numbered from medial to lateral.

- **Origin:** (Except the first) Adjacent sides of the tendons of the flexor digitorum longus muscle. The first arises from the medial side of the tendon fro the second digit. The others lie between the flexor tendons. They pass forward below the deep transverse metatarsal ligament that separates them from the interosseous muscles.

Each lumbrical enters the corresponding toe from the side of the hallux (cf palmar lumbricals enter from the side of the pollex).

- **Insertion:** Medial side of the dorsal digital expansions.
- **Action:** Flexes proximal phalanx extends middle and distal phalanges i.e. they prevent the terminal phalanges from buckling under the pull of flexor digitorum longus (straighten the toes).

- **Nerve supply:** Medial plantar nerve to the first lumbrical and lateral plantar nerve to the second, third, and fourth.

The lateral three lumbricals are bipennate and arise from adjacent tendons but the medial lumbrical is unipennate and arises from the medial side of the most medial tendon (to the second toe). The three lateral lumbricals are supplied by the lateral plantar nerve and the medial lumbrical is supplied by the medial plantar nerve.

Flexor hallucis longus (tendon) enters the sole from the back of the leg after descending in the most lateral osseofascial compartment of the flexor retinaculum. It crosses deep to the tendon of the flexor digitorum longus and continues forward to enter the fibrous flexor sheath of the big toe and inserts into the base of its distal phalanx.

- **Nerve supply:** Tibial nerve (S1 and S2).
- **Action:** Flexes the big toe. With the foot fixed, it raises the heel in walking. It also maintains the longitudinal arch of the foot.

THIRD LAYER

Deep (third) layer contains a special adductor of the hallux and short special flexor on each side.

Flexor hallucis brevis a two bellied muscle on the plantar surface of the first metatarsal.

- **Origin:** Medial part of the plantar surface of the cuboid bone and from slips of tibialis posterior muscle that reach the cuboid and lateral cuneiforms.

- **Insertion:** Medial belly: Medial sesamoid and medial side of base of the proximal phalanx of the hallux. Lateral belly: Lateral side of the base of the distal phalanx of the hallux.

Flexor hallucis longus passes between the two bellies to reach the big toe.

- **Action:** Flexion of the proximal phalanx of the big toe. It helps to maintain both medial and lateral longitudinal arches of the foot.
- **Nerve supply:** Medial plantar nerve (S1 and S2).

Adductor hallucis has two heads of origin (cf adductor pollicis in the hand) fuse before they insert.

- **Origin:** *Oblique head*: Bases of second, third and fourth metatarsals and the sheath of the tendon of fibularis longus (distal attachment of the long plantar ligament); *Transverse head*: Plantar surface of the deep transverse metatarsal ligament.
- **Insertion:** The heads converge and insert together into the lateral side of the base of the proximal phalanx of the big toe.
- **Action:** Adducts all of the toes together and flexes the big toe. It aids in maintaining arches of the foot.
- **Nerve supply:** Lateral plantar nerve (deep branch, S1 and S2).

Flexor digiti minimi brevis covers the plantar surface of the fifth metatarsal.

- **Origin:** Base of the fifth metatarsal and sheath of the tendon of fibularis longus.
- **Insertion:** Lateral side of the base of the proximal phalanx of the fifth toe.
- **Action:** Flexes the proximal phalanx of the little toe.
- **Nerve supply:** Lateral plantar nerve (superficial branch, S1 and S2).

DEEPEST LAYER

Deepest (fourth) layer contains a dorsal four and plantar three interossei and

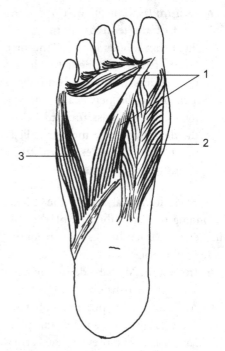

FIG. 52-3 Sole of foot — deep layer of muscles. (1) Adductor hallucis m. transverse and oblique heads; (2) Flexor hallucis brevis m; (3) Flexor digiti minimi brevis m.

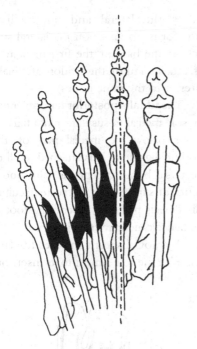

FIG. 52-4 Dorsal interossei muscles.

- **Origin:** Two heads from adjacent sides of the metatarsals between which they lie. The tendons pass above (deep to) the deep transverse metatarsal ligament.
- **Insertion:** Medial side of the base of the proximal phalanx of the second toe, lateral side of the second, third, and fourth toes (the base of the respective proximal phalanx on the side appropriate for abduction).
- **Action:** Abduct the middle three digits from the longitudinal axis of the second toe (the hallux and little toe have special abductors). Their attachment to the extensor expansion is too flimsy to enable them to extend the toes at the interphalangeal joints.

Plantar Interossei

Plantar interossei slightly more superficial. There are only three muscles that adduct the lateral three toes toward the long axis of the second toe.

- **Origin:** Medial side of the (lateral) third, fourth, and fifth metatarsals.
- **Insertion:** Lateral sides of the base of the proximal phalanges of the same toes.
- **Nerve supply:** Lateral plantar nerve (S1 and S2).
- **Action:** Adduct lateral three digits towards the longitudinal axis of the second toe (none are inserted into the second toe). They hold the metatarsals together and strengthen the metarsal arch. Their attachment to the extensor expansion is too

tendons of fibularis longus and tibialis posterior.

There are seven interossei muscles: four dorsal and three plantar. Their tendons pass above the deep transverse ligaments and insert into the bases of the proximal phalanges. They are innervated by the lateral plantar nerve.

Interossei of this layer are similar to those of the fourth layer of the hand but their action is relative to the second toe contrasting action relative to the third finger in the hand. Four dorsal interossei abduct from the line of the second toe.

Dorsal Interossei

Dorsal interossei are bipennate muscles.

flimsy so that they do not extend the toes at interphalangeal joints.

At the ankle, the **tendon of fibularis longus** lies posterior to that of fibularis brevis. It curves from the lateral side of the leg behind the lateral malleolus then below it and then below the fibular trochlea. Its tendon turns around the lateral side of the foot in a groove on the cuboid and passes medially forward in the fibrous canal formed by the groove on the plantar surface of the cuboid and a "sheath" derived from the long plantar ligament. It continues across the lateral and intermediate cuneiforms to insert on the lateral surface of the base of the first metatarsal (cf attachment of the tendon of tibialis anterior to the same bones).

The **tibialis posterior tendon** enters the sole medial to sustentaculum tali and passes below the medial part of the "spring" ligament. It is attached to all of the tarsal bones except the talus and a recurrent slip reaches the sustentaculum tali. Tibialis posterior inverts the foot at the talocalcaneonavicular joint.

A synovial sheath encloses fibularis longus in the sole almost to its insertion.

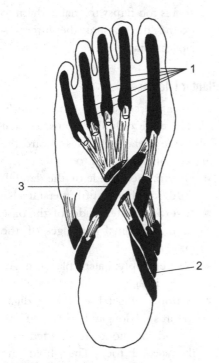

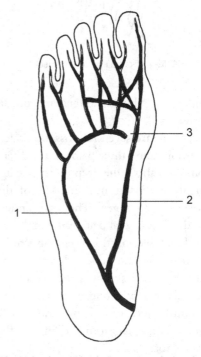

FIG. 52-5 Synovial sheaths in the sole of the foot. (1) Digital sheaths; (2) Sheath of flexor hallucis longus m; (3) Sheath of fibularis longus m.

FIG. 52-6 Arteries of the sole of the foot. (1) Lateral plantar a; (2) Medial plantar a; (3) Deep plantar branch of dorsalis pedis a.

On the medial side of the sole, there are three sheaths behind the medial malleolus, one each for tibialis posterior, flexor digitorum longus and flexor hallucis longus. In the toes, sheaths enclose the long and short flexor tendons beginning near the middle of the metatarsals (distal to the attachment of the lumbricals) and extending to the bases of the terminal phalanges where the tendons insert.

Vinculae, are vascular folds of synovial sheaths (mesotendons) connecting the visceral and parietal layers between the tendons and the proximal phalanges. The triangular vincula breve are located near the insertion of each tendon. Vinculae longa connect the layers opposite the proximal phalanges. Blood vessels run in the vinculae to supply the tendons.

3. Arteries

The arteries of the sole are terminal branches of the **posterior tibial artery**.

The medial and lateral plantar arteries course between the muscles of the first and second layers (the superficial neurovascular plane). The deep branches of the lateral plantar artery course medially between the muscles of the third and fourth layer (the deep neurovascular plane).

Medial plantar artery is the smaller terminal branch of the posterior tibial artery that passes forward (with the medial plantar nerve) on the medial side of the foot to the base of the first metatarsal where it anastomoses with the digital branch of the first metatarsal artery. It then follows the medial side of the hallux anastomosing with the first plantar metatarsal artery.

The artery provides three superficial digital branches, that accompany corresponding branches of the medial plantar nerve between the three medial toes.

Muscular branches supply the muscles of the medial side of the sole.

The medial plantar artery does not usually form an arch analogous to the superficial arterial arch of the palm.

Lateral plantar artery is the larger terminal branch of the posterior tibial artery passes obliquely forward from its point of origin under the flexor retinaculum to the base of the fifth metatarsal bone (cf ulnar artery in the palm). It curves medially at the base of the fifth metatarsal to a deeper plane (between the third and fourth muscle layers) as the plantar arterial arch with the deep branch of the lateral plantar nerve and terminates at the base of the first metatarsal. It anastomoses with the deep plantar branch of the dorsalis pedis artery.

Branches of the plantar arterial arch are muscular branches supply adjacent muscles and articular (recurrent) branches that supply tarsal joints.

Three posterior perforating branches ascend through the proximal end of the lateral three interosseous spaces. They anastomose with the corresponding branches from the dorsal metatarsal arteries of the arcuate artery on the dorsum of the foot.

Four plantar metatarsal arteries arise from the plantar arterial arch. The second, third and fourth arteries pass forward on the plantar surface of the interossei between the metatarsal bones. Each provides an anterior perforating branch and divides into a pair of plantar digital arteries. The anterior perforating branches ascend through the distal end of the intermetatarsal spaces of the lateral four digits to anastomose with the corresponding branch from the dorsal metatarsal artery.

Plantar digital branches supply adjacent sides of the lateral four toes.

Proper plantar digital branches to the lateral side of the fifth toe, arise where the lateral plantar artery becomes the plantar arch.

First plantar metatarsal artery arises at the junction of the deep plantar branch of the dorsalis pedis artery and the plantar arch. At the first cleft, it provides a plantar digital artery to the medial side of the digit and divides into a pair of plantar digital arteries for the adjacent sides of the first and second toes.

All the digital arterial branches lie dorsal to their corresponding nerves. They supply the plantar surfaces and sides of the digits and form plexuses in their digital pads.

4. Nerves

Medial plantar nerve is the larger terminal branch of the tibial nerve. Its distribution to the skin and nail beds and distribution to four muscles, two deep (the first lumbrical and flexor hallucis brevis) and two superficial (flexor digitorum brevis and abductor hallucis) is similar to the distribution to analogous structures in the palm by the median nerve.

The medial plantar nerve arises beneath the flexor retinaculum then deep to abductor hallucis and passes forward between abductor hallucis and flexor digitorum brevis divide opposite the bases of the metatarsals into a proper plantar digital nerve to the hallux and three common digital nerves.

Cutaneous branches innervate the skin on the medial side of the sole.

Muscular branches innervate the abductor hallucis and flexor digitorum brevis.

Articular branches innervate the tarsal and metatarsal joints.

Proper digital nerve supplies a muscular branch to the flexor hallucis brevis muscle, pierces the plantar aponeurosis, and descends to the medial side of the foot and big toe to supply the skin.

Three common plantar digital nerves pass between the processes of the plantar aponeurosis. Each splits into two proper digital nerves.

First common digital nerve provides a muscular branch to the first lumbrical and supplies the skin on adjacent sides of the first and second toes.

Second common digital nerve supplies the skin on the adjacent sides of the second and third toes.

Third common digital nerve gives off a communicating branch to the lateral plantar nerve and supplies the skin on the adjacent sides of the third and fourth toes.

Lateral plantar nerve is the other smaller terminal branch of the tibial nerve. Its cutaneous distribution to the lateral one and a half digits and all of the muscles supplied by the median nerve including all seven interossei and adductor hallucis. It resembles the distribution of the ulnar nerve to analogous structures in the palm.

The lateral plantar nerve enters the sole passes obliquely forward, between the first and second muscle layers, between flexor digitorum brevis and quadratus plantae (between first and second muscular layers), and terminates by dividing into superficial and deep branches.

Cutaneous branches innervate the skin on the lateral side of the sole.

Muscular branches innervate the quadratus plantae and abductor digiti minimi muscles.

Articular branches innervate the calcaneocuboid joint.

Superficial branch splits into two rami:

Medial branch (the fourth common digital nerve) provides a communicating branch to the medial plantar nerve and in turn divides into two proper digital nerves that supply the adjacent sides of the fourth and fifth toes.

Lateral branch provides a proper digital branch to the skin of the lateral side of the fifth toe and muscular branches to the flexor digiti minimi, third plantar, and fourth dorsal interossei.

Deep (muscular) branch curves across the base of the metatarsal bones together with the plantar arch (between the third and fourth muscular layers). They supply the three lateral lumbricals, interossei of three medial spaces, and both the heads of adductor hallucis.

The articular branches supplies the tarsometatarsal and intermetatarsal joints.

Chapter **53**

Bones and Joints
of Lower Extermity

Chapter Outline

1. Hip Bone
2. Thigh
3. Leg
4. Foot

1. Hip Bone

See Chapter 42.

2. Thigh

FEMUR

The femur or thigh bone is the longest and strongest bone in the body. It consists of a shaft with two ends, an upper, proximal end, and lower or distal end.

Proximal End

Proximal end consists of a smooth, rounded, head connected with the

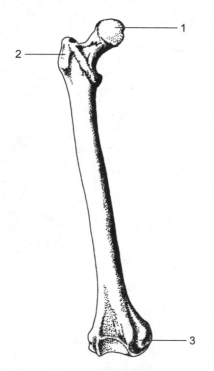

FIG. 53-1   Right femur — anterior aspect. (1) Head; (2) Greater trochanter; (3) Medial epicondyle.

shaft by a neck prominent greater and lesser trochanters (Gk, a runner) at the junction of the neck and shaft are associated with muscular attachments.

Head is smooth forming about two thirds of a sphere and faces upward, medially and slightly forward. It articulates with the acetabulum of the hip bone to form the hip joint and has a fovea below its center for attachment of the ligament of the head of the femur.

Neck is a thick bar connecting the head with the shaft in the region of the trochanters. It has numerous pits on its surface for the entrance of blood vessels. The neck subtends an angle of from 115° to 140° with the shaft and it points forward at about 8°.

Greater trochanter is a rough quadrilateral eminence that projects from the junction of the neck and the proximal end of the shaft. On its lateral surface, an oblique ridge runs diagonally from posterosuperior to anteroinferior. Gluteus medius is attached to the ridge and bursae beneath the tendon of gluteus medius and beneath gluteus maximus are related to anterior and posterior smooth areas on either side of the ridge. From the superior border of the greater trochanter, piriformis is attached to the upper superior border and obturator internus and the gemelli to a smaller depression in front. The greater trochanter is a traction epiphysis of its attached muscles.

Trochanteric fossa is a roughened depression on the inner (medial) surface line," below). The intertrochanteric line receives the anterior part of the capsule of the hip joint and the iliofemoral ligament.

The intertrochanteric crest is a thick, rough ridge on the posterior surface of the femur that joins the back of the greater trochanter with the lesser trochanter. It gives attachment to quadratus femoris (at the quadrate tubercle, a rounded prominence at about the center of the intertrochanteric crest).

Shaft

Shaft is smooth and cylindrical above but compressed from before backward in its lower third. It has is a gradual convex curvature forwards.

Linea aspera a thickened ridge (a buttress) of bone on the back of the middle third of the femoral shaft. It has medial, intermediate and lateral lips. Adductor longus, intermuscular septa and the short head of biceps are attached together to linea aspera.

Lateral lip extends from the gluteal tuberosity above to the lateral supracondylar line below. It gives origin for vastus lateralis, vastus intermedius and short head of biceps femoris are attached to the lateral lip.

Medial lip extends from the spiral line above to the medial supracondylar line below. Vastus medialis, adductor longus and adductor magnus are attached to the medial lip (and the spiral line).

Gluteal tuberosity is the upward prolongation from the lateral lip of linea aspera to the base of the greater

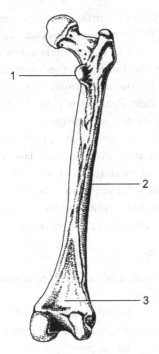

FIG. 53-2 Right femur — posterior aspect. (1) Lesser trochanter; (2) Linea aspera; (3) Popliteal surface.

of the greater trochanter. It gives attachment to obturator externus.

Lesser trochanter is a conical process that projects medially backwards from the junction of the medial part of the neck and the shaft. It gives attachment to the tendon of iliopsoas (it is a traction epiphysis).

Intertrochanteric line is a ridge on the anterior surface at the junction of the neck with the shaft. It descends obliquely from the greater trochanter into the linea aspera. The upper part of the line is rough, the lower part is less prominent (see "pectinate or spiral trochanter. It gives partial attachment to gluteus maximus.

Pectineal line is the upward prolongation of the intermediate lip of the linea aspera to the base of the lesser trochanter. It gives attachment to pectineus.

Spiral (pectineal) line is the upward continuation of the medial lip of the linea aspera. It winds around below the lesser trochanter to become continuous anteriorly with the intertrochanteric line. It gives attachment to vastus medialis.

Medial supracondylar line extends from the medial lip of linea aspera to the adductor tubercle. It is smooth above by femoral vessels passing between bone and adductor magnus to enter the popliteal fossa. It gives attachment to adductor magnus.

Lateral supracondylar line extends from the lateral epicondyle to the linea aspera. It gives attachment to short head of biceps and lateral intermuscular septum and distally, a small area for origin of plantaris.

Popliteal surface of the femur is the flattened, triangular area between the lateral and medial supracondylar lines. It forms the upper part of the floor of the popliteal fossa and is in contact with the popliteal artery. The medial head of gastrocnemius is attached to the popliteal surface just above the medial condyle and plantaris laterally above the lateral condyle.

Nutrient foramen is located on the linea aspera in the middle of the shaft. It is directed upward.

Anterior surface of the femoral shaft is smooth and convex. It gives attachment to vastus intermedius on its upper two thirds and the lower third is related to the suprapatellar pouch of the knee joint. Articularis genu is attached above the articular margin.

Lateral surface is separated from the anterior surface by the lateral border. It gives attachment to vastus intermedius in its upper two thirds.

Medial surface is covered by but not attached to vastus medialis.

Distal End

Distal end is broadened for articulation with the tibia.

Medial and lateral condyles are united anteriorly and laterally by a shallow articular depression, the patellar surface. The condyles bulge beyond the plane of the shaft posteriorly and are separated by the intercondylar notch. The lateral condyle is broader than the medial and the medial condyle is longer than the lateral condyle.

Intercondylar line is a ridge on the posterior surface that separates the popliteal surface of the shaft from the intercondylar fossa.

Medial and lateral epicondyles are the most prominent points that project from the sides of the condyles. The medial epicondyle gives attachment to the tibial collateral ligament and the lateral epicondyle gives attachment to the fibular collateral ligament.

Adductor tubercle is a pointed projection on the medial epicondyle at the termination of the medial lip of the linea aspera. It provides an attachment for the tendon of adductor magnus.

PATELLA

The patella is a large sesamoid bone developed in the tendon of quadriceps femoris.

Anterior surface is a convex surface roughened by vertical ridges from the insertion of tendon fibers.

Posterior (articular) surface is a smooth area subdivided by a vertical ridge into a large lateral and a smaller medial facet. The ridge occupies the groove on the patellar surface of the femur.

Base is the thick, superior border and gives attachment to the tendinous fibers of rectus femoris and vastus intermedius.

Apex is the inferior end of the patella giving attachment to the patellar ligament.

3. Leg

TIBIA

The tibia is the larger, medial weight bearing bone of the leg.

Proximal End

Proximal end is large and expanded more from side to side than from before backwards. The upper surface has two smooth, articular facets separated from each other by a wide irregular intercondylar area.

Medial condyle is an oval superior articular surface for articulation with the medial condyle of the femur. Its

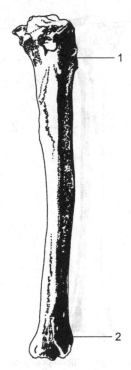

FIG. 53-3 Right tibia — lateral aspect. (1) Tuberosity; (2) Medial malleolus.

central area is slightly concave with a crescentic margin on which lies the medial semilunar cartilage. In front, medially and behind is a rough margin for attachment of the capsular ligament. The medial posterior margin of the condyle gives attachment to the oblique posterior ligament of the knee. The attachment of semimembranosus grooves the posterior and medial condyle.

Lateral condyle has a broader, shorter and more circular articular surface than the medial condyle. The lateral semilunar cartilage rests on the margin of the

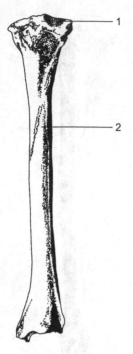

FIG. 53-4 Right tibia–posterior aspect.
(1) Intercondylar eminence; (2) Soleal line.

articular surface and the margin is rough
for attachment of the capsular ligament.

Anterior intercondylar area gives
attachment to the anterior extremities
of the medial and lateral menisci and
the anterior cruciate ligament.

Posterior intercondylar area is a
broad groove separating the posterior
aspects of the condyles. It gives attachment to the posterior cruciate ligament
and behind the eminence, it gives
attachment to the posterior extremities
of the medial and lateral menisci.

Intercondylar eminence is a projection between the articular surfaces divided
into two intercondylar tubercles (lateral

and medial) by a median notch. The
medial tubercle is the larger and more
prominent. Nothing is attached to the
tubercles.

Tibial tuberosity is situated anteriorly below the lower margins of the
condyles. Its upper part is smooth lying
deep to the subcutaneous infrapatellar
bursa and its lower part is rough for
attachment of the ligamentum patellae.

Shaft

Shaft is thick above and narrows distally.
It presents three borders and three
surfaces.

Anterior border is subcutaneous
and prominent and extending from the
tuberosity down to the anterior margin
of the malleolus.

Medial border is a blunt border
extending from the medial condyle to
the back of the medial malleolus below.

Interosseous border is a sharp,
prominent border extending from in
front of the facet for the head of the
fibula to the apex of the fibular notch.
The border gives attachment to the
interosseous membrane.

Medial surface is smooth and convex and bounded by the anterior and
medial borders. It is subcutaneous
except near its upper end where sartorius, gracilis, semitendinosus and medial
ligament of the knee joint are attached.

Lateral surface is bounded by the
medial and interosseous borders. It is
slightly concave above but becomes
convex below. There is a shallow groove
in its upper two thirds for attachment of

tibialis anterior. The lower part of the bone is covered in sequence from the medial side by tendons of tibialis anterior, extensor hallucis longus and extensor digitorum longus.

Posterior surface is smooth and flat and bounded by the anterior and interosseous borders.

Soleal line crosses the upper third of the posterior surface and extends obliquely down from the facet for the head of the fibula to the medial border. The triangular area above the line gives attachment to popliteus. The line gives attachment to soleus and popliteal fascia. Below the line, flexor digitorum longus and tibialis posterior are attached. A large nutrient foramen, directed downward, lies below the soleal line.

Vertical line extends distally from the soleal line, dividing the middle third of the posterior surface into lateral and medial parts. Tibialis posterior is attached lateral to the vertical line and flexor digitorum longus medial to the line.

Distal End

Distal end of the tibia projects medially and downward as the medial malleolus.

Medial malleolus forms the subcutaneous prominence on the medial side of the ankle. It is grooved on its posterior surface by the malleolar sulcus by the tendons of tibialis posterior and flexor digitorum longus (and flexor hallucis longus may produce a shallow groove lateral to the malleolar sulcus).

Fibular notch is a triangular groove on the lateral surface that receives the distal end of the shaft of the fibula. It is roughened by attachment of the interosseous ligaments uniting the tibia and fibula. The borders of the notch are sharp from attachment of the anterior and posterior tibiofibular ligaments.

Anterior surface is smooth and rounded. It is covered by extensor tendons above and rough for attachment of the anterior ligament of the ankle below.

Posterior surface is a rounded surface that is marked by a shallow oblique groove passing down and medially to the back of the medial malleolus. The groove accommodates the tendon of tibialis posterior.

Inferior articular surface is quadrilateral, wider in front than behind and concave anteroposteriorly. It is continuous medially with the malleolar articular surface and articulates with the body of the talus.

FIBULA

Fibula is the smaller lateral leg bone. The head of the fibula articulates with the lateral condyle of the tibia and the lower end of the fibula articulates with the fibular notch of the tibia to form the inferior tibiofibular joint. The lateral malleolus of the fibula articulates with the talus to form part of the ankle joint.

Proximal End

Proximal end has three major parts.

The head is the irregular, rounded upper aspect of the fibula. It has an articular facet superiorly for articulation with the lateral condyle of the tibia to form the superior tibiofibular joint. The posterior surface of the head is rough for attachment of soleus muscle.

Styloid process is the pointed process that projects upward from the posterolateral aspect of the head. It is the site of attachment of biceps femoris that splits to enclose the lateral ligament of the knee joint.

Neck is the constricted part of the shaft below the head.

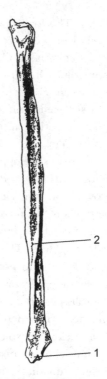

FIG. 53-6   Right fibula — lateral aspect. (1) Lateral malleolus; (2) Lateral surface.

FIG. 53-5   Right fibula — posterior aspect. (1) Apex; (2) Lateral malleolus.

Shaft

Shaft has three borders and three surfaces.

Interosseous border ascends from the apex of a triangular articular area. In the lower shaft, the line divides into an anterior line (the interosseous) border for attachment of the interosseous membrane. The posterior line is the medial crest that ends on the medial side of the head and gives attachment to a layer of deep fascia.

Anterior border is a sharp border that descends from the neck. It gives attachment to the superior extensor retinaculum at its lower end.

Posterior border is a blunt border that descends from the medial side of the head to the medial margin of the posterior surface of the malleolus.

Anterior (extensor) surface is the area between the anterior border and interosseous border. Extensor digitorum longus it attached to the upper three quarters, fibularis longus below and extensor hallucis longus to the middle half of the surface.

Lateral (fibular) surface is the continuation of the anterior surface of the shaft lying between the anterior and posterior borders. It may be grooved by the tendons of fibularis longus and brevis.

Posterior (flexor) surface between the interosseous and posterior borders begins at the neck and passes below to join the interosseous border. It is associated with flexor muscles.

Distal End

Lateral malleolus is the distal end of the fibula. It is more prominent than the medial malleolus of the tibia. It articulates with the lateral surface of the talus and a groove for fibularis longus and brevis muscle tendons as they pass into the foot.

Articular facet on the medial surface of the lateral malleolus is triangular for articulation with the lateral surface of the talus.

Malleolar fossa located behind the articular facet provides attachment for part of the lateral ligament of the ankle

joint (inferior transverse tibiofibular and posterior talofibular ligaments).

4. Foot

TARSUS

The posterior half of the foot consists of seven irregularly cuboidal tarsal bones arranged in two rows.

The talus lies immediately distal to the tibia and fibula by means of three surfaces of the trochlea, a part of the body of the talus. The inferior surface of the talus rests in part on the calcaneus, the bone of the heel. The head of the talus articulates with the concave posterior surface of the navicular.

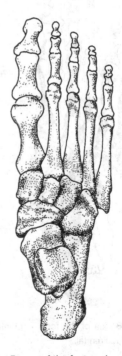

FIG. 53-7   Bones of the foot — dorsal aspect.

The navicular articulates with the anterior end of the talus and the three cuneiforms (medial, intermediate and lateral).

The *cuboid* articulates with the anterior surface of the calcaneus.

TALUS

Talus or ankle bone lies between the lower end of the tibia and upper surface of the calcaneus. No muscles or tendons are attached to the talus.

Head is the anterior, rounded end of the body of the talus that articulates with the navicular.

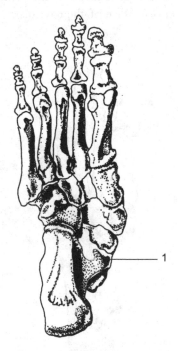

FIG. 53-8 Bones of the foot — plantar aspect; (1) Sustentaculum tali.

Neck is the constricted portion between the head and body. The anterior ligament of the ankle joint is attached to the anterior part of the neck.

Body of the talus is the cuboidal posterior part of the talus. It presents the *trochlea.*

Trochlear (dorsal) surface is the convex upper aspect of the body that articulates with the tibia. It is wider in front than behind, convex from before backwards and slightly concave from side to side. The posterior ligament of the ankle joint is attached just behind the trochlear surface.

Inferior (plantar) surface carries anterior, middle and posterior articular facets for articulation with the calcaneus to form the talocalcaneal joints.

Sulcus tali is the deep groove between the middle and posterior articular facets. It gives attachment to the interosseous talocalcaneal ligament.

CALCANEUS

Calcaneus is the largest of the tarsal bones. It is situated inferiorly and slightly laterally to the talus. It carries most of the body weight in the erect position.

FIG. 53-9 Bones of the right foot — medial aspect.

Superior (proximal) surface presents, from behind forward, a rough area covered by a fat pad; an oval posterior facet for articulation with the body of the talus; the sulcus calcanei; a long, narrow articular facet for articulation with the head of the talus; and a rough tubercle anterolaterally for attachment of the extensor digitorum brevis. The latter also gives attachment to the bifurcated ligament, inferior extensor retinaculum and inferior fibular retinaculum.

Inferior (plantar) surface is slightly concave with medial and lateral tubercles at its posterior end. The plantar aponeurosis, abductor hallucis and flexor digitorum brevis are attached to the medial tubercle. Flexor accessorius is attached in front of the lateral tubercle and abductor digiti minimi to both tubercles.

Posterior surface large and convex broader below. The lower half is rough providing insertion for the tendocalcaneus. The upper half is smooth contacting a bursa between it and the tendon.

Medial surface is concave. Sustentaculum tali projects from the anterosuperior aspect. It has a groove on its inferior surface for the tendon of flexor hallucis longus. The plantar calcaneonavicular (spring) ligament is attached to the anterior part of sustentaculum tali. The flexor retinaculum is attached to the edges of the groove for flexor hallucis longus and the medial ligament of the ankle is attached to the medial surface of the sustentaculum tali.

Anterior surface is small and articulates with the cuboid.

Sulcus calcanei is a rough elongated depression between the middle and posterior articular facets. It gives attachment to the interosseous talocalcaneal ligament.

Lateral surface is rough and subcutaneous. It has a small elevation for the calcaneofibular ligament. In front is the fibular tubercle separating a groove for fibularis brevis tendon above and fibularis longus tendon below. The superior and inferior fibular retinaculae are attached to the calcaneus near the fibular tubercle.

NAVICULAR

Navicular is roughly disc shaped and situated on the medial side of the foot between the talus posteriorly and the three cuneiform bones anteriorly.

Proximal surface carries a large concave facet for articulation with the head of the talus.

Distal surface has a slightly crescent facet divided by two vertical ridges into three facets for articulation with the medial, intermediate and lateral cuneiforms.

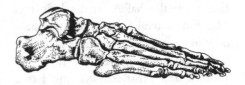

FIG. **53-10** Bones of the right foot — lateral aspect.

CUNEIFORM BONES

Cuneiform bones are wedge-shaped bones (medial, intermediate and lateral) situated between the navicular proximally and the three medial metatarsal bones distally. The lateral and intermediate bones have the broad surface of the wedge superiorly and the medial cuneiform is the largest, it has a rough and narrow dorsal surface. The medial and lateral cuneiforms project distally further than the intermediate cuneiform and form a socket for the base of the second metatarsal.

Cuboid lies on the lateral side of the foot between the calcaneus behind and fourth and fifth metatarsals in front. The plantar surface has a deep groove anteriorly for the tendon of fibularis longus (a sesamoid bone may develop in the tendon and slide on the facet). The long plantar ligament attaches to both lips of the groove converting it into a canal. The short plantar ligament attaches to the rough tuberosity behind the groove.

METATARSAL BONES

Five metatarsal bones are set side by side between the tarsal bones and the digits. They are numbered from one to five, from the hallux to the little toe. The four lateral metatarsals resemble one another but the metatarsal to the hallux differs considerably. It is the shortest and most massive and has a single facet for the medial cuneiform. Tibialis anterior is attached to the medial side and fibularis longus to the

lateral side of the base. It has two depressions on either side of the base for sesamoids in the tendon of flexor hallucis brevis and the first dorsal interosseous is attached to its lateral surface.

Head of metatarsals articulate distally with the phalanges to form the metatarsophalangeal joints. They are smaller than the bases of the metatarsals, compressed laterally and have a rounded articular surface that extends further on the plantar than on the dorsal side. On each side is a groove for the capsule and a tubercle for a collateral ligament. On the plantar side is a groove for the flexor tendons.

Shaft is concave in its plantar surface and convex on its dorsal surface.

Base of metatarsals of the lateral four are wedge shaped. The base of the fifth metatarsal has a projecting tuberosity for attachment of fibularis brevis.

PHALANGES

Phalanges of the foot are similar in arrangement to those in the hand, however the hallux has only two phalanges (proximal and distal) while the lateral four toes each have three (proximal, middle and distal).

Base of phalanx has a concave articular facet for the head of the corresponding metatarsal and on its plantar surface, a groove tor the flexor tendons.

Head has a pulley shaped articular surface for the base of the middle phalanx.

Shaft curved slightly with the concavity toward the sole.

Joints of Lower Extremity

Chapter Outline

The joints of the lower extremity are classified into five groups.

1. Hip Joint

Type

The hip joint is a spheroidal, synovial ball-and-socket joint.

Articulating Element

The globular shaped head of the femur articulates with the acetabulum of the hip bone. The depth of the socket is increased by the acetabular labrum, a fibrocartilaginous ring triangular in cross section that is attached to the rim of the acetabulum and the transverse ligament of the acetabulum.

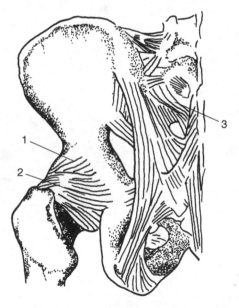

Ligaments

Transverse ligament of the acetabulum extends across the acetabular notch below, forming a foramen. Articular vessels and nerves enter the fossa under the ligament.

Articular capsule extends proximally from the rim of the acetabulum, and the transverse acetabular ligament to the intertrochanteric line anteriorly. Laterally, it reaches the middle of the neck of the femur. It is strong but loose below to allow movement. The capsule has both longitudinal and circular fibers. The circular fibers are deeper and form a collar, the zona orbicularis that clasp the free capsule close to the neck.

Synovial membrane lines the capsule forming a tubular sheath around the ligament of the head of the femur (see below) and covers the intracapsular part of the neck of the femur enveloping it up to the articular head. It also continues over both surfaces of the labrum and the pad of fat in the acetabular fossa.

Supporting Ligaments

Iliofemoral ligament is an inverted Y-shaped band lying in front of the hip joint with its apex at the anterior inferior iliac spine. Its two divergent limbs that are attached below to the intertrochanteric line. It becomes taut in full extension of the hip and helps to maintain upright posture.

Pubofemoral ligament lies on the medial and inferior capsule of the hip joint. It extends anteriorly from the pubic part of the acetabular rim and blends

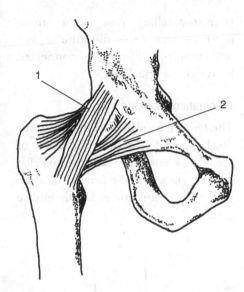

FIG. 53-12 Hip joint — anterior aspect. (1) Upper and lower bands of iliofemoral ligament; (2) Pubofemoral ligament.

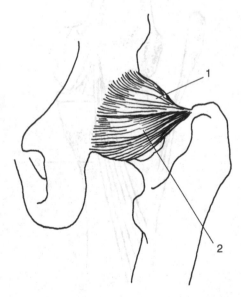

FIG. 53-13 Hip joint ligaments — posterior aspect. (1) Iliofemoral ligament; (2) Ischiofemoral ligament.

below with the fibrous capsule and iliofemoral ligament. The pubofemoral ligament becomes tight in extension and limits abduction of the hip joint.

Ischiofemoral ligament lies on the posterior aspect of the capsule. It extends posteriorly from the ischial part of the acetabulum. It spirals laterally and upward behind the neck of the femur and inserts into the superior part of the neck of the femur.

Ligament of head of femur is a flattened, triangular intracapsular ligament extending from its apex at the fovea of the head of the femur to its base at the transverse ligament and margins of the acetabular notch inferiorly. It is covered by a sleeve of synovial membranea and is taut in adduction of the hip. The ligament carries a branch of the obturator artery.

Movements

The hip joint allows movement in all possible directions.

Flexion: Psoas major, pectineus and rectus femoris (checked by contact with abdominal wall when knee is flexed or tension in hamstrings if knee is extended).

Extension: Gluteus maximus, hamstrings (checked by capsule, iliofemoral, pubofemoral and ischiofemoral ligaments and iliopsoas).

Abduction: Gluteus medius, gluteus minimus and tensor fascia lata (checked by tension in adductors).

Adduction: Adductors longus, brevis and magnus, pectineus and sartorius (checked by contact with opposite limb, tension in abductors when thigh is flexed).

Medial rotation: Tensor fascia lata, gluteus medius and minimus and adductors longus, brevis and magnus (checked by tension in lateral rotators).

Lateral rotators: Piriformis, quadratus femoris, obturators internus and externus, and gemelli (checked by tension in medial rotators).

Circumduction: A composite movement.

Blood Supply

Deep division of the superior and inferior gluteal arteries, medial and lateral femoral circumflex arteries. Metaphysial and epiphysial branches of the medial femoral circumflex artery are carried in retinacula of the neck to the head and neck of the femur. The posterior division of the obturator artery gives rise to the artery of the ligament of the head.

Nerve Supply

Branches from the femoral nerve (nerve to rectus femoris), anterior division of the obturator nerve, superior gluteal nerve, and nerve to quadratus femoris.

2. Knee Joint

Type

A synovial, condylar hinge joint composed of medial and lateral (tibiofemoral) condylar joints and a semiplane (femoropatellar) joint. The three joint cavities (two tibiofibular and one femoropatellar) are connected by restricted openings.

Articulating elements

The femoral condyles articulate with the tibial condyles. The posterior aspect of the patella articulates with the patellar articular surface of the femur. The articular surfaces are covered with hyaline cartilage.

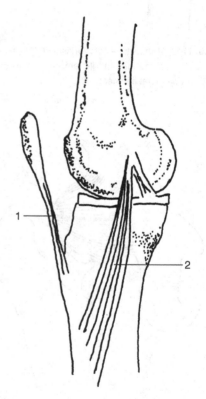

FIG. 53-14 Knee joint — medial aspect. (1) Ligamentum patellae; (2) Tibial collateral ligament.

Extracapsular (Collateral) Ligaments

Tibial collateral ligament a flat band extends from the medial epicondyle of the femur to the medial surface of the shaft of the tibia below the medial condyle. The points of attachment are slightly behind the vertical axis of the bones so that they become taut in extension and prevent hyperextension.

Fibular collateral ligament, a cordlike band, extends from a tubercle on the lateral condyle of the femur behind the groove for popliteus to the lateral surface of the head of the fibula. The tendon of popliteus passes deep to the ligament and the tendon of biceps splits on each side of its fibular attachment.

Medial and lateral patellar retinacula reinforce the capsule of the knee anteriorly. They are expansions from the medial and lateral vasti and insert into the front of the tibial condyles as far as the collateral ligaments.

Oblique popliteal ligament is a flat band on the lateral side of the knee joint. It is an expansion of the tendon of semimembranosus and inserts into the lateral condyle of the femur and reinforces the joint posteriorly.

Patellar ligament is a strong, thick band that extends from the inferior aspect of the patella to the tuberosity of the tibia. It is a continuation of the tendon of quadriceps femoris.

The posterior surface of the ligament is separated from the synovial membrane of the joint by the infrapatellar fat pad, and from the upper end

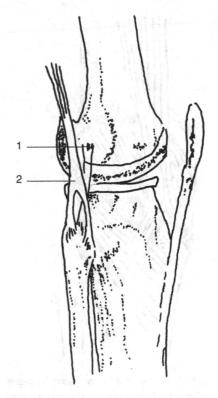

FIG. 53-15 Knee joint — lateral aspect. (1) Fibular collateral ligament; (2) Tendon of biceps femoris m.

of the tibia by the deep infrapatellar bursa.

Arcuate popliteal ligament strengthens the lower lateral part of the knee posteriorly. It passes from the back of the head of the fibula upward and medially over the tendon of popliteus and spreads over the back of the capsule of the joint.

Intracapsular Ligaments and Cartilages

The cruciate ligaments are placed in the vertical plane between the condyles.

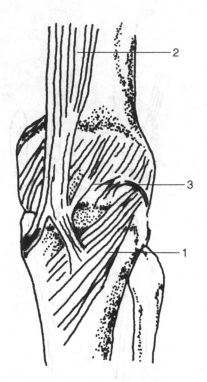

FIG. 53-16 Knee joint — posterior ligaments. (1) Tendon of popliteus m; (2) Semimembranosus m; (3) Oblique posterior ligament.

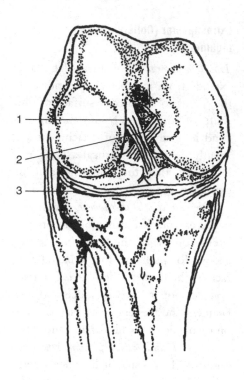

FIG. 53-17 Knee joint — cruciate ligaments. (1) Anterior cruciate ligament; (2) Posterior cruciate ligament; (3) Lateral meniscus.

They are strong rounded cords crossing one another like the limbs of an X. Their designation is taken from their relation to the intercondylar eminence of the tibia. They prevent anterior displacement (anterior ligament) or posterior displacement (posterior ligament) of the tibia and aid in the distribution of synovial fluid.

Anterior cruciate ligament extends from the anterior part of the intercondylar fossa of the tibia upward, backward, and laterally to the posterior part of the medial aspect of the lateral condyle of the femur.

Posterior cruciate ligament extends from the posterior part of the intercondylar fossa of the tibia upward, forward, and medially to the anterior part of the lateral surface of the medial condyle of the femur.

Medial and lateral menisci (or semilunar cartilages).

The menisci are crescentic fibrocartilages with flat lower surfaces and concave upper surfaces that deepen the articulating surface of the condyles and act as shock absorbers. Their peripheral margins are thick, convex, and attached

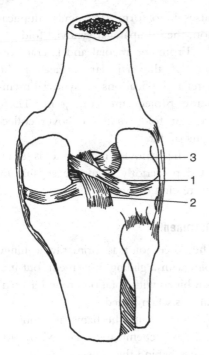

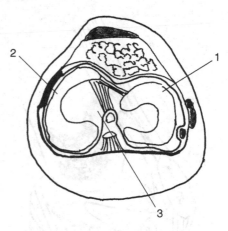

FIG. 53-19 Knee joint — horizontal section. (1) Lateral meniscus; (2) Medial meniscus; (3) Cruciate ligaments.

FIG. 53-18 Knee joint — posterior aspect. (1) Posterior meniscofemoral ligament; (2) Posterior cruciate ligament; (3) Anterior cruciate ligament.

to the capsule. The inner margins are thin, concave, and free. Their horns or fibrous ends, are attached to the anterior intercondylar area of the tibia in front and to the posterior end of the intercondylar eminence behind.

Medial meniscus is situated on the medial aspect of the upper end of the tibia. It is larger than the lateral meniscus, nearly oval in shape and is thin and pointed at its attachment. The peripheral border is attached to the capsule.

Lateral meniscus is situated on the lateral aspect of the upper end of the tibia. It is almost circular in shape and weakly attached to the margin of the lateral tibial condyle and is not attached where it crosses the tendon of popliteus.

Posterior meniscofemoral ligament is a fibrous band that extends from the posterior part of the lateral meniscus behind the posterior cruciate ligament to be inserted on the medial condyle of the femur. A comparable anterior ligament passing in front of the posterior cruciate ligament is sometimes present.

Transverse genicular ligament connects the anterior convex margins of the two menisci.

Articular capsule is a thin, strong, membrane attached just outside the margins of the femoral condyles laterally and above. It is attached to the intercondylar notch posteriorly. It is deficient anteriorly, where it is replaced by quadriceps femoris, the patella, and the patellar ligament.

Synovial membrane lines the articular capsule and is reflected onto the

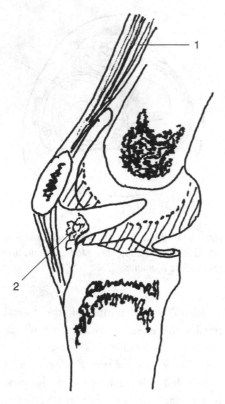

FIG. 53-20 Knee joint — sagittal section.
(1) Quadriceps tendon; (2) Infrapatellar fat pad.

bones, and extends as far as their articular margins. It leaves the capsule posteriorly to pass forward around the
cruciate ligaments that are outside of
the articular cavity between the synovial membrane and the back of the
capsule. It extends a short distance
upward onto the deep surface of the
quadriceps tendon from behind the
patella to line the suprapatellar bursa.
Below the patella, it covers the deep
surface of the infrapatellar fat pad that

raises it to form a triangular duplication, the infrapatellar synovial fold.

From the medial and lateral borders of the articular surface of the
patella, duplications of synovial membrane project into the joint. These
are alar folds that also cover collections of fat.

The synovial membrane is separated posteriorly from the capsule by
the tendon of popliteus.

Movements

The knee joint is primarily a hinge
joint with a gliding movement, but it is
combined with rotation around a vertical axis when flexed.

Flexors are the hamstring muscles
and gastrocnemius and limited by soft
tissues behind the knees.

Extension is performed by quadriceps and limited by tension in the
anterior cruciate ligaments, oblique
popliteal ligament and medial and
lateral ligaments. The terminal part of
extension is accompanied by medial
rotation of the femur on the tibia. As
the knee comes into full extension, the
anterior cruciate ligament becomes
taut terminating the extension of the
lateral condyle. The medial condyle
which has a longer and more curved
articular surface, continues to extend
and the lateral condyle rotates
forward.

Medial rotation of the femur on the
tibia tightens the oblique popliteal ligament. The lateral and medial ligaments

tighten simultaneously. This "screw home" movement "locks" the joint in a slightly hyperextended position. To unlock the knee into flexion, contraction of popliteus laterally rotates the femur before flexion can begin.

The patella maintains a shifting contact with the femur in all positions of the knee. It protects the front of the joint and increases the traction angle of the tendon of quadriceps.

Blood Supply

Blood supply to the knee joint is provided by the vessels that form the genicular anastomosis, chiefly the superior genicular branch of the femoral artery, genicular branches of the popliteal artery, recurrent branches of the anterior tibial artery and descending branch of the lateral femoral circumflex artery.

Nerve supply of the knee is provided by branches of the femoral nerve (to the vasti) and the saphenous nerve; branches of the tibial and common fibular nerves; and the genicular branch of the obturator nerve (Hilton's law).

Bursae communicating with the knee joint

- Suprapatellar (quadriceps) bursa.
- Popliteus bursa that often opens into the superior tibiofibular joint.
- Medial gastrocnemius bursa between the medial head of gasctrocnemius and the medial femoral condyle.

Bursae surrounding but not opening into the knee joint

- Prepatellar bursa, subcutaneous over the patella.
- Infrapatellar bursae, one subcutaneous over the patellar ligament and the other deep to the patellar ligament in front of the tibia.
- Beneath the insertions of sartorius, gracilis and semitendinosus (anserine bursa).
- Beneath semimembranosus tendon (often communicating with the knee through medial gastrocnemius bursa).
- Beneath lateral head of gastrocnemius.
- Between lateral ligament and biceps tendon.

3. Tibiofibular Joints

SUPERIOR TIBIOFIBULAR JOINT
Type

Synovial plane joint

Articulating Elements

Head of the fibula articulates with the lateral condyle of the femur.

Ligaments

Articular capsule is thicker in front and attached to the margins of the facets on the tibia and fibula. It is lined with a synovial membrane.

Anterior ligament of the head of the fibula is a few bands pass obliquely

from the front of the head of the fibula
to the front of the lateral condyle of the
femur.

Posterior ligament of the head of
the fibula is a broad band passes
obliquely from the back of the head of
the fibula to the back of the lateral
condyle of the femur.

Movement

The joint provides for a slight gliding
movement allowing some flexibility
between tibia and fibula in movements
at the ankle joint.

Blood Supply

It is provided by the lateral inferior
genicular and anterior recunent tibial
arteries (from the anterior tibial
artery).

Nerve Supply

Nerve to popliteus (from tibial) and the
recurrent (genicular) branch of the
common fibular nerve.

Interosseous membrane (middle
tibiofibular syndesmosis) extends
between the interosseous borders of
the tibia and fibula with fibers directed
down and laterally. Its upper margin has
a concave border and does not reach the
superior tibiofibular joint allowing passage of the anterior tibial vessels. Below,
it is continuous with the interosseous
ligament of the inferior tibiofibular
joint and is perforated by numerous
small vessels.

INFERIOR TIBIOFIBULAR JOINT

Type

Fibrous syndesmosis

Articulating Elements

Lower end of the fibula articulates with
the fibular notch of the tibia.

Ligaments

Anterior tibiofibuar ligament is a flat
band that passes obliquely from the
anterior border of the fibular notch of
the tibia to the lateral malleolus.

Posterior tibiofibular ligament is a
flat band that passes obliquely from the
posterior border of the fibular notch of
the tibia to the lateral malleolus.

Transverse tibiofibular ligament is a
wide band that crosses the back of the
joint. It extends from the inferior border of the posterior surface of the tibia
to the upper end of the malleolar fossa
of the fibula.

Interosseous ligament is a continuation of the interosseous membrane
consists of short thick bands extending
from the floor of the fibular notch of
the tibia to above the articular surface of
the lateral malleolus.

Movement

This joint accommodates the talus during ankle movement.

Blood Supply

Branches of the fibular artery and anterior and posterior tibial artery.

Nerve Supply

Deep fibular, tibial and saphenous nerves.

4. Ankle (Talocrural) Joint

Type

Synovial hinge joint

Articulating Elements

The malleoli at the distal end of the tibia articulate with the talus to form a mortise and tenon like joint (the trochlea of the talus is the tenon and the distal end of the medial malleolus of the tibia and lateral aspect of the medial malleolus of the fibula form the mortise). The transverse tibiofibular ligament completes the socket posteriorly.

Ligaments

Articular capsule is attached to the tibia and fibula above the malleoli, posterior and lateral margins of the articular surfaces of the talus and anteriorly to the neck of the talus. It is thin in front and behind but strengthened laterally by ligaments.

The synovial membrane lines the deep surfaces of the ligaments. The synovial cavity extends upward between the distal ends of the tibia and fibula into the inferior tibiofibular joint as far as the interosseous ligament.

Deltoid (medial collateral) ligament is a strong triangular ligament on the medial side of the joint attached by its blunted apex to the margins and tip of the medial malleolus above. Its broadened inferior attachment stretches from the neck of the talus along the tuberosity of the navicular and sustentaculum tali to the medial side of the body of the talus.

Three Collateral Ligaments

Anterior talofibular ligament is a short ligament extends forwards from the anterior border of the lateral malleolus to the lateral side of the neck of the talus.

Calcaneofibular ligament is a rounded cord extends downward and backward from the tip of the lateral malleolus to the middle of the lateral surface of the calcaneus.

Posterior talofibular ligament is a strong horizontal band extends from the floor of the malleolar fossa of the fibula to the posterior tubercle of the talus.

Movements

This joint provides for dorsiflexion (tibialis anterior) and plantar flexion (soleus and gastrocnemius). In dorsiflexion, the broader anterior part of the trochlea of the talus completely fills the mortise of the joint and in this position, stability of the joint is greatest.

Blood Supply

Anterior and posterior tibial and fibular arteries.

Nerve Supply

Tibial and lateral branch of the deep fibular nerves.

5. Joints of the Foot

INTERTARSAL JOINTS (7)

Intertarsal joints are all plane joints and have dorsal, plantar, and interosseous ligaments. They are lined with a synovial membrane and permit limited gliding. Their blood supply is from branches of the dorsalis pedis, medial and lateral plantar arteries. Their innervation is derived from the deep fibular, medial and lateral plantar nerves.

SUBTALAR (TALOCALCANEAL) JOINT

Subtalar is a clinical term for the composite of the anterior and posterior joints between calcaneus and talus (the talocalcanel and talocalcaneonavicular joints) where movement occurs between the talus and the remainder of the foot. It is more commonly referred to the posterior articulation behind the tarsal canal.

Articulating Elements

Concave facets on the inferior surface of the talus articulates with the convex posterior facet on the superior surface of the calcaneus.

Ligaments

Capsule attached to the margins of the articulating surfaces.

Medial talocalcaneal ligament connects the medial tubercle of the talus to the posterior part of sustentaculum tali.

Lateral talocalcaneal ligament connects the lateral process of the talus to the lateral surface of the calcaneus (and calcaneofibular ligament).

Posterior talocalcaneal ligament connects the lateral tubercle of the talus and upper medial part of calcaneus.

Interosseous talocalcaneal ligament is a transverse bundle located in the tarsal canal separating the anterior and posterior parts of the talocalcaneal joints. It connects the sulcus of the talus and the sulcus of the calcaneus.

Movements

The subtalar axis, passing through the long axis of the talus and about which the calcaneus rotates relative to the talus is about 45° to the floor and 16° medial to the long axis of the second metatarsal. Three types of movement occur in combination around this axis:

(a) *inversion* — elevation of the medial border of the foot and depression of the lateral border about the longitudinal axis and eversion — depression of the medial border of the foot and elevation of the lateral border about the longitudinal axis.

(b) *abduction* — outward rotation about a vertical axis through the tibia and adduction, inward rotation.

(c) *dorsiflexion and plantar flexion* — about a transverse axis but significantly less than between the tibia and the talus.

When these movements occur simultaneously, they result in supination of the foot (a combination of inversion, adduction and plantar flexion) and pronation (a combination of eversion, abduction and dorsiflexion).

All talocalcaneal ligaments are directed down and backwards and prevent backward displacement of the foot.

TALOCALCANEONAVICULAR JOINT
Type

Synovial ball and socket joint

Articulating Elements

The head of the talus (ball) articulates with a socket formed by two bones and two ligaments. The bones are the posterior surface of the navicular and the anterior articular surface of the calcaneus. The ligaments are the upper surface of the plantar calcaneonavicular ligament (spring ligament) and the bifurcate ligament.

Ligaments

Capsule is formed by the spring ligament below, deltoid ligament medially, interosseous ligament posteriorly, bifurcate ligament laterally and dorsal talonavicular ligament above.

Plantar calcaneonavicular ligament (spring ligament) bridges the gap between the sustentaculum tali and navicular on the plantar aspect of the talocalcaneonavicular joint. It extends from the sustentaculum tali and distal surface of calcaneus to the entire width of the inferior surface of the navicular. Medially, it blends with the deltoid ligament of the ankle and laterally with the lower part of the bifurcate ligament. Its upper surface is smooth and contains a fibrocartilaginous plate. It is

an important support for the head of the talus.

Calcaneocuboid part of bifurcate ligament is a thick dense fibroelastic ligament the medial limb of which extends from the sustentaculum tali to the navicular.

CALCANEOCUBOID JOINT
Articulating Elements

Facets on the anterior surface of the calcaneus articulate with a facet on the posterior surface of the cuboid.

Ligaments

The capsule is completely surrounds the joint cavity.

Plantar calcaneocuboid (short plantar) ligament is a short, wide and strong ligament attached to the tubercle on the anterior end of the calcaneus and adjoining cuboid.

Long plantar ligament is the longest tarsal ligament. It extends between the plantar surface of the calcaneus in front of the tuberosity to the tuberosity of the cuboid. Its superficial fibers spread forward to the bases of the third, fourth, and fifth metatarsals.

Bifurcated ligament is a strong Y-shaped ligament band. The stem of is attached to the dorsal surface of the calcaneus lateral to the facet for the head of the talus. The calcaneocuboid part is attached to the medial surface of the cuboid. The calcaneonavicular part extends to the lateral surface of the navicular.

Transverse tarsal joint is the irregular articular plane that extends across the width of the foot. It is composed of the talonavicular joint medially, and the calcaneocuboid joint laterally. Inversion and eversion of the foot take place primarily at this joint.

Additional ligaments unite the cuboid and navicular. These are the dorsal, plantar, and interosseous cuboideonavicular ligaments.

Three distal intertarsal joints are:

CUNEONAVICULAR JOINT
Articulating Elements

Facets on the anterior surface of the navicular articulate with facets on the posterior surface of the three cuneiform bones.

Ligaments

Dorsal and plantar cuneonavicular ligaments.

CUNEOCUBOID JOINT
Articulating Elements

A facet on the medial surface of the cuboid articulates with the posterior facet on the lateral surface of the lateral cuneiform.

Ligaments

Dorsal, plantar, and interosseous cuneocuboid ligaments.

INTERCUNEIFORM JOINTS

Intercuneiform joints provide for two articulations.

Articulating Elements

A facet on the medial side of the lateral cuneiform articulates with a facet on the lateral side of the middle cuneiform. A facet on the lateral side of the second cuneiform articulates with a facet on the medial side of the medial cuneiform.

Ligaments

Dorsal, plantar, and interosseous intercuneiform ligaments.

TARSOMETATARSAL JOINTS
Type

Plane type

Articulating Elements

The base of the first metatarsal articulates with the medial cuneiform; of the second metatarsal with all three cuneiforms; of the third metatarsal with the lateral cuneiform; of the fourth metatarsal with the cuboid and lateral cuneiform and of the fifth metatarsal with the distal surface of the cuboid.

Ligaments

Dorsal, plantar, and interosseous tarsometatarsal ligaments. The latter are strong and help to maintain the transverse arch of the foot.

Movement

Slight gliding.

INTERMETATARSAL JOINTS
Type

Plane type

Articulating Elements

Bases of the metatarsal bones.

Ligaments

Dorsal, plantar, and interosseous intermetatarsal ligaments. The first and second metatarsals are interconnected by interosseous fibers only.

Movements

Slight gliding that contributes to the flexibility of the foot.

METATARSOPHALANGEAL JOINTS
Type

Condyloid type

Articular Elements

Rounded heads of the metatarsals articulate with the cupped bases of the proximal phalanges.

Ligaments

Capsule is loose and attached closer to the articular borders dorsally than on the plantar side. Synovial membrane lines the non-articular surface.

Plantar ligaments are dense fibrocartilaginous plate attached firmly to the proximal base of each phalanx. All plantar ligaments are interconnected by the deep transverse metatarsal ligament at the heads and joint capsules of the metatarsals.

The plantar ligament of the first metatarsophalangeal joint is replaced by sesamoid bones in the tendon of the flexor hallucis brevis muscle.

Collateral ligaments extend from the tubercles on each side of the head of a metatarsal bone to the side of the base of the phalanx and to the margins of its plantar ligament. The extensor expansions on the dorsal surface take the place of the dorsal ligaments.

Movements

The joints permit dorsiflexion, plantar flexion, abduction, and adduction.

INTERPHALANGEAL JOINTS
Type

Hinge type

Articulating Elements

Heads of the proximal and middle phalanges articulate with the shallow cavities at the bases of the middle and distal phalanges.

Ligaments

The *capsule* is attached at the borders of the articular surfaces and completed dorsally by the extensor tendons.

The *plantar ligaments* (or plates) are thickened plantar segments of the capsule where the flexor tendons cross the joint. These plates are attached to the plantar edge of the base of the phalanx and slides like a visor over the joint in flexion.

The *collateral ligaments* are arranged like those of the metatarsophalangeal joints and strengthen the joint at the sides. They are attached eccentrically from a tubercle on the sides of the head of the proximal phalanx to a palmar plate and plantar edge of the base of the next distal phalanx. The collateral ligaments are therefore slack in extension but taut in flexion.

Movements

Trochlear surfaces permit only dorsi and plantar flexion.

Chapter 54

Notes on Locomotion

Chapter Outline

1. Arches of the Foot
2. Movements Within the Foot
3. Standing
4. The Walking Foot
5. Important Ingredients of Walking
6. Gait Parameters
7. Swing Phase
8. Stance Phase
9. Locomotion

The morphology of the foot is adapted for two purposes:

1. As a support of the weight of the body
2. As a lever for propulsion, i.e. locomotion — walking and running

The calcaneus supports the talus and has a prominent backward prolongation to receive the attachment of the gastrocnemius and soleus muscles that produce a large part of the forward thrust, raising the heel and thereby the body upon the toes, with the heads of the metatarsal bones as the fulcrum. The action distributes the weight along the tarsals, metatarsals and digits. Flexor hallucis longus and flexor digitorum longus act on the toes and serve to keep their plantar surfaces against the ground.

1. Arches of the Foot

A characteristic feature of the human foot is that the bones do not lie flat on the ground but are modified to make an arched structure for supporting and propelling the weight of the body that falls on the talus. Arches of the foot are medial longitudinal arch, lateral longitudinal arch and transverse arch.

MEDIAL LONGITUDINAL ARCH

Bones: Calcaneus, talus, navicular, three cunieforms and first, second and third metatarsals.

Support: Plantar calcaneonavicular ligament, tibialis posterior, flexor hallucis longus, flexor digitorum longus, intrinsic muscles of first toe and plantar aponeurosis.

LATERAL LONGITUDINAL ARCH

Bones: Calcaneus, cuboid, fourth and fifth metatarsals.

Support: Long plantar and plantar calcaneocuboid ligaments, tibularis longus and intrinsic muscles of the fifth toe.

TRANSVERSE ARCH

Bones: Anterior part of tarsus and bases of metatarsals.

Support: Fibularis longus and transverse head of adductor hallucis.

Arches are maintained by:
1. Shape of bones
2. Ligamentous tension that holds the bones together
3. Long and short muscles
4. Plantar aponeurosis — central part.

The arches of the foot form a half dome by which the body weight is distributed all around from the talus to the ground. The ligaments and muscles on the sole of the foot are stronger and more powerful than those on the dorsum. It is the ligaments and muscle action that are the main support of the arches. When the muscles are weak, the arches tend to collapse, the ligaments are strained and give rise to flat foot.

2. Movements Within the Foot

Most of the movement takes place in the distal part of the foot between the metatarsals and phalanges but some important movements take place between the tarsal bones especially at the subtalar, talocalcaneonavicular and calcaneocuboid joints. These joints are:
  (a) Inversion — turning the sole of the foot medially.
  (b) Eversion — turning the sole of the foot laterally.

INVERSION

Produced by: Tibialis anterior and tibialis posterior.

Limited by: Tension in fibular muscles and lateral part of interosseous talocalcanen ligament (a common way of spraining the ankle).

EVERSION (is more limited)

Produced by: Fibular muscles.

Limited by: Tension in tibialis anterior, tibialis posterior and deltoid ligament.

Slight gliding movements also occur in the other joints of the foot during inversion and eversion and when the weight falls on the foot.

3. Standing

The center of gravity of the body is about 1 cm behind the sacral promontory.

A perpendicular from the center of gravity meets the ground half way between the heels behind and metatarsals in front. It passes posterior to the hip joints, anterior to the knees and anterior to the ankles.

The hip joint is "close packed." The iliofemoral, ischiofemoral and pubofemoral ligaments resist further extension.

The knee joint is "close packed." The femur is medially rotated on the tibia, collateral and cruciate ligaments are taut.

There is little or no muscular activity required at the hip or knee.

There is little activity in the intrinsic muscles of the foot. Support is maintained by ligaments linking the plantar surfaces of the bones. When the foot is used to apply force to the ground, the intrinsic muscles become active and tension in the tendons of the long flexor muscles ties the extremities of the arches together taking the strain off ligaments.

4. The Walking Foot

Locomotion is the translation of the body from one point to another. The center of gravity is located just in front of the second sacral vertebra midway between the hip joints. Walking is initiated by inclining the body ahead of its center of gravity. To regain balance, one leg must be brought forward ahead of the shifting center of gravity. The leg that remains weight bearing is the "stance" leg and must be in contact with the ground. The leg that moves to regain balance is the "swing" limb. Walking is divided into a "stance" phase (60% of the gait cycle) in which the leg is weight bearing and a "swing" phase (40% of the gait cycle) in which the leg is moved to another point of contact.

Several manoeuvres (determinants of gait) together increase efficiency and decrease energy expenditure. They aim to decrease the amplitude of vertical displacement of the pelvis and to decrease the degree of undulation. In the vertical plane, the center of gravity is displaced twice in the vertical axis. The lowest points occur at "heel strike" and at the "toe-off."

Pelvic tilt involves a drop of the pelvis on the swing side that decreases vertical undulation slightly.

Pelvic rotation refers to the oscillation of the pelvis about an axis of the spine. This lessens the angle between the pelvis and thigh and decreases the vertical undulation of the center of gravity.

Knee flexion during the stance phase: As the body moves over its center of gravity, the knee flexes approximately

15° until the foot is flat on the ground. This decreases the amount of pelvic undulation slightly.

Knee ankle relationship: At heel strike, the ankle is dersiflexed and gradually plantar flexes. This rotation occurs around the ankle joint and as the ankle joint passes over the heel, the knee flattens out the two arcs of motion.

Pelvic shift is the movement of the pelvis laterally to maintain body balance as one leg is lifted from the ground.

5. Important Ingredients of Walking

1. Balance (equilibrium) — the person must remain upright during gait and in standing.
2. Transference of weight — the walker transfers weight from one limb to the other.
3. Clearance and leg movement — the leg moved must miss or slide lightly over the walking surface.

Each person has a walking signature.

6. Gait Parameters

1. *Step length* is the distance from the heel contact point of one foot with the walking surface to the point of heel contact with the other foot. Step length increases with walking speed and subject height etc. Typically, it is approximately 0.75 meters.

2. *Stride length* is the distance from heel contact to the point of heel contact with the same foot again. Stride length was the basis of measurement of the mile (a Roman mile was 1000 steps by a Roman soldier with a stride length of 5.2 feet).

3. *Cadence* (is from kad — to fall as in cadaver — fallen one) the number of steps taken per minute. Typically, it is 110 steps per minute.

4. *Walking speed* is 82.5 (110 × 0.75) meters per minute (1.4 meters per second). Walking speeds above two meters per second are considered fast and the transition to running usually occurs at about four meters per second. Maximum running speeds are usually about ten meters per second.

7. Swing Phase

The swing phase accomplishes:
1. Forward placement of the foot
2. Toe clearance
3. Preparation for foot contact

The swing of the limb from the pelvis is much like the swing of a pendulum that has two links (thigh and shank). The swing should require little energy expenditure in normal walking. It lasts from toe off to heel contact and takes up about 38% of the gait cycle (stride). Muscle activity in this stage is relatively low.

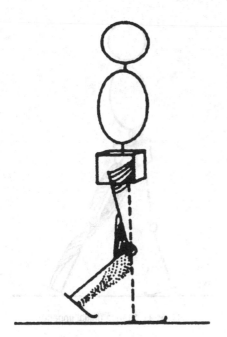

FIG. 54-1 Initial swing.

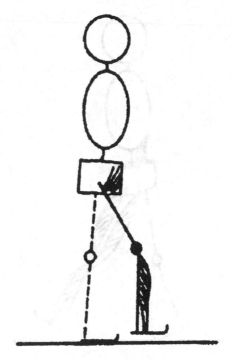

FIG. 54-2 Mid swing.

INITIAL (EARLY) SWING

This phase involves lifting of the limb from the floor and initial advancement of the thigh. Critical events are hip flexion to 20° and knee flexion to 60°.

Rectus femoris, ihopsoas, gracilis, sartorius, and tensor fascia lata acting at the hip: advance the thigh.

Biceps femoris (short head), gracilis, and sartorius acting at the knee insures toe clearance.

Pretibial muscles that cross the ankle become active but provide no critical function at this time.

MID SWING

Iliopsoas, gracilis, and sartorius advance the thigh at the hip. Pretibial muscles acting at the ankle insure toe clearance.

TERMINAL SWING

Deep extensors maintain the trunk erect. Hamstrings acting at the hip decelerate the limb for heel contact. Quadriceps femoris acting at the knee create step length for heel contact. Pre-tibial muscles crossing the ankle position foot for heel contact.

8. Stance Phase

The stance phase takes up about 62% of the gait cycle of which around

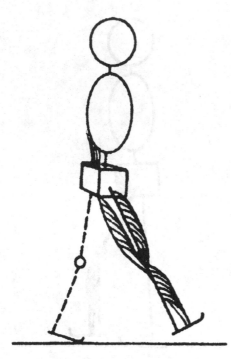

FIG. 54-3   Terminal swing.

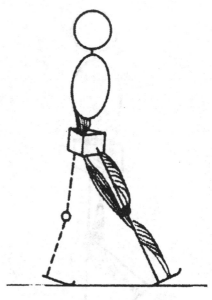

FIG. 54-4   Stance (initial impact).

12% is double support. During normal walking, many muscles function
eccentrically.

The stance phase provides:

1. Structural stability for the mass of
   the torso to move over (cf an
   inverted pendulum).
2. Roll over on the stance foot.
3. Shock absorption from the foot,
   ankle, knee and pelvis.
4. Preparation for swing.

INITIAL CONTACT (HEEL STRIKE)

Deep extensors stabilize the trunk.
Hamstrings and gluteus maximus extend
the hip propelling the trunk forward.

Quadriceps femoris stabilizes the
slightly flexed knee. Pretibial muscles
acting across the ankle relax allowing
lowering of the foot to midstance and
transferring weight from the heel to
the plantar arch.

LOADING RESPONSE

Gluteus maximus and hamstrings
restrain hip flexion. Quadriceps femoris
restrains knee flexion. Pretibial muscles
acting at the ankle restrain plantar flexion of the foot.

MID-STANCE

Abdominal muscles maintain the trunk
erect. Gluteus medius, gluteus minimus,
tensor fascia lata maintain lateral hip
stability.

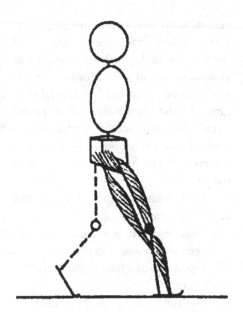

FIG. 54-5 Stance (loading response).

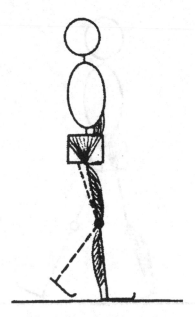

FIG. 54-6 Early mid-stance.

EARLY MID-STANCE

Quadriceps femoris restrain knee flexion. Soleus acting at the ankle controls
tibial advancement.

LATE MID-STANCE

Soleus and gastrocnemius acting at the
ankle control tibial advancement.

TERMINAL STANCE

Lipsilateral extensors control weight
shift of the trunk. Gluteus medius and
gluteus minimus maintain lateral hip
stability. Soleus and gastrocnemius
maintain a neutral ankle.

SUMMARY

In the stance phase, extensor muscles
are active to support the body weight.

FIG. 54-7 Late mid-stance.

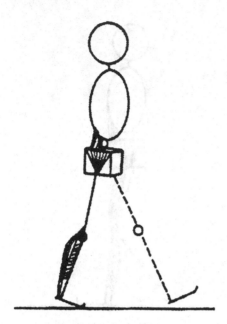

FIG. 54-8 Terminal stance.

In the swing phase, flexors contribute to floor clearance of the foot.

Intrinsic muscles of the foot prevent stretching of plantar ligaments when a load is applied to the foot.

Dorsiflexors raise the toes to clear the ground during the swing phase. Plantar flexors and hip extensors thrust the whole body forward. Hip muscles position the trunk over the foot that contacts the ground.

9. Locomotion

Locomotion is controlled by a central program.

The basic rhythmic pattern of neural activity underlying locomotion is generated by neurons intrinsic to the central nervous system (pattern generators). Walking movements have been shown not to be reflex in origin but generated by neurons located exclusively in the spinal cord. A central program is the expression of a neural circuit that produces a particular pattern of motor output that does not require afferent feedback for its essential pattern.

Descending systems moderate the central program to:

1. Provide tonic excitatory bias to extensor motor systems.
2. Open and close spinal reflex circuits.

Neurons giving rise to the rubrospinal tract, vestibulospinal tract, reticulospinal tract and the descending noradrenergic system are rhythmically active in phase with locomotor movements.

Ascending information from the spinal cord related to locomotion is carried to the cerebellum via the dorsal and ventral spinocerebellar tracts. The dorsal tract carries information about muscle activity while the ventral tract informs the cerebellum of active processes (pattern generation for locomotion) within the spinal cord.

Afferent information is important in:

1. Switching the motor program from one phase to another (swing phase to or from stance phase).
2. Opening and closing reflex pathways in different parts of a step cycle channeling reflex activity to compensate for changing terrain.

Back

The back is the posterior part of the trunk, extending from the neck to the pelvis. It supports the head, acts as a central pillar of the body and connects the upper and lower segments of the trunk. In addition, it provides attachments for the ribs, reduces the shock of impact in walking and running and protects the spinal cord. The back is a composite of 33 bones (vertebrae) and their discs (symphyses), ligaments, fascia and muscles.

Chapter 55

Fascia and Muscles of the Back

Chapter Outline

1. Prevertebral (Anterior Vertebral) and Psoas Muscles

These muscles form a bundle arising (and inserting) from the vertebral bodies and discs and are supplied by the ventral rami of spinal nerves.

Longus Colli

Longus colli consists of three parts.

Vertical part: Comprises muscular slips that pass between the bodies of upper thoracic vertebrae to the front of the bodies of lower cervical vertebrae.

Lower oblique part: Comprises slips from the bodies of upper thoracic vertebrae upward and laterally to transverse processes (anterior tubercles) of lower cervical vertebrae.

Upper oblique part: Comprises slips from the transverse processes (anterior tubercles) of upper cervical vertebrae upward and medially to the anterior tubercle of the atlas.

- **Nerve supply:** Venral ramus of C2, C3 and C4.
- **Action:** Flexion and lateral flexion of the neck. It is active in talking, swallowing and coughing.

Longus Capitis

- **Origin:** Anterior tubercles of transverse processes of 3–6 cervical vertebrae.
- **Insertion:** Base of the skull (inferior surface of basioccipital bone).

- **Nerve supply:** Branches of cervical plexus C1–C4.
- **Action:** Flexes the head.

Rectus Capitis Anterior

- **Origin:** Front of the lateral mass of the atlas.
- **Insertion:** Basioccipital bone close to the condyle.
- **Action:** Flexes the head at the atlanto-occipital joint.

Rectus Capitis Lateralis

- **Origin:** Transverse process of the atlas.
- **Insertion:** Jugular process of the occipital bone.
- **Nerve supply:** Ventral rami of C1–C2.
- **Action:** Lateral flexion of the head to the same side. It stabilizes the atlanto-occipital joint.

These short prevertebral muscles, together with sternomastoid and upper deep muscles of the back, adjust and stabilize the skull on the spine.

2. Lateral Vertebral Muscles (Scalenes)

Three muscles connect the anterior tubercles (costal elements) of the cervical vertebral to the first and second ribs. They are innervated by ventral rami of cervical nerves.

Scalenus Anterior

- **Origin:** Anterior tubercles of the transverse processes of lower cervical vertebrae.
- **Insertion:** By a narrow tendon to the scalene tubercle on the first rib.
- **Nerve supply:** Branches from the cervical plexus (C4–C7 ventral rami).

Scalenus Medius

- **Origin:** Posterior tubercles of transverse processes of upper cervical vertebrae.
- **Insertion:** Large area of the upper surface of the first rib behind the subclavian groove.
- **Nerve supply:** Branches of the cervical nerves (ventral rami of C3–C7).

The brachial plexus and subclavian artery emerge between scalenus anterior and medius.

Scalenus Posterior

Scalenus posterior may be regarded as a part of scalenus medius crossing the first rib to reach the second rib.

- **Origin:** Posterior tubercles of transverse processes of the lower four to six cervical vertebrae.
- **Insertion:** Posterior outer surface of the second rib.
- **Nerve supply:** Ventral rami of C5–C7.
- **Action:** Lateral flexion of the neck and elevation of the second rib.

Scalene muscles together elevate the first and second ribs during respiration.

3. Dorsal Muscles of the Vertebral Column

FASCIA

Nuchal fascia is the cervical part of the fascia of the back. It is attached to the skull beneath the superior nuchal line, to the ligamentum nuchae, and to the spines of C7 and T1 to T6, and is continuous with the thoracolumbar fascia below. It is the posterior part of the prevertebral layer of cervical fascia covering splenius capitus.

Thoracolumbar fascia is the downward prolongation of the nuchal fascia in the neck. In the thoracic and lumbar regions, it is formed by the fusion of posterior aponeuroses of internal oblique and transversus abdominis. It splits into three layers that form two muscular compartments.

In the midline, it is attached to the vertebral spines below T7, their supraspinous ligaments, and the medial crest of the sacrum. In the thoracic region, it is attached to the angles of the ribs.

Below, it is attached to the iliac crest.

Laterally, it is thin in the thoracic region and attaches to the angles of the ribs.

In the lumbar region, it is thick and split into layers that enclose muscles continuous with the aponeurotic origin of transversus abdominis. The aponeurosis splits into two layers to enclose the deep vertebral muscles.

Posterior layer is attached to the vertebral spines and gives origin to latissimus dorsi.

Middle layer inserts into the tips of the vertebral transverse processes. With the posterior layer, it encloses the erector spinae.

Anterior layer is a strong, thin layer that inserts midway along the transverse processes ane with the middle layer, enclosing quadratus lumborum.

MUSCLES

The muscles of the back may be divided into three groups

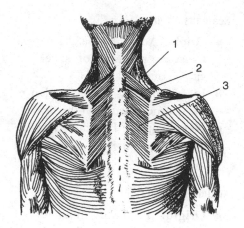

FIG. 55-2 Second layer of superficial back muscles. (1) Levator scapulae; (2) Rhombiod minor; (3) Rhombiod major.

Superficial group — acting on limbs
    Trapezius
    Latissimus dorsi
    Levator scapulae
    Rhomboideus major and minor

Intermediate group — respiratory
    Serratus posterior superior
    Serratus posterior inferior

Deep group — intrinsic (supplied by dorsal rami)
    Splenius
    Suboccipital muscles
    Longitudinal muscles: iliocostalis, longissimus, spinalis
    Oblique muscles: semispinalis, multifidus, rotatores
    Minor muscles: interapinales, intertransversarii, levatores costarum

Superficial Group

The superficial group include trapezius, latissimus dorsi, levator scapulae and

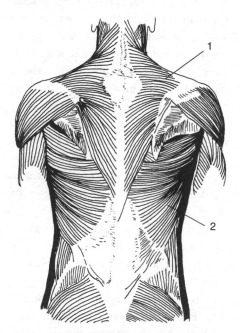

FIG. 55-1 Superficial back muscles. (1) Trapezius; (2) Latissimus dorsi.

the rhomboids that act on the upper limbs.

Intermediate Group

The intermediate group consists of two flat quadrilateral muscles that have spread across the deep group (indicated by their nerve supply by ventral rami of spinal nerves).

Serratus Posterior Superior

This muscle is thin, quadrilateral muscle and aponeurosis.

- **Origin:** Ligamentum nuchae and spines of last cervical and several upper thoracic vertebrae deep to splenius. Fibers are directed down and laterally.
- **Insertion:** Outer surfaces of the second to fifth ribs beyond the angles.
- **Action:** A weak elevator of the upper ribs.
- **Nerve supply:** Branches of the ventral rami of T1 to T4.

Serratus Posterior Inferior

This muscle is a flat quadrilateral muscle, situated at the junction of the thoracic and lumbar regions deep to latissimus dorsi.

- **Origin:** Spines of the lower thoracic and upper lumbar vertebrae.
- **Insertion:** Fibers pass laterally and upwards to the outer surfaces of the ninth to twelfth ribs beyond their angles.
- **Action:** Depresses or fixes the lower ribs holding them against the upper pull of the diaphragm.

- **Nerve supply:** Branches of ventral rami of T9 to T12.

Deep Muscles

Splenius is the detached part of the deep group of vertebral muscles that acts as a strap holding the deeper muscles of the neck. It has a single origin but divides into two parts, splenius capitis that inserts into the skull and splenius cervicis that inserts into the cervical vertebrae.

Splenius

- **Origin:** Lower half of the nuchal ligament and spines of the lower cervical and upper six thoracic spines.
- **Insertion:**
  Capitis: Superior nuchal line and mastoid process.
  Cervicis: Posterior tubercles of C1 to C3.
- **Action:** Draws head and neck back and rotates the face towards the side on which the muscle is acting.
- **Nerve supply:** Lateral branches of the dorsal rami of C2 to C5 or C6.

Longitudinal muscles is a broad thick muscle mass, the erector spinae attached from the pelvis to the skull by aponeurotic origin from spines and fleshy origin from interosseous sacroiliac ligament and the iliac crest. A little below the last rib, it splits into three columns.

Iliocostalis

Iliocostalis is the most lateral column consisting of the iliocostalis lumborum, iliocostalis thoracis and iliocostalis cervicis.

- **Origin:** Crest of the ilium.
- **Insertion:** Angles of ribs. Its upper fascicles arise from lower ribs and insert in upper ribs and at cervical levels end in the posterior tubercles of cervical transverse processes.

Longissimus

Longissimus is the intermediate column consisting of the thoracis, cervicis and capitis muscles.

- **Origin:** Transverse processes of lower level.
- **Insertion:** Transverse processes of higher level. Longissimus capitis inserts into the posterior margin of the mastoid process (beneath splenius and sternoclieidomastoid).

Spinalis

Spinalis is the narrow, most medial column consisting of thoracis, cervicis, and capitis.

- **Origin:** Upper lumbar spines.
- **Insertion:** Upper thoracic spines.

Erector spinae extend the vertebral column and, acting on one side, bend the column to that side. Longissimus capitis bends the head and rotates the face toward the same side.

Erector spinae muscles are innervated serially by dorsal rami of spinal nerves.

Oblique muscles are covered by the erector spinae muscle. They extend obliquely from the transverse processes to the spines (the transversospinal group).

Semispinalis

Semispinalis derives its name from covering half of the length of the vertebral column thoracis, cervicis, capitis lies superficially and spans from four to six segments from the tips of the thoracic transverse processes. Semispinalis capitis is the largest muscle mass on the back of the neck.

- **Origin:** Tips of the transverse processes from T12 and upward.
- **Insertion:** Vertebral spines as far as the occipital bone.
  Throracis: Spines of C6 or C7 and T1 to T4.
  Cervicis: Spines of C2–C5.
  Capitis: Occipital bone between superior and inferior nuchal lines immediately beneath trapezius.

Multifidus

Multifidus extends throughout the length of the vertebral column but individual fascicles span three segments. It is a transversospinal muscle (to transverse processes in the thoracic region, articular processes in the cervical level and mamillary processes in the lumbar region).

- **Origin:** Posterior aspects of the sacrum to the transverse process of C4.
- **Insertion:** Lower border of every vertebral spine to C2.

Rotatores

Rotatores consists the cervicis, thoracis, lumborum. The shortest and deepest of the transvesrospinal group spaning one segment. They are best developed in the throacic region.

- **Origin:** Root of one transverse process.
- **Insertion:** Lamina of the vertebra immediately above it.

Interspinales

Interspinales consists the cervicis, thoracis, lumborum.

- **Origin:** Spines of the cervical and lumbar vertebrae.
- **Insertion:** Spine of the vertebra immediately above.

Intertransversarii

Intertransversarii consists the lumborum, cervicis. They are absent in the thoracic region.

- **Origin:** Transverse processes of the cervical and lumbar vertebrae.
- **Insertion:** Transverse process above it.

Levatores Costarum

They are present only in the thoracic region.

NERVES OF THE DEEP BACK MUSCLES

The dorsal rami of spinal nerves pass backward between transverse processes and supply muscles, joints and skin. They are "mixed" nerves and divide into medial and lateral branches (except C1, S4 and S5 and Co1). The muscular component supplies longitudinal muscles occupying the space between the vertebral spines and angles of the ribs.

CUTANEOUS BRANCHES OF DORSAL RAMI

The dorsal ramus of C1, C6 and C7 have no cutaneous branches.

The dorsal ramus of C2 has a large medial branch that ascends to the vertex as the greater occipital nerve.

The dorsal ramus of C3 supplies the upper medial part of the neck as the third occipital nerve.

The dorsal rami of the other cervical nerves supply the back of the neck (except C7 and C8).

The dorsal ramus of T2 is large and reaches the point of the shoulder.

Medial branches of the upper T6 or T7 nerves supply the skin of the scapular region.

Lateral branches of the T7 nerves and lower supply the skin of the lower chest and loin. Those of the first three lumbar nerves become the superior cluneal nerves that supply the gluteal region to the greater trochanter.

The dorsal rami of L4 and L5 have no cutaneous branches.

Lateral branches of the dorsal rami of S1–S5 form the middle cluneal nerves that supply the skin over the back of the sacrum and adjacent gluteal region.

Dorsal rami of S4, and Co1 do not divide into medial and lateral branches but form a cutaneous nerve supplying the skin over the coccyx.

Chapter 56

Suboccipital Triangle

Chapter Outline

1. Boundaries
2. Suboccipital Muscles
3. Contents

The suboccipital triangle is located at the upper portion of the back of the neck beneath the occipital bone and adjacent muscles of the back.

1. Boundaries

The suboccipital triangle has the following boundaries.

Superior and medial boundary: Rectus capitis posterior major.

Superior and lateral boundary: Obliquus capitis superior.

Inferior and lateral boundary: Obliquus capitis inferior.

Roof: Semispinalis capitis and longissimus capitis.

Floor: Contains the posterior arch of the atlas and posterior occipitoatlantal membrane (being pierced by the vertebral artery).

Contents: The four suboccipital muscles (rectus capitis posterior major, rectus capitis posterior minor, obliquus capitis inferior and obliquus capitis superiror), two nerves, (suboccipital and greater occipital nerve) and two arteries (vertebral and occipital).

2. Suboccipital Muscles

These are a group of small muscles in the same plane as the deep (cf oblique/transversospinal) vertebral muscles and connect the atlas and axis to one another or to the head. They are supplied by the suboccipital nerve (dorsal ramus of C1).

Obliquus Capitis Inferior

- **Origin:** Spine of the axis.
- **Insertion:** Tip of the transverse process of the atlas.

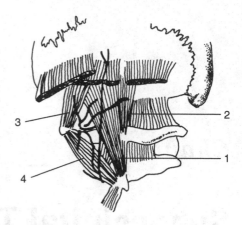

FIG. 56-1 Suboccipital triangle. (1) Rectus capitis posterior major m; (2) Rectus capitis posterior minor m; (3) Obliquus capitis superior m; (4) Obliquus capitis inferior m.

- **Action:** Rotates the atlas and thus the head (turns the head toward the same side).
- **Nerve supply:** Branch of the dorsal ramus of the suboccipital nerve (C1).

Obliquus Capitis Superior

- **Origin:** Tip of the transverse process of the atlas.
- **Insertion:** Between the superior and inferior nuchal lines on the occipital bone.
- **Action:** Extends the head (pulls the head to the same side).
- **Nerve supply:** Branch of the dorsal ramus of the suboccipital nerve (C1).

Rectus Capitis Posterior Minor

The rectus capitis posterior minor lies medial to rectus capitis posterior major and is more closely applied to the base of the skull.

- **Origin:** Posterior tubercle of the arch of the atlas. It widens as it ascends.
- **Insertion:** Occipital bone between the inferior nuchal line and the foramen magnum.
- **Action:** Extends the head.
- **Nerve supply:** Branch of the dorsal ramus of the suboccipital nerve (C1).

Rectus Capitis Posterior Major

- **Origin:** Spine of the axis.
- **Insertion:** Occipital bone between the inferior nuchal line and the foramen magnum.
- **Action:** Extends the head.
- **Nerve supply:** Branch of the dorsal ramus of the suboccipital nerve (C1).

The two recti fix the skull on the atlas during rotation.

3. Contents

Occipital artery is a posterior branch of the external carotid artery in the upper part of the neck.

In the anterior triangle of the neck, the occipital artery passes upwards and back beneath the digastric muscle and mastoid process and attached muscles then backward in the occipital groove. It lies on obliquus capitis superior and semispinalis capitis then pierces trapezius, and ascends with the greater occipital nerve to the scalp.

A descending branch arising on obliquus capitis superior is part of a collateral circulation after ligation or blockage of the external carotid or subclavian artery.

Vertebral artery supplies the posterior part of the brain. It arises in the neck from the first part of the subclavian artery and ascends through the foramina transversaria of C1 to C6. It winds behind the lateral mass of the atlas and passes the lateral edge of the posterior atlantooccipital membrane to enter the skull through the foramen magnum. In the suboccipital triangle, it gives branches to the suboccipital muscles and meningeal branches in the posterior cranial fossa.

Suboccipital nerve is the dorsal ramus of C1 that emerges above the posterior arch of the atlas below the vertebral artery and innervates the suboccipital muscles and the semispinalis muscle. It has no cutaneous branches.

Greater occipital nerve is the dorsal ramus of C2. It emerges below the obliquus capitis inferior muscle supplying it and divides.

The lateral branch innervates splenius and semispinalis capitis while.

The larger medial branch, known as the greater occipital nerve pierces semispinalis capitis and trapezius, ascends to the vertex of the skull with the occipital artery and supplies the skin of the scalp as far forward as the vertex.

Chapter 57

Vertebral Column

Chapter Outline

1. Components of a Vertebra
2. Cervical Vertebrae
3. Thoracic Vertebrae
4. Lumbar Vertebrae
5. Sacrum
6. Coccyx

The vertebral column consists of 33 vertebrae: seven cervical, 12 thoracic, five lumbar, five fused sacral and four fused coccygeal vertebrae. Intervertebral discs binding the vertebrae together, distribute the forces over the entire surface of the vertebral bodies, allow limited movement between adjacent vertebrae and comprise 25% of the total length of the vertebral column.

There are seven cervical vertebrae but eight cervical nerves. The first nerve passes between the occipital bone and the first vertebra and so on until the eighth nerve passes between the seventh cervical and first thoracic vertebrae.

There are three curvatures of the vertebral column in the sagittal plane. Before birth, the column has a single primary curvature or flexure anteriorly. Two secondary curvatures concave anteriorly develop in the cervical and lumbar regions. The secondary cervical curvature develops as the child holds the head upright and the secondary lumbar curvature develops as the child learns to walk upright and balance on two feet.

1. Components of a Vertebra

Each vertebra consists of three parts.

Body is a short cylindrical weight bearing part with a smooth rim. The vertebral bodies increase in size caudally (with increase in weight from above). They consist of spongy bone covered by compact bone and have transverse and vertical trabeculae that resist tensile and compressive forces.

Vertebral arch located behind the body, encloses and protects the spinal cord. It consists of pedicles that project backward from the posterolateral aspects of the body and have indented borders that form the vertebral notches. The superior notches on one vertebra combine with the inferior notches on the adjacent vertebra to complete an intervertebral foramen that transmit spinal nerves.

The vertebral arch is completed by laminae, flat, sloping plates that project back from the ends of the pedicles and fuse in the midline. The vertebral foramen is formed by the posterior surface of the vertebral body and the vertebral arch. Successive foramina form the vertebral canal that lodges the spinal cord, its meninges and blood vessels embedded in adipose tissue.

PROCESSES

Spinous process projects backward from the junction of the laminae. In the cervical region, spines are short and bifid, in the thoracic region, they are long and in the lumbar region, they are thick.

Transverse processes project laterally on each side from the junction of the pedicle and the lamina. They serve as levers for attachment of vertebral muscles and ligaments and provide points of articulation for the necks of most ribs. In the cervical region, transverse processes are short and bifid with a foramen transversarium, in the thoracic region, they are thick and in the lumbar region, they bear an accessory process.

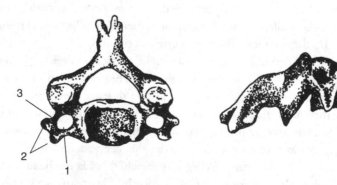

FIG. 57-1   Typical cervical vertebra. (1) Foramen transversarium; (2) Tubercles; (3) Costotransverse bar.

Articular (zygapophyseal) processes project above and below from the junction of the pedicle and the lamina.

Superior articular processes (zygapohyses) are projections on the superior aspect of the vertebral arch at the junction of pedicle and lamina. They carry the superior articular facets that face posteriorly in the cervical region, posteromedially in the thoracic region, medially in the lumbar region and posteriorly in the sacrum.

Inferior articular processes (zygapophyses) project from the inferior aspect of the vertebral arch at the junction of the pedicle and lamina that carry the inferior articular facet for articulation with the superior articular process of the vertebra below.

Superior and inferior vertebral notches are formed between the pedicles of adjacent vertebrae. They transmit the spinal nerves.

2. Cervical Vertebrae

The vertebrae of the neck are relatively small but increase in size below toward the thoracic region. They are characterised by the foramen transversarium in the transverse processes on both sides that transmit the vertebral arteries. The two parts of the transverse processes are united by a costotransverse bar and terminate in anterior and posterior tubercles. Superior articular processes face posteriorly and inferior; articular processes face anteriorly permitting lateral bending.

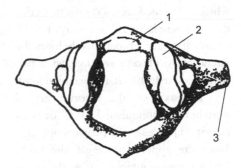

FIG. 57-2 Atlas from above. (1) Facet for dens; (2) Superior articular process; (3) Transverse process.

The atypical cervical vertebrae are the first cervical vertebra, second cervical vertebra and the seventh cervical vertebra.

First cervical vertebra or **atlas** has no body and a rudimentary spine. It consists of an anterior arch, that bears a facet for articulation with the odontoid process of the second cervical vertebra and an anterior tubercle for the attachment of the anterior longitudinal ligament. Lateral masses, bear superior and inferior articular facets for articulation with the occipital condyles above and superior facets of the axis respectively. Rectus capitis anterior is attached to the anterior surface of the lateral mass.

Transverse processes are long. Rectus capitis lateralis is attached to the front of the transverse process and obliquus capitis superior and inferior and some deep muscles of the back are attached to the back of the process.

The posterior arch (cf laminae of other vertebrae) bears a posterior tubercle which is a rudimentary spine to which is attached rectus capitis posterior minor. There is a shallow groove for the vertebral artery immediatly behind the lateral mass. Flexion and extension primarily occur at the atlantooccipital joint.

The **axis** of the second cervical vertebra is distinguished by its dens or odontoid process, that projects upwards from the upper surface of the body. The apex gives attachment to the apical

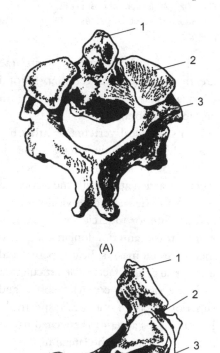

(A)

(B)

FIG. 57-3 Axis (A) from above and (B) from the side. (1) Dens; (2) Superior articular facet; (3) Foramen transversarium.

ligament and lateral tubercles give attachment to the alar ligaments. On the anterior aspect of the dens is a facet for articulation with the anterior arch of the atlas, and on the posterior aspect is a facet for a bursa separating it from the transverse ligament of the atlas. Rotation primarly occurs at the atlantoaxial joint.

Seventh cervical vertebra or vertebra prominens is characterized by a very long (not bifid) spine, for which it is named (although it may not be the most prominent spine) and gives attachment to the ligamentum nuchae. The foramen transversarium is small and transmits small accessory veins.

The combination of flexion-extension and lateral flexion permits circumduction of the head.

3. Thoracic Vertebrae

The second to the eighth thoracic vertebrae (of 12) are considered "typical" thoracic vertebrae. The body of the vertebrae is kidney shaped and there are demifacets on the upper and lower borders of the junction of the body and arch for articulation with the head of the corresponding rib. There are also facets on most of the transverse processes for articulaton with the tubercles of the ribs.

Articular surfaces of thoracic zygapophyses lie in a plane on the arc of a circle (center of rotation located in the vertebral body) thereby permitting a

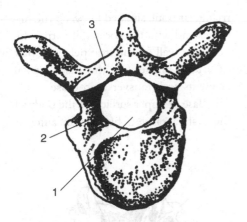

FIG. 57-4 Thoracic vertebra — superior aspect. (1) Vertebral foramen; (2) Pedicle; (3) Lamina.

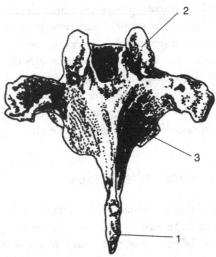

FIG. 57-6 Thoracic vertebra from behind. (1) Spine; (2) Superior articular process; (3) Inferior articular process.

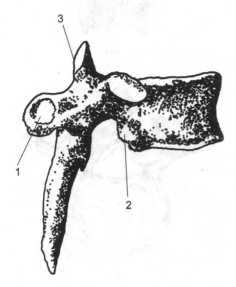

FIG. 57-5 Thoracic vertebra — lateral aspect. (1) Facet for costal tubercle; (2) Costal facet; (3) Superior articular process.

possible (limited by the ribs and their attachment to the sternum).

The atypical thoracic vertebrae are the first thoracic vertebra, eleventh thoracic vertebra and the twelfth thoracic vertebra.

First thoracic vertebra resembles a cervical vertebra but has an entire facet on its superior surface for the first rib and a demifacet inferiorly. The spine is more nearly horizontal and may be more prominent than that of the seventh cervical vertebra.

Eleventh thoracic vertebra resembles a lumbar vertebra. It has an entire facet on the pedicle and none on the transverse process.

Twelfth thoracic vertebra resembles a lumbar vertebra but has a complete costal facet mostly on the pedicle.

small degree of rotation increased by having 12 members. Some flexion and extension and lateral flexion is also

Three articular processes or tubercles are present in place of a transverse process. Superior (mamillary) process on the posterior border of the superior articular process, lateral (costal) tubercles homologous to ribs and inferior (accessory) articular processes, the true transverse process of the vertebra, are present.

4. Lumbar Vertebrae

Lumbar vertebrae are characterized by the large size of the body (increasing caudally), absence of costal facets and foramina in the transverse process. The bodies are kidney shaped and pedicles and laminae are short and thick. The spines are quadrilateral in shape project horizontally backwards. The orientation of the articular processes limit rotation but allows a considerable amount of flexion and extension and a moderate amount of lateral flexion.

Fifth lumbar vertebra is the largest of all the vertebrae. It has a small spine and a large, thick transverse process. The joints formed between the inferior articular process and superior articular facets of the sacrum lie in the frontal plane.

5. Sacrum

The sacrum is formed by the fusion of five embryonic sacral vertebrae. In the adult, the sacrum is a flattened triangular bone situated between the hip bones and with the coccyx, it forms the posterior wall of the true pelvis. A pair of lateral masses are the fused costal elements and transverse processes.

Base or upper surface of the body of the first sacral vertebra has an anterior

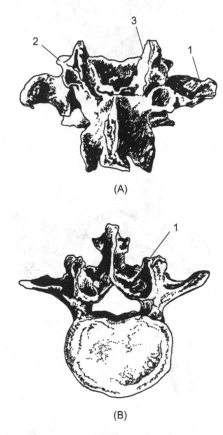

(A)

(B)

FIG. 57-7 (A) Typical Lumbar vertebra from behind. (1) Transverse process; (2) Mamillary process; (3) Superior articular process. (B) Lumbar vertebra from above. (1) Mamillary process.

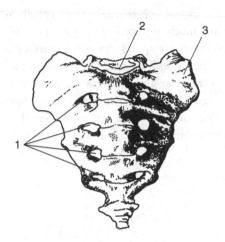

FIG. 57-8 Sacrum — anterior view. (1) Pelvic sacral foramina; (2) Promontory; (3) Ala.

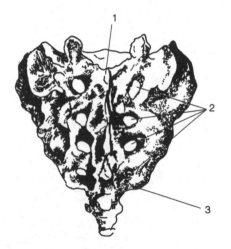

FIG. 57-9 Sacrum — posterior view. (1) Median sacral crest; (2) Dorsal sacral foramina; (3) Sacral canal.

margin that projects forward as the promontory. Posteriorly, the triangular vertebral foramen marks the beginning

of the sacral canal. The right and left lateral parts or alae of the sacrum are extends from the side of the body and represent fused costal and transverse elements. Each ala is crossed by the lumbosacral trunk, iliolumbar artery, obturator nerve and psoas major.

Apex is the oval inferior surface of the body of the fifth sacral vertebra that articulates with the first coccygeal vertebra.

Pelvic surface is concave and shows the bodies and ossified discs where fusion took place and are now represented by four transverse ridges. The pelvic sacral foramina lateral to the ridges are smooth and rounded laterally from the emergence of the upper four sacral nerves.

Dorsal surface is convex and marked by two vertical rows of four dorsal sacral foramina. In the midline, the median sacral crest is formed from the reduced spinous processes. The lateral sacral crest lies lateral to the sacral foraminae. The crests are formed from the transverse processes.

Sacral hiatus is an oblique entrance into the vertebral canal formed by the failure of the laminae of the fifth (sometimes fourth) sacral vertebrae to fuse.

Sacral horns are tubercles of the inferior articular processes of the fifth sacral vertebrae and are connected to the horns of the coccyx.

Lateral surface is a narrow edge below but broadens above. Its upper part contains the auricular surfaces covered with cartilage and articulates with the ilium.

6. Coccyx

The coccyx is frequently a single bone resulting from the fusion of four coccygeal vertebrae. The vertebrae are much reduced having no pedicles, laminae or spines. Its pelvic surface is concave and smooth. The dorsal surface is irregular and presents a pair of coccygeal horns, the larger and most superior of the articular processes that articulate with the horns of the sacrum.

Chapter 58

Vertebral Joints

Chapter Outline

The joints of the vertebral column are cartilaginous joints between the vertebral bodies; synovial joints between the vertebral arches; the articulation between axis and atlas; and between the atlas and the skull.

1. Joints Between the Vertebral Bodies

- **Type:** Cartilaginous symphyses.
- **Articulating elements:** The bodies of adjacent vertebrae articulate with one another.
- **Ligaments:**

Anterior longitudinal ligament is a collection of longer and shorter fibers that extends along the anterior surface of the bodies of the vertebrae binding them together. It extends from the anterior tubercle of the atlas to the pelvic surface of the sacrum.

Posterior longitudinal ligament is a similar band that extends in the vertebral canal on the posterior border of the vertebral bodies from the axis (where it is continuous with the tectorial membrane) to the sacrum. It is broader over the intervertebral discs and narrower over the vertebral bodies.

Intervertebral discs are fibrocartilaginous pads adherent to the hyaline cartilage covering the surfaces of vertebral bodies. They vary in thickness in different parts of the column together constituting about one quarter of its length. Each consists of an outer, firm annulus fibrosus, a fibrous ring of concentrically and spirally arranged fibre bundles and an inner, soft nucleus pulposus, a soft plastic material containing bundles of collagen fibres, connective tissue cells and amorphous intercellular material.

With advancing age, the entire disc may become fibrocartilaginous.

2. Joints Between the Vertebral Arches

Arches are connected by synovial joints between articular processes and accessory ligaments that connect laminae, transverse processes and spines.

- **Type:** Plane synovial joints allowing some gliding between facets.
- **Articulating elements:** Articular processes of adjacent vertebrae.
- **Ligaments:**

Articular capsules are loose capsules that enclose the synovial joints between the articular processes.

Ligamenta flava are a pair of elastic ligaments that connects the laminae of adjacent vertebrae. It extends laterally to the capsule of the joints between facets contributing to the posterior boundary of the intervertebral foramen. Its content of elastic fibers prevents buckling in flexion and compression of the spinal cord.

Interspinous ligaments between adjacent spines from the base to the tip are best developed in the lumbar region.

Supraspinous ligament connects the tips of all the spinous processes. In the neck, it merges with the ligamentum nuchae that extends in the midline from the exterior occipital protuberance of the skull to the spine of the seventh cervical vertebra.

Intertransverse ligament extends between the transverse processes of adjacent vertebrae. Best developed in the lumbar region.

3. Atlantoaxial Joint

Atlantoaxial joint consists of three articulations, two lateral atlantoaxial joints and one median atlantoaxial joint.

- **Lateral atlantoaxial joints (2)**
- **Type:** Synovial pivot type.
- **Articular elements:** Reciprocal articular processes of the atlas and axis.
- **Median atlantoaxial joint (1)** comprises two separate joints, one between the anterior surface of the dens and the inner aspect of the anterior arch of the atlas and one between the posterior surface of the dens and the transverse ligament of the atlas.

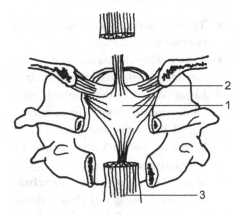

FIG. 58-1 Joints between the occipital bone, the atlas and the axis. (1) Cruciate ligament; (2) Alar ligament; (3) Posterior longitudinal ligament.

- **Ligaments:**

Articular capsule is thin, extending between the lateral masses of the atlas and the margins of the articular surface of the axis.

Anterior atlantoaxial membrane is a strong membrane that extends between the anterior arch of the atlas and the front of the body of the axis between the lateral joints. The anterior longitudinal ligament overlies the membrane in the median plane.

Posterior atlantoaxial membrane is a broad, thin membrane between the lower border of the posterior arch of the atlas and the upper border of the laminae of the axis.

Transverse ligament of atlas is a strong band extending behind the dens between the lateral masses. The superior crus extends from this ligament to the basilar part of the occipital bone and inferior crus descends to the body of the axis. The crura and the transverse ligament constitute the cruciform ligament of the atlas.

- **Accessory ligaments**

These ligaments are occipitoaxial. Tectorial membrane extends between the basilar part of the occipital bone and the body of axis where it is a continuation of the posterior longitudinal ligament. It covers the dens and its ligaments.

Alar ligaments are short strong bands that extend between sides of the apex of the dens and the anterior margin of foramen magnum.

Apical ligament of the dens extends from the apex of the dens to the middle of the anterior margin of the foramen magnum.

- **Movement:** The atlantoaxial joint provides for rotation of the skull and atlas together. The dens acts as a pivot enclosed in the ring formed by the anterior arch of the atlas and transverse ligament. Articular surfaces of the atlas glide forward and backward. Alar ligaments are the chief check ligaments in rotation.

4. Atlanto-Occipital Joint

- **Type:** Synovial elipsoid.
- **Articulating elements:** Superior articular facets of the lateral masses of the atlas and the condyle of the occipital bone.
- **Ligaments:**
Articular capsule is a thin and loose capsule envelops each joint.

Anterior atlanto-occipital membrane is densely woven fibres extend between the anterior margin of the foramen magnum and the upper border of the anterior arch of the atlas.

Posterior atlanto-occipital membrane extends from the posterior margin of foramen magnum to the upper border of the posterior arch of the atlas.

- **Movements:** The atlanto-occipital joint permits flexion, extension, and lateral bending. No rotary movement is possible.

5. Lumbosacral Joint

- **Type:** cartilaginous, symphyseal.
- **Articulating elements:** Fifth lumbar vertebra and the first segment of the sacrum.
- **Ligaments:**
Anterior longitudinal ligament attached to the intervertebral discs and margins of the vertebral bodies anteriorly and posterior longitudinal ligament attached to the intervertebral discs and margins of the vertebral bodies posteriorly. There are synovial joints between the articular processes, ligamenta flava and interspinal and supraspinal ligaments.

The strong iliolumbar ligament passes laterally from the transverse process of the fiftn lumbar vertebra to the posterior part of the inner lip of the iiac crest.

6. Sacrococcygeal Joint

- **Type:** Cartilaginous through an intervertebral disc.
- **Articulating elements:** Apex of the sacrum and the base of the coccyx.
- **Ligaments:** Intervertebral disc reinforced on all sides by sacrococcygeal ligaments and interarticular ligaments between the cornua of the sacrum and coccyx.
- **Movements of the vertebral column:** Motion takes place at the intervertebral disks and joints of the articular processes. Movements of the column include flexion,

extension, lateral bending, and rotation and the axis of each movement is through the nucleus pulposus. The extent of movement between adjacent vertebrae is small but the entire vertebral column has a considerable range of motion. The range of motion is limited by associated ligaments and muscles and the compression and extension of intervertebral discs.

In relation to movements of the head on the vertebral column, flexion and extension occur mostly at the atlanto-occipital joint (extension is freer than flexion) and rotation occurs primarily as the skull and atlas rotate around the dens of the axis.

Vertebral canal is a vertebral arch and the posterior surface of its body enclose a vertebral foramen. The vertebral foraminae and the associated ligaments comprise the vertebral (spinal) canal. The canal is triangular in the cervical and lumbar regions and circular in the thoracic region.

The vertebral canal is lined with extradural fat in which lies the internal vertebral plexus of veins that drains the marrow of vertebral bodies and drains into intervertebral veins. The internal vertebral plexus has no valves and venous blood can partly bypass the inferior vena cava (to the superior vena cava) in coughing or abdominal straining. The vertebral canal also contains the meninges, spinal cord and its vessels, and the roots of the spinal nerves and their sheaths.